CNC Programming Guide

Rick Calverley
Director of Education
Lincoln College of Technology
Grand Prairie, Texas

Publisher
The Goodheart-Willcox Company, Inc.
Tinley Park, IL
www.g-w.com

Copyright © 2024
by
The Goodheart-Willcox Company, Inc.

All rights reserved. No part of this work may be reproduced, stored, or transmitted in any form or by any electronic or mechanical means, including information storage and retrieval systems, without the prior written permission of
The Goodheart-Willcox Company, Inc.

Library of Congress Control Number: 2022942616

ISBN 978-1-63776-702-3

1 2 3 4 5 6 7 8 9 – 24 – 27 26 25 24 23 22

The Goodheart-Willcox Company, Inc. Brand Disclaimer: Brand names, company names, and illustrations for products and services included in this text are provided for educational purposes only and do not represent or imply endorsement or recommendation by the author or the publisher.

The Goodheart-Willcox Company, Inc. Safety Notice: The reader is expressly advised to carefully read, understand, and apply all safety precautions and warnings described in this book or that might also be indicated in undertaking the activities and exercises described herein to minimize risk of personal injury or injury to others. Common sense and good judgment should also be exercised and applied to help avoid all potential hazards. The reader should always refer to the appropriate manufacturer's technical information, directions, and recommendations; then proceed with care to follow specific equipment operating instructions. The reader should understand these notices and cautions are not exhaustive.

The publisher makes no warranty or representation whatsoever, either expressed or implied, including but not limited to equipment, procedures, and applications described or referred to herein, their quality, performance, merchantability, or fitness for a particular purpose. The publisher assumes no responsibility for any changes, errors, or omissions in this book. The publisher specifically disclaims any liability whatsoever, including any direct, indirect, incidental, consequential, special, or exemplary damages resulting, in whole or in part, from the reader's use or reliance upon the information, instructions, procedures, warnings, cautions, applications, or other matter contained in this book. The publisher assumes no responsibility for the activities of the reader.

The Goodheart-Willcox Company, Inc. Internet Disclaimer: The Internet resources and listings in this Goodheart-Willcox Publisher product are provided solely as a convenience to you. These resources and listings were reviewed at the time of publication to provide you with accurate, safe, and appropriate information. Goodheart-Willcox Publisher has no control over the referenced websites and, due to the dynamic nature of the Internet, is not responsible or liable for the content, products, or performance of links to other websites or resources. Goodheart-Willcox Publisher makes no representation, either expressed or implied, regarding the content of these websites, and such references do not constitute an endorsement or recommendation of the information or content presented. It is your responsibility to take all protective measures to guard against inappropriate content, viruses, or other destructive elements.

Image Credits. Front cover: Pixel B/Shutterstock.com; Back cover: Pixel B/Shutterstock.com

Preface

Technology in the manufacturing industry is progressing at a pace that has not been seen before. As the manufacturing industry grows and previous generations of workers continue to retire, companies are experiencing difficulties filling skilled positions. This is a great time to be entering the field of manufacturing!

In the past, manufacturing facilities were filled with manually operated mills and lathes, a skilled machinist crafting parts at each machine. Today, manufacturing facilities are filled with computer numerical control (CNC) machines, and CNC programmers and operators use these machines to produce parts at a higher level of quality, a higher rate of production, and a higher level of consistency than was attainable in the past.

CNC programmers are highly skilled professionals. They have a fundamental understanding of machine technology, machining operations, cutting tools, and part workholding methods. Along with this knowledge, CNC programmers possess a detailed understanding of programming commands and functions used to control the machine. CNC programmers are also able to work efficiently with machining personnel in today's fast-paced production environment.

Advancing technology in the manufacturing industry requires a new educational approach to prepare tomorrow's workforce. Students must be able to develop the relevant skills needed for career success. *CNC Programmer's Guide* has been designed to meet these needs. This text will help students prepare for a career in the exciting field of CNC machining.

The content of this text is organized in three sections. The first section, *CNC Mill Programming*, covers the commands, functions, and programming techniques used in CNC milling operations. The second section, *CNC Lathe Programming*, covers the commands, functions, and programming techniques used in CNC lathe operations. The third section, *Subprogramming, Probe Programming, and Macros*, covers subprogramming applications, probe programming, macro programming, and advanced topics. This content organization is designed for different course structures and flexibility in training programs. The three sections contain comprehensive coverage focused on specific machine processes and programming applications.

CNC Programmer's Guide features an extensive number of programs throughout the text to illustrate examples of how CNC operations are programmed. These are practical examples with detailed explanations to help students understand how codes work and when they are used. Depending on the CNC machine and controller type, programming commands and functions can vary significantly. This text explains different programming formats where appropriate, with an emphasis on Haas and Fanuc controllers.

CNC Programmer's Guide employs a building-block approach to present fundamental concepts before progressing to more advanced topics. Programming commands and formats are introduced before specific types of operations are covered. Coverage includes processes involved in program planning, machine setup, and part verification.

The primary focus of *CNC Programmer's Guide* is CNC programming techniques and practices. For more in-depth coverage of CNC manufacturing, including CNC machining technology, machine types, tooling, and advanced topics, refer to *CNC Manufacturing Technology*, also published by Goodheart-Willcox.

About the Author

Rick Calverley is the Director of Education at Lincoln College of Technology in Grand Prairie, Texas, where he designed and implemented Lincoln's first program in CNC Machining and Manufacturing Technology. As a third-generation machinist, trained mold maker, tool and die maker, and CNC multiaxis programmer, Mr. Calverley has seen this industry go through radical changes. He started his career in 1982, running manual machinery in his father's shop. During his 30+ years of making parts, he has produced parts used to make turbochargers, transmissions, plastic injection molds, aircraft parts, and even parts for the International Space Station. Before joining Lincoln, Mr. Calverley worked as a CNC programmer for several manufacturing firms, including Solidiform, Inc., a defense contractor in the aerospace industry.

Mr. Calverley is the author of **CNC Manufacturing Technology**. He holds an AS degree in marine technology from the College of Oceaneering, and has completed additional college-level industry and academic programs. He holds all 12 NIMS Machining Level 1 certifications, as well as three Mastercam Associate Level certifications.

Mr. Calverley has served on the national councils for the Fabricators and Manufacturers Association, the Mastercam Educators Alliance, and the Haas Technical Education Community. He also established the first National Tooling and Machining Association student chapter.

Acknowledgments

The author and publisher wish to thank the following companies, organizations, and individuals for their contribution of resource material, images, or other support in the development of **CNC Programmer's Guide**.

The author and publisher extend special appreciation to Madalyn Belle Photography for substantial contribution to the photography program.

CNC Software, Inc.

Haas Automation, Inc.

Iscar

Rick Calverley thanks his wife, Gina, and his five sons—Jacob, Zech, Sam, Gabe, and Israel—for the support they provided throughout the process of writing this textbook.

Reviewers

The author and publisher wish to thank the following industry and teaching professionals for their valuable input into the development of *CNC Programmer's Guide*.

Danny R. Adkins
Ivy Tech Community College
Evansville, IN

Brian Aiken
Pickens County Career and Technology Center
Liberty, SC

Brendan Anderson
SUNY Alfred State College
Alfred, NY

David Black
Anderson W. Clark Magnet High School
Glendale, CA

Daniel Colquitt
Baker College
Flint, MI

Ed Doherty
Suncoast Technical College
Sarasota, FL

Jeremy Dutton
Gateway Technical College
Kenosha, WI

Joel Eisele
Monroe Community College
Rochester, NY

William Gelches
Delaware County Community College
Media, PA

Paul Gorsky
Lincoln Technical Institute
Mahwah, NJ

Douglas Green
Solano Community College
Fairfield, CA

Eugene L. Horst Jr.
SUNY Morrisville
Morrisville, NY

Michael E. Jones
York County Community College
Wells, ME

Jack Krikorian
Technology and Manufacturing Association
Schaumburg, IL

Angel Madiedo
Palm Beach State College
Lake Worth, FL

Carrie Marsico
Cuyahoga Community College
Cleveland, OH

Eric McKell
Brigham Young University
Provo, UT

Alister McLeod
Indiana State University
Terre Haute, IN

Terry Morse
Delta College
University Center, MI

Mathieu Ordiway-Thiem
Lockheed Martin High Speed Wind Tunnel
Dallas, TX

Todd Sanders
Danville Community College
Danville, VA

Kevin Schmidt
Camden County College
Blackwood, NJ

Derek Seeke
Guilford Technical Community College
Jamestown, NC

Rich Shouse
Gateway Technical College
Kenosha, WI

Vincent Stadler
Monroe Community College
Rochester, NY

Jason Taylor
Shelton State Community College
Tuscaloosa, AL

John Templeton
NTMA Training Centers of Southern California
Santa Fe Springs, CA

Steve B. Tornero
Stark State College
North Canton, OH

Jacob Tucker
Pasadena City College
Pasadena, CA

Leonard Walsh
Goodwin University
East Hartford, CT

Chris Waterworth
Greater Lawrence Technical School
Andover, MA

Brian Wickham
Hudson Valley Community College
Troy, NY

Features of the Textbook

The instructional design of this textbook includes student-focused learning tools to help you succeed. This visual guide highlights these features.

Chapter Opening Materials

Each chapter begins with a chapter outline, a list of learning objectives, and a list of key technical terms. The **Chapter Outline** summarizes the topics that will be covered in the chapter. The **Learning Objectives** clearly identify the knowledge and skills to be gained when the chapter is completed. The **Key Terms** list the key technical terms to be learned in the chapter. When key terms are introduced, they are highlighted in *bold italic* type.

Additional Features

Additional features are used throughout the body of each chapter to further learning and knowledge. **Safety Notes** alert you to potentially dangerous practices and conditions. **From the Programmer** features provide practical advice and guidance that is especially applicable on the job. **Debugging the Code** features explain proper use of programming commands and help you develop troubleshooting skills.

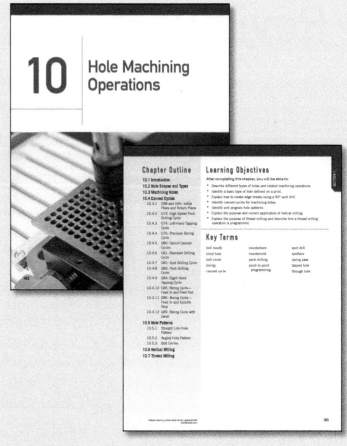

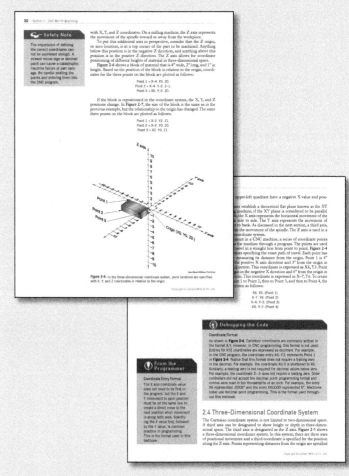

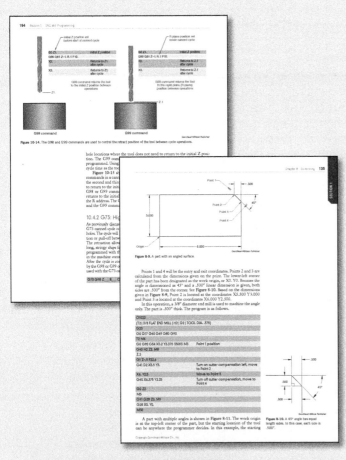
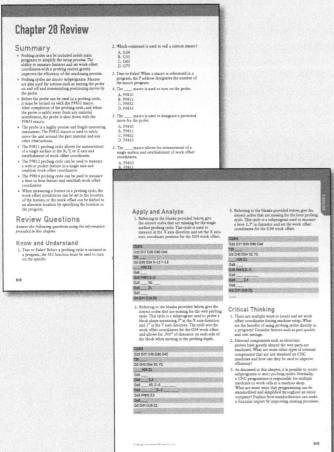

Illustrations

Illustrations have been designed to clearly and simply communicate the specific topic. The text makes extensive use of detailed illustrations and photographs to support explanations of concepts.

Program Examples

Numerous **Program Examples** show students the programmed code corresponding to discussions in the text. These examples are set off from the main text with color highlighting. Explanations of specific codes appear to the right of the lines in the program.

End-of-Chapter Content

End-of-chapter material provides an opportunity for review and application of concepts. A concise **Summary** provides an additional review tool and reinforces key learning objectives. This helps you focus on important concepts presented in the text. **Know and Understand** questions enable you to demonstrate comprehension of the chapter material. **Apply and Analyze** questions extend learning and allow you to apply knowledge and skills. **Critical Thinking** questions help you develop higher-order thinking and problem-solving skills.

TOOLS FOR STUDENT AND INSTRUCTOR SUCCESS

Student Tools

Student Text
CNC Programmer's Guide is a comprehensive text that focuses on the techniques, processes, and procedures used by CNC programmers.

Lab Workbook
- Hands-on practice includes review questions corresponding to each chapter in the textbook.
- Lab activities offer students opportunities to reinforce knowledge and apply programming skills.

G-W Digital Companion
For digital users, e-flash cards and vocabulary exercises allow interaction with content to reinforce knowledge of key terms and topics.

Instructor Tools

LMS Integration
Integrate Goodheart-Willcox content within your Learning Management System for a seamless user experience for both you and your students. EduHub® LMS–ready content in Common Cartridge® format facilitates single sign-on integration and gives you control of student enrollment and data. With a Common Cartridge integration, you can access the LMS features and tools you are accustomed to using and G-W course resources in one convenient location—your LMS.

G-W Common Cartridge provides a complete learning package for you and your students. The included digital resources help your students remain engaged and learn effectively:

- **Digital Textbook**
- Online **Lab Workbook** content
- **Drill and Practice** vocabulary activities

When you incorporate G-W content into your courses via Common Cartridge, you have the flexibility to customize and structure the content to meet the educational needs of your students. You may also choose to add your own content to the course.

For instructors, the Common Cartridge includes the Online Instructor Resources. QTI® question banks are available within the Online Instructor Resources for import into your LMS. These prebuilt assessments help you measure student knowledge and track results in your LMS gradebook. Questions and tests can be customized to meet your assessment needs.

Online Instructor Resources
- The **Instructor Resources** provide instructors with time-saving preparation tools such as answer keys, editable lesson plans, and other teaching aids.
- **Instructor's Presentations for PowerPoint**® are fully customizable, richly illustrated slides that help you teach and visually reinforce the key concepts from each chapter.
- Administer and manage assessments to meet your classroom needs using **Assessment Software with Question Banks**, which include hundreds of matching, completion, multiple choice, and short answer questions to assess student knowledge of the content in each chapter.

See www.g-w.com/cnc-programmers-guide-2024 for a list of all available resources.

Professional Development
- Expert content specialists
- Research-based pedagogy and instructional practices
- Options for virtual and in-person Professional Development

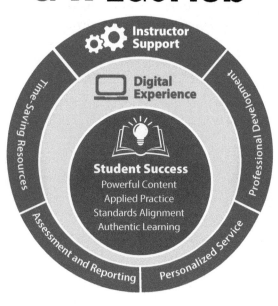

Brief Contents

Section 1—CNC Mill Programming

1. Machining Mathematics for Milling ... 2
2. Cartesian Coordinate System and Machine Axes for Milling ... 24
3. Preparatory Commands: G-Codes ... 44
4. Miscellaneous Functions: M-Codes ... 60
5. Address Codes for Mill Programming ... 78
6. Steps in Program Planning ... 94
7. Mill Program Format ... 110
8. Contouring ... 124
9. Pockets and Slots ... 152
10. Hole Machining Operations ... 182
11. Facing and Island Machining ... 218
12. Setup Sheets ... 238
13. Machine Setup ... 260

Section 2—CNC Lathe Programming

14. Machining Mathematics for Turning ... 288
15. Cartesian Coordinate System and Machine Axes for Turning ... 314
16. Preparatory Commands: Lathe G-Codes ... 334
17. Miscellaneous Functions and Address Codes for Lathe Programming ... 350
18. Lathe Program Planning ... 376
19. Lathe Contour Programming ... 396
20. Hole Machining on a Lathe ... 430
21. Programming Grooves and Parting Off Operations ... 452
22. Threading ... 480
23. Live Tooling ... 502
24. Lathe Setup ... 524

Section 3—Subprogramming, Probe Programming, and Macros

25. Main Programs and Subprograms ... 544
26. Subprogramming Techniques ... 562
27. Probing for Work Offsets ... 588
28. Probing Inside of the Program ... 604
29. On Machine Verification ... 620
30. Macros ... 638
31. Additional Tips and Tricks ... 652

Contents

Section 1
CNC Mill Programming

Chapter 1
Machining Mathematics for Milling 2
   1.1 Introduction 4
   1.2 Converting Fractions to Decimals 4
   1.3 Geometric Shapes 6
   1.4 Angles 11
   1.5 Trigonometry. 12
   1.6 Bolt Circles 16
   1.7 Milling Speeds and Feeds 19

Chapter 2
Cartesian Coordinate System and Machine
Axes for Milling 24
   2.1 Introduction 26
   2.2 Number Line. 26
   2.3 Two-Dimensional Coordinate System 27
   2.4 Three-Dimensional Coordinate System 28
   2.5 Absolute and Incremental Positioning. 31
   2.6 Machine Home and Work Origin. 32
   2.7 Four-Axis Machines. 33
   2.8 Five-Axis Machines 36
   2.9 Polar Coordinates. 38

Chapter 3
Preparatory Commands: G-Codes 44
   3.1 Introduction 46
   3.2 Using G-Codes in a Program 46
   3.3 G-Code Commands 46
   3.4 Startup Blocks 56

Chapter 4
Miscellaneous Functions: M-Codes 60
   4.1 Introduction 62
   4.2 Use of M-Codes 62
   4.3 M-Codes as Program Functions 64
   4.4 M-Codes as Machine Functions 68

Chapter 5
Address Codes for Mill Programming. 78
   5.1 Introduction 80
   5.2 Address Codes 80

Chapter 6
Steps in Program Planning 94
   6.1 Introduction 96
   6.2 Print Review 97
   6.3 Part Workholding 99
   6.4 Tool Selection 101
   6.5 Order of Operations 103

Chapter 7
Mill Program Format 110
   7.1 Introduction 112
   7.2 Opening Statement. 113
   7.3 Program Body. 117
   7.4 Program Closing Statement 120

Chapter 8
Contouring 124
   8.1 Introduction 126
   8.2 Point-to-Point Programming 126
   8.3 Cutter Compensation 128
   8.4 Calculating Angular Moves 134
   8.5 Calculating Radial Moves. 138
   8.6 Chamfering 145

Chapter 9
Pockets and Slots 152
   9.1 Introduction 154
   9.2 Pocket Milling 154
   9.3 Pocket Finishing 166
   9.4 Slots 167

Chapter 10
Hole Machining Operations 182
   10.1 Introduction 184
   10.2 Hole Shapes and Types 184
   10.3 Machining Holes 190
   10.4 Canned Cycles. 192
   10.5 Hole Patterns. 203
   10.6 Helical Milling 210
   10.7 Thread Milling 210

Chapter 11
Facing and Island Machining. 218
   11.1 Introduction to Facing 220
   11.2 Facing Tools. 220
   11.3 Facing Program Strategy. 223

11.4 Creating the Facing Program 226
11.5 Island Machining 227

Chapter 12
Setup Sheets 238
   12.1 Introduction 240
   12.2 Exchange of Information 240
   12.3 Setup Sheets 241
   12.4 Setup Sheet Formats 243
   12.5 Efficiency in Production 245

Chapter 13
Machine Setup 260
   13.1 Introduction 262
   13.2 Setting the Workholding 262
   13.3 Establishing Work Coordinates 264
   13.4 Establishing Tool Length Offsets 268
   13.5 Tool Diameter Offsets 274
   13.6 Running the First Piece 278

Section 2
CNC Lathe Programming

Chapter 14
Machining Mathematics for Turning 288
   14.1 Introduction 290
   14.2 Converting Fractions to Decimals 290
   14.3 Geometric Shapes 292
   14.4 Angles 296
   14.5 Trigonometry 298
   14.6 Tapers 303
   14.7 Thread Measurements 306
   14.8 Turning Speeds and Feeds 308

Chapter 15
Cartesian Coordinate System and Machine
Axes for Turning 314
   15.1 Introduction 316
   15.2 Number Line 316
   15.3 Two-Axis Coordinate System 317
   15.4 Machine Home and Part Origin 325
   15.5 C Axis Coordinate Programming 326

Chapter 16
Preparatory Commands: Lathe G-Codes 334
   16.1 Introduction 336
   16.2 Using G-Codes in a Program 336
   16.3 G-Code Commands 338
   16.4 Startup Blocks 346

Chapter 17
Miscellaneous Functions and Address Codes
for Lathe Programming 350
   17.1 Introduction 352
   17.2 Use of M-Codes 352
   17.3 M-Codes as Program Functions 354
   17.4 M-Codes as Machine Functions 359
   17.5 Address Codes 367

Chapter 18
Lathe Program Planning 376
   18.1 Introduction 378
   18.2 Print Review 379
   18.3 Part Workholding 381
   18.4 Tool Selection 385
   18.5 Order of Operations 389

Chapter 19
Lathe Contour Programming 396
   19.1 Introduction 398
   19.2 Point-to-Point Programming 398
   19.3 Tool Nose Radius Compensation 400
   19.4 Programming Radial Moves 402
   19.5 Programming Angular Moves 406
   19.6 Turning Canned Cycles 408

Chapter 20
Hole Machining on a Lathe 430
   20.1 Introduction 432
   20.2 Hole Drilling Cycles 432
   20.3 Machining Holes with Live Tooling 439

Chapter 21
Programming Grooves and Parting Off Operations . . . 452
   21.1 Introduction 454
   21.2 Planning Grooving Programs 454
   21.3 Programming Straight Wall Grooves 458
   21.4 Programming Chamfers 461
   21.5 G75 Grooving Cycle 463
   21.6 Tapered Wall Grooves 466
   21.7 Full Radius Grooves 470
   21.8 Parting Off 473

Chapter 22
Threading 480
   22.1 Introduction 482
   22.2 Thread Terminology 482
   22.3 Programming Threads 484
   22.4 Tapered Threads 493
   22.5 Multi-Start Threads 496

Chapter 23
Live Tooling 502
- 23.1 Introduction 504
- 23.2 Special Considerations with Live Tooling 505
- 23.3 Radial Hole Machining Canned Cycles....... 507
- 23.4 Slot Milling............................. 512
- 23.5 CAM Software Programs 516

Chapter 24
Lathe Setup 524
- 24.1 Introduction 526
- 24.2 Tool Installation 526
- 24.3 Tool Offsets 527
- 24.4 Work Offsets............................ 531
- 24.5 Setup Sheets............................ 533
- 24.6 Setup Sheet Formats..................... 534
- 24.7 Efficiency in Production.................. 537

Section 3
Subprogramming, Probe Programming, and Macros

Chapter 25
Main Programs and Subprograms 544
- 25.1 Introduction 546
- 25.2 Main Programs and Subprograms 546
- 25.3 Mill Subprogramming 547
- 25.4 Lathe Subprogramming 554

Chapter 26
Subprogramming Techniques................... 562
- 26.1 Introduction 564
- 26.2 The Work Coordinate System 564
- 26.3 Using Work Offsets for Subprograms....... 564
- 26.4 Lathe Subprogramming 569
- 26.5 Using Subprograms for Multiple Parts and Fixtures 572
- 26.6 Layering Subprograms 574
- 26.7 Contouring Subprograms 578
- 26.8 Using Subprograms in Multiaxis Machining .. 582

Chapter 27
Probing for Work Offsets...................... 588
- 27.1 Introduction 590
- 27.2 Bore Probing Cycle....................... 591
- 27.3 Boss Probing Cycle....................... 592
- 27.4 Rectangular Pocket Probing Cycle......... 592
- 27.5 Rectangular Block Probing Cycle.......... 593
- 27.6 Pocket X Axis Probing Cycle 594
- 27.7 Pocket Y Axis Probing Cycle 595
- 27.8 Web X Axis Probing Cycle 595
- 27.9 Web Y Axis Probing Cycle 596
- 27.10 Outside Corner Probing Cycle 598
- 27.11 Inside Corner Probing Cycle 599
- 27.12 Single Surface Probing Cycle 600

Chapter 28
Probing Inside of the Program.................. 604
- 28.1 Introduction 606
- 28.2 Using the Probe as a Tool 606
- 28.3 Starting Probe Position and Protected Moves .. 607
- 28.4 Single Surface Measurement.............. 608
- 28.5 Web/Pocket Measurement 610
- 28.6 Bore/Boss Measurement 614

Chapter 29
On Machine Verification 620
- 29.1 Introduction 622
- 29.2 Pros and Cons of On Machine Verification.... 622
- 29.3 Tool Offset Adjustment and Tolerance Verification............................. 623

Chapter 30
Macros 638
- 30.1 Introduction 640
- 30.2 Using Variables in Macros................. 641
- 30.3 Local Variables.......................... 641
- 30.4 Global Variables 641
- 30.5 System Variables 644
- 30.6 Macro Programming Example 648

Chapter 31
Additional Tips and Tricks..................... 652
- 31.1 Introduction 654
- 31.2 Verifying the Program 654
- 31.3 Establishing Tool Offsets 654
- 31.4 Running the First Part 656
- 31.5 CAM Programming....................... 656
- 31.6 Lathe Taper 657
- 31.7 Second-to-Last Cut and Finish Pass 659
- 31.8 Programming Feed Rates for Arcs 660
- 31.9 Subprogramming Applications............ 660
- 31.10 Metric Programming..................... 661
- 31.11 Scaling 662
- 31.12 Mirror Imaging.......................... 663

Reference Section 666
Glossary 680
Index 686

Feature Contents

From the Programmer

Trigonometric Calculations	13
Speeds and Feeds	19
Calculating Spindle Speeds	20
Coordinate Entry Format	28
Positioning Mode	32
G-Code Entry Format	50
G41: Left-Hand Cutter Compensation	55
Sequence of Commands	63
Notes in Programs	65
Tool and Offset Numbers	84
Print Specifications	97
Machining Inside Corners	99
Machining Slots	102
Program Entry Format	103
General Rules for Programming	113
Startup Block	116
G53: Machine Coordinate System	117
Program Building and Planning	121
Climb Milling	128
D Offset Number	132
Coordinate Calculations	136
Climb Milling	156
CAM and High-Speed Machining	165
Drill Point Calculation	185
Drilling Pilot Holes	186
Using the G76 Canned Cycle	197
G74 and G84 Canned Cycles	201
Canned Cycle Programming	203
Hole Programming	213
Efficiency in Machining	223
Teamwork in Machining	245
Repeatable Workholding	266
Trigonometric Calculations	299
Speeds and Feeds	309
Calculating Spindle Speeds	309
Coordinate Entry Format	322
Part Orientation	325
Work Offsets	326
Multiaxis Turning	328
G-Code Entry Format	340
Sequence of Commands	353
Notes in Programs	355
Automatic Chamfering Addresses	369
Spindle Speed Commands	371
Print Specifications	379
Roughing and Finishing Passes	380
Interpreting Dimensions	381
Identifying Workholding Requirements	384
Cutting Insert Data	388
Programming Efficiency	398
Walking the Path	401
Arc Programming Methods	406
Type 1 and Type 2 Cycles	415
Boring Bar Size	417
Tapping on a Lathe	437
Hole Machining	439
Multiaxis Programming Formats	445
Becoming a Great Programmer	463
Grooving Operations	472
Efficiency in Machining	474
Ball Screw Assemblies	482
Tapered Thread Cutting	487
Thread Forms	495
Hole Orientation	508
Contour Programming	512
Multiaxis Programming	519
Part Origin and Tool Setup	527
Setup Sheets in Machining	537
Programming Modes	550
Using Subprograms	558
Programming Modes	565
Development of Work Offsets	569
Developing Subprograms	575
Subprogram Applications	582
Parameter Entry	590
Probe Position	591
Identifying Probing Cycles	600
Benefits of Probing Systems	601
Initial Work Offset Setting	606
Macros and Variables	609
Using Probing Cycles in Subprograms	617

Single Surface Z Axis Measurement 625
Advancing Probing Technology. 635
Using Global Variables. 644
Reading Variables. 645
Applying Macros 649
Tool Offset Adjustment. 655

Debugging the Code

Coordinate Format 28
Canned Cycle Loop Count 205
Coordinate Format 321
Machine Alarms 402
G96 and G97 Commands 434
Safe Movement 462
Thread Chamfer Pullout 491
Interpreting Dimensions 509
Offset Measurement. 530
Work Offset Designation 567
Programming Modes 578
Feed Rate. 608
Tool Offset Adjustment. 624
Tool Offset Settings. 647

Section 1
CNC Mill Programming

1. Machining Mathematics for Milling
2. Cartesian Coordinate System and Machine Axes for Milling
3. Preparatory Commands: G-Codes
4. Miscellaneous Functions: M-Codes
5. Address Codes for Mill Programming
6. Steps in Program Planning
7. Mill Program Format
8. Contouring
9. Pockets and Slots
10. Hole Machining Operations
11. Facing and Island Machining
12. Setup Sheets
13. Machine Setup

Itsanan/Shutterstock.com

1 Machining Mathematics for Milling

Chapter Outline

1.1 Introduction
1.2 Converting Fractions to Decimals
   1.2.1 Communicating Precision
1.3 Geometric Shapes
   1.3.1 Circles
   1.3.2 Polygons
   1.3.3 Lines
1.4 Angles
   1.4.1 Supplementary Angles
   1.4.2 Complementary Angles
1.5 Trigonometry
1.6 Bolt Circles
1.7 Milling Speeds and Feeds

Learning Objectives

After completing this chapter, you will be able to:

- Convert fractions to decimals.
- Explain how required precision is communicated in machining.
- Identify two-dimensional and three-dimensional geometric shapes encountered in machining.
- Name the parts of a right triangle.
- Use the Pythagorean theorem to solve for sides of triangles.
- Differentiate between supplementary and complementary angles.
- Use basic trigonometry to solve for sides and angles in a right triangle.
- Calculate speeds and feeds in milling.

Key Terms

acute angle	geometric shape	rhombus
angle	hypotenuse	right angle
arc	isosceles triangle	right triangle
bolt circle	line	scalene triangle
chip load	line segment	sine
circle	numerator	square
complementary angles	obtuse angle	straight angle
cosine	parallelogram	supplementary angles
curved line	polygon	tangent
denominator	Pythagorean theorem	tangent line
diameter	quadrilateral	triangle
equilateral triangle	radius	trigonometry
flute	rectangle	

Chapter opening photo credit: Sujin Nimdee/Shutterstock.com

1.1 Introduction

Math is essential in the machining and manufacturing trades. Machinists must have a good working knowledge of math and must be able to make calculations for numerous tasks. This chapter explains the basic math principles used in creating computer numerical control (CNC) milling programs. Math is used throughout production, from program planning and machine setup to part inspection. As you will learn in this chapter, machinists must be able to make calculations based on information provided on prints. Often, machinists use their knowledge of common geometric shapes to determine unknown dimensions. This chapter explains how to calculate linear distances and angles and how to determine locations of features. This chapter also covers trigonometric functions and calculations for cutting speeds and feeds.

When reading prints, it is important to identify the system of measurement used for dimensions. In the United States, most prints are dimensioned in decimal inches or fractional inches. The inch is the basic unit of linear measurement in the US Customary system. All measurements that you will encounter in this text are based on this system. In most countries outside the United States, prints are dimensioned using the International System of Units (abbreviated SI), known as the metric system. Metric prints used in the manufacturing industry are commonly dimensioned in millimeters. The system of measurement used on a print is normally indicated in or near the title block. Most CNC machines can read programs prepared in inches or millimeters. In Chapter 3, you will learn how to set the correct mode for working units when writing a CNC program.

When making unit conversions and other machining calculations, a calculator is often useful. This text assumes access to a simple scientific calculator, such as the one shown in **Figure 1-1**. However, any calculator with a fraction key and sine, cosine, and tangent functions is sufficient to perform machining calculations. Consider the calculator as a tool, just like a file or wrench, to support your work in the shop.

In this text, green buttons are used to display exactly which buttons on the calculator to press and in which order. The prompt will look like this:

Goodheart-Willcox Publisher
Figure 1-1. A scientific calculator is a useful tool for completing mathematical calculations.

1.2 Converting Fractions to Decimals

Most prints used in manufacturing today are dimensioned in decimal inches. However, you will also encounter prints that use fractions to specify dimensions. It is common to see fractional dimensions on older prints and drawings of simple parts. The use of fractions normally signifies a noncritical measurement or a feature that is not designed to meet a high degree of precision in manufacturing. See **Figure 1-2**.

Fractions cannot be used in CNC programs, however. Instead, they must be converted to decimal values. Knowing how to convert fractions is essential for machinists. Having this knowledge is also important when making measurements because many measuring tools make readings in decimal units.

Fractions have a numerator and a denominator. The *numerator* is the number on top of the fraction and indicates the number of parts in the fraction. The *denominator* is the number on the bottom of the fraction. It

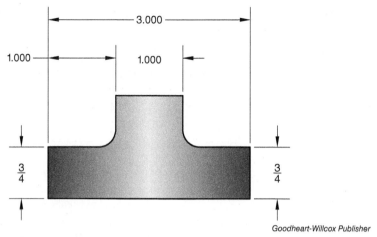

Figure 1-2. A basic part with dimensions in decimal inches and fractional inches.

indicates the whole number quantity into which the parts are divided. For example, in the fraction 3/4, the numerator is 3 and the denominator is 4.

The process of converting fractions to decimals is a straightforward division problem. To convert a fraction into a decimal, divide the denominator into the numerator. The result is a decimal.

Take a look at this sample problem:

$$\tfrac{3}{4} = 4\overline{)3} \text{ or } 3 \div 4$$

$$\tfrac{3}{4} = .750$$

The process is even easier using a calculator. Use the following sequence:

$$3 \div 4 =$$

1.2.1 Communicating Precision

The answer to the problem in the previous example is .750. Mathematically, it would also be correct to express the answer as .75, or even .7500. The trailing zero after the decimal point does not hold any value. Similarly, any leading zero on the leftmost side of the decimal point does not hold any value.

$$00.75 = 0.75 = .750 = 0.750000 = 0000.7500000$$

All of these numbers are equal. However, machinists have a specific way of communicating. To prevent confusion, machinists *talk* or *work* to three places or four places to the right of the decimal point, depending on the required precision. The third place to the right of the decimal point, for example, is the thousandths place:

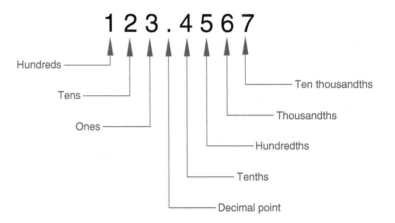

As an example of how machinists communicate, a print dimension of .750 is read out as *seven hundred fifty thousandths* (.750). If you were measuring a drill bit that was .5 in diameter, you would say *five hundred thousandths* (.500). This system provides an efficient and effective method of communication.

A similar method is used when the required precision is in ten thousandths of an inch. As shown previously, the fourth place to the right of the decimal point is the ten thousandths place. This place is referred to by machinists using the abbreviation *tenths*. For example, a print dimension of .0005 is read out as *five tenths*. In actual measurement, the referenced dimension is five ten-thousandths of an inch.

1.3 Geometric Shapes

Geometric shapes include two-dimensional objects, such as circles, arcs, triangles, and polygons; and three-dimensional objects, such as spheres, pyramids, cubes, and polyhedrons. See **Figure 1-3**. Two-dimensional geometric shapes are flat. Imagine drawing a shape on a piece of paper. You can draw left and right or up and down, but only in two dimensions. A three-dimensional shape has depth. Consider the difference between a circle and a sphere. A circle is two-dimensional, and spheres are three-dimensional. You should know how to recognize basic geometric shapes and how the geometry of a part can be evaluated to determine dimensional information.

1.3.1 Circles

Circles are two-dimensional objects often encountered in machining, such as when drilling holes or cutting round shafts. A *circle* is defined as a closed plane curve that is an equal distance at all points from its center point. A circle can be thought of as a closed loop with a center point. If the distance from that center point to every point on the outside is the same, the closed loop is a circle, **Figure 1-4**. An *arc* is a portion of a circle. An arc has a start point, end point, center point, and defined direction (clockwise or

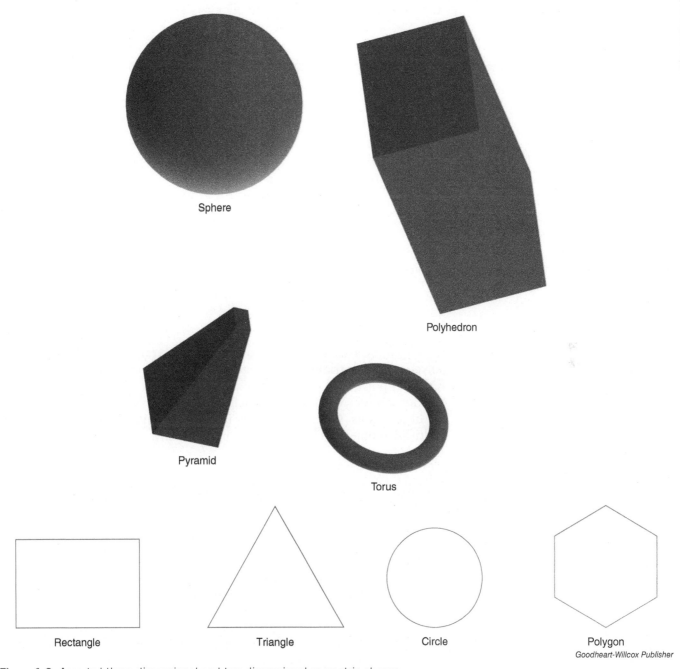

Figure 1-3. Assorted three-dimensional and two-dimensional geometric shapes.

counterclockwise). Many shapes encountered in machining are formed by arcs. For example, a 90° round corner blend represents one-quarter of a circle.

The distance from the center point to the edge of a circle or arc is defined as the *radius*. The radius is frequently used in creating CNC programs and calculating locations. The distance across a circle through the center point is the *diameter*. The radius of a circle is one-half the diameter. See **Figure 1-5**. Hole sizes are often defined on a print in terms of diameter, whereas corner blends and other partial circles are defined by radius. When you encounter a circular feature on a print, pay close attention to specifications and verify whether the dimension defines a radius or a diameter.

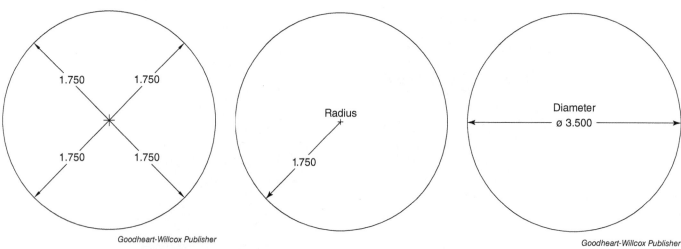

Figure 1-4. A circle is a closed plane curve defined by a center point that is equidistant from all points on the curve.

Figure 1-5. The radius of a circle is the distance from the center point to the edge of the circle. The diameter is the distance from edge to edge through the center point. The radius is one-half the diameter.

1.3.2 Polygons

A *polygon* is a two-dimensional shape with straight sides. Examples of polygons include triangles, rectangles, pentagons, and hexagons, **Figure 1-6.** A regular polygon is made up of equal-length sides with equal interior and exterior angles. Many of the machining projects in later chapters make use of polygons. Often, machining involves cutting rectangular blocks or multi-sided objects. The prefix *poly* means many.

Triangles

Triangles are particularly important polygons in machining applications. A *triangle* is made up of three sides that form three angles. There are several types of triangles that you will commonly encounter in machining work. The following types are defined by the relationship between the sides and angles.

- A *scalene triangle*, **Figure 1-7,** has no equal sides or equal angles.

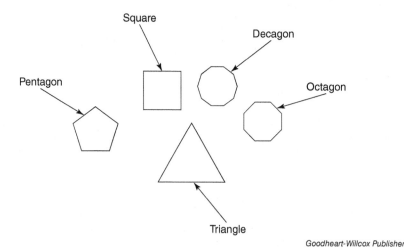

Figure 1-6. Examples of polygonal geometric shapes. As seen here, polygons can have an unlimited number of sides.

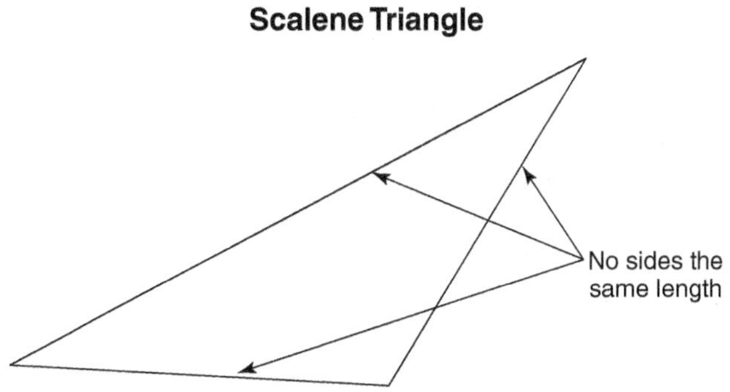

Scalene Triangle

Figure 1-7. A scalene triangle has no equal sides or angles.

- An *isosceles triangle*, Figure 1-8, has two equal sides and two equal angles.
- An *equilateral triangle*, Figure 1-9, has all equal sides and all equal angles.
- A *right triangle*, Figure 1-10, has one 90° angle. A 90° angle is defined as two lines that are exactly perpendicular to each other. On a print, a right triangle is designated by a square symbol in the 90° corner of the triangle. The longest side in a right triangle, called the *hypotenuse*, is always across from the 90° angle. The right triangle has special mathematical properties that make it advantageous for use in machining.

All triangles share the same mathematical property: the sum of all three angles must equal 180°.

$$180° = \text{Angle 1} + \text{Angle 2} + \text{Angle 3}$$

The ability to calculate angles without all of the angles defined is often necessary in print reading. Using the previous formula, you can calculate any angle of a triangular shape if you know the other two angles. If you

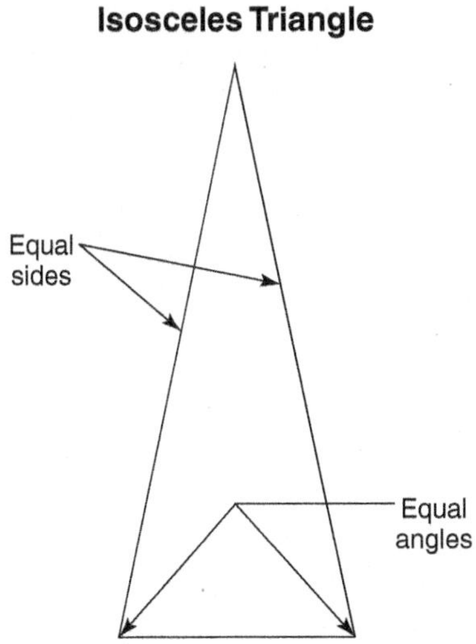

Isosceles Triangle

Figure 1-8. An isosceles triangle has two equal sides and two equal angles.

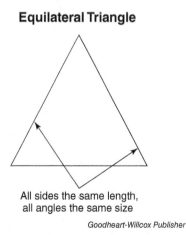

Figure 1-9. An equilateral triangle has all equal sides and all equal angles.

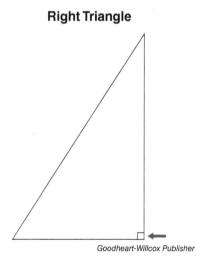

Figure 1-10. A right triangle, denoted by a right angle symbol.

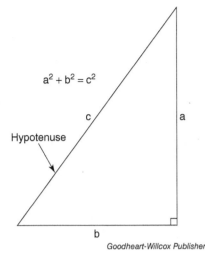

Figure 1-11. The sides of a right triangle can be calculated using the Pythagorean theorem.

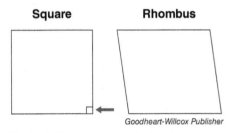

Figure 1-12. Squares and rhombuses have four equal sides. A square has four right angles, while a rhombus has no right angles. Because all four sides of the square are equal, the presence of one right angle, indicated by the right angle symbol, implies that all four angles measure 90°.

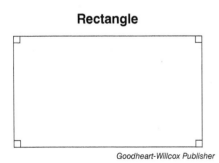

Figure 1-13. A rectangle has four right angles and two sets of equal sides.

have a triangle where you know one angle is 45° and another angle is 55°, calculate the third angle as follows:

$$180° - \text{Angle 1} - \text{Angle 2} = \text{Angle 3}$$
$$180° - 45° - 55° = \text{Angle 3}$$
$$80° = \text{Angle 3}$$

Right triangles have a unique characteristic defined by the Pythagorean theorem. The *Pythagorean theorem* states that, in a right triangle, the square of one side plus the square of the second side is equal to the square of the hypotenuse, **Figure 1-11**:

$$a^2 + b^2 = c^2$$

Consider the following example. In the triangle shown in **Figure 1-11**, side *a* measures 2″ and side *b* measures 1.5″. What is the length of side *c*?

$$a^2 + b^2 = c^2$$
$$(2″)^2 + (1.5″)^2 = c^2$$
$$4″ + 2.25″ = c^2$$
$$c = \sqrt{6.25} = 2.5″$$

To solve this problem on your calculator, press the following keys:

$$\boxed{2} \; \boxed{x^2} \; \boxed{+} \; \boxed{1.5} \; \boxed{X^2} \; \boxed{=} \; \boxed{\sqrt{x}}$$

The last input key takes the square root. This last step changes c^2 to the desired solution, *c* or 2.5″.

As shown in these examples, using standard algebra, you can calculate any side of a right triangle if you know the other two sides. Additional mathematical methods for working with right triangles and finding all of their sides and angles are covered later in this chapter.

Quadrilaterals

A *quadrilateral* is a four-sided, two-dimensional polygon. Note that the prefix *quad* means four and indicates the number of sides. Quadrilaterals are among the most commonly machined shapes and include squares, rectangles, and rhombuses. Mathematically, or geometrically, these three different quadrilaterals all have specific properties.

- A *square* is a quadrilateral with four equal sides and four right (90°) angles.
- A *rhombus* has four equal sides but no right angles. Compare a square with a rhombus, **Figure 1-12**. Notice the right angle symbol in the square.
- A *rectangle* is also a four-sided object with four right angles, **Figure 1-13**. However, rectangles do not have four equal sides. Instead, they have two pairs of equal sides.

Notice that each of these shapes has parallel sides and that the opposite sides are equal in length. Squares, rhombuses, and rectangles are all *parallelograms*. A parallelogram is a quadrilateral in which the opposite sides are both parallel and equal.

1.3.3 Lines

Lines are important geometric features in machining. With the exception of commands for machining circular features, most CNC machining commands are along a line. A *line* is a continuous, straight, one-dimensional geometric element with no end. This means a line is infinite. It is impossible to machine an infinitely long part. When the term *line* is used in machining, it actually refers to a line segment. A *line segment* is a line with a definitive beginning and end. Line segments, like lines, do not curve or waver. A *curved line* is an arc, partial circle, or spline.

A line that touches an arc or circle at exactly one point is referred to as a *tangent line*. If a line crosses or touches an arc at more than one point, it is not tangent, but intersecting, **Figure 1-14**.

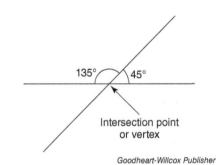

Figure 1-14. A circle displaying a tangent and nontangent (intersecting) line.

1.4 Angles

Most machining requires a good working knowledge of angles. An *angle* measures the rotational distance between two intersecting lines or line segments from their point of intersection (or vertex). Angles are usually given in degrees, **Figure 1-15**.

There are 360° in a circle (one degree is 1/360 of a circle). A straight line passing through the center point of a circle divides the circle in half and forms two semicircles, each representing 180°. See **Figure 1-16**.

There are four common types of angles:

- A *right angle* measures exactly 90° between two intersecting lines. When two lines intersect and form a right angle, **Figure 1-17**, all other angles of intersection are also 90°. Notice that the sum of the angles is 360°.
- An *acute angle* measures less than 90°. Anything less than a right angle is an acute angle, **Figure 1-18**.
- An *obtuse angle* measures more than 90°, **Figure 1-19**. Notice that the second leg of the angle intersects the first at an angle greater than a right angle.

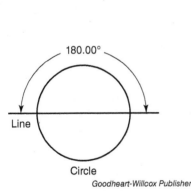

Figure 1-16. A circle contains 360°. A line passing through the center point of the circle divides the circle in half and forms two semicircles. Each semicircle is a 180° arc.

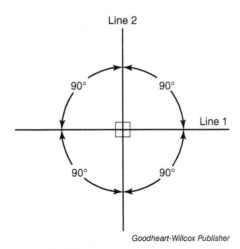

Figure 1-17. Two lines intersecting at 90° angles.

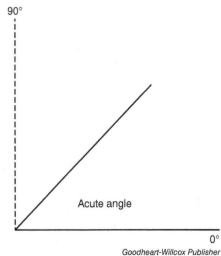

Figure 1-18. An acute angle is one that is less than 90°.

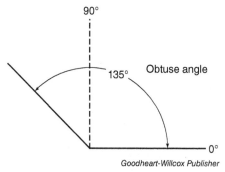

Figure 1-19. An obtuse angle is one that is more than 90°.

- A *straight angle* measures exactly 180°. A straight line represents a straight angle. Refer to the horizontal line shown in **Figure 1-16**.

1.4.1 Supplementary Angles

Supplementary angles are two angles that add up to 180°. Recall that a straight angle is 180°. When one straight line intersects another at an angle, the angles on the opposite sides of the vertex are *supplementary*. If one angle is known, the other can be calculated. For example, if one line intersects another at 50°, calculate the second angle as follows. See **Figure 1-20**.

$$50° + x = 180°$$
$$x = 180° - 50°$$
$$x = 130°$$

The prints used in a normal machining environment do not define every feature. In fact, it is incorrect to overdimension an object on a print. Often, one angle is defined, and you must calculate the second or supplementary angle.

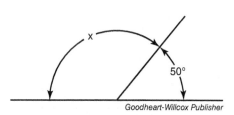

Figure 1-20. When one supplementary angle is known, the other supplementary angle can be calculated by subtracting the known angle from 180°.

1.4.2 Complementary Angles

Complementary angles are two angles that add up to 90°. Problems involving complementary angles normally start with a 90° angle that has been intersected by a line. You can find the missing angle that completes the 90° angle if the other angle is known, **Figure 1-21**.

In this example, the given angle is 25°. What is the complementary angle?

$$90° = 25° + x$$
$$90° - 25° = x$$
$$x = 65°$$

In **Figure 1-22**, the right angle symbol in the lower-left corner indicates a 90° angle. If the known angle is 34°, what is the complementary angle?

$$90° = 34° + x$$
$$90° - 34° = x$$
$$x = 56°$$

As you move forward with print reading, you will see prints that have right angles or right triangles. Often, the print may define only one angle. You can calculate unknown angles using the principles previously discussed as well as basic trigonometry.

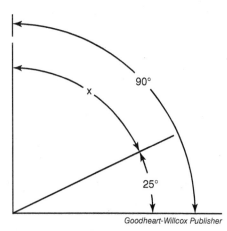

Figure 1-21. An unknown angle can be calculated by subtracting its known complementary angle from 90°.

1.5 Trigonometry

Trigonometry is a branch of mathematics that deals with the relationships between the sides and angles of triangles and with functions of angles. Essentially, trigonometry is the mathematics of triangles and angles. This text focuses on right triangles.

The three fundamental principles for working with right triangles are:

- Right triangles always have one angle of 90°.
- The sum of all three angles in a triangle is 180°.
- The Pythagorean theorem states that $a^2 + b^2 = c^2$.

Start by reviewing the parts of right triangles, **Figure 1-23**. Immediately, you should identify each triangle as a right triangle because it has the right angle symbol. The hypotenuse, or longest side, is directly across from the right angle and is designated by the letter H.

The other two sides are named based on their relation to a given angle or for the angle for which you are trying to solve. They are called the *opposite side* and *adjacent side*, indicated by the letters O and A, respectively, **Figure 1-24**.

The mathematical relationships between sides and angles are given by three trigonometric functions:

- The *sine* of a given angle is the ratio of the opposite side to the hypotenuse.
- The *cosine* of a given angle is the ratio of the adjacent side to the hypotenuse.
- The *tangent* of a given angle is the ratio of the opposite side to the adjacent side.

"SOHCAHTOA" is a helpful mnemonic for remembering the definitions of sine, cosine, and tangent and the mathematical calculations to solve for right triangle sides and angles.

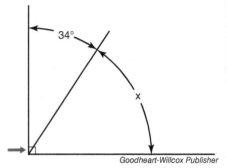

Figure 1-22. A complementary angle is easily calculated when a right angle is given (as shown by the right angle symbol) and one of the complementary angles is known.

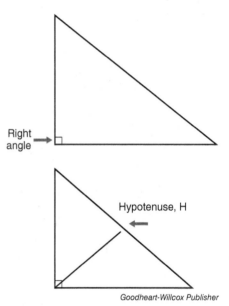

Figure 1-23. The right angle symbol denotes a right angle. The side directly across from the 90° angle in a right triangle is the hypotenuse.

S = sine
O = opposite
H = hypotenuse
C = cosine
A = adjacent
H = hypotenuse
T = tangent
O = opposite
A = adjacent

$$\text{Sine (sin)} = \frac{\text{Opposite (O)}}{\text{Hypotenuse (H)}}$$

$$\text{Cosine (cos)} = \frac{\text{Adjacent (A)}}{\text{Hypotenuse (H)}}$$

$$\text{Tangent (tan)} = \frac{\text{Opposite (O)}}{\text{Adjacent (A)}}$$

Trigonometric functions can be used to determine all of the sides and angles of a right triangle. Sine, cosine, and tangent calculations can be made using reference tables or a scientific calculator. Although reference tables are helpful tools, the probability of error increases with manual calculations. The quickest and most accurate calculations are made with a calculator. First, identify what sides and angles you know and what you are trying to find. Then, choose the appropriate formula and use the correct calculator inputs.

The following is an example of finding the length of a side in a right triangle when one angle and one side are known. The first step is to identify the sides. The side opposite the right angle is always the hypotenuse.

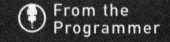

Trigonometric Calculations

On a scientific calculator, the sine of a given angle is calculated by entering the angle and then pressing the SIN key. The COS and TAN keys are used in the same manner to calculate cosine and tangent.

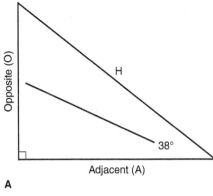

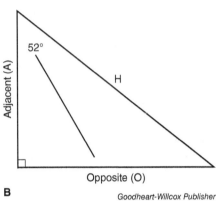

Figure 1-24. The opposite and adjacent sides of a right triangle are determined relative to the known angle. A—This right triangle has one angle defined as 38°. The side directly opposite that angle is known as the opposite side. The side of the triangle that is next to or touching the 38° angle is called the adjacent side. B—This is the same triangle, but the known angle is now 52° and the opposite and adjacent sides change positions.

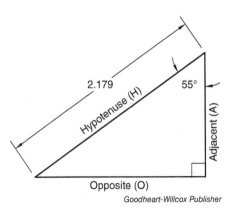

Figure 1-25. A right triangle with one known side and one known angle.

The side opposite the known angle is the *opposite side*. The side touching the known angle is the *adjacent side*. In **Figure 1-25**, the length of the hypotenuse and one angle are known. To calculate the opposite side, use the sine function. The known angle (55°) is represented as *a* in the formula:

$$\sin a = \frac{\text{Opposite (O)}}{\text{Hypotenuse (H)}}$$

$$\sin 55 \times \text{Hypotenuse (H)} = \text{Opposite (O)}$$
$$.8192 \times 2.179 = \text{O}$$
$$1.785'' = \text{O}$$

To solve this problem using a calculator, press the following keys:

| 55 | SIN | × | 2.179 | = |

The adjacent side in **Figure 1-25** can be calculated using the cosine function. The known angle (55°) is represented as *a* in the formula:

$$\cos a = \frac{\text{Adjacent (A)}}{\text{Hypotenuse (H)}}$$

$$\cos 55 \times \text{Hypotenuse (H)} = \text{Adjacent (A)}$$
$$.5736 \times 2.179 = \text{A}$$
$$1.250'' = \text{A}$$

To solve this problem using a calculator, press the following keys:

| 55 | COS | × | 2.179 | = |

The following summarizes the calculator inputs used for sine, cosine, and tangent formulas when solving for one side of a right triangle. These formulas are used when one angle and one side are known. The first two formulas were used in the previous examples. Use the appropriate input based on the side and angle you know.

Other trigonometric functions are used when you know two sides of a right triangle and need to find an angle. The inverse sine, cosine, and tangent functions are used to find an unknown angle relative to two known sides. On a typical calculator, the inverse trigonometric functions are labeled SIN^{-1}, COS^{-1}, and TAN^{-1}. These are typically second functions located on or above the SIN, COS, and TAN keys. Second functions on a calculator are accessed by pressing the second function key before pressing the appropriate key. The second function key may be color coded and is typically labeled 2nd or Shift.

The inverse trigonometric functions are also known as arcsine, arccosine, and arctangent. In the following formulas, –1 denotes an inverse function. The angle to be calculated is represented as a.

$$\text{Angle } a = \sin^{-1}\left(\frac{\text{Opposite (O)}}{\text{Hypotenuse (H)}}\right)$$

$$\text{Angle } a = \cos^{-1}\left(\frac{\text{Adjacent (A)}}{\text{Hypotenuse (H)}}\right)$$

$$\text{Angle } a = \tan^{-1}\left(\frac{\text{Opposite (O)}}{\text{Adjacent (A)}}\right)$$

The first step in solving for an unknown angle in a right triangle is to identify the two known sides. In **Figure 1-26**, there are two known sides. Since the known sides are the opposite and adjacent sides, the inverse tangent function is used to solve for the unknown angle:

$$\text{Angle } a = \tan^{-1}\left(\frac{\text{Opposite (O)}}{\text{Adjacent (A)}}\right)$$

$$\text{Angle } a = \tan^{-1}\left(\frac{1.750}{2.500}\right)$$

$$\text{Angle } a = \tan^{-1}(0.7)$$

$$\text{Angle } a = 34.992°$$

To solve this problem using a calculator, press the following keys:

| 1.75 | ÷ | 2.5 | = | 2nd | Tan |

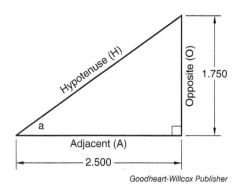

Figure 1-26. A right triangle with two known sides and an unknown angle to solve.

The following summarizes the calculator inputs used for inverse sine, cosine, and tangent formulas when solving for an unknown angle. These formulas are used when two sides are known. Use the appropriate input based on the sides you know.

1. Opp ÷ Hyp = 2nd SIN

2. Adj ÷ Hyp = 2nd COS

3. Opp ÷ Adj = 2nd TAN

As previously discussed in this chapter, the Pythagorean theorem can be used to calculate a side of a right triangle when you know two sides. First, identify the sides you know. Then, select one of the following formulas and calculate the unknown side. Use the appropriate calculator inputs based on the given information for the triangle.

1. Adj x^2 + Opp x^2 = $\sqrt{x}$ (Hyp)

2. Hyp x^2 − Opp x^2 = $\sqrt{x}$ (Adj)

3. Hyp x^2 − Adj x^2 = $\sqrt{x}$ (Opp)

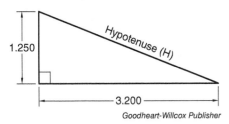

Figure 1-27. A right triangle with two known sides and one unknown side to solve. The unknown side is the hypotenuse.

In **Figure 1-27**, two sides are known and the hypotenuse must be calculated. When considering how to label the two known sides, it is irrelevant which side is defined as opposite or adjacent. To understand why, look at the Pythagorean theorem equation:

$$a^2 + b^2 = c^2$$

Side c is always the longest side, or the hypotenuse. The hypotenuse must always be identified in this manner. In comparison, sides a and b can have an interchangeable order in the equation. Just as 6 + 3 = 9 and 3 + 6 = 9, the designation of opposite and adjacent will not affect the final answer. Since the unknown side in this example is the hypotenuse, use the following calculator formula and key entry.

Calculator formula:

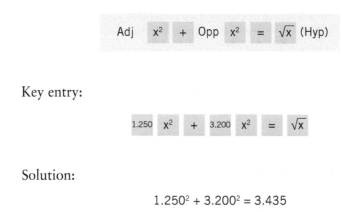

Key entry:

Solution:

$$1.250^2 + 3.200^2 = 3.435$$

1.6 Bolt Circles

A common feature found on a machining print is the ***bolt circle***. A bolt circle is a theoretical circle on which the center points of holes lie in a circular pattern of holes. The holes in the pattern are equally spaced, with equal angles between holes. Common examples are the bolt circle for the lugs on a car's wheel and the holes on a bolted joint between pipe flanges. **Figure 1-28** illustrates a circular pattern of six holes on an 8″ diameter bolt circle. Notice that each hole is positioned at the same distance from the theoretical arc center and each hole is 60° apart. A set of six holes equally spaced in a 360° circular pattern creates 60° spacing.

To create a CNC program, you will need to calculate each hole position from the center of the theoretical arc. In the example shown in **Figure 1-28**, the theoretical arc is defined by an 8″ diameter bolt circle. Each hole position can be determined by making right triangle calculations. First, identify the right triangles that can be created for use in calculations. **Figure 1-29** shows the triangle for one of the hole positions.

It is mathematically possible to calculate the position, or two unknown sides, for each triangle using the trigonometry formulas. However, there is an easier way to make the calculations using a calculator. As you will learn in Chapter 2, the calculations can be used to plot X and Y coordinates for the hole locations. Chapter 2 explains how to use coordinates in the Cartesian coordinate system. For now, making calculations is sufficient. **Figure 1-29** shows the X and Y coordinate distances corresponding to the sides of the right triangle.

Chapter 1 Machining Mathematics for Milling 17

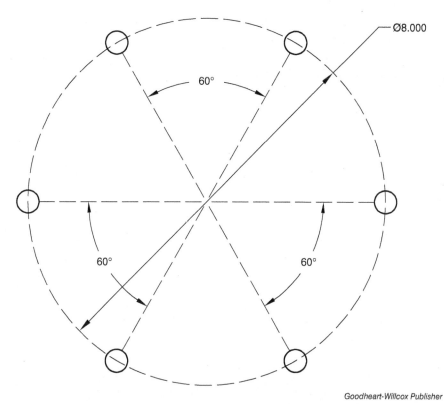

Figure 1-28. An 8″ diameter bolt circle locating six holes in a 360° circular pattern. The holes are the same distance from the theoretical arc center and equally spaced. This creates 60° separation between holes.

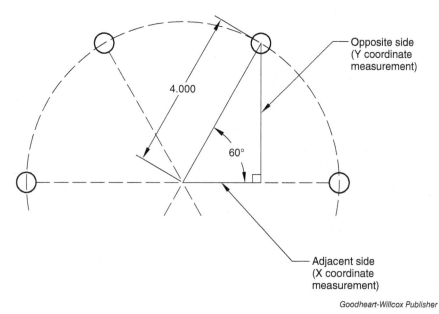

Figure 1-29. A right triangle used to calculate the position of a hole in a circular hole pattern.

There are two numbers that must be known before you start your calculations. The first number is the radius of the theoretical arc the holes are centered on. In this example, the radius is 4″, or half the diameter of the 8″ bolt circle. The second number is the angular position of each hole that is being calculated.

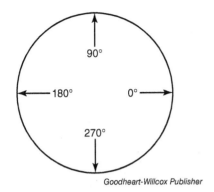

Figure 1-30. Angle positions on a circle with 0° oriented at 3 o'clock.

There is a standard in CNC programming that also applies for these calculations. As previously discussed, there are 360° in a circle. **Figure 1-30** shows the 0°, 90°, 180°, and 270° angle positions on a circle in a default orientation. The zero angle position (0°) is oriented at 3 o'clock. The angle position rotates counterclockwise, with 12 o'clock representing 90°, 9 o'clock representing 180°, and 6 o'clock representing 270°. With this orientation, the following bolt circle formulas can be utilized.

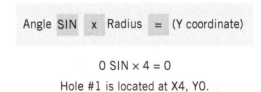

With this information, you can now calculate the X and Y coordinate positions of the holes. See **Figure 1-31**. The following calculations are for the holes identified as Hole #1 and Hole #2. These coordinate calculations are based on measuring from the center of the bolt circle. The radius is 4″. The angular position of Hole #1 is 0° and the angular position of Hole #2 is 60°.

Hole #1 X coordinate:

Angle COS x Radius = (X coordinate)

0 COS × 4 = 4

Hole #1 Y coordinate:

Angle SIN x Radius = (Y coordinate)

0 SIN × 4 = 0

Hole #1 is located at X4, Y0.

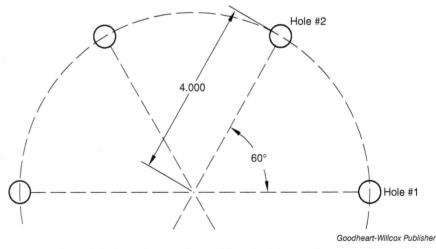

Figure 1-31. Bolt circle formulas can be used to calculate coordinates for the hole centers. The coordinate position of Hole #1 is located at 0° and Hole #2 is located at 60°.

Hole #2 X coordinate:

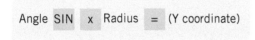

$$60 \cos \times 4 = 2$$

Hole #2 Y coordinate:

$$60 \sin \times 4 = 3.464$$

Hole #2 is located at X2, Y3.464.

1.7 Milling Speeds and Feeds

When cutting material on a CNC milling machine, the relationship between how fast the spindle turns and how fast the cutting tool moves across the material is vitally important. Machine shops try to get parts cut as quickly as possible. In addition, tooling reacts differently at different speeds and feeds. For example, if you turn an end mill very slowly and feed it into the material rapidly, the tool may suffer a catastrophic failure and shatter. Conversely, if you run an end mill at a high speed but feed it slowly, it takes excessive time to cut the material and can dull the cutting edges of the tool.

Calculating speeds and feeds in the modern machining environment is one of the most overlooked and miscalculated aspects of machining. Fortunately, there is a relationship between the material, the cutting tool, and the machine that can be easily calculated. Every CNC programmer should be able to accurately calculate a tool's proper speed and feed.

The first factor in calculating speeds and feeds is the cutting tool. The machining industry today primarily uses tools made from solid carbide, so this is the material assumed in this text. When working with other cutter materials, consult manufacturer specifications. Milling calculations also depend on the diameter of the cutter and how many cutting edges it has.

The next important factor is the type of material being cut. Stainless steels do not cut as easily as aluminum, for example. Different materials have different rates of material removal and require careful consideration of cutting speeds. Cutting speed is a measure of the tool's movement in feet per minute. It is expressed in surface feet per minute (sfm) and is commonly referred to as surface footage. In simple terms, surface footage refers to the number of linear feet a point on a rotating tool travels in one minute. Tooling manufacturers provide charts with recommended cutting speeds for common materials. On a chart, recommended speeds are expressed in ranges of values. These values are obtained through empirical testing. Cutting speeds can vary based on the tooling manufacturer, so it is best to get the range of recommended values directly from the manufacturer. See the Reference Section in this text for information on recommended cutting speeds for common materials.

Surface footage is used in calculating the spindle speed. The rotating speed of the spindle is measured in revolutions per minute (rpm).

From the Programmer

Speeds and Feeds

Calculating the correct spindle speeds and cutting feeds is the best way to make your tools last as long as possible. Too slow of feed means the tool is cutting for longer amounts of time to do the same amount of work. Too fast of feed means the tool does not have enough time to cut, and can fail on the cutting edge. Often the best resource for speed and feed settings can be the tool distributor or a tooling engineer. Always calculate speeds and feeds to remain as efficient as possible.

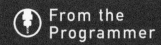

From the Programmer

Calculating Spindle Speeds

The mathematical constant 3.82 is used in the calculation for spindle speed. This value is the result of dividing the number of inches per foot by pi: 12/3.14 = 3.82.

The following examples are used to calculate the speed of two different end mills with two different types of materials. Consider a 1/2″ end mill cutting aluminum at 700 sfm and a 3/8″ end mill cutting 304 stainless steel at 450 sfm. The formula for calculating spindle speed, measured in rpm, is given below:

$$\text{rpm} = \frac{3.82 \times \text{sfm}}{\text{Diameter of tool (in inches)}}$$

Calculate the speed for the 1/2″ (.500″) end mill for cutting aluminum.

$$\frac{3.82 \times 700}{.500} = 5348 \text{ rpm}$$

The speed for this operation is 5348 rpm. Remember this number for use in calculating the feed.

Now, calculate the speed for the 3/8″ end mill for cutting stainless steel.

$$\frac{3.82 \times 450}{.375} = 4584 \text{ rpm}$$

Now that you know the speed, you can calculate feed rate. The formula for feed rate, measured in inches per minute (ipm), is given below:

Feed Rate (ipm) = rpm × Chip Load × Number of Cutting Edges (Flutes)

Feed rate depends on chip load and the number of cutting edges as well as spindle speed. *Chip load* refers to the actual thickness of the chip being cut or the depth of each cutting edge as it passes through the material. Chip load is measured in inches per tooth and varies with each cutter, but will generally range from .005″ to .010″. Any end mill, whether solid carbide or inserted, will have multiple cutting edges, or flutes. A *flute* is the recessed groove along a cutting edge allowing for chip removal. Tools with more flutes can feed faster, but have less clearance space for chip removal and a smaller internal web, making them weaker in heavy cuts.

Return to the previous example and calculate the feed rate for each cutter. The 1/2″ end mill used for cutting aluminum has a speed of 5348 rpm and three flutes. Calculate the feed rate as follows:

5348 rpm × .007 × 3 = 112.3 ipm

Now calculate the feed rate for the 3/8″ end mill used for cutting stainless steel. This cutter has two flutes.

4584 rpm × .007 × 2 = 64.2 ipm

The feed rate for this operation is 64.2 ipm at a speed of 4584 rpm.

Chapter 1 Review

Summary

- A good working knowledge of math principles is needed to create CNC milling programs.
- A scientific calculator is a useful tool for making machining calculations.
- To convert a fraction to decimal format, divide the denominator into the numerator.
- To ensure a proper level of accuracy and ease of language, machinists express decimals in the thousandths and ten thousandths place.
- When evaluating a print, machinists identify basic geometric shapes and evaluate part geometry to determine dimensional information.
- Four common types of triangles encountered in machining work are scalene, isosceles, equilateral, and right triangles.
- The Pythagorean theorem states that, for a right triangle, the square of one side plus the square of the second side is equal to the square of the hypotenuse.
- Right triangles can be solved for all sides and all angles using standard formulas.
- Supplementary angles are two angles that add up to 180°. Complementary angles are two angles that add up to 90°.
- The hypotenuse is the longest side of a right triangle and is always across from the 90° angle.
- Basic trigonometry is used to solve for sides and angles in a right triangle.
- Cutting speed is a measure of a tool's movement in feet per minute and is expressed in surface feet per minute (sfm). Surface footage is used in calculating the spindle speed, which is measured in revolutions per minute (rpm).
- Feed rate is calculated based on the spindle speed, the chip load, and the number of cutting edges on the tool.

Review Questions

Answer the following questions using the information provided in this chapter.

Know and Understand

1. Most prints used in manufacturing today are dimensioned in ____.
   A. fractional inches
   B. decimal inches
   C. millimeters
   D. centimeters

2. The position three places to the right of the decimal point is called the ____ place.
   A. hundredths
   B. tenths
   C. ten thousandths
   D. thousandths

3. *True or False?* A circle is a closed plane curve that is an equal distance from its center point to all points.

4. The distance across a circle through the center point is the ____.
   A. length
   B. width
   C. diameter
   D. radius

5. Examples of ____ include triangles, rectangles, pentagons, and hexagons.
   A. pyramids
   B. polygons
   C. rhombuses
   D. trapezoids

6. The ____ triangle has no equal sides or equal angles.
   A. isosceles
   B. equilateral
   C. obtuse
   D. scalene

7. A right triangle must contain one ____ angle.
   A. 30°
   B. 45°
   C. 60°
   D. 90°

8. *True or False?* The sum of all angles in any triangle is 90°.

9. The Pythagorean theorem is written as ____.
   A. $a^2 \times b^2 = c^2$
   B. $a^2 + b^2 = c^2$
   C. $a^2 - b^2 = c^2$
   D. $b^2 - a^2 = c^2$

10. An acute angle is defined as ____.
    A. greater than 180°
    B. equal to 90°
    C. greater than 0°
    D. less than 90°

11. *True or False?* Two angles are complementary if they add up to 180°.

12. ____ is defined as a branch of mathematics that deals with the relationships between the sides and angles of triangles and with functions of angles.
    A. Solid modeling
    B. Plane geometry
    C. Pythagorean theory
    D. Trigonometry

13. The longest side in a right triangle, or the side directly across from the right angle, is called the ____.
    A. apex
    B. adjacent side
    C. hypotenuse
    D. opposite side

14. *True or False?* Besides the hypotenuse, the other two sides in a right triangle are called the opposite and the adjacent.

15. ____ is a measure of a tool's movement in feet per minute and is expressed in surface feet per minute (sfm).
    A. Chip load
    B. Feed rate
    C. Spindle speed
    D. Surface footage

3. Identify the following geometric shape. Assume all sides are equal.

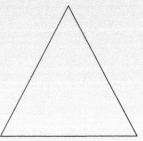

4. Identify the following geometric shape. Assume all sides are equal.

5. List and briefly describe the four types of triangles discussed in this chapter.

6. Use the Pythagorean theorem to solve for the unknown side of this right triangle.

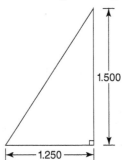

7. Use the Pythagorean theorem to solve for the unknown side of this right triangle.

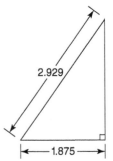

Apply and Analyze

1. Convert 3/32 to its decimal equivalent. Round to the nearest thousandths.

2. Convert 1 3/16 to its decimal equivalent. Round to the nearest thousandths.

8. Identify the unnamed side of this right triangle.

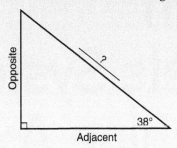

9. Label all of the sides of this right triangle.

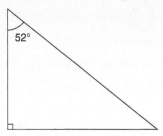

10. Use a calculator to solve for the missing angle of this right triangle.

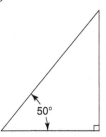

11. Use a calculator to solve for the missing side of this right triangle. Give answer to the nearest thousandths.

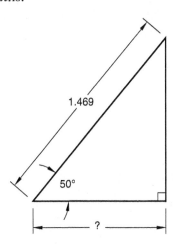

12. Use a calculator to solve for the two missing angles of this right triangle. Round answers to the nearest whole angle.

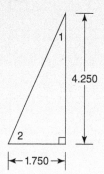

13. Using a 5/8" diameter end mill, what is the correct speed, in rpm, for cutting aluminum at 750 surface feet per minute (sfm)?

14. When cutting mild steel in a milling machine with a calculated speed of 3726 rpm, what is the feed rate in inches per minute (ipm)? Assume .005" chip load with a 3-flute end mill.

Critical Thinking

1. Take a look around your home, school, or workplace and identify various geometric shapes. Can you measure and calculate any irregularities in these shapes? How do these irregularities affect the construction of these shapes?

2. Consider how right triangles are used in everyday life. Consider the construction of your home, the school building, or other built structures, such as a storage shed. Determine the approximate angle of a roof or calculate the length of a diagonal between two walls that meet at 90°.

3. Depending on metal composition and cutting tools used, recommended cutting speeds can vary. How does this impact your machining time? Why does this make it so important to analyze materials and cutting tools?

2 Cartesian Coordinate System and Machine Axes for Milling

Chapter Outline

2.1 Introduction
2.2 Number Line
2.3 Two-Dimensional Coordinate System
2.4 Three-Dimensional Coordinate System
2.5 Absolute and Incremental Positioning
2.6 Machine Home and Work Origin
2.7 Four-Axis Machines
2.8 Five-Axis Machines
2.9 Polar Coordinates

Learning Objectives

After completing this chapter, you will be able to:

- Describe how the Cartesian coordinate system is applied in CNC milling.
- Identify the four quadrants in a two-dimensional coordinate system.
- Plot X and Y coordinate positions in a two-dimensional coordinate system.
- Plot coordinates in three-dimensional space using the X, Y, and Z axes.
- Explain the relationship between machine home and the work origin.
- Describe the machining capabilities and configurations of four-axis milling machines.
- Describe the machining capabilities and configurations of five-axis milling machines.
- Identify the axes defining rotational movement on four-axis and five-axis milling machines.
- Use polar coordinates to plot points defining angular relationships between features.

Key Terms

3 + 1 milling
absolute positioning
Cartesian coordinate system
full four-axis milling

incremental positioning
machine home
origin
polar coordinates

rotary
trunnion
work envelope
work origin

Chapter opening photo credit: teh_z1b/Shutterstock.com

arsa35/Shutterstock.com

Figure 2-1. The organization of rows and columns in a vending machine is a common example of the Cartesian coordinate system.

2.1 Introduction

The basic principle of creating any CNC program is positioning a tool at a known location and moving that tool to a series of defined locations. The system used to plot and calculate machine positions and movement is known as the Cartesian coordinate system. The *Cartesian coordinate system* specifies each point uniquely in a plane with a pair of alphanumeric coordinates. This system is named for its developer, French scientist and mathematician René Descartes. An influential philosopher, Descartes is considered the father of analytic geometry. His developments led to the coordinate plotting system that is widely used today.

An example of the Cartesian coordinate system in use is a vending machine. See **Figure 2-1**. When you purchase a snack or beverage from a vending machine, you select the row of the item you want and then the number in that row. Specifying two points of intersection tells the machine where to locate your item.

2.2 Number Line

To understand how points are located in the Cartesian coordinate system, look at the simple number line shown in **Figure 2-2**. A number line is a straight line with numbered divisions spaced equally along the line. The zero mark in the middle of the line establishes the starting point and is known as the *origin*. The numbered divisions along the line are located in relationship to the origin. The numbers can have any defined value and unit format. For example, each number can represent one inch, one foot, or one mile. In this example, the numbers represent whole number inches.

The numbers to the right of the origin are positive numbers. The numbers to the left of the origin are negative numbers. Movement in the right direction is positive and movement in the left direction is negative. The numbers define the amount of linear movement in a given direction. For example, moving five units to the right of the origin represents a movement of 5″ in the positive direction. Moving three units to the left of the origin represents a movement of 3″ in the negative direction. These distances are represented by Point 1 and Point 2 in **Figure 2-2**. Specifying distances in relation to a known origin is a basic way to describe tool movement.

The horizontal number line shown in **Figure 2-2** establishes a single axis of movement. In CNC programming, additional axes are used to define the direction of the cutting tool.

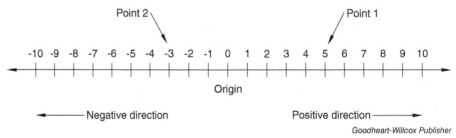

Goodheart-Willcox Publisher

Figure 2-2. A number line measures distance and direction. Point 1 represents a movement of 5″ from the origin in the positive direction. Point 2 represents a movement of 3″ from the origin in the negative direction.

2.3 Two-Dimensional Coordinate System

A number line represents a single axis of travel. However, all CNC machines have at least two axes of travel. Adding a second number line establishes the second coordinate axis in the Cartesian coordinate system. See **Figure 2-3**. In this system, the horizontal axis is designated as the X axis and the vertical axis is designated as the Y axis. The X and Y axes intersect at the origin and are perpendicular to each other (oriented at 90°). Points representing distances from the origin are specified with X and Y coordinates. The coordinate system origin is designated as X0, Y0 and can be located anywhere in space. This system allows the programmer to define the location of the workpiece and the direction of tool movement.

As shown in **Figure 2-3**, the coordinate system is divided into four quadrants. Coordinates are specified as positive or negative based on their location from the origin. A coordinate to the right of the origin has a positive X value and a coordinate to the left of the origin has a negative X value. In **Figure 2-3**, notice that each quadrant has positive or negative X and Y values based on its location in relation to the origin. The quadrants are numbered 1–4 in a counterclockwise direction. Coordinates in the upper-right quadrant have a positive X value and positive Y value.

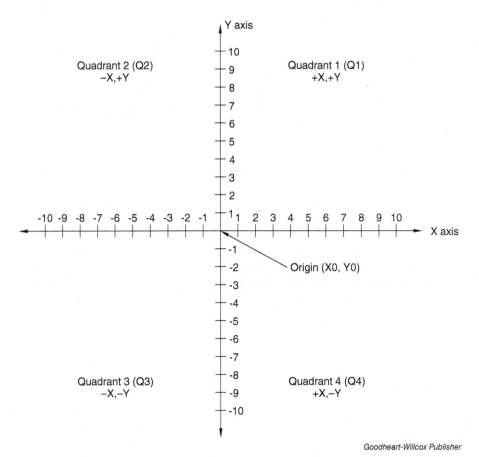

Goodheart-Willcox Publisher

Figure 2-3. A two-dimensional coordinate system has two axes designated as the X axis and Y axis. The axes intersect at the origin and divide the system into four quadrants. Points are located in relation to the origin and have positive or negative values based on their location from the origin.

Coordinates in the upper-left quadrant have a negative X value and positive Y value.

The X and Y axes establish a theoretical flat plane known as the XY plane. On a milling machine, if the XY plane is considered to be parallel to the machine table, the X axis represents the horizontal movement of the machine table from side to side. The Y axis represents the movement of the table from front to back. As discussed in the next section, a third axis, the Z axis, represents the movement of the spindle. The Z axis is used in a three-dimensional coordinate system.

To define movement in a CNC machine, a series of coordinate points is communicated to the machine through a program. The points are used by the machine to travel in a straight line from point to point. **Figure 2-4** shows a series of points specifying the exact path of travel. Each point has an X,Y coordinate measuring its distance from the origin. Point 1 is 6″ from the origin in the positive X axis direction and 3″ from the origin in the positive Y axis direction. This coordinate is expressed as X6, Y3. Point 2 is 7″ from the origin in the negative X direction and 6″ from the origin in the positive Y direction. This coordinate is expressed as X–7, Y6. To create movement from Point 1 to Point 2, then to Point 3, and then to Point 4, the CNC program is written as follows:

> X6. Y3. (Point 1)
> X–7. Y6. (Point 2)
> X–4. Y–3. (Point 3)
> X5. Y–7. (Point 4)

 Debugging the Code

Coordinate Format

As shown in **Figure 2-4**, Cartesian coordinates are commonly written in the format X,Y. However, in CNC programming, this format is not used. Entries for XYZ coordinates are expressed as decimals. For example, in the CNC program, the coordinate entry X6. Y3. represents Point 1 in **Figure 2-4**. Notice that this format does not require a trailing zero in the decimal. For example, the coordinate X6.0 is shortened to X6. Similarly, a leading zero is not required for decimal values below zero. For example, the coordinate Z–.5 does not require a leading zero. Older controllers did not accept the decimal point programming format and entries were read in ten thousandths of an inch. For example, the entry X6 represented .0006″ and the entry X60000 represented 6″. Machines today use decimal point programming. This is the format used throughout this textbook.

From the Programmer

Coordinate Entry Format

The X axis coordinate value does not need to be first in the program, but the X and Y movement to each position must be on the same line to create a direct move to the next position when movement is along both axes. Specifying the X value first, followed by the Y value, is common practice in programming. This is the format used in this textbook.

2.4 Three-Dimensional Coordinate System

The Cartesian coordinate system is not limited to two-dimensional space. A third axis can be designated to show height or depth in three-dimensional space. The third axis is designated as the Z axis. **Figure 2-5** shows a three-dimensional coordinate system. In this system, there are three axes of positional movement and a third coordinate is specified for the position along the Z axis. Points representing distances from the origin are specified

Chapter 2 Cartesian Coordinate System and Machine Axes for Milling 29

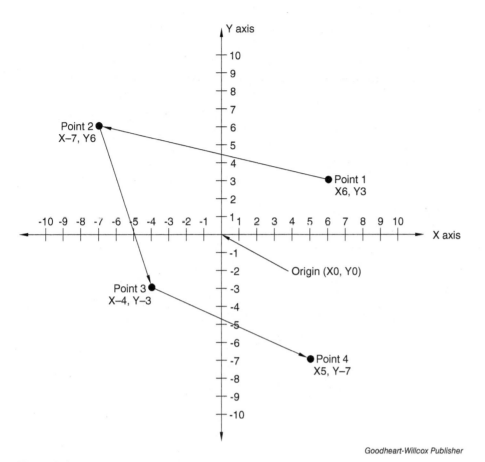

Figure 2-4. Plotted coordinates representing the path of travel in a CNC program.

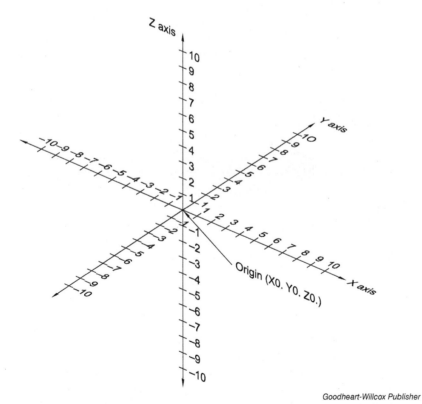

Figure 2-5. A third axis, designated as the Z axis, is used in the three-dimensional coordinate system.

> **Safety Note**
>
> The importance of defining the correct coordinates cannot be expressed enough. A missed minus sign or decimal point can cause a catastrophic machine failure or part damage. Be careful plotting the points and entering them into the CNC program.

with X, Y, and Z coordinates. On a milling machine, the Z axis represents the movement of the spindle toward or away from the workpiece.

To put this additional axis in perspective, consider that the Z origin, or zero location, is at a top corner of the part to be machined. Anything below this position is in the negative Z direction, and anything above this position is in the positive Z direction. The Z axis allows for coordinate positioning of different heights of material in three-dimensional space.

Figure 2-6 shows a block of material that is 4″ wide, 2″ long, and 1″ in height. Based on the position of the block in relation to the origin, coordinates for the three points on the block are plotted as follows:

Point 1 = X–4. Y0. Z0.
Point 2 = X–4. Y–2. Z–1.
Point 3 = X0. Y–2. Z0.

If the block is repositioned in the coordinate system, the X, Y, and Z positions change. In **Figure 2-7**, the size of the block is the same as in the previous example, but the relationship to the origin has changed. The same three points on the block are plotted as follows:

Point 1 = X–2. Y2. Z1.
Point 2 = X–2. Y0. Z0.
Point 3 = X2. Y0. Z1.

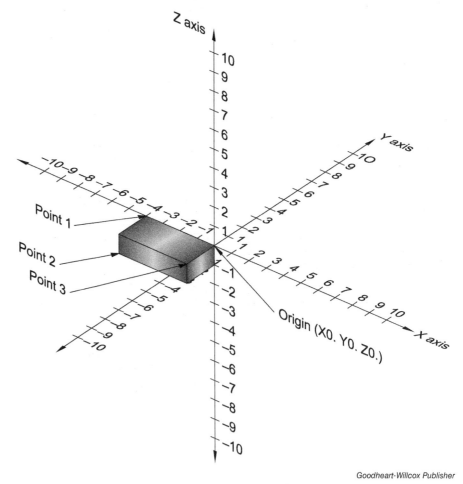

Goodheart-Willcox Publisher

Figure 2-6. In the three-dimensional coordinate system, point locations are specified with X, Y, and Z coordinates in relation to the origin.

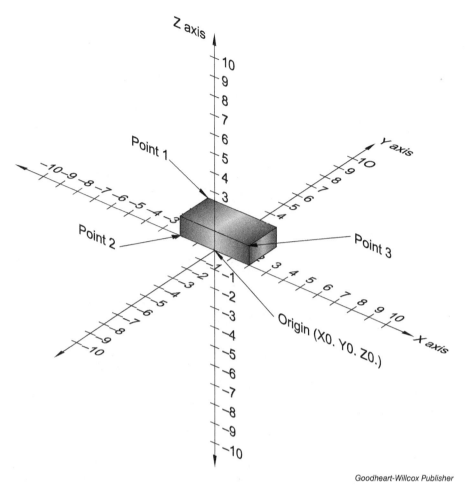

Goodheart-Willcox Publisher

Figure 2-7. The block of material has different X, Y, and Z coordinates after moving it to a new position.

2.5 Absolute and Incremental Positioning

The coordinates specified for point locations in the previous examples are based on absolute coordinate entry. Absolute coordinates are measured in relation to the origin (X0, Y0, Z0). In a CNC program, moves specified with absolute coordinates are referred to as ***absolute positioning*** moves, meaning that they represent absolute measurements in reference to the origin. Absolute positioning is the appropriate method of movement for most purposes. A second way to specify movement is to use incremental positioning. ***Incremental positioning*** refers to measuring from the current location to a second location. Incremental coordinates represent a relative measurement from the current position of the tool, not from the origin. Incremental coordinates are also called *relative coordinates* because they are located relative to the previous coordinate. See **Figure 2-8**. The example shows the same path of travel defined in **Figure 2-4**, but with coordinates specified using incremental positioning. Compare the two location methods. Referring to the incremental coordinate for Point 1, the previous location is considered to be the origin. Note the positive and negative values corresponding to the entries for Points 2, 3, and 4. Point 2 is 13″ from Point 1 in the negative X direction and 3″ from Point 1 in the positive Y direction. Coordinates must be entered carefully in the program when using incremental positioning because if an incorrect position is specified, each successive move will be incorrect.

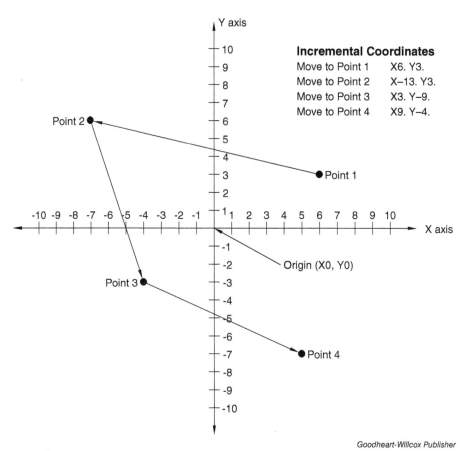

Figure 2-8. Coordinates defining a path of travel using incremental positioning. Incremental coordinates are measured relative to the previous coordinate location.

> **From the Programmer**
>
> **Positioning Mode**
>
> Although it is possible, and sometimes preferred, to program in incremental moves, most CNC programs are written using absolute positioning. An origin, or zero point, is established, and all machine or tool movements are made relative to that zero point. Making sure you know where the machine is located and where it is headed is critical in CNC programming.

Incremental positioning is sometimes used when a print uses a chain of dimensions to define linear distances between features. In addition, incremental positioning can be used with special programming methods to create arcs. Chapter 3 covers the programming commands used to establish absolute and incremental movement.

2.6 Machine Home and Work Origin

In a CNC milling operation, the machine controller must know where the material, or part, is located. This is achieved by measuring the distance from the machine's home position to the part origin location. Every milling machine has a fixed location called *machine home*. This position varies based on the machine type and manufacturer and can be anywhere within the *work envelope* of the machine. On a three-axis milling machine, this position is most often located at the limit of the positive X axis travel, the positive Y axis travel, and the positive Z axis travel. This would position the machine table to the far left and completely forward with the spindle at its highest position. A typical three-axis milling machine home position is shown in **Figure 2-9**.

In a part setup, the part material is secured, and the distance from machine home to the *work origin* is then measured. The work origin represents the X0. Y0. location on the part and is positioned at a suitable point based on the part geometry and dimensions. The distance from machine

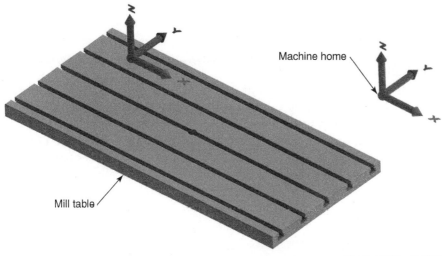

Figure 2-9. Machine home is a fixed position usually located at the limits of machine travel along the positive X axis, positive Y axis, and positive Z axis.

home to the work origin is entered into the controller as the work coordinate system origin. The CNC program is written using this same origin position. **Figure 2-10** shows a typical machine setup with a vise used to hold a piece of material to be machined.

2.7 Four-Axis Machines

A three-axis milling machine has three axes of movement designated as the X, Y, and Z axes. A four-axis milling machine adds another axis for use in CNC milling. A *rotary* is a rotating unit that establishes a fourth axis and

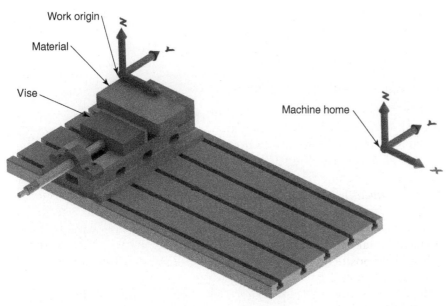

Figure 2-10. A workpiece secured to a milling table using a vise. The work origin defines the X0, Y0 location on the part and is located in relation to the machine home position.

allows for rotation of the workpiece. In a program, the amount of rotational movement is expressed in degrees.

There are a number of types of rotary devices in the manufacturing industry. A rotary can be integrated directly into the CNC mill, or it can be a stand-alone unit that can be removed from the machine. A rotary table is shown in **Figure 2-11**. This unit can stand up vertically or be mounted horizontally on the milling table. The T-slots on the circumference of the table are used for mounting work.

Another common type of rotary is a collet indexer, **Figure 2-12**. The collet allows direct holding of round stock on the outside of the material. An indexer can be mounted horizontally or vertically in most cases, depending on the machining required.

Haas Automation Inc.

Figure 2-11. A platter-style rotary table provides a fourth axis for machining.

Goodheart-Willcox Publisher

Figure 2-12. A collet indexer mounted on a milling machine.

On a milling machine, an axis of rotation parallel to the X axis is commonly designated as the A axis. Rotation about the A axis is in a positive or negative direction. The point of rotation is through the center point of the spindle on the rotary device. The positive or negative rotational direction is based on the mounting orientation of the device. When looking along the positive X axis, positive rotation is clockwise. **Figure 2-13** shows a platter-style rotary table and its direction of rotation.

The process of positioning a rotary axis at a fixed angular position and completing three-axis milling operations is known as *3 + 1 milling*. In a typical 3 + 1 milling operation, the rotary device is programmed to rotate the part to a fixed position. Then, the cutting tool is engaged to machine the part. Additional sides of the part can be machined in the same manner by rotating the work to different positions. In this type of machining, the axis of rotation is fixed while movement of the cutting tool occurs along the X, Y, and Z axes. In *full four-axis milling*, the rotary axis is in motion at the same time tool movement occurs along the X, Y, and Z axes. Continuous linear and rotational movement allows for machining more complex parts in a single setup.

Figure 2-14 shows a cylindrical part held in place by a collet rotary device and the rotational direction around the A axis. This part has a slot machined on the outer surface. Simultaneous linear and rotary motion is used in machining the slot because it is rotated 10° from the center axis of the part. This work setup is an example of full four-axis milling. The setup permits synchronous rotation of the part and linear movement of the cutting tool.

When programming a rotary movement, an address code is used to specify the machine axis. Rotation about the A axis is defined with the A letter code. The rotational movement can be specified individually or in combination with other axis movements. For example:

- **A90.** Rotates the A axis in a positive 90° rotation.
- **A–60.** Rotates the A axis in a negative 60° rotation.
- **A40. Y1. X2.** Rotates the A axis in a positive 40° rotation while simultaneously moving in the X and Y axis directions.

Goodheart-Willcox Publisher

Figure 2-13. A rotary table mounted to a milling table permits rotation of the workpiece. The axis of rotation is designated as the A axis.

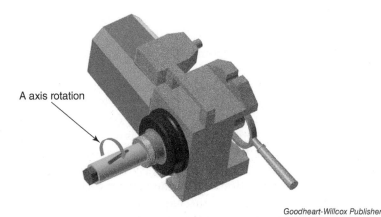

Goodheart-Willcox Publisher

Figure 2-14. A part requiring rotational slot machining.

2.8 Five-Axis Machines

A five-axis milling machine has five axes of movement. Multiaxis machines can have different configurations, but five-axis milling usually refers to machines with three axes of linear movement and two axes of rotary movement. The five-axis milling machine is rapidly becoming the most utilized tool in the modern machine shop. It provides versatility and speed and has the ability to machine complex parts with fewer operations, thus reducing costs. It also adds complexity to programming and setup.

There are a number of ways a five-axis machine can be configured. The rotary axes can be fully integrated into the machine, or they can be part of an add-on device that is removed when not in use. **Figure 2-15** shows a small add-on unit that saves space and can be easily removed when not needed. This is a side-mounted, two-axis rotary unit with a tilting rotary table.

Other machines can be factory built with multiaxis capability. A five-axis CNC milling machine is shown in **Figure 2-16**.

In five-axis milling, the two axes defining rotational movement are typically designated as the A axis and B axis. However, whether an axis is

Haas Automation Inc.

Figure 2-15. An add-on rotary unit mounted to a milling machine. The unit provides two rotary axes for five-axis milling.

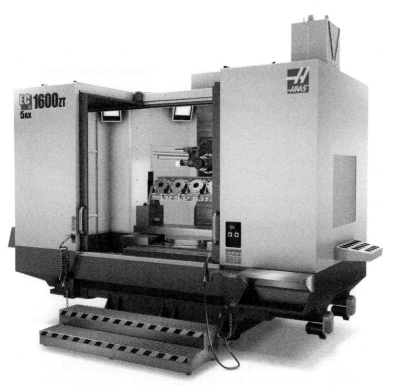

Haas Automation Inc.

Figure 2-16. A five-axis machining center.

designated as A or B may depend on the setup or factory controller configuration. For specific details on the machine configuration, refer to the operator's manual.

Figure 2-17 shows an add-on unit used in a five-axis milling configuration. This unit is a side-mounted rotary with a tilting rotary table. This configuration allows the workpiece to be rotated to different positions along two axes and reduces the amount of setups required.

Figure 2-18 shows another add-on rotary unit called a *trunnion*. A trunnion is a rotary table that allows for rotation in one or two axis directions. The trunnion in **Figure 2-18** allows for rotation around the A axis and B axis. This unit contains a platter for mounting work. The saddle

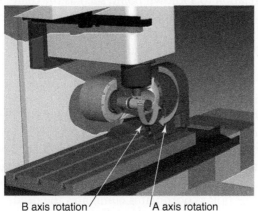

Goodheart-Willcox Publisher

Figure 2-17. A compact rotary unit mounted on a milling machine for five-axis machining. The unit is light and easy to install and remove as required.

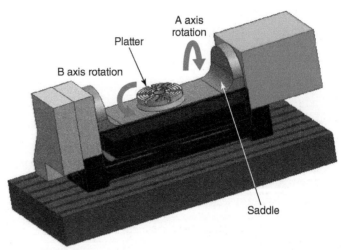

Figure 2-18. A trunnion table with two rotary axes mounted to a conventional machining table.

rotates from a horizontal to a vertical position around the A axis and the platter rotates independently around the B axis.

In a program, address codes are used to specify motion for each machine axis. Each of the five axes can be programmed for individual movement or combined movement with other axes. Rotational movement is expressed in degrees. For example:

- **A90. B45.** Rotates the A axis and B axis simultaneously.
- **B45. X.5.** Rotates the B axis 45° while moving .5″ in the X axis direction.
- **A10. B−20. X.2 Y−.5 Z−.25.** Moves all axes synchronously.

2.9 Polar Coordinates

Some prints may specify part dimensions based on angular relationships between features. For example, the center points of holes in a bolt circle are defined by a radius and an angle. Coordinate locations for hole centers in circular patterns are commonly defined with polar coordinates. *Polar coordinates* are coordinates that reference a linear distance and an angle. Polar coordinates are specified in relation to a known point, such as the center point of a circle.

Figure 2-19 shows an example of polar coordinate dimensioning used to locate a hole. A linear dimension specifies the distance from the first hole to the second hole and an angular dimension specifies the polar angle from the zero angle position. As discussed in Chapter 1, the zero angle position is oriented at the 3 o'clock position. Angles are measured in a counterclockwise direction from this position.

Figure 2-20 shows an example of a print where a 1″ diameter bolt circle is defined for four holes in a circular pattern. In a program, polar coordinates are used to specify the hole locations for drilling. The center point of the pattern is used as the origin for coordinates. The holes are located by specifying a radius value and an angular position relative to the center of the pattern. The use of polar coordinates eliminates the need to mathematically calculate each hole position, thus simplifying the programming process and alleviating the possibility of calculation errors. Using Cartesian coordinates would require manual calculations to be made with trigonometric functions.

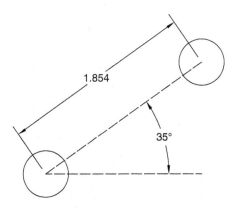

Goodheart-Willcox Publisher

Figure 2-19. Polar coordinates define the location of a feature by specifying a linear distance and angular direction.

In this example, the work coordinate system origin is located at the top-left corner of the workpiece. But the center of the bolt circle is located at X2, Y–2.5. By programming that position and using specialized codes for polar coordinate programming, the four holes in the pattern can be drilled by giving the radius and angular position.

Different types of controllers use different codes for polar coordinate programming. Although bolt circles are a common use for polar coordinate positioning, there are other instances where this technique could prove useful. Polar coordinate programming and other methods used for machining holes in milling operations are covered in Chapter 10.

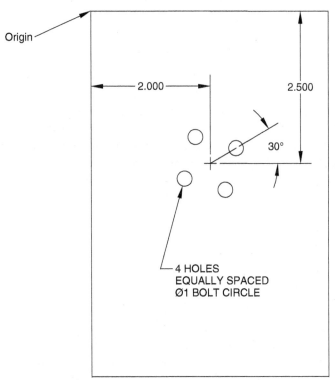

Goodheart-Willcox Publisher

Figure 2-20. A part with a circular pattern of holes located on a 1″ diameter bolt circle. Polar coordinates are used in programming to machine the holes.

Chapter 2 Review

Summary

- The Cartesian coordinate system is used to locate points representing X and Y values relative to the origin.
- A two-dimensional coordinate system is made up of X and Y axes and is divided into four quadrants.
- A three-dimensional coordinate system has a third axis designated as the Z axis.
- Absolute positioning refers to specifying moves with absolute coordinates measured in relation to the origin.
- Incremental positioning refers to specifying moves with coordinates measured in relation to the current position.
- Machine home is a fixed location on a milling machine that usually represents the limit of the positive X axis travel, the positive Y axis travel, and the positive Z axis travel.
- The work origin represents the X0. Y0. location on the part and is measured relative to the machine home position.
- A rotary is a unit that establishes a fourth axis on a milling machine and allows for rotation of the workpiece.
- On a milling machine, an axis of rotation parallel to the X axis is commonly designated as the A axis.
- In full four-axis milling, the rotary axis is in motion at the same time tool movement occurs along the X, Y, and Z axes.
- In five-axis milling, the two rotary axes are in motion while tool movement along the X, Y, and Z axes occurs simultaneously.
- Polar coordinates are commonly used to define coordinate locations for hole centers in circular patterns.

Review Questions

Answer the following questions using the information provided in this chapter.

Know and Understand

1. The Cartesian coordinate system was developed by ____.
   A. Ray Cartesian
   B. Carte Desysteme
   C. René Descartes
   D. Remy Descarme

2. On a number line, the starting point is known as ____.
   A. the origin
   B. point zero
   C. coordinate home
   D. machine home

3. *True or False?* The most common method used to program a CNC machine is incremental positioning.

4. A two-dimensional coordinate system is made up of the ____ axes.
   A. X and Z
   B. A and Z
   C. A and B
   D. X and Y

5. In the figure below, in which quadrant is the solid box located?

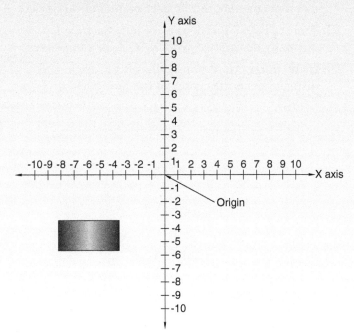

A. Quadrant 1
B. Quadrant 2
C. Quadrant 3
D. Quadrant 4

6. In the figure below, what is the coordinate for the center point of the solid circle?

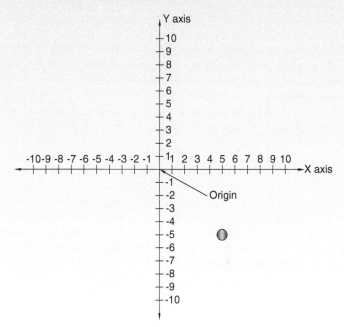

A. X5. Y–5.
B. X–5. Y–5.
C. X5. Y5.
D. X–5. Y5.

7. On a milling machine, which axis represents the movement of the spindle in relationship to the workpiece?

A. A axis
B. B axis
C. Y axis
D. Z axis

8. In the figure below, what is the coordinate of Point 1 on the block of material?

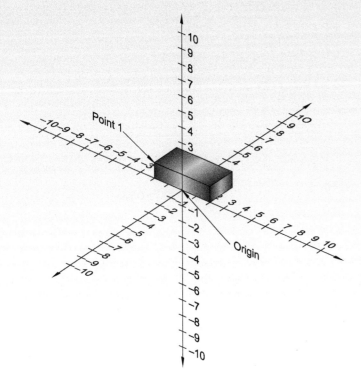

A. X0. Y2. Z1.
B. X–2. Y0. Z1.
C. X2. Y2. Z0.
D. X–2. Y0. Z0.

9. *True or False?* A rotary unit on a four-axis milling machine establishes an axis of rotation.

10. On a milling machine, the axis that establishes rotation about the X axis is commonly designated as the _____ axis.

   A. B axis
   B. Z axis
   C. A axis
   D. X axis

11. *True or False?* The rotary axes on a five-axis milling machine can be fully integrated into the machine, or they can be part of an add-on device.

12. *True or False?* A five-axis milling machine is normally designated with X, Y, Z, B, and C axes.

13. Using polar coordinates, what is the angle and distance from the center of the bolt circle to Point 1?

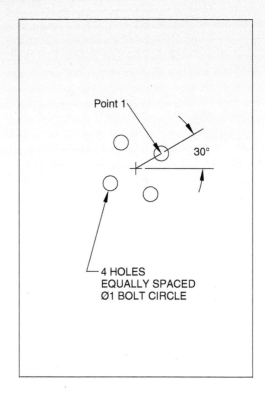

A. 30° and 0″
B. 30° and .5″
C. 30° and 1″
D. 60° and 1″

Apply and Analyze

1. What is the purpose of using the Cartesian coordinate system in CNC programming?
2. What are the three designated axes of movement on a three-axis milling machine?
3. Explain the additional benefit provided by a four-axis milling machine in comparison to a three-axis milling machine.
4. Why are five-axis milling machines becoming more widely used, and what benefit do they provide to a machine shop?
5. What is the benefit of using polar coordinates in a program to specify locations of holes in a circular pattern defined by a bolt circle?

Critical Thinking

1. The Cartesian coordinate system has important uses in the machining industry and other industries, including mechanical design and engineering. What are some examples in everyday life where coordinate systems are used to designate distance and direction?
2. Sketch a map of your home. Designate one direction as the X axis direction and designate a second direction at 90° to establish the Y axis direction. Start from an origin position, such as a point near the front door, and map out the distance and direction to different rooms, such as the kitchen and dining room. Select one of the rooms and plot coordinates to locate items such as furniture and appliances. What is the most logical location for the origin of coordinates based on the layout of the room?
3. Describe the benefits of four- and five-axis machining. Why is the ability to rotate work a benefit? What are some examples of common parts that might benefit from four- or five-axis machining?

3 Preparatory Commands: G-Codes

Chapter Outline

3.1 Introduction
3.2 Using G-Codes in a Program
3.3 G-Code Commands
    3.3.1 G00: Rapid Positioning
    3.3.2 G01: Linear Interpolation
    3.3.3 G02 and G03: Circular Interpolation
    3.3.4 G20 and G21: Measurement Modes
    3.3.5 G90 and G91: Absolute and Incremental Positioning
    3.3.6 G40, G41, and G42: Cutter Compensation
    3.3.7 G54: Work Offset

3.4 Startup Blocks

Learning Objectives

After completing this chapter, you will be able to:

- Explain the purpose of preparatory commands.
- Identify the most common G-code commands for milling operations.
- Identify G-code groups and their purpose.
- Understand the difference between modal and nonmodal commands.
- Describe the purpose of the G00 and G01 movement commands.
- Identify coordinates specified in absolute positioning and incremental positioning mode.
- Explain the use of cutter compensation commands.

Key Terms

absolute positioning	incremental positioning	preparatory command
address	linear interpolation	startup block
block	modal command	word
G-code	nonmodal command	work offset

3.1 Introduction

A CNC program is a series of letters and numbers that, when combined, create action from the milling machine. A letter and number grouping used to execute a command in a program is called a *word*. The letter preceding the number in a word is called an *address*. An address is a single-letter character that defines what a machine should do with the numerical data that follows. Addresses are used in programs to designate commands and machine functions. The most commonly used program address is the letter G, or *G-code*. A G-code identifies a *preparatory command*. The purpose of the preparatory command is to prepare the machine controller, or preset the machine, into a specific state of operation. This chapter covers the most commonly used preparatory commands in CNC mill programming.

A G-code positioning command by itself will not create any motion or movement, but it places the machine in the operational mode for the program entries that follow. Conversely, the machine will not operate without the operational mode being set. For example, the line of machine code below, by itself, will not create any movement from the machine.

X5. Y3.

Although there is a designated set of coordinates to a target position, the controller has not been put into an operational mode. Is this a rapid move or a move at a feed rate? Are these coordinates in inches or in metric units? Are the coordinates specified in absolute or incremental positioning mode? These questions cannot be answered without an accompanying G-code address.

3.2 Using G-Codes in a Program

Once the machine is preset into a specific operating mode with the appropriate G-code command, a motion or movement can occur. For example, the G01 command creates straight-line movement. The programming example shown in **Figure 3-1** moves the machine from its current location, the origin, to the X5. Y3. position. The G01 code commands a linear move in a straight line.

The single line of code shown in **Figure 3-1** is called a *block*. A *block* is a single word or a series of words forming a complete line of CNC code. Notice that F20. appears at the end of the block as the last word. The address code F designates the feed rate. The programmed feed rate is used by the G01 command and must be specified. The feed rate is the rate at which the cutting tool moves into the material in inches per minute (ipm). This is discussed in more detail in Chapter 5.

Most G-code commands are *modal commands*, meaning once they are turned on, the machine stays in that condition until the mode is canceled or until a subsequent command changes the machine's condition. **Figure 3-2** shows a series of straight-line movements programmed with the G01 command. All of the movements are in a linear direction and occur in succession because the G01 command is modal and presets the mode of operation to straight-line movement.

3.3 G-Code Commands

The table in **Figure 3-3** lists the most commonly used G-code commands. It is not a complete list, and some codes are specific to certain controller

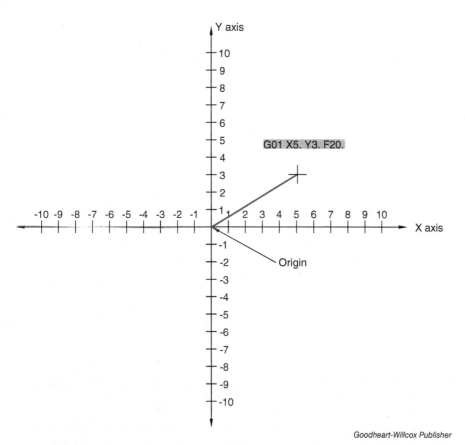

Figure 3-1. The G01 command initiates straight-line movement. It is used in the line of code shown to move the machine from its current location to the X5. Y3. location.

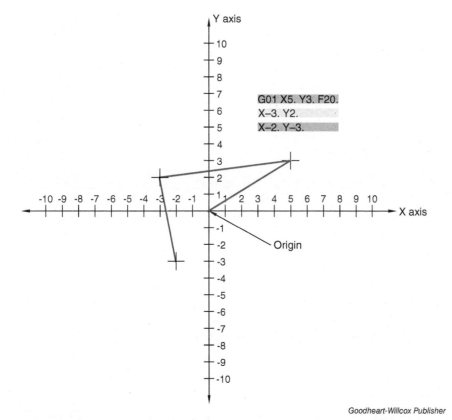

Figure 3-2. Using the G01 command to carry out a series of straight-line movements. The G01 command is a modal command and remains in effect until canceled.

Commonly Used G-Codes

G-Code	Function	Mode
G00	Rapid positioning	Modal
G01	Linear interpolation in feed mode	Modal
G02	Circular interpolation clockwise (CW)	Modal
G03	Circular interpolation counterclockwise (CCW)	Modal
G04	Dwell (in milliseconds)	Nonmodal
G09	Exact stop check	Nonmodal
G10	Programmable offset setting mode	Nonmodal
G11	Cancels G10—Programmable offset setting mode cancel	Nonmodal
G15	Polar coordinate command cancel (Fanuc)	Modal
G16	Polar coordinate command (Fanuc)	Modal
G17	XY plane designation	Modal
G18	XZ plane designation	Modal
G19	YZ plane designation	Modal
G20	US Customary units of input	Modal
G21	Metric units of input	Modal
G28	Machine zero return	Nonmodal
G29	Return from machine zero	Nonmodal
G30	Machine zero return	Nonmodal
G31	Skip function	Nonmodal
G40	Cutter compensation cancel	Modal
G41	Cutter compensation left	Modal
G42	Cutter compensation right	Modal
G43	Tool length offset—positive	Modal
G44	Tool length offset—negative	Modal
G49	Tool length offset cancel	Modal
G50	Scaling function cancel	Modal
G51	Scaling function	Modal
G52	Local coordinate system setting	Nonmodal
G53	Machine coordinate system	Nonmodal
G54	Work coordinate offset 1	Modal
G55	Work coordinate offset 2	Modal
G56	Work coordinate offset 3	Modal
G57	Work coordinate offset 4	Modal
G58	Work coordinate offset 5	Modal
G59	Work coordinate offset 6	Modal
G60	Unidirectional positioning	Nonmodal
G61	Exact stop mode	Modal
G62	Automatic corner override mode (Fanuc)	Modal
G63	Tapping mode (Fanuc)	Modal
G64	Cutting mode (Fanuc); cancels G61 (Haas)	Modal
G65	Custom macro call	Nonmodal
G66	Custom macro modal call (Fanuc)	Modal
G67	Custom macro modal call cancel (Fanuc)	Modal

Continued

Figure 3-3. G-code commands used in programming. Some machine-specific codes are not listed.

G-Code	Function	Mode
G68	Coordinate system rotation	Modal
G69	Coordinate system rotation cancel	Modal
G70	Bolt hole circle (Haas)	Nonmodal
G71	Bolt hole arc (Haas)	Nonmodal
G72	Bolt holes along an angle (Haas)	Nonmodal
G73	High speed peck drilling cycle	Modal
G74	Left hand threading cycle	Modal
G76	Fine boring cycle	Modal
G80	Canned cycle cancel	Modal
G81	Drilling cycle	Modal
G82	Spot drilling cycle	Modal
G83	Full retract peck drilling cycle	Modal
G84	Right hand thread drilling cycle	Modal
G85	Boring cycle—bore in, bore out	Modal
G86	Boring cycle—bore in, rapid out	Modal
G87	Back boring cycle (Fanuc)	Modal
G88	Boring cycle—bore in, dwell, rapid out (Fanuc)	Modal
G89	Boring cycle—bore in, dwell, bore out (Fanuc)	Modal
G90	Absolute positioning mode	Modal
G91	Incremental positioning mode	Modal
G92	Set work coordinate shift amount	Nonmodal
G98	Return to initial level in canned cycle	Modal
G99	Return to R level in canned cycle	Modal
G110–G129	Additional work coordinate system locations	Modal

Goodheart-Willcox Publisher

Figure 3-3. *(Continued)*

types. Consult your machine manufacturer's technical manual or website for exact G-codes that are available. Many of these codes will be discussed in greater detail in subsequent chapters.

In a CNC program, multiple G-code commands can be used in the same block of code, as long as they do not have conflicting functions. For example, a block written as **G90 G54 G00 X12. Y–6.** is a valid line of code. Referring to the G-code commands in **Figure 3-3**, this line is easily deciphered:

- **G90.** Establishes absolute positioning mode.
- **G54.** Designates work coordinate offset 1 as the work coordinate system.
- **G00.** Places the machine in rapid positioning mode.

This line will instruct the machine to move to the absolute coordinate position X12. Y–6. in the G54 work coordinate offset in full rapid mode. The G-code commands in this block define the positioning mode for how coordinates are located, establish the work offset, and define the mode of movement for the cutting tool. The functions of these commands do not conflict with each other.

In CNC programming, G-codes are organized into specific groups by function. The group designation identifies modal G-code commands that

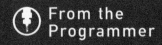

G-Code Entry Format

On many machines, G-code commands beginning with a zero in the digit portion can be shortened to a single digit following the letter G. For example, G00 can be shortened to G0 and G01 can be shortened to G1. This alternate format may not be recognized on older machines.

cannot be in effect at the same time. Multiple G-codes from different groups can be placed on the same block of code, but no two G-codes from the same group can be specified on the same block. G-codes belonging to the same group are associated with the same function. For example, a block that states **G01 G40 G41 D1 X–1** is invalid. The G40 and G41 commands are both related to the same function: cutter compensation. Cutter compensation commands are used to compensate for the tool diameter by offsetting the tool to the left or right of the programmed cutting path. The G40 and G41 commands have conflicting functions. The G40 command cancels cutter compensation and G41 initiates cutter compensation left. If conflicting G-codes are used on a single block of code, the machine's alarm will activate.

G-code groups are listed in the table in **Figure 3-4**. Not all of the groups are listed, as some pertain only to lathe operations. The number of groups and the group numbers vary by controller type. Generally, G-code groups are numbered from Group 00 to Group 25. Group 00 consists of *nonmodal commands*. A nonmodal command is only active in the block in which it appears and terminates as soon as the function is complete. All other G-code commands are modal and remain in effect until they are canceled, or a new G-code command places the machine in a different mode of operation.

3.3.1 G00: Rapid Positioning

The G00 (or G0) command is only used when the tool is not engaged into the part. This command will move the machine in full rapid movements. This command is commonly used when going to the home position or

G-Code Groups

Group Number	Description	G-codes
00	Nonmodal G-codes	G04, G09, G10, G11, G28, G29, G30, G31, G52, G53, G60, G65, G70, G71, G72, G92
01	Motion commands	G00, G01, G02, G03
02	Plane selection	G17, G18, G19
03	Dimensioning mode	G90, G91
06	Unit input	G20, G21
07	Cutter radius offset	G40, G41, G42
08	Tool length offset	G43, G44, G49
09	Canned cycles	G73, G74, G76, G80, G81, G82, G83, G84, G85, G86, G87, G88, G89
10	Return mode	G98, G99
11	Scaling cancel	G50
12	Coordinate system	G54–G59, G110–G129
13	Cutting modes	G61, G62, G63, G64
14	Macro mode	G66, G67
16	Coordinate rotation	G68, G69
18	Polar input	G15, G16

Goodheart-Willcox Publisher

Figure 3-4. G-code groups.

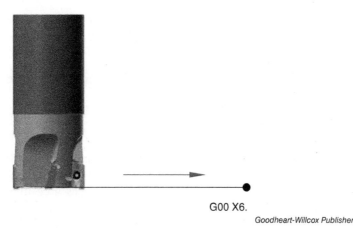

Figure 3-5. The G00 command sets the machine to rapid movement. Shown is a programmed rapid move to the X6. position.

coming away from the home position. **Figure 3-5** shows a rapid move to the X6. position.

Caution should be used when programming movement along multiple axes simultaneously. When a machine is moving in full rapid mode, a block of code with both X and Y coordinates may not necessarily move on a straight line to that position, depending on the machine. Older machines may move at the same rate along each axis to the nearest position first. See **Figure 3-6**.

3.3.2 G01: Linear Interpolation

The G01 (or G1) command is used when entering a piece of material while cutting, and when exiting a cut. This command creates linear movement from the current position to a specified position. *Linear interpolation* refers to determining a straight-line distance by calculating intermediate points between a start point and end point. The G01 command is followed by a

> **Safety Note**
>
> When using the G00 command, be cautious of any possible collisions with this movement. G00 is a modal command, so all movements following it will be in full rapid mode until it is canceled.

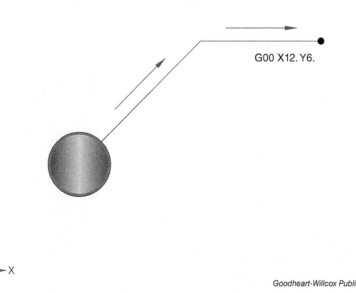

Figure 3-6. Depending on the machine, rapid movement programmed along two axes with the G00 command may not occur along a straight line. Shown is a programmed move to X12. Y6. Notice that the movement of the machine was to the closest axis position first.

feed rate command (F command) to designate the rate at which the cutting tool moves into the material in inches per minute (ipm). See **Figure 3-7**.

When more than one axis of movement is programmed on a single block of code, the G01 command will make a direct linear move to the specified coordinates. The example in **Figure 3-8** shows a move to X12. Y6.

3.3.3 G02 and G03: Circular Interpolation

The G02 (G2) and G03 (G3) commands produce radial or circular tool movement at a specified feed rate. The G02 command produces a clockwise rotation, and the G03 command produces a counterclockwise rotation. See **Figure 3-9**.

When programming a G02 or G03 command, it is common to use a G01 command first to position the tool at the start point of the radius or arc, then use the G02 or G03 command to complete the arc. The arcs generated can be partial arcs or full circles, depending on the code. Since these are modal commands, the G01 command will need to follow the G02 or G03 command to return the machine condition to linear interpolation mode. Programming methods using the G02 and G03 commands are explained later in this text.

3.3.4 G20 and G21: Measurement Modes

The G20 and G21 commands set the type of working units used by the machine. CNC machines can read programs in US Customary units or metric units. In the US Customary system, also known as the English (or Imperial) system, the basic unit of linear measurement is the inch. Most prints for manufacturing in the United States are dimensioned in decimal inches. On metric prints, dimensions are commonly specified in millimeters. It can be beneficial to program in one unit format over another. For example, if the print is dimensioned in metric units, it is logical to write the CNC code in the same units.

The G20 command sets the machine to use US Customary units. The G21 command sets the machine to use metric units. Using the G21 command means that the lines of code and specified offsets will be based on

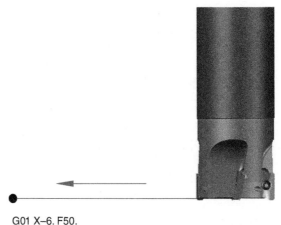

G01 X–6. F50.

Goodheart-Willcox Publisher

Figure 3-7. The G01 command creates linear movement. Shown is a programmed move to the X–6. position. A feed rate command (F command) follows the G01 command and specifies the rate at which the cutting tool moves into the material.

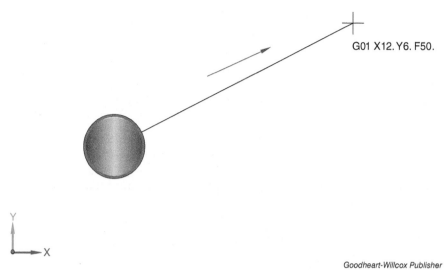

Figure 3-8. The G01 command creates straight-line movement to the specified coordinates.

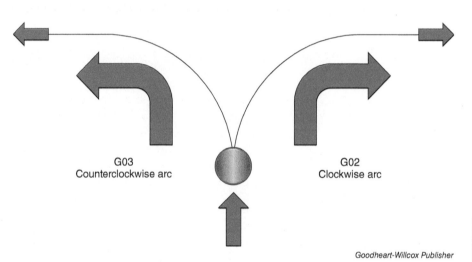

Figure 3-9. The G02 and G03 commands produce motion in a circular path. The G02 command generates a clockwise rotation, and the G03 command generates a counterclockwise rotation.

metric units. Always make sure to specify the correct code for the unit format. On some machines, the G20 and G21 commands check that the machine control has been set to the correct mode for the program and will produce an alarm if the units are not set correctly.

3.3.5 G90 and G91: Absolute and Incremental Positioning

There are two basic ways to define coordinates for locating machine positions and movement. The G90 and G91 commands set the positioning mode used to locate coordinates. The G90 command sets the current positioning mode to *absolute positioning*. In this mode, coordinate values are measured from the coordinate system origin (X0. Y0.). The origin represents a fixed point from which coordinates are located. This method

> **Safety Note**
>
> Be aware that if tool offsets are specified while the machine is in inch mode and then the program uses a G21 command, the offsets will not automatically convert. For example, the machine will read a 15″ offset as a 15 mm offset. This is potentially dangerous and can result in a machine crash. Always specify the appropriate unit format for machining and do not use more than one unit format in a program.

locates absolute coordinates and is the appropriate positioning method for most purposes.

The G91 command sets the current positioning mode to *incremental positioning*. In this mode, coordinate values are measured from the current point to the next point. Incremental coordinates represent a relative measurement from the current position of the tool, not from the coordinate system origin. Incremental coordinates are also called relative coordinates because they are located relative to the previous coordinate.

Figure 3-10 shows coordinates located with each positioning method. The coordinates are represented as Point 1 and Point 2. Each point represents a possible destination position for the cutting tool. The tool start position is the coordinate X7. Y5.

A block of code written as **G90 G01 X–6. Y–5. F50.** will take the tool to Point 1. The G90 command specifies absolute positioning and locates the coordinate in relation to the coordinate system origin. Point 1 is the absolute coordinate X–6. Y–5.

A block of code written as **G91 G01 X–6. Y–5. F50.** will take the tool to Point 2. The G91 command specifies incremental positioning. The same X–6. Y–5. coordinate is used, but the coordinate is an incremental

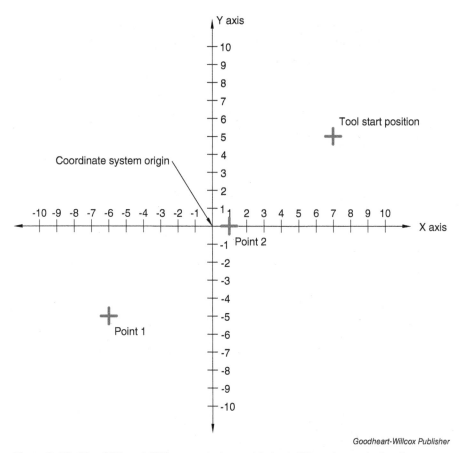

Goodheart-Willcox Publisher

Figure 3-10. The G90 and G91 commands provide two different ways to locate coordinates for tool moves. The G90 command activates absolute positioning mode and moves the tool from its current position (the start position) to an absolute position measured from the origin. The coordinate entry X–6. Y–5. moves the tool to Point 1. The G91 command activates incremental positioning mode and moves the tool from its current position an incremental amount in the specified axis direction. The coordinate entry X–6. Y–5. locates the tool at Point 2. This changes the cutter path and results in a different finish location for the move.

coordinate located in relation to the current position of the tool. The current position is the tool start location, not the origin. This locates Point 2 six units in the negative X axis direction and five units in the negative Y axis direction from the X7. Y5. tool start location. Notice that using the G91 command results in an entirely different cutter path and finish location.

Moves defined with absolute positioning are used for most programming applications. Incremental positioning moves must be programmed carefully because if an incorrect coordinate position is specified, all subsequent moves will be incorrect. Incremental moves are used when it is appropriate to specify coordinates relative to the current position. A practical use for incremental positioning in a program is drilling a pattern of holes aligned along a straight path, as explained in Chapter 10.

3.3.6 G40, G41, and G42: Cutter Compensation

The G41 and G42 commands are used to offset the tool to the left or right of the programmed cutting path. When programming coordinate positions in a CNC program, the centerline of the spindle is being moved to those points. The machine does not consider the diameter of the cutting tool being used. To offset the center of the tool and allow the tool edge to follow the correct path, the G41 or G42 command is used. Each command uses the stored radius offset setting for the cutting tool. The G41 command activates left-hand cutter compensation and the G42 command activates right-hand cutter compensation. If the tool is cutting on the left-hand side of the material, the G41 command is used. The G42 command will offset the tool to the right-hand side of the part or piece of material. The offset direction of the tool is determined by viewing the tool from behind as it moves in the cutting direction. See **Figure 3-11**.

The G41 and G42 commands are modal commands and will continue to offset the tool for any additional moves until canceled. The G40 command cancels the G41 or G42 command.

Additional explanations and examples of the G41 and G42 commands and other G-code commands executed for specific programs are provided later in this text.

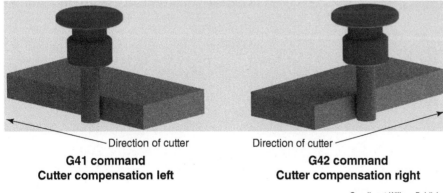

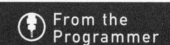

From the Programmer

G41: Left-Hand Cutter Compensation

In most contour milling operations, the cutting tool is positioned to cut material on the left side of the part. Most milling programs created for these operations use the G41 command to establish left-hand cutter compensation.

Figure 3-11. The G41 command offsets the tool to the left of the programmed cutting path. The G42 command offsets the tool to the right of the programmed cutting path. The offset direction corresponds to the side of the part machined when viewing the tool from behind in the direction of the cutting path.

3.3.7 G54: Work Offset

As previously discussed, the work coordinate system (WCS) defines the X0. Y0. origin location of the part. In CNC programming, the work coordinate system can be defined as a *work offset*. A work offset is a coordinate setting that establishes the location of the work origin relative to the machine home position. Coordinates for a work offset are entered by the operator into the machine during setup. The appropriate work offset command is used in the program to designate the work coordinate system. For example, the G54 command is used to activate work coordinate offset 1 as the work coordinate system. This command is commonly used when machining a single part. On most machines, it is possible to set a number of work offsets referenced to specific programming commands. The G55, G56, G57, G58, and G59 commands are additional work offset commands used for this purpose. Multiple work offsets are defined when several parts are to be machined in one program. During setup, each workpiece is secured and work offsets are defined to establish a different work origin for each part.

3.4 Startup Blocks

Most CNC programmers will begin new programs with a *startup block*. This block, also referred to as a default block or safety block, cancels any machine conditions left from the previous program or establishes a new starting condition for the current program. Canceling modes that may have been previously active prevents errors and helps ensure the machine will start and operate safely. A startup block usually contains the appropriate commands to set the unit format and positioning mode. While this practice can vary from programmer to programmer, a typical startup block could be written as **G90 G80 G17 G40 G20**. The commands are read as follows:

- **G90.** Establishes absolute positioning mode.
- **G80.** Cancels a previously active canned cycle.
- **G17.** Designates the XY plane for machining.
- **G40.** Cancels cutter compensation.
- **G20.** Places the machine in inch mode.

Chapter 3 Review

Summary

- The program address G identifies a preparatory command.
- The purpose of a preparatory command is to prepare the machine controller, or preset the machine, into a specific state of operation.
- The G00 command places the machine in rapid positioning mode.
- The G01, G02, and G03 commands are interpolation commands used to move the machine to specific coordinate positions.
- Multiple G-code commands can be used on a single block in a program as long as they do not conflict with each other.
- A modal command stays active until it is canceled or until a subsequent command changes the machine's condition.
- A nonmodal command is only active in the block in which it appears.
- G-code commands are grouped by their use and purpose.
- CNC machines can operate using US Customary or metric units. The G20 or G21 command is used to set the units used by the machine.
- The G90 command creates absolute movements measured from the work coordinate system origin. The G91 command creates incremental movements measured from the machine's current position.
- The G41 command offsets the tool to the left-hand side of the part. The G42 command offsets the tool to the right-hand side of the part. The G40 command cancels any cutter compensation in the controller.

Review Questions

Answer the following questions using the information provided in this chapter.

Know and Understand

1. The letter used in a word in CNC programming is referred to as a(n) ____.
   A. symbol
   B. coordinate
   C. address
   D. designator

2. G-code commands are considered ____ commands.
   A. address
   B. compensation
   C. incremental
   D. preparatory

3. The ____ command will initiate full rapid machine travel.
   A. G00
   B. G01
   C. G02
   D. G03

4. *True or False?* To program an angled move while cutting, use the G01 command.

5. To produce radial tool movement in a counterclockwise rotation, the ____ command is used.
   A. G00
   B. G01
   C. G02
   D. G03

6. To move a tool to the absolute position of X5. Y–3. in relationship to the origin, use the ____ command.
   A. G41
   B. G90
   C. G42
   D. G91

7. To move a tool incrementally from its current position to X5. Y–3., use the _____ command.
   A. G41
   B. G90
   C. G42
   D. G91

8. *True or False?* The G90 command is used to cancel cutter compensation.

9. To position a tool cutting on the left-hand side of a piece of material, the correct cutter compensation command is _____.
   A. G40
   B. G41
   C. G42
   D. G43

10. To position a tool cutting on the right-hand side of a piece of material, the correct cutter compensation command is _____.
    A. G40
    B. G41
    C. G42
    D. G43

11. *True or False?* The G43 command cancels the G41 or G42 command.

12. The G43 command sets the _____ in a positive direction.
    A. tool length offset
    B. cutter compensation
    C. positioning mode
    D. coordinate system

Apply and Analyze

1. What is the correct code to set absolute positioning mode current and move the tool in rapid travel from Point 1 to Point 2?

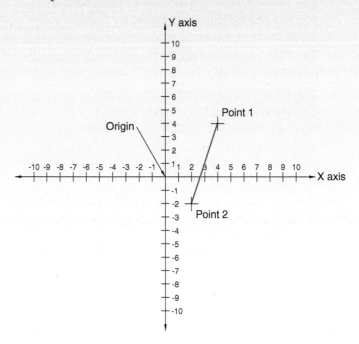

2. What is the correct code to set incremental positioning mode current and move the tool in rapid travel from Point 1 to Point 2?

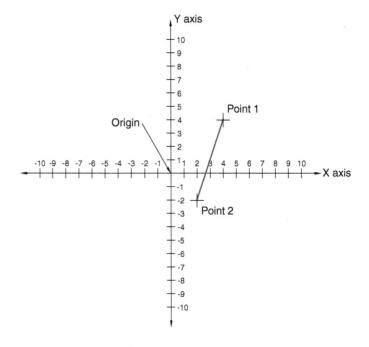

3. Identify the G-code command to set the correct cutter compensation for the tool shown based on the tool orientation and cutting direction.

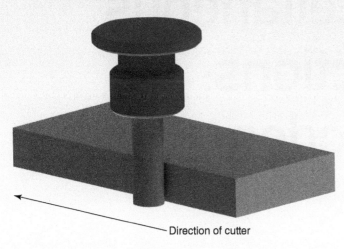

Direction of cutter

4. Identify and briefly describe the two types of working unit formats and corresponding commands used in CNC programming.
5. When can the programmer use multiple G-code commands in the same block of code?
6. Explain the difference between modal and nonmodal commands.
7. Why are G-code commands organized into groups?
8. What is the purpose of a startup block? Briefly explain the types of commands that are commonly included at the beginning of a program.

Critical Thinking

1. Think about the movement that takes place in your daily routine. How many different types of modes do you enter? Consider walking, running, sitting, standing, riding in a vehicle or an elevator, and other types of movement. If you were to assign a G-code to represent each type of movement, how many different codes would it take?
2. Consider the use of G-codes by different machines. Why is it important for G-codes to be standardized for different machine types? How does this simplify machining? Explain the difficulty of using different machine controllers with different G-codes.
3. Depending on the type of machining and programming you may attempt in a machine shop, the most common G-codes used may vary. What are some common G-codes you might use if you were machining? How does the type of machining influence the selection of G-codes?

4 Miscellaneous Functions: M-Codes

Chapter Outline

4.1 Introduction
4.2 Use of M-Codes
4.3 M-Codes as Program Functions
    4.3.1 M00: Program Stop
    4.3.2 M01: Optional Stop
    4.3.3 M02, M30, and M99: Program End
4.4 M-Codes as Machine Functions
    4.4.1 Coolant Functions
    4.4.2 Spindle Functions

Learning Objectives

After completing this chapter, you will be able to:

- Explain the purpose of miscellaneous functions.
- Identify machine functions and program functions.
- Identify the most common M-codes for milling operations.
- Describe the functions of the stop codes M00 and M01.
- Describe the functions of the end codes M02, M30, and M99.
- Explain the difference between modal and nonmodal M-codes.
- Apply techniques for initiating the coolant codes M07, M08, and M09.
- Use the spindle control functions M03, M04, and M05.

Key Terms

flood coolant
miscellaneous function
mist coolant
program loop
program rewind

4.1 Introduction

The CNC program address *M* identifies a ***miscellaneous function***. Miscellaneous functions are referred to as M-functions or M-codes. Sometimes, M-codes are referred to as machine functions. However, not all M-codes control machine functions, so this is not an accurate definition. Some M-codes relate only to controlling the program process and not a machine function. In this text, M-codes are referred to as miscellaneous functions.

In a CNC program, the programmer often needs to control machine functions or the program operation outside of machine movements or cutting commands. This is most often done with the functions of M-codes. Examples of machine control functions include turning the spindle on, turning the coolant on, and operating a pallet changer. Examples of program control functions include program stops, program rewinds, and program optional stops.

Since some M-codes are used to control machine functions, they can vary greatly by manufacturer and control type. Even the same type of milling machine can use different M-codes, depending on the functions and features of the machine. This chapter covers the most commonly used M-codes in mill programming. Not all M-codes are discussed. The table in **Figure 4-1** lists the most commonly used M-codes for milling operations.

4.2 Use of M-Codes

Unlike G-codes, where multiple codes can be placed on a single block of program code, there can only be one M-code on a single block of code. The M-code can exist on its own line and perform a function. For example:

G00 G90 G54 X–2.375 Y3.85	
M08	Coolant on
G43 H1 Z2.	

In this example, the machine moves in rapid mode to an absolute X and Y coordinate, and then the coolant is initiated with the M08 code. On the last line, the tool length offset command is initiated. An alternative to these lines of programming could be the following:

G00 G90 G54 X–2.375 Y3.85 M08
G43 H1 Z2.

Is there a difference in these two programming approaches? Yes—the difference is *when* the coolant is turned on. In the first example, the X and Y positioning is fully completed, and then the coolant is activated. In the second example, the coolant is initiated as the X and Y axis moves are happening.

Not all M-codes can be used in this manner. M-codes that control program processing functions should not be on a line with movement commands. M00, M01, M02, M30, M98, and M99 are examples of functions that should reside either on their own line or on lines with no movement. For example:

G01 X3.25 Y–12.6 F55. M00

> **Safety Note**
>
> It is always essential to know how your machine operates and how the controller carries out functions. On many machines, coolant will initiate with axis movement when the M08 function is read by the controller on the same line. However, this is not true of every machine. The Haas controller will complete motion first and then initiate coolant, even if these actions are on the same line of code. Be cautious and verify that the coolant is coming on before the tool starts the cut. If this is not the case, edit the program as needed.

Commonly Used M-Codes

M-Code	Description	Controller
M00	Program stop	Fanuc/Haas
M01	Optional program stop	Fanuc/Haas
M02	Program end	Fanuc/Haas
M03	Spindle on clockwise	Fanuc/Haas
M04	Spindle on counterclockwise	Fanuc/Haas
M05	Spindle stop	Fanuc/Haas
M06	Tool change	Fanuc/Haas
M07	Coolant on—mist coolant Coolant on—shower coolant	Fanuc/Haas
M08	Coolant on	Fanuc/Haas
M09	Coolant off	Fanuc/Haas
M10	4th axis brake on	Fanuc/Haas
M11	4th axis brake release	Fanuc/Haas
M12	5th axis brake on	Fanuc/Haas
M13	5th axis brake release	Fanuc/Haas
M17	Automatic pallet change (APC) pallet unclamp and open APC door	Haas specific
M18	Automatic pallet change (APC) pallet clamp and close APC door	Haas specific
M19	Orient spindle	Fanuc/Haas
M30	Program end and rewind	Fanuc/Haas
M31	Chip auger forward	Haas specific
M33	Chip auger stop	Haas specific
M41	Spindle low gear override	Fanuc/Haas
M42	Spindle high gear override	Fanuc/Haas
M80	Automatic door open	Haas specific
M81	Automatic door close	Haas specific
M83	Auto air jet on	Haas specific
M84	Auto air jet off	Haas specific
M88	Coolant through spindle on	Haas specific
M89	Coolant through spindle off	Haas specific
M97	Local subprogram call	Fanuc/Haas
M98	Subprogram call	Fanuc/Haas
M99	Subprogram return or loop	Fanuc/Haas

Goodheart-Willcox Publisher

Figure 4-1. M-codes commonly used in mill programming.

From the Programmer

Sequence of Commands

CNC program lines are not read from left to right, one block at a time, like you would read a sentence. The full line of code is processed at the same time. That is why an X and Y coordinate move on the same line creates a straight-line angle and not two straight lines. The X and Y coordinates are being processed simultaneously. Likewise, depending on the machine controller, coolant is turned on simultaneously when tool motion initiates. This occurs regardless of where the M08 code is placed on the line. The order in which commands are placed within a single line of code does not make a difference. However, most programmers will follow a certain pattern in their methodology and specify commands in a logical order that represents a normal sequence of functions.

This block of code has a feed movement command to an X and Y location, but the M00 code specifies a program stop. Although this block of code will function, it is not recommended or considered the best programming practice. The correct lines of programming should read as follows:

```
G01 X3.25 Y-12.6 F55.
M00
```

This code allows the machine to complete the X and Y axis movement command, and then perform the program stop function.

M-codes can be modal or nonmodal. Nonmodal M-codes are only active in one block of the program. Modal M-codes stay active until they are canceled or altered. **Figure 4-2** shows partial listings of modal and nonmodal M-codes.

For example, in a line of code that states **M03 S3500**, the spindle is turned on clockwise at 3500 revolutions per minute (rpm). The spindle will remain rotating until the program ends or an M05 (spindle stop) code is initiated in the program body. The M04 function can also change the direction of the spindle to counterclockwise, thus canceling the M03 function. However, it is best practice to use the M05 code to cancel the spindle rotation to prevent possible damage to the spindle.

4.3 M-Codes as Program Functions

As previously discussed, miscellaneous functions can be used to control either program processing functions or machine functions. Program processing functions can be used to stop, pause, or end the program processing sequence. There are multiple applications for these functions, and there are specific places in programs to use them.

4.3.1 M00: Program Stop

The M00 function is defined as a *mandatory program stop*. Each time the machine controller reads the M00 code, the spindle will stop, any machine motion will stop, and the part program will stop. No control settings will change and the program functions will resume with a cycle start button push. Once the cycle start is engaged, the previous feed rate, coolant condition, and coordinate system will still be active. The spindle will need to be

Modal M-Codes

M-Code	Description	Codes that Cancel
M03	Spindle on clockwise	M05
M04	Spindle on counterclockwise	M05
M05	Spindle stop	M03, M04
M07	Mist coolant or shower coolant on	M09, M08
M08	Coolant on	M09, M07
M09	Coolant off	M07, M08

Nonmodal M-Codes

M-Code	Description
M00	Program stop
M01	Optional program stop
M02	Program end
M06	Tool change
M30	Program end and rewind

Goodheart-Willcox Publisher

Figure 4-2. Examples of modal and nonmodal M-codes.

restarted with an M03 or M04 code, but the spindle speed set prior to the M00 code will be restored.

The M00 code is used any time the program needs to completely stop. For example, the program can be stopped to perform an inspection check of a feature that was just machined. See **Figure 4-3**. In this example, the 4.000 linear dimension has a tight tolerance. This dimension can be checked at a specific point in the program. To machine the outer surfaces of the part, the cutter travels around the rectangular boundary. Before proceeding to other machining, the program is stopped with the M00 code to check the 4.000±.001 dimension.

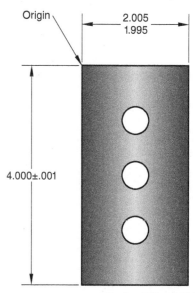

Goodheart-Willcox Publisher

Figure 4-3. To machine this part, the cutter travels around the outside of the rectangular area. To allow for inspection of the 4.000 linear dimension before the holes are machined, the program is stopped with the M00 code. The program includes a note to inform the operator when the part is to be checked.

T1 M06 (1/2" End mill)	Tool change to Tool 1
G00 G90 G54 X-.75 Y.25 S2674 M03	Rapid mode, absolute mode, G54 work offset, go to XY position, spindle on clockwise to 2674 rpm
G43 H1 Z.25 M08	Tool length offset 1, move to Z.25, coolant on
G01 Z-.5 F6.42	Feed move to Z-.5 at 6.42 ipm
G41 D1 X0. F53.48	Turn on cutter compensation left Tool 1, feed move to absolute X position at 53.48 ipm
X2.00	Single axis move to X position
Y-4.00	Single axis move to Y position
X0.	Single axis move to X position
Y0.	Single axis move to Y position
G40 Y.75	Turn off cutter compensation, move to Y.75
Z.2 F6.42	Move to Z.2 at 6.42 ipm
G00 Z.25	Rapid move to Z.25
M00 (Inspect 4.00±.001 dimension)	Machine and spindle stop to inspect feature
G91 G28 Z0. M09	Incremental mode, machine home in Z, coolant off
T2 M06 (1/2" Spot drill)	Tool change to Tool 2
G00 G90 G54 X1. Y-1. S4584 M03	Rapid mode, absolute mode, G54 work offset, go to XY position, spindle on clockwise to 4584 rpm
G43 H2 Z2. M08	Tool length offset 2, move to Z2., coolant on
G98 G81 Z-.25 R.1 F35.84	Return to reference position, turn on drilling cycle, Z-.25 hole depth, drilling feed rate at 35.84 ipm
Y-2.	Second hole
Y-3.	Third hole

After verifying the dimension, the program is restarted and the holes are machined using the G81 canned cycle command. This is a drilling cycle used to machine the three holes in the part. Study the portions of the program that appear before and after the block containing the M00 code. The additional text to the right of each block explains the meaning of the related commands and functions. The specific operations indicated are explained in more detail in later chapters.

> **From the Programmer**
>
> **Notes in Programs**
>
> Anything written in a program and enclosed by parentheses ( ) is a note and is not read as code. Referring to the previous example, there is a note to inspect the 4.000±.001 dimension on the same line as the M00 code. There are also notes used to identify the 1/2" end mill and 1/2" spot drill. You can add a note at any place in a program. It is good practice to note part numbers, tool references, and any special machining information for the operator.

Another example of using the M00 code is to move or rotate a part in a fixture or vise at a designated point in the program. It is good practice to create a note in the program that explains the purpose of the M00 code when it appears.

4.3.2 M01: Optional Stop

The M01 function is defined as an *optional program stop*. Similar to the M00 code, the M01 code is added to a program to create a full program stop. The difference is that the function of the M01 code can be turned on or off using the Optional Stop switch on the control panel. When this switch is turned on, the program will stop when an M01 code is read. If the switch is off, the program will proceed without stopping.

The most common use of the M01 code is with tool change commands. Operators and programmers prefer to have the ability to pause at tool changes and ensure the cutting tool is free from chips or debris. An optional program stop also provides the opportunity to make sure the tool is not broken or damaged. The following is the same program from the previous example, but the M01 code is used instead of the M00 code and is specified after the machine is sent to the home position.

| G91 G28 Z0. M09 |
| M01 (Prepare to tool change) |
| T2 M06 (1/2" Spot drill) |
| G00 G90 G54 X1. Y–1. S4584 M03 |
| G43 H2 Z2. M08 |
| G98 G81 Z–.25 R.1 F45.84 |

In this case, the operator can stop at all tool changes to inspect the tool if needed. However, if the Optional Stop switch is in the off position, the program will continue without interruption in machining. Although pausing the program is a good practice, the operator should be aware that this adds time to machining and the program should only be stopped if there is a need for such an inspection.

Programmers will also use the M01 code for verifying the program on the machine during setup. The program can be stopped during tool changes or after critical features are machined to make sure the program is operating as expected. Once the program is verified and running as needed, the Optional Stop switch is turned off and production is allowed to take place.

4.3.3 M02, M30, and M99: Program End

At the end of every program, a function is needed to tell the machine that the program is complete. Without this notification, the operator would need to complete at least two resets on the controller to restart the program and continue the machining process.

Older machines often used the M02 code to define the end of a program. This was common when programs were run from 3.5" floppy drives or external memory sources. At the end of each program, the machine controller had to be reloaded with the file and reactivated. The following example uses the M02 function to terminate the program.

O0930	Program number
(T2 | 1/2 SPOT DRILL | H2)	
(T3 | 1/2 DRILL | H3)	
G20	
G00 G17 G40 G49 G80 G90	
T2 M06	Tool change to Tool 2
G00 G90 G54 X1. Y-1. S3500 M03	Spindle on clockwise to 3500 rpm
G43 H2 Z2. M08	Coolant on
G98 G81 Z-.25 R.1 F45.84	
Y-2.	
Y-3.	
G80	
M05	Spindle off
G91 G28 Z0. M09	Coolant off
M01	Optional stop
T3 M06	Tool change to Tool 3
G00 G90 G54 X1. Y-1. S2292 M03	Spindle on clockwise to 2292 rpm
G43 H3 Z2. M08	Coolant on
G98 G83 Z-.625 R.1 Q.1 F18.36	
Y-2.	
Y-3.	
G80	
M05	Spindle off
G91 G28 Z0. M09	Coolant off
G28 X0. Y0.	
M02	Program stops here

The M30 function ends the program and executes a ***program rewind***. The term *rewind* is a reference to the past when CNC programs were loaded onto tapes. The M30 code would stop the program and rewind the tape back to the beginning. Using the M30 function is the most common way to end a CNC program. It stops the program at the end and then resets it back to the beginning. By comparison, the M02 function ends the program but does not reset it to the beginning. After the M30 code is executed, the Cycle Start button can be pushed to start the program again. The next program is the same program from the previous example, but the M30 function is used instead of the M02 function to return to the beginning.

```
O0930
( T2 | 1/2 SPOT DRILL | H2 )
( T3 | 1/2 DRILL | H3 )
G20
G00 G17 G40 G49 G80 G90
T2 M06
G00 G90 G54 X1. Y-1. S3500 M03
G43 H2 Z2. M08
G98 G81 Z-.25 R.1 F45.84
Y-2.
Y-3.
G80
M05
G91 G28 Z0. M09
M01
T3 M06
G00 G90 G54 X1. Y-1. S2292 M03
G43 H3 Z2. M08
G98 G83 Z-.625 R.1 Q.1 F18.36
Y-2.
Y-3.
G80
M05
G91 G28 Z0. M09
G28 X0. Y0.
M30
```

Program stops here and rewinds to beginning

The M99 function executes a *program loop*. This code is used at the end of a program, but it does not actually stop a program, it loops the program. As the CNC program reaches the end and the M99 code is processed, the program loops back to the start and continues to run. The M99 code is most often used in subprogramming. It is commonly used at the end of a subprogram to loop back into the main program, or at the end of a program with an M01 or M00 code.

4.4 M-Codes as Machine Functions

Miscellaneous functions that control machine operations are much like electrical switches. One M-code turns on a specific machine function and another M-code is used to turn off that function. Most machines come equipped with spare M-codes for special functions. The spare M-codes can be used to operate optional equipment or electrical sources for add-on features.

4.4.1 Coolant Functions

Many machining functions require the use of coolant to reduce friction and heat during the machining process. Machine coolants have become more advanced over time, and delivery methods from the machine have also evolved. Two common delivery methods for coolant are flood delivery and mist delivery. *Flood coolant* is liquid coolant that floods the cutting area. *Mist coolant* is a mix of fluid and air sprayed at the cutting area. Machines can be equipped with a variety of coolant delivery systems. Special cutting tools are available for machines that inject high-pressure coolant through the spindle. This functionality enables coolant to directly reach the cutting area on deep cuts and remove chips through pressure. Some machines are equipped with a programmable coolant nozzle system. The nozzle can be programmed to automatically move to different positions during machining so that coolant targets the cutting area and flushes away chips. In addition to using liquid coolant to clear chips and debris, chip removal can be accomplished with compressed air. Controllers can use different M-codes to control these delivery methods, but the codes most often used are M07 (mist coolant) and M08 (flood coolant).

M07: Mist Coolant

The M07 code activates the mist coolant function. Mist cooling is becoming more utilized in machining as coolants improve and cutter coatings provide more lubricity at the cutting edge. See **Figure 4-4**. Mist cooling combines a light air blast with microparticles of coolant and provides a direct spray at the cutting edge. This method minimizes consumption of coolant and reduces condensation loss, making it cost effective.

The M07 function can be activated at any point in the program. Where it is programmed determines the timing of the mist arriving at the cutter or material, as shown in the following example programs.

oYOo/Shutterstock.com

Figure 4-4. Mist coolant is a spray of fluid and air targeted at the cutting edge. It minimizes the volume of coolant used in machining.

Example 1:

G20	
G00 G17 G40 G49 G80 G90	
T2 M06	
G00 G90 G54 X1. Y–1. S3500 M03	
G43 H2 Z2. M07	Mist coolant turns on during move to Z2.
G98 G81 Z–.25 R.1 F45.84	
Y–2.	
Y–3.	
G80	
M05	

Example 2:

G20	
G00 G17 G40 G49 G80 G90	
T2 M06	
M07	Mist coolant turns on after tool change
G00 G90 G54 X1. Y–1. S3500 M03	
G43 H2 Z2.	
G98 G81 Z–.25 R.1 F45.84	
Y–2.	
Y–3.	
G80	
M05	

Example 3:

G20	
G00 G17 G40 G49 G80 G90	
T2 M06	
G00 G90 G54 X1. Y–1. S3500 M03	
G43 H2 Z2.	
G98 G81 Z–.25 R.1 F45.84 M07	Mist coolant turns on during drill cycle
Y–2.	
Y–3.	
G80	
M05	

M08: Flood Coolant

The M08 code is the most widely used coolant function. It floods coolant at full pressure. See **Figure 4-5**. The benefit of flooding the coolant is it provides massive coolant supply to the cutter and material, while flushing any chips or debris away from the part being machined.

Flood delivery of coolant can be a useful tool, especially in pocketing operations and on vertical milling machines. Flood cooling supplies a high volume of coolant, provides lubrication, removes heat, and clears chips from surfaces. Chips and debris can cause tool breakage or inferior surface finishes.

Figure 4-5. Flood cooling floods the cutting area with coolant at full pressure.

Flood cooling requires machines with larger coolant tanks and has additional maintenance costs compared to other coolant delivery methods. Coolant loss occurs in flood cooling due to carryout on parts, accumulation on chips, and evaporation. A flood coolant system must be cleaned regularly and machine coolant must be monitored for contaminants such as bacteria and tramp oil. Tramp oil is unwanted oil that migrates into coolant from an outside source, such as machine lubricants and oil-based protective coatings on materials.

The M08 code is used in the same manner as the M07 code in a CNC program. If the M07 code is active and the M08 code is initiated, the mist cooling function will cease and the flood cooling function will activate.

Delivering coolant through the spindle is an option on some machines. Fanuc controls often use the M08 code for this option, while Haas machines use the M88 code. Other controllers also have functions such as M13, which turns on the spindle and coolant simultaneously. Always check the machine's technical manual to verify specific uses of codes in a machine.

M09: Coolant Off

The M09 code is used to turn coolant off. Any code that was used to initiate coolant will be canceled with the M09 code. This code is often found at the end of programs or at tool change locations, although it is not completely necessary. The coolant will "pause" at tool changes and will be shut off at the end of the program with the M30 or M02 code. The following example uses the M09 function before a tool change.

Z.2 F6.42	
G00 Z.25	
G91 G28 Z0. M09	Coolant is turned off
T2 M06 (1/2" Spot drill)	
G00 G90 G54 X1. Y-1. S4584 M03	
G43 H2 Z2. M08	Coolant is restarted

The following example shows what happens at the tool change without the M09 code.

Z.2 F6.42	
G00 Z.25	
G91 G28 Z0.	
T2 M06 (1/2" Spot drill)	Coolant pauses during tool change
G00 G90 G54 X1. Y–1. S4584 M03	Coolant restarts after tool change
G43 H2 Z2.	

The following example shows what happens at the end of a program without the M09 code.

T1 M06	
G00 G90 G54 X–.75 Y.25 S3820 M03	
G43 H1 Z2. M08	Coolant is turned on
Z.2	
G01 Z–.5 F20.	
X–.25 F45.84	
X1.25	
Y–1.25	
X–.25	
Y.25	
Y.75	
Z.2 F6.42	
G00 Z2.	
M05	
G91 G28 Z0.	
G28 X0. Y0.	
M30	Coolant is turned off at M30

4.4.2 Spindle Functions

Spindle on forward, spindle on reverse, and spindle stop are all M-code machine functions. For virtually every CNC operation, the spindle requires activation to begin rotation of the cutting tool. All spindles can rotate in either a clockwise or counterclockwise direction. On a CNC milling machine, the direction of rotation is based on the standard view looking through the center of the tool from the spindle side of the machine. On a vertical milling machine, clockwise rotation is viewed by looking down at the center axis of the spindle from above the machine.

M03: Spindle Forward or Spindle Clockwise

In most cases, the M03 code is used to start the spindle. The M03 code initiates forward (clockwise) rotation of the spindle. The M03 code is accompanied by the spindle speed address code (S) and a numerical value to designate the spindle speed in revolutions per minute (rpm). The following block of code will initiate the spindle to rotate clockwise at 3500 rpm. See **Figure 4-6.**

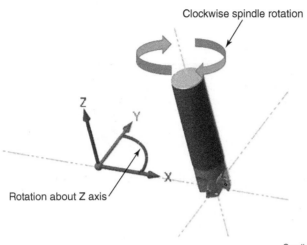

Figure 4-6. The M03 code initiates forward (clockwise) rotation of the spindle. On a vertical milling machine, the spindle rotation is about the Z axis of the machine.

```
M03 S3500
```

Figure 4-6 shows the rotation of the spindle in a clockwise direction. Rotation is about the Z axis of the machine. Notice the faces of the yellow inserts at the cutting edges. They are rotated into the material to create the cut. Standard drills, taps, and end mills are used to cut in this direction. The following is an example of using the M03 function to initiate rotation of the spindle in a program. If the M03 function is given on its own block, the machine reverts to the previous spindle speed.

G00 G90 G54 X-.75 Y.25 S3820 M03	Spindle on clockwise to 3820 rpm
G43 H1 Z2. M08	
Z.2	
G01 Z-.5 F20.	
X-.25 F45.84	
X1.25	
Y-1.25	
X-.25	
G00 Z2.	
M05	Spindle off
G91 G28 Z0.	
G28 Y0.	
M03	M03 function initiated with no speed; will revert to previous speed (3820 rpm)
G43 H3 Z2. M08	

M04: Spindle Reverse or Spindle Counterclockwise

The M04 code has some uses, but it is not a common function in milling programs. The M04 code turns the spindle in reverse, or in a counterclockwise rotation. There are left-hand drills, taps, and end mills that require the use of the M04 code. The M04 code is utilized in the same manner as the

M03 code, and it must also be accompanied by a spindle speed specification to designate the rpm of the spindle.

T4 M06 (Left hand drill)	
G00 G90 G54 X–.75 Y.25 S3820 M04	Spindle on counterclockwise to 3820 rpm
G43 H1 Z2. M08	
Z.2	
G01 Z–.5 F20.	

Left-hand taps also require the use of the M04 code in programming and may be the most common of the left-hand tools used in milling operations. Specific tapping cycles and correct procedures for tapping operations are covered in more detail later in the text.

M05: Spindle Stop

The M05 spindle stop function is used to stop all spindle motion. It is most commonly used before a tool change or at the end of the program. It is not mandatory to stop at these locations, because an M06 tool change and an M30 or M02 program stop will stop the spindle from turning. Most programmers add this code to their normal programming code to shut the spindle off early and safely, before any doors on the machine would open and cause a possible safety concern.

The following program examples show common uses for the M05 function.

Example 1:

Z.2 F6.42	
G00 Z2.	
M05	Stops spindle before machine returns to home position
G91 G28 Z0.	
G28 X0. Y0.	
M30	

Example 2:

Z.2 F6.42	
G00 Z.25	
M05	Stops spindle before a tool change
G91 G28 Z0.	
T2 M06 (1/2" Spot drill)	
G00 G90 G54 X1. Y–1. S4584 M03	
G43 H2 Z2.	

Remember that different machines and controllers can have a multitude of M-codes. Many machines come with optional M-code functions specifically designed to work as switches to turn external devices on or off. As machines become increasingly sophisticated, the need for additional M-code functions will also increase.

Chapter 4 Review

Summary

- The program address M identifies a miscellaneous function.
- Miscellaneous functions can be used to control machine functions or program processing functions.
- Only one M-code can be placed on a single line of code in a CNC program.
- Miscellaneous codes that control program processing functions can be used to stop, pause, or end the program processing sequence.
- Miscellaneous codes can be used for a variety of machine functions. They act much like electrical switches to turn functions on and off.
- The M00 and M01 codes can be used to create program stops. The M00 code forces the program to stop each time it is encountered. The M01 program stop can be bypassed using a selector switch on the machine control.
- The M02 and M30 codes are used to end a program. The M02 code ends the program while the M30 code stops the program and rewinds to the program beginning.
- The M99 code is used primarily for subprograms to loop back into the main program after the subprogram is complete.
- Coolant functions are controlled with the M07, M08, and M09 codes. The M07 code turns on mist coolant, while the M08 code turns on flood coolant. The M09 code is used to turn off coolant.
- Spindle control functions are commanded with the M03, M04, and M05 codes. The M03 code initiates forward (clockwise) spindle motion. The M04 code initiates reverse (counterclockwise) spindle motion. The M05 code is used to cease all spindle motion.

Review Questions

Answer the following questions using the information provided in this chapter.

Know and Understand

1. M-codes are used to control _____.
   A. machine movements
   B. cutting commands
   C. machine functions and program functions
   D. work offsets

2. M-codes are considered _____ functions.
   A. preparatory
   B. compensation
   C. interpolation
   D. miscellaneous

3. The _____ command will initiate a full program stop.
   A. M03
   B. M00
   C. M04
   D. M05

4. *True or False?* Multiple M-codes can be used on a single line of a program.

5. The _____ code must be programmed in conjunction with a selector switch on the controller.
   A. M03
   B. M00
   C. M01
   D. M99

6. Which of the following codes is used to loop a program back to the main program or to the program start?
   A. M02
   B. M30
   C. M09
   D. M99

7. The M07 code will activate which of the following functions?
   A. spindle forward
   B. mist coolant on
   C. flood coolant on
   D. coolant off

8. *True or False?* The M03 function initiates spindle forward motion.

9. For the spindle rotation to be stopped, the _____ code is used.
   A. M08
   B. M11
   C. M05
   D. M50

10. For a tool change, which M-code must be utilized?
    A. M09
    B. M16
    C. M07
    D. M06

11. *True or False?* The M03 code is accompanied by a spindle speed specification.

12. If using a left-handed cutting tool, which M-code would be used specifically for that tool?
    A. M04
    B. M03
    C. M08
    D. M99

13. On Haas machines, a specific M-code is used for delivering coolant through the spindle. Which code is used?
    A. M09
    B. M07
    C. M08
    D. M88

Apply and Analyze

1. Why is it good practice to include notes in a program?

2. Referring to the blanks provided in the following program, give the correct M-codes to turn the spindle on clockwise and turn the spindle off before the machine returns to the home position.

| G00 G90 G54 X-.75 Y.25 S3820 _____ |
| G43 H1 Z2. M08 |
| Z.2 |
| G01 Z-.5 F20. |
| X-.25 F45.84 |
| G00 Z2. |
| _____ |
| G91 G28 Z0. M09 |
| M01 |

3. Referring to the blank provided in the following program, give the correct M-code to initiate mist coolant after the spindle is turned on and before the machine travels down the Z axis.

| G00 G90 G54 X-.75 Y.25 S3820 M03 |
| _____ |
| G43 H1 Z2. |
| Z.2 |
| G01 Z-.5 F20. |
| X-.25 F45.84 |
| G00 Z2. |

4. Referring to the blank provided, give the correct M-code to initiate an optional stop before the tool change.

| G00 G90 G54 X1. Y–1. S4584 M03 |
| G43 H2 Z2. M08 |
| G98 G81 Z–.25 R.1 F45.84 |
| Y–2. |
| Y–3. |
| G80 |
| M05 |
| G91 G28 Z0. M09 |
| ____ |
| T3 M06 |
| G00 G90 G54 X1. Y–1. S2292 M03 |
| G43 H3 Z2. M08 |

5. Referring to the blanks provided, give the correct M-codes to turn on the spindle counterclockwise, initiate flood coolant before the machine travels down the Z axis, turn off the spindle before the machine returns to the home position, and end and rewind the program.

| T2 M06 (Left-hand drill) |
| G00 G90 G54 X1. Y–1. S4584 ____ |
| |
| G43 H2 Z2. |
| G98 G81 Z–.25 R.1 F45.84 |
| Y–2. |
| Y–3. |
| G80 |
| |
| G91 G28 Z0. M09 |
| |

Critical Thinking

1. Think about the functions performed by the appliances in your home. If the appliance functions were controlled by M-codes, which functions would be modal and which would be nonmodal? For example, the different cycles of a washing machine might be considered to be modal because they are active for a period of time and end before the start of the next cycle. Which appliances have both modal and nonmodal functions? Which appliances have only nonmodal functions?

2. Why is it important to have spare M-codes available in a CNC machine? What are some add-on equipment features that could be used in a CNC machining operation? Research CNC door openers online and determine how door-opening systems might assist in high-production manufacturing.

3. Consider the placement of M-codes in a CNC program. What are some possible benefits to starting these functions at different locations inside the program? What are some possible machining failures that could result from specifying an M-code in the wrong location in a program?

5 | Address Codes for Mill Programming

Chapter Outline

5.1 Introduction
5.2 Address Codes
   5.2.1 A: Rotary Axis Movement—X Axis
   5.2.2 B: Rotary Axis Movement—Y Axis
   5.2.3 C: Rotary Axis Movement—Z Axis
   5.2.4 D: Tool Diameter Offset
   5.2.5 F: Feed Rate
   5.2.6 G: Preparatory Command
   5.2.7 H: Tool Length Offset
   5.2.8 I: Arc Center Location
   5.2.9 J: Arc Center Location
   5.2.10 K: Arc Center Location
   5.2.11 L: Loop Count
   5.2.12 M: Miscellaneous Function
   5.2.13 N: Block Number
   5.2.14 O: Program Number
   5.2.15 P: Program Number and Dwell
   5.2.16 Q: Repeat Depth
   5.2.17 R: Rapid Plane Position and Arc Radius
   5.2.18 S: Spindle Speed
   5.2.19 T: Tool Number
   5.2.20 U, V, and W: Alternate Axis Designations
   5.2.21 X: Axis Movement
   5.2.22 Y: Axis Movement
   5.2.23 Z: Axis Movement

Learning Objectives

After completing this chapter, you will be able to:

- Differentiate between the terms *word*, *address*, and *block* in CNC programming.
- Explain the purpose of address codes.
- Identify common address codes in CNC milling operations.
- Use address codes in a milling program.

Key Terms

address
address code
block
canned cycle
dwell
end-of-block (EOB)
program number
word

Chapter opening photo credit: Dmitry Kalinovsky/Shutterstock.com

5.1 Introduction

CNC programs are constructed using *words* to create *blocks* of code. As you learned in Chapter 3, a *word* is a letter and number grouping used to execute a command in a program. The letter preceding the number is called an *address*. This chapter looks specifically at the letter addresses that define what the machine controller should do with the numerical data that follows in a block of code.

The word M03, for example, consists of the address *M* and the number *03*. This word signals the machine controller to turn on the spindle clockwise. The letter M designates a miscellaneous function and the number 03 designates the spindle forward function. This letter and number combination instructs the machine to execute a specific function. You will not find individual letters or numbers in a CNC program, only *words* that fully define the function or command.

A *block* is a complete line of CNC code. A block represents a single line in a program and can be a single word or a series of words. The following lines represent three blocks of code:

M08	Coolant on
M01	Optional stop
G00 G90 G54 X–3.2 Y1.8 M03 S5000	Rapid mode, absolute mode, G54 work offset, go to XY position, spindle on clockwise to 5000 rpm

A block is often referred to simply as a *line of code*. A line of code is much like a sentence. Each block in a program ends with a semicolon (;). The semicolon character is used to separate each line and is known as the *end-of-block (EOB)* character. In software, this character is automatically generated by pressing the Enter key on the keyboard at the end of a block. The semicolon character is used in processing and is not entered by the programmer. It is displayed on the control screen to represent the end of each line. On the machine, the blocks of code in the previous program would look like this:

| M08; |
| M01; |
| G00 G90 G54 X–3.2 Y1.8 M03 S5000; |

The semicolon is read by the machine controller as the EOB character. The CNC machine control panel will have an EOB key. This key is pressed at the end of a block when entering program code or making edits manually at the control.

5.2 Address Codes

An *address code* is a letter code that is coupled with a number to create a word. The address code defines what the machine controller should do with the numerical data that follows. Address codes are fairly standard between machine controllers and even between lathe and mill applications. Address codes used in lathe programming are covered in later chapters. **Figure 5-1** lists commonly used address codes for milling operations.

Commonly Used Address Codes

Code	Description
A	Rotary or indexing axis about X axis
B	Rotary or indexing axis about Y axis
C	Rotary or indexing axis about Z axis
D	Tool diameter offset number used in cutter compensation
E	Second feed rate function
F	Feed rate function
G	Preparatory command
H	Tool length offset
I	X axis distance in circular interpolation or alternate X axis designation
J	Y axis distance in circular interpolation or alternate Y axis designation
K	Z axis distance in circular interpolation or alternate Z axis designation
L	Loop count for canned cycles
M	Miscellaneous function
N	Block number
O	Program number
P	Program designator in subprogramming or dwell time
Q	Canned cycle repeat depth
R	Arc radius in circular interpolation or rapid plane reference in canned cycles
S	Spindle speed function
T	Tool number designator
U	Axis motion for optional external U axis
V	Axis motion for optional external V axis
W	Axis motion for optional external W axis
X	X axis movement
Y	Y axis movement
Z	Z axis movement

Goodheart-Willcox Publisher

Figure 5-1. Address codes commonly used in mill programming.

5.2.1 A: Rotary Axis Movement—X Axis

The A address is used to specify rotary axis movement where rotation is around the machine's X axis. It specifies angular rotation and is followed by a positive or negative number that will delineate the distance and direction of rotation.

```
G00 G90 G54 X0. Y0. A–45.0
```

In this block, the machine will complete a rapid move to X0. Y0. and a rotary axis move to an angle of –45°. The angle can be expressed to a maximum of three decimal places, or in thousandths of a degree.

5.2.2 B: Rotary Axis Movement—Y Axis

The B address is used to specify rotary axis movement where rotation is around the machine's Y axis. It specifies angular rotation and is followed by a positive or negative number that will delineate the distance and direction of rotation.

```
G01 X-3. B12.5 F38.5
```

In this block, the machine will complete a feed move to X–3. and a simultaneous rotary axis move to an angle of 12.5°.

5.2.3 C: Rotary Axis Movement—Z Axis

The C address is used to specify rotary axis movement where rotation is around the machine's Z axis. It specifies angular rotation and is followed by a positive or negative number that will delineate the distance and direction of rotation.

```
G00 Z10.
C100.
```

In this block, the machine will complete a rapid move to Z10. and a rotary axis move to an angle of 100°.

5.2.4 D: Tool Diameter Offset

The D address is used in generating cutter compensation in a milling program. The D address and following number designates the diameter offset that the controller will use to calculate the proper cutting position. The D address references the diameter or radius setting of the tool offset stored in the machine controller.

```
G01 G41 D6 X0. Y-5. F62.
```

In this block, the cutter will move to X0. Y–5. while initiating cutter compensation left. The control will use the diameter or radius that is input into the tool #6 offset value.

5.2.5 F: Feed Rate

The F address is used to command the feed rate, in inches per minute (ipm), that the cutter will travel. The feed rate specified in the program is modal. The last feed rate commanded will stay active until a new feed rate is designated.

```
G01 Z-2.5 F35.
X10. Y-3.
```

In the first block, the machine will travel to Z–2.5 at a 35 ipm feed rate. It will stay at a feed rate of 35 ipm to travel to X10. Y–3.

5.2.6 G: Preparatory Command

The G address is used for a preparatory command. The purpose of the preparatory command is to prepare the machine controller, or preset the machine, into some condition of operation. A G-code command can be modal or nonmodal. Multiple G-code commands can be used on the same block in a program, as long as they do not have conflicting functions.

```
G00 G90 G54 X9.
```

In this block, the machine executes a rapid move to the absolute position X9. in the first work offset.

5.2.7 H: Tool Length Offset

The H address is used to designate the tool length offset for a cutting tool in a milling program. The tool length offset, sometimes referred to as the tool height offset, is a measurement used to compensate for different lengths of tools used in machining. The tool length offset represents the distance from the machine home Z axis position to the work coordinate system Z0. position. Because each tool has a different length, the machine controller must be able to reference a measurement for the Z axis position of the tool relative to the machine home position. See **Figure 5-2**. During machine setup, the tool length offset value is entered by the operator on the tool offset page in the machine controller. The H address and the number that follows in a program references the stored value.

There are different methods and equipment used to establish tool length offsets in machining. On some setups, the tool length offset represents the

IndustrialVibes/Shutterstock.com

Figure 5-2. Tools of various lengths require accurate tool length offsets to be specified in the machine controller.

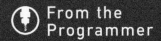

From the Programmer

Tool and Offset Numbers

Most programmers will use the tool number as the same number for the D offset and H offset designation. For example, tool #1 uses D1 and H1 for the tool diameter and length offsets, which is easy to remember. But you are not required to use the same number. Tool #1 can use any D or H offset number, and in some cases you can use more than one D or H offset for the same tool. However, be careful. If you use the wrong offset for a tool, a crash is bound to happen.

true length of the tool measured from the gage line of the spindle to the tool tip. The gage line is an established reference point on the spindle.

In most cases, the G43 command is used with an H address in a program to specify the tool length offset. The following is an example of designating the tool length offset after a tool change.

| T2 M06 |
| G54 X0. Y0. M03 S3500 |
| G43 H2 Z1. |

In the third block in this example, the G43 command and H2 tool offset designation instructs the control to reference the tool #2 offset value.

5.2.8 I: Arc Center Location

The I address is used to designate the location of the arc center in programmed radial moves. The I address value designates the distance and direction from the arc start point to the arc center point along the X axis. The specified I value represents an incremental distance measured from the arc start point.

The I address is used with circular interpolation commands in arc programming. In the following example, the G03 command is used to create radial tool movement in a counterclockwise direction. See **Figure 5-3**.

Y0.	
G03 X0. Y2. I–2.0	The I value represents the distance from the arc start point to the arc center point
G01 X–2.5	

5.2.9 J: Arc Center Location

The J address is used to designate the location of the arc center in programmed radial moves. The J address value designates the distance and

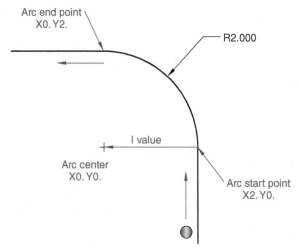

Goodheart-Willcox Publisher

Figure 5-3. The I address is used in arc programming to specify the distance and direction from the arc start point to the arc center point along the X axis.

direction from the arc start point to the arc center point along the Y axis. The specified J value represents an incremental distance measured from the arc start point.

In the following example, the G02 circular interpolation command is used with the J address to create radial tool movement in a clockwise direction. See **Figure 5-4**.

X0.	
G02 X2. Y0. J–2.0	The J value represents the distance from the arc start point to the arc center point
G01 Y–2.	

5.2.10 K: Arc Center Location

The K address is used to designate the location of the arc center when programming radial moves in the XZ or YZ plane. The K address value designates the distance and direction from the arc start point to the arc center point along the Z axis. To program circular interpolation moves in the XZ or YZ plane, the G18 or G19 command is used to activate the corresponding plane. In the following example, the G02 circular interpolation command is used with a K address value to create radial tool movement in a clockwise direction in the XZ plane.

X0. Z0.	
G02 X2. Z–2. K–2.0	The K value represents the distance from the arc start point to the arc center point
G01 Z–2.5	

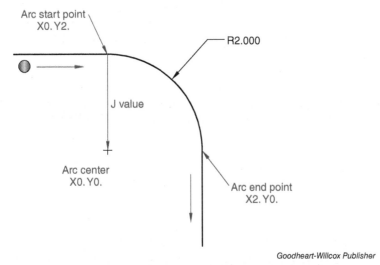

Goodheart-Willcox Publisher

Figure 5-4. The J address is used in arc programming to specify the distance and direction from the arc start point to the arc center point along the Y axis.

5.2.11 L: Loop Count

The L address is used to create loops (repetitive cycles) in a CNC program. The L address can be used in incremental programming, such as a drilling cycle that repeats at equal distances apart. The most common use is in subprogramming to repeat the same program multiple times. The following is an example of using the L address to repeat a *canned cycle* for drilling holes at equal increments.

G98 G81 R.5 Z–1. F10.	
G91 X.5 L10	The L10 word repeats the cycle 10 times

The following is an example of using the L address in subprogramming to repeat a program.

M98 P1020 L3	The L3 word repeats the O1020 program three times

5.2.12 M: Miscellaneous Function

The M address is used to designate a miscellaneous function. M-codes control machine functions and program functions. Only one M-code can be used on a single block of code.

G00 G90 G54 X0. Y.5 M03 S5000	M03 initiates forward rotation of the spindle
M08	M08 initiates flood delivery of coolant

5.2.13 N: Block Number

The N address is used for block numbering in a program. A block number, or sequence number, identifies a block of code with a sequential number. An N address number can be placed in front of any block for reference purposes and searching capabilities.

In older controllers, block numbers provided the easiest way to search through blocks to find a specific place in the program. Many programmers will still place block numbers at tool changes or strategic locations to aid in searching for content, but this is optional. Block numbers are primarily used for organization and may be omitted to reduce program length. In most cases, block numbers have no purpose other than numbering the lines. However, some canned cycles in lathe programming require certain blocks to be numbered in order to be referenced.

The following example shows a typical way to number blocks in a program. Blocks may be numbered in increments as shown so that additional lines can be inserted between existing blocks if needed.

N10 T10 M06
N20 G00 G90 G54 X–.5 Y1. M03 S5000
N30 G43 H10 Z2.0

5.2.14 O: Program Number

The O address is used to designate the *program number*. It is only used in the opening block of a program. Each program begins with the letter O followed by a four- or five-digit number. It is good practice to use a number that represents a part number or engineering number so that the part created by the program can be easily identified.

It is common to include a program name or description on the same line as the program number. The name or description is enclosed in parentheses. Sometimes, a revision number is included as part of the program name, as shown in the following example.

```
O42378 (Part #13-42378 Rev 1)
G90 G17 G20 G80 G40
T4 M06
```

The numbering format used by the programmer should follow the naming standards of the company or organization. On some machines, certain program numbers are reserved for special use. For example, program numbers in the 9000 series (such as O9001, O9002, and so on) are reserved for macro programs on some machines and should not be used for general purpose part programs.

5.2.15 P: Program Number and Dwell

The P address is used with several different commands and functions in programming. One application is in subprogramming to designate a program number. In the main program, a programming function to exit that program and enter into another program must be specified. The M98 function is used for this purpose. The M98 function is followed by a P address with the appropriate subprogram number, as shown in the following example.

```
M98 P1356
```

In this block, the M98 function is the subprogram call into program O1356 (referenced by P1356).

Another application for the P address is in canned cycles to call a dwell function. *Dwell* allows for a pause when the tool is at full depth. The P address value designates the amount of time for the pause. A number with a decimal point represents seconds and a number without a decimal point represents milliseconds.

Example 1:

```
G82 Z–0.720 P0.3 R0.1 F15.     Specified dwell is .3 seconds
```

Example 2:

```
G82 Z–0.720 P5 R0.1 F15.       Specified dwell is 5 milliseconds
```

5.2.16 Q: Repeat Depth

The Q address is used with canned cycle commands for hole machining operations. The Q address is used with a designated value to specify incremental depth positions in a drilling cycle.

```
G83 Z–0.875 Q0.1 R0.1 F15.
```

In this block, the Q0.1 value designates a .100″ incremental peck in a peck drilling cycle. Peck drilling refers to drilling a hole to a partial depth and retracting the tool before drilling deeper. The Q address is also used in the G73 canned cycle for high-speed peck drilling.

5.2.17 R: Rapid Plane Position and Arc Radius

The R address is used with canned cycle commands and with circular interpolation commands in arc programming. In one application, the R address is used to designate a rapid plane defining the retract position of the tool in a drilling cycle.

```
G83 Z–0.875 Q0.1 R0.1 F15.
```

In this block, the R address references the rapid plane position above the part where the tool moves to begin the cycle and then rapids out to end the cycle. The R0.1 address value specifies the Z position of the rapid plane (Z.1).

The R address is also used with circular interpolation commands to identify the size of a radius being machined. In this application, the R address value defines the radius of the arc in radial movement. This programming method for arcs specifies the arc radius rather than the distance from the arc start point to the arc center point.

```
G03 X1.25 Y2.625 R.75
```

In this block, the R.75 value indicates a .750″ (3/4″) radius.

5.2.18 S: Spindle Speed

The S address is used to set the spindle speed in revolutions per minute (rpm). In the following example, the spindle is turned on clockwise and the spindle speed is set at 3500 rpm.

```
M03 S3500
```

5.2.19 T: Tool Number

The T address is used to designate the tool number for a tool change. The numerical value following the T address specifies the number of the tool to be loaded by the machine in the spindle when a tool change function is called. The T address only designates the number of the tool and does not control the tool change function. The M06 machine function instructs the machine to perform a tool change.

On a CNC machining center, an automatic tool changer is used to store tools and perform tool changes automatically. Different machines have different types of equipment for this purpose. On some machines, the tool is brought into a ready position in the tool carousel when the corresponding tool number is referenced with the T address. Placing the T address in advance on a block earlier in the program allows the tool change to occur instantly when the M06 machine function is read by the machine.

On a machine equipped with an umbrella-style tool changer, each tool must be placed in a numbered pocket corresponding to the number of the tool. See **Figure 5-5.** When a tool change occurs on this type of machine, the tool currently in use must first be placed in its pocket on the umbrella carousel. Then, the carousel is rotated to the subsequent tool called in the program. This adds time to each tool change. On this type of machine, the tool designation and tool change function are programmed as follows.

```
T4 M06
```

When this block is processed, the machine places the current tool in the spindle back into its correct location and then rotates the umbrella to the #4 position. Once rotation is complete, tool #4 will be inserted into the spindle.

A machine with a side-mount tool changer allows the tool to be prepositioned for a faster tool change. See **Figure 5-6.** On this type of machine, the T address and tool change function can be written on the same block or the T address can be programmed several lines ahead of the tool change as shown below.

```
T4
...
...
...
M06
```

Lutsenko_Oleksandr/Shutterstock.com

Figure 5-5. A machining center with an umbrella-style tool changer. Each tool is stored in a numbered pocket. During a tool change, the tool currently in use must be returned to its corresponding pocket before the tool called in the program is located and inserted in the spindle.

Goodheart-Willcox Publisher

Figure 5-6. A side-mount tool carousel with 24-tool storage capacity. The machine moves the tool to the ready position when the T address is initiated in the program. This function allows for tool changes to occur in less than one second.

In this program, the side-mount tool carousel positions itself in advance to the #4 position, and the tool change occurs immediately when the M06 function is initiated.

5.2.20 U, V, and W: Alternate Axis Designations

The U address is used to designate movement on an alternate axis parallel to the X axis. Just as the I, J, and K addresses are alternate designations for the X, Y, and Z axes in circular interpolation moves, the U, V, and W axis designations can be substituted for X, Y, and Z. Although this is not common in milling, the U, V, and W addresses can be used for incremental positioning on the alternate axes.

5.2.21 X: Axis Movement

The X address is used as a primary axis designation. The X axis is the machine axis that designates movement from left to right, in relationship to the cutting tool. It is a standard axis in CNC mill programming. In the following example, the second block contains an X coordinate position of –.5, as measured in the coordinate system defined by the G54 work offset.

```
T10 M06
G00 G90 G54 X-.5 Y1. M03 S5000
G43 H10 Z2.0
```

5.2.22 Y: Axis Movement

The Y address is used as a primary axis designation. The Y axis is the machine axis that designates movement from back to front, in relationship to the cutting tool. It is a standard axis in CNC mill programming. In the following example, the second block contains a Y coordinate position of 1., as measured in the coordinate system defined by the G54 work offset.

```
T10 M06
G00 G90 G54 X-.5 Y1. M03 S5000
G43 H10 Z2.0
```

5.2.23 Z: Axis Movement

The Z address is used as a primary axis designation. The Z axis is the machine axis that designates vertical movement through the centerline of the cutting tool. It is a standard axis in CNC mill programming. In the following example, the third block contains a Z coordinate position of 2.0, as measured in the coordinate system defined by the G54 work offset. The Z position of the tool is relative to the tool length offset specified for tool #10.

```
T10 M06
G00 G90 G54 X-.5 Y1. M03 S5000
G43 H10 Z2.0
```

Chapter 5 Review

Summary

- A word is a letter and number grouping used to execute a command in a program. The letter preceding the number is called an address.
- A block is a complete line of CNC code. A block represents a single line in a program and can be a single word or a series of words.
- An address code is a letter code that is coupled with a number to create a word. The address code defines what the machine controller should do with the numerical data that follows.
- The A address is used to specify rotary axis movement where rotation is around the machine's X axis. The B address is used to specify rotary axis movement where rotation is around the machine's Y axis. The C address is used to specify rotary axis movement where rotation is around the machine's Z axis.
- The D address is used in generating cutter compensation in a milling program.
- The F address is used to command the feed rate, in inches per minute (ipm), that the cutter will travel.
- The H address is used to designate the tool length offset for a cutting tool in a milling program.
- The I, J, and K addresses are used to designate the location of the arc center in programmed radial moves.
- The N address is used for block numbering in a program. The O address is used to designate the program number.
- The S address is used to set the spindle speed in revolutions per minute (rpm).
- The T address is used to designate the tool number for a tool change.

Review Questions

Answer the following questions using the information provided in this chapter.

Know and Understand

1. The letter preceding the number in a word is called a(n) ____.
   A. symbol
   B. block
   C. address
   D. designator

2. Single lines of CNC code are referred to as ____.
   A. addresses
   B. words
   C. sentences
   D. blocks

3. An address code is a letter code that is coupled with a number to create a ____.
   A. sentence
   B. word
   C. program
   D. block

4. *True or False?* Each block in a program ends with a period.

5. *True or False?* An address code defines what the machine controller should do with the numerical data that follows.

6. On a CNC mill, the C address is used to specify rotary axis movement around which axis?
   A. W
   B. X
   C. Y
   D. Z

7. On a CNC mill, the H address is used to designate the ____.
   A. spindle speed
   B. feed rate
   C. tool diameter offset
   D. tool length offset

8. On a CNC mill, the P address is used to designate the ____.
   A. spindle speed
   B. subprogram number
   C. feed rate
   D. rapid plane

9. *True or False?* The C address is used in generating cutter compensation.

10. The S address is used to set the ____.
    A. tool length offset
    B. feed rate
    C. spindle speed
    D. tool number

11. The O address is used to designate the _____ number.
    A. block
    B. sequence
    C. program
    D. tool

12. *True or False?* The R address can be used to designate the size of the arc radius in arc programming.

13. The _____ address is used in arc programming to designate the incremental distance from the arc start point to the arc center point along the X axis.
    A. I
    B. J
    C. K
    D. L

14. The _____ address is used in arc programming to designate the incremental distance from the arc start point to the arc center point along the Y axis.
    A. I
    B. J
    C. K
    D. L

15. The _____ address is used to designate the tool number for a tool change.
    A. M
    B. T
    C. H
    D. D

Apply and Analyze

1. Describe the purpose of the H address code and identify the command with which it is typically used in a program.

2. Describe two different applications for the R address code in canned cycles and arc programming.

3. Referring to the blanks provided, give the correct address codes to designate tool #1 as the tool length offset and set the initial feed rate for cutting to 20 ipm.

| G00 G90 G54 X-.75 Y.25 S3820 M03 |
| G43 _____ Z2. M08 |
| Z.2 |
| G01 Z-.5 _____ |
| X-.25 F45.84 |
| G00 Z2. |

4. Referring to the blanks provided, give the correct address codes to set the feed rate for cutting to 10 ipm and designate the incremental distance to complete the circular interpolation move shown.

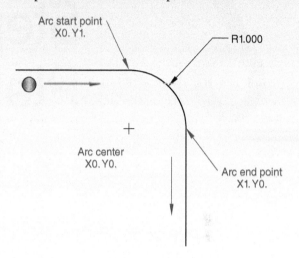

| G01 X0. Y1. _____ |
| G02 X1. Y0. _____ |
| G01 Y-2. |

Critical Thinking

1. Address codes vary in other types of CNC equipment in comparison to milling machines. Research other types of CNC machines and describe some of the alternative equipment that they use. Is there a need for specific address codes for other types of machines, and what purpose do they serve?

2. A wide variety of different letters and numbers can be used as entries in a single block in a CNC program. Each letter and number combination represents something specific to the machine. What are some other software programming languages? How are they different in comparison to CNC programming?

3. Explain the importance of designating address codes for tool offsets correctly. What might happen if the wrong tool length offset value is entered for a tool in a program? How would a part be affected in machining if the wrong offset value is specified for cutter compensation?

6 Steps in Program Planning

Chapter Outline

6.1 Introduction
6.2 Print Review
6.3 Part Workholding
   6.3.1 Vises
   6.3.2 Fixtures
   6.3.3 Vacuum Tables
   6.3.4 Magnetic Chucks
6.4 Tool Selection
6.5 Order of Operations

Learning Objectives

After completing this chapter, you will be able to:

- Explain the importance of preplanning a CNC mill program.
- Review a print to determine critical features and machining strategy.
- Explain how to establish an appropriate workholding setup based on the final part and program.
- Identify tools needed to create a CNC program.
- Identify features that may reduce the total number of tools required to produce a part.
- Define the order of operations in a program.

Key Terms

bull nose end mill	fixture	spot drill
design intent	order of operations	tapping
end mill	reamer	workholding

6.1 Introduction

In this chapter, you will learn how a print can tell you exactly how to machine a part. You will learn how complex operations can be managed by planning the production sequence step-by-step. You will also learn how workholding impacts part production.

Think of the many aspects of machining a part that must be planned and communicated to the machine. What tools are available? How do you control where each tool goes and how fast it gets there, and how do you determine how fast you can machine a piece of material? These are things that are all controlled by the CNC program—and, ultimately, by the programmer.

The most critical, and often overlooked, steps in creating a successful CNC part program occur during the planning stage. Beginning programmers may grab a print and start producing code, adding tools that make sense to them. This approach overlooks the complex issues of part workholding, machine availability, and the part geometry to be machined. An experienced programmer will spend time considering the machines and tooling available, the most efficient workholding method, and the customer's requirements.

It is always a good idea to assemble a team when preparing to program a part and move it to production. A customer representative (perhaps a salesperson or manager), someone from the engineering staff, a setup specialist, and an operator could all provide helpful feedback in the planning process to make a quality finished part. Every program produced and every part manufactured should give end users a product that fits their needs in a cost-effective manner, while giving your company a profitable solution. This type of customer service and smart planning takes a team of highly skilled and organized professionals.

In a machine shop environment, there can be a perception or culture that leaves all of the part production decisions to the programmer. Maybe someone even says, "There just isn't enough time to meet with all these people." Maybe the programmer believes it is easier to work out all of these decisions alone and thinks programs can be created without other input to save time. Consider this in a different way. There is not enough time or money to make scrap parts or pay for the cost of scrap parts. Parts made incorrectly represent wasted time and can lead to missed deadlines or unhappy customers. Mistakes in part production are extremely expensive and can lead to the demise of a machine shop.

Developing a CNC program requires effective planning and decision making. The following general steps are used in the program planning stage to improve efficiency and eliminate unnecessary loss of time. The order of these steps may vary depending on company practice, but this is a typical sequence.

1. Review the print.
2. Establish the part workholding.
3. Identify the required tools.
4. Determine the order of operations.

6.2 Print Review

Interpreting design intent is an important first step in reviewing a print. The term *design intent* refers to the way in which the part is dimensioned and noted to define critical features of the part and the part function. For example, if a print is dimensioned exclusively from one point, that location must be critical to the function and orientation of the part. If a tolerance on a feature is ±.001″, that feature is critical. Conversely, if a feature's tolerance is ±.020″, that feature can be considered noncritical. By examining the tolerances of features and the originating points from which tolerances are established, a print reader can gain an understanding of how a part functions, what features are critical in manufacturing, and how the part can be machined.

The part shown in **Figure 6-1** is dimensioned as 6″ × 3″, with a total tolerance of .020″ for the 6″ dimension. The print reader can determine that this part could be manufactured from 3″ wide stock. In order to meet tolerance requirements, the 6″ length cannot be left as saw cut, but the part is not intended for a close fit into another assembly. To meet these requirements, the machinist can saw the 3″ wide material about 6.1″ long, and then machine both ends to 6.00″. For the hole size and location dimensions, the tolerances are a little tighter. The location tolerances reference the top-left corner of the part. A logical location for the work coordinate system (WCS) origin—the point where both the X and Y coordinate values are zero—is at the same corner on the top-left edge. In the Cartesian coordinate system, this would locate the part in quadrant 4, or the X+, Y– quadrant. This is a typical work setup that corresponds to using a vise to secure the workpiece so that the edge along the X axis is aligned against the fixed jaw. The coordinates for the part origin relative to machine home are established by the machinist during setup and entered into the work offset designated in the program. If multiple parts are to be machined in the same program using multiple work offsets, the program planning should reflect the quantity of parts required.

An important feature to this part's design is the hole diameter. The hole has a limit dimension of .500″–.501″. This dimension specifies a very close size tolerance of .001″. The tolerance specification actually tells you how

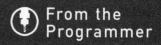

Print Specifications

The information on a print provides the specifications needed to manufacture a part. The machinist should make the part to meet the specifications and dimensions provided. Never attempt to measure or scale a drawing to determine a missing dimension or make an assumption about the designer's intent. If necessary, consult with the designer for clarification.

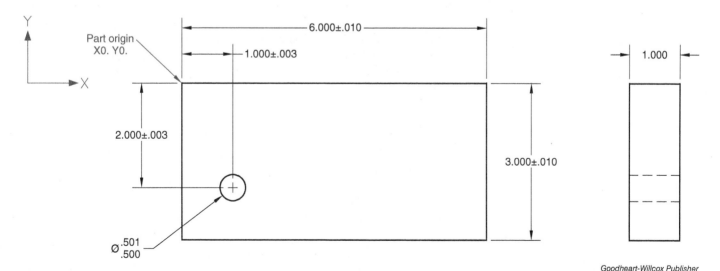

Goodheart-Willcox Publisher

Figure 6-1. A drawing of a 6″ × 3″ rectangular part with a hole feature. The dimensions and tolerances specified are used to determine how to machine the part.

to machine this feature. There are a variety of ways to make a hole in a part, including drilling, milling, reaming, and boring. A drilled hole must be spot drilled first and then drilled, producing a hole that is no closer than ±.003″ of a specified size. The resulting hole is not perfectly round or accurately positioned. A milled hole is accurate in position and roundness, but due to machining technique, milled holes can have some taper from top to bottom. A third option is a reamed hole, which must be spot drilled, drilled undersize, and then reamed. A *reamer* is a tool that only cuts on the sides and must enter through a previously drilled hole. For the hole shown in **Figure 6-1,** an adequate machining technique would be to drill the hole .485″ in diameter and then ream with a .5005″ reamer. The last option is to bore the hole. This is accomplished by drilling a hole and then using a boring head to machine a round, close-diameter hole. Boring achieves the best results—producing holes that are accurate in position and diameter—but it is slower than the other operations. Boring this hole is not the most cost-effective operation, but it is sometimes required to meet print specifications.

In communicating where dimensions are established and which features are most important, the print shown in **Figure 6-1** tells us how to machine this part. In summary, establish the top-left corner as the WCS origin, use an *end mill* to cut the part length to 6.00″, and then *spot drill*, drill, and ream a .5005″ hole. This part can be finished in four machining operations, using four tools. See **Figure 6-2.**

The part shown in **Figure 6-3** has a hole with a .500±.005″ diameter dimension. The size tolerance for this hole is .01″. This hole can be simply drilled. In addition, there is a pocket feature. The right-side view on the print shows this pocket is .500″ deep. The pocket also has .188″ radii in the corners. This means that a tool measuring 3/8″ in diameter or less must be used to create the correct corner radius.

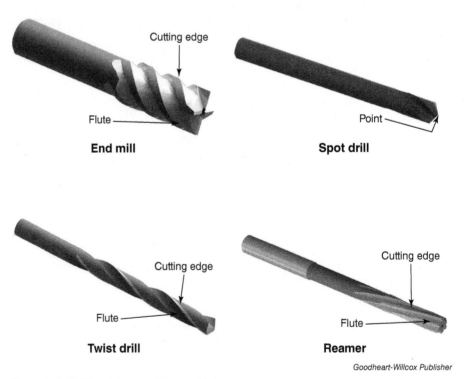

Goodheart-Willcox Publisher

Figure 6-2. Cutting tools used in machining.

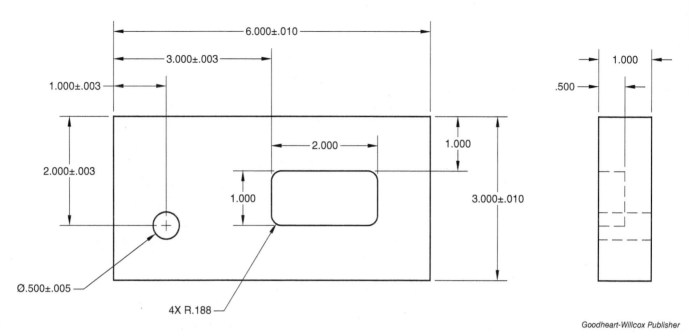

Figure 6-3. A drawing of a 6″ × 3″ rectangular part with a hole and a pocket feature. Based on tolerance specifications, the hole can be spot drilled and then drilled with a 1/2″ diameter drill.

A review of this print determines that the same size stock used for the previous part is suitable. The 3″ wide stock will be cut to 6.1″ long and then machined to an overall length of 6.00″. The .500″ hole will be spot drilled and then drilled with a 1/2″ diameter drill. The pocket should be roughed and then completed with a finish contour operation. By carefully examining the print, the print reader gains a much better understanding of the part, which can then inform the best machining process.

6.3 Part Workholding

Once the requirements of the print are fully understood, the programmer must decide how the part will be held on the machine. **Workholding** refers to any device that is used to secure a workpiece against the forces of machining. Holding the part securely during the machining process is critical for part accuracy and safety, but machinists also need to have access to machine as many features as possible in a single workholding configuration. This reduces setup and overall cycle time.

There are a variety of workholding methods and devices used to mount work in milling operations. Determining which workholding method to use is a critical decision that must be made in the preprogramming phase. If you do not know how you are going to hold the part, there is no way to determine how to machine it.

6.3.1 Vises

For some projects, workholding will be as simple as placing a workpiece in a vise. See **Figure 6-4.** A machine vise has a fixed jaw and a movable jaw for securing work. Bolting a vise on a mill table can be traced back to the earliest manual milling machine operations.

> **From the Programmer**
>
> **Machining Inside Corners**
> An inside corner should not be cut with a tool that has the same size as the corner radius. For example, if the corner radius is .188″, then a 3/8″ end mill will fit into that corner. However, using the same size tool will create gouging, or a poor finish, in that corner. It is better to rough with a 3/8″ end mill, leaving .010″ stock, and then program a finish contour with a 5/16″ end mill to create a good finish.

Goodheart-Willcox Publisher

Figure 6-4. A machine vise used for workholding. On a T-slot mill table, the vise is bolted directly to the table with T-slot nuts and studs or T-bolts.

Vises have improved greatly over the years. Vises can be mounted in a vertical or horizontal orientation on the mill table, and can hold a single piece of material or multiple pieces. In many operations, the fixed jaw of a vise is used as a reference surface aligned with the machine's X or Y axis to establish the work coordinate system position.

Often, soft jaws made from aluminum are used for holding work in a vise. Soft aluminum jaws are used to assist in holding odd-shaped parts and to protect finished work surfaces.

Multiple vises can be used on a milling machine to run several parts at once, or to perform multiple operations on the same part. The cost, ease of use, and versatility make vise workholding setups simple and reliable.

6.3.2 Fixtures

More complicated parts, such as castings, may require a specialty fixture to be constructed for workholding. A *fixture* is a custom workholding device used to position and secure an irregular-shaped workpiece for machining. A good fixture should have repeatability on workpiece placement. It should provide ease of use and full access to all areas that require machining. See **Figure 6-5**.

The use of specialty fixtures can greatly improve multiaxis milling operations. As the workpiece is rotated to machine on multiple sides, the fixture can secure the workpiece, while still providing access to machine.

6.3.3 Vacuum Tables

Some workpieces require specialized workholding devices. A vacuum table uses a pump to generate vacuum from below and hold down the workpiece. See **Figure 6-6**. Vacuum tables are often used to secure thin materials or large shapes that require cutting the entire periphery.

Although vacuum tables are not common in most machining scenarios, they do provide a great alternative when the need arises. Generally, only light machining passes are performed on a vacuum table to prevent the workpiece from becoming dislodged from the vacuum.

Aumm graphixphoto/Shutterstock.com

Figure 6-5. A fixture is used to secure complex workpieces and provide full access to machining.

Figure 6-6. A vacuum table is used in conjunction with a vacuum pump. The pump generates negative air pressure to create a strong vacuum and secure the workpiece.

6.3.4 Magnetic Chucks

Another alternative workholding solution is a magnetic chuck. Magnetic chucks are not very common in most machine shops, but they can be used successfully in the right situation. A magnetic chuck has a strong electromagnetic surface that can hold ferrous metals. The most common application for a magnetic chuck is in grinding operations where very small amounts of material are removed and very high tolerances are maintained. **Figure 6-7** shows a magnetic chuck with the ability to tilt at different angles.

6.4 Tool Selection

The next important decision in program planning is determining what tools to use. Start by making a list of tools and identify any part surfaces and

Figure 6-7. A magnetic chuck is used to secure workpieces made from ferrous metals. This device is a magnetic sine plate.

> **From the Programmer**
>
> **Machining Slots**
>
> A slot should not be cut with a tool that has the same size as the slot width. If machining a .625" (5/8") wide slot, using a 5/8" diameter end mill will leave a rough finish. Slot milling normally requires a roughing operation followed by a finishing operation. Using a tool with the same size as the slot width will leave no space for chip evacuation and cause the chips to drag through the cut, leaving a rough surface finish. Use a smaller end mill (1/2"), rough down the middle, and then finish each side individually.

features that may require a special or uncommon tool. **Figure 6-8** shows a part with a boss. The transition between the boss and base surface has a .100" radius. This transition would require a *bull nose end mill*. See **Figure 6-9**. A bull nose end mill does not have a square edge, but rather a rounded edge to create a radius transition like the one in **Figure 6-8**. The tool selected for this operation is a 1/2" bull nose end mill with a .100" radius.

To reduce the number of tool changes and cycle time, programmers aim to minimize the number of tools used. Always evaluate the possibility of using the same tools for multiple features. For example, if cutting a .625" wide slot and a .750" wide slot on one part, a 1/2" end mill can accomplish both cuts without using a second tool. See **Figure 6-10**.

In most cases, it is also best practice to spot drill every hole before drilling. If any specialized tooling is required, it is best to identify this in the planning phase so that it can be accounted for while programming.

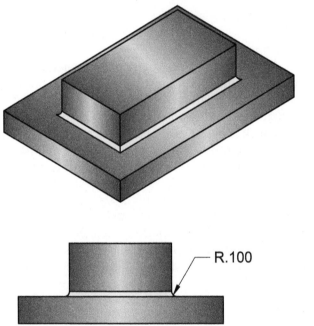

Goodheart-Willcox Publisher

Figure 6-8. A rectangular part with a radius transition between the boss and base surface.

Goodheart-Willcox Publisher

Figure 6-9. A bull nose end mill has a rounded edge to create a radius transition.

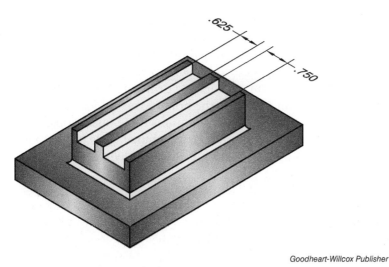

Goodheart-Willcox Publisher

Figure 6-10. A rectangular part with slots of different widths cut with a 1/2″ end mill.

6.5 Order of Operations

To this point in the planning, the print has been reviewed, workholding has been established, and tooling has been selected. These procedures are normally followed specifically in that order, although there might be cases when a slightly different order is taken. The last decision to be made is the sequence in which machining operations will occur, or the *order of operations*. Determining the order of operations is normally straightforward, but it is often possible to make smart decisions with the machining sequence to translate into reduced cycle times and higher-quality parts. The following examples show how the order of operations can make a significant difference.

Figure 6-11 shows a .5″ thick plate with two machined pockets and 12 tapped holes. *Tapping* forms internal threads in a drilled hole through use of a tap. The two pockets are different sizes. It is common to think of each pocket separately and rough one pocket and finish it before moving to the next pocket. If both pockets are roughed with one tool first, and then

From the Programmer

Program Entry Format

As discussed in previous chapters, on most machines, programming commands with a zero in the digit portion can be shortened to a single digit following the code letter. For example, G00 can be shortened to G0 and G01 can be shortened to G1. This format is used in programs in this chapter and in remaining chapters of this textbook.

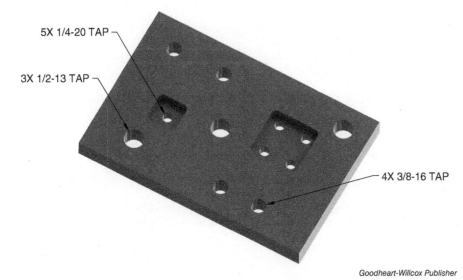

Goodheart-Willcox Publisher

Figure 6-11. A plate with 12 tapped holes. The part also includes two pockets. The sample program is used for machining the four 3/8–16 tapped holes.

finished with a second tool, two tool changes are eliminated. Tool changes require time and reduce the efficiency of the machining.

The real opportunity for gaining improved efficiencies through the order of operations comes from the tapped holes. Before a hole is tapped, it must be drilled with the appropriate drill size. Most drilled holes require a spot drill operation to ensure the hole starts on location and straight. This means that each of the 12 holes requires three operations. In the following program, the four 3/8–16 holes are machined. This program is used to spot drill, drill, and tap each of the four holes individually.

```
O0425
(DATE: 10-18-20)
(PROGRAMMER: REC)
(MATERIAL - STEEL INCH - 1030 - 200 BHN)
(T1 | 1/2 SPOT DRILL)
(T2 | 5/16 DRILL)
(T3 | 3/8–16 TAP)
G20
G0 G17 G40 G49 G80 G90
T1 M6 (T1 | 1/2 SPOT DRILL)
G0 G90 G54 X-5.1951 Y3.2958 S3500 M3
G43 H1 Z2. M8
G98 G81 Z-.125 R.1 F45.84
G80
M5
G91 G28 Z0. M9
M01
T2 M6 (T2 | 5/16 DRILL)
G0 G90 G54 X-5.1951 Y3.2958 S855 M3
G43 H2 Z2. M8
G98 G81 Z-.6 R.1 F4.24
G80
M5
G91 G28 Z0. M9
M01
T3 M6 (T3 | 3/8–16 TAP)
G0 G90 G54 X-5.1951 Y3.2958 S713 M3
G43 H3 Z2. M8
G98 G84 Z-.6 R.1 F44.5625
G80
M5
G91 G28 Z0. M9
M01
T1 M6 (T1 | 1/2 SPOT DRILL)
G0 G90 G54 X-3.8481 Y3.2958 S3500 M3
G43 H1 Z2. M8
G98 G81 Z-.125 R.1 F45.84
G80
```

Continued

```
M5
G91 G28 Z0. M9
M01
T2 M6 (T2 | 5/16 DRILL)
G0 G90 G54 X-3.8481 Y3.2958 S855 M3
G43 H2 Z2. M8
G98 G81 Z-.6 R.1 F4.24
G80
M5
G91 G28 Z0. M9
M01
T3 M6 (T3 | 3/8-16 TAP)
G0 G90 G54 X-3.8481 Y3.2958 S713 M3
G43 H3 Z2. M8
G98 G84 Z-.6 R.1 F44.5625
G80
M5
G91 G28 Z0. M9
M01
T1 M6 (T1 | 1/2 SPOT DRILL)
G0 G90 G54 X-1.4471 Y.6313 S3500 M3
G43 H1 Z2. M8
G98 G81 Z-.125 R.1 F45.84
G80
M5
G91 G28 Z0. M9
M01
T2 M6 (T2 | 5/16 DRILL)
G0 G90 G54 X-1.4471 Y.6313 S855 M3
G43 H2 Z2. M8
G98 G81 Z-.6 R.1 F4.24
G80
M5
G91 G28 Z0. M9
M01
T3 M6 (T3 | 3/8-16 TAP)
G0 G90 G54 X-1.4471 Y.6313 S713 M3
G43 H3 Z2. M8
G98 G84 Z-.6 R.1 F44.5625
G80
M5
G91 G28 Z0. M9
M01
T1 M6 (T1 | 1/2 SPOT DRILL)
G0 G90 G54 X-2.4719 Y.6313 S3500 M3
G43 H1 Z2. M8
G98 G81 Z-.125 R.1 F45.84
```

Continued

```
G80
M5
G91 G28 Z0. M9
M01
T2 M6 (T2 | 5/16 DRILL)
G0 G90 G54 X-2.4719 Y.6313 S855 M3
G43 H2 Z2. M8
G98 G81 Z-.6 R.1 F4.24
G80
M5
G91 G28 Z0. M9
M01
T3 M6 (T3 | 3/8-16 TAP)
G0 G90 G54 X-2.4719 Y.6313 S713 M3
G43 H3 Z2. M8
G98 G84 Z-.6 R.1 F44.5625
G80
M5
G91 G28 Z0. M9
G28 Y0.
M30
```

This program consists of 106 lines of programming with 12 tool changes. Review the four 3/8–16 tapped holes in **Figure 6-11** and determine if any operations can be adjusted.

If the holes were all spot drilled, then all drilled, and then all tapped, the program could be reduced. The revised program is as follows.

```
O0425
(DATE: 10-18-20)
(PROGRAMMER: REC)
(MATERIAL - STEEL INCH - 1030 - 200 BHN)
(T1 | 1/2 SPOT DRILL)
(T2 | 5/16 DRILL)
(T3 | 3/8-16 TAP)
G20
G0 G17 G40 G49 G80 G90
T1 M6 (T1 | 1/2 SPOT DRILL)
G0 G90 G54 X-5.1951 Y3.2958 S3500 M3
G43 H1 Z2. M8
G98 G81 Z-.125 R.1 F45.84
X-3.8481
X-1.4471 Y.6313
X-2.4719
G80
M5
G91 G28 Z0. M9
M01
```

Continued

```
T2 M6 (T2 | 5/16 DRILL)
G0 G90 G54 X-5.1951 Y3.2958 S855 M3
G43 H2 Z2. M8
G98 G81 Z-.6 R.1 F4.24
X-3.8481
X-1.4471 Y.6313
X-2.4719
G80
M5
G91 G28 Z0. M9
M01
T3 M6 (T3 | 3/8-16 TAP)
G0 G90 G54 X-5.1951 Y3.2958 S713 M3
G43 H3 Z2. M8
G98 G84 Z-.6 R.1 F44.5625
X-3.8481
X-1.4471 Y.6313
X-2.4719
G80
M5
G91 G28 Z0. M9
G28 Y0.
M30
```

The change in the order of operations reduces the program to 43 lines and only three tool changes. Program length does not necessarily result in reduced machining time, but reducing tool changes does reduce overall cycle times. When planning this part in its entirety, including all of the holes and pockets, the difference between poor planning and a solid order of operations could be 38 tool changes or 10 tool changes.

Efficiency gains in machining should be considered before a program is started. The time saved from preplanning will save in machine setup, operation, and cycle times. Although it can be an often overlooked step in the programming process, take the extra time and consider all of these factors before beginning a program.

Chapter 6 Review

Summary

- The planning that happens before a CNC mill program is created is vital to its success. It is necessary to understand all of the customer's requirements and the capabilities of the machinery and available tools.
- Design intent refers to the way in which the part is dimensioned and noted to define critical features of the part and the part function.
- By examining a print, important information on machining procedures can be derived. The dimensioning and tolerancing on the print identify critical features and can help in determining factors such as the work coordinate system origin and workholding.
- Workholding is a critical factor in machining and programming. The material must be securely held while allowing full access to all features that need machining. A program cannot be created without first determining how the part will be held in place.
- Tooling is an important factor to consider before starting the CNC program. Utilizing the same tool for multiple features can reduce cycle times. Specialty tooling needs to be identified before the program goes to the machine for production. When possible, minimize the amount of tools used and the tool changes needed.
- Identify features and operations that can be consolidated to reduce cycle times. Often, the same tool can be used to produce multiple features and improve efficiency.
- Before programming begins, consider the order of operations. Having a clear, definitive list of operations in a logical order can greatly reduce cycle times and eliminate mistakes in program operation.
- A clear idea of customer requirements, tooling needed, workholding, and order of operations is likely to reduce the total machine time for any product.

Review Questions

Answer the following questions using the information provided in this chapter.

Know and Understand

1. What is an important factor to consider before writing a CNC mill program?
    A. Customer requirements
    B. Workholding
    C. Tooling
    D. All of the above.

2. *True or False?* A typical sequence in program planning begins with reviewing the print.

3. *True or False?* Examining tolerances on a print can help determine what features are critical in manufacturing.

4. *True or False?* A reamer is used to form threads in a drilled hole.

5. A _____ has a fixed jaw and a movable jaw for securing work.
    A. magnetic chuck
    B. mill table
    C. vise
    D. vacuum table

6. *True or False?* A magnetic chuck is used for holding nonferrous metals.

7. A _____ is a custom workholding device used to position and secure an irregular-shaped workpiece for machining.
    A. casting
    B. spot drill
    C. reamer
    D. fixture

8. *True or False?* Tool selection does not impact machine cycle time.

9. A _____ end mill has a rounded edge on the end.
    A. square nose
    B. square end
    C. bull nose
    D. tapered

10. *True or False?* It is best practice to spot drill prior to a drill operation.

11. Which of the following is impacted by the order of operations in a program?

    A. Cycle time
    B. Tool usage
    C. Program length
    D. All of the above.

12. Usually, the last step in program planning is to determine _____.

    A. workholding
    B. tolerances
    C. the order of operations
    D. tooling

Apply and Analyze

1. When determining the workholding method to be used in machining, explain the factors that must be considered to allow for machining accuracy and efficiency.

2. Study the drawing below. Which dimensions need careful consideration during the program planning phase?

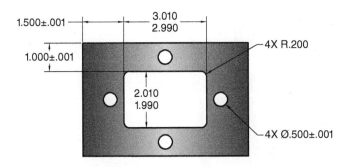

3. Study the drawing below. What is the most appropriate location on the part for establishing the work coordinate system origin? Explain why.

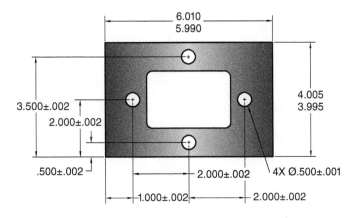

4. Study the drawing below. What is the most efficient order of operations for machining the tapped holes?

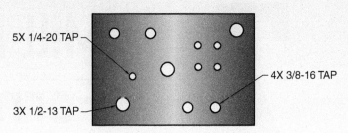

Critical Thinking

1. Explain what each team member in production can contribute to program planning. Other than the personnel discussed in this chapter, who else might need to be included? Why is the customer a valuable contributor? What can an operator or setup person contribute? What about sales personnel?

2. Preplanning is an important consideration in most projects. Consider a project you may have in your personal life—perhaps painting a room, repairing a vehicle, or even taking a trip. Write out an explanation of how you might plan for the project. Who else might need to be involved in the planning? What supplies or tools are needed? What are the steps required and in what order should they be completed to make the project successful?

3. This chapter discussed the importance of efficiency and cycle times. What is the impact of not having machines running production to the shop personnel? How does this affect profitability? How can planning prevent inefficiencies? Conduct research and find out what the average hourly shop rates are for machine shops in your area.

7 | Mill Program Format

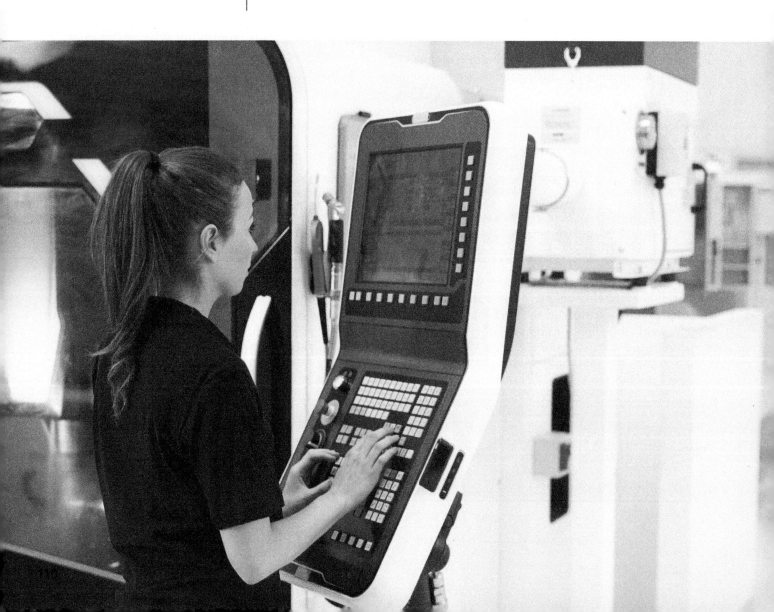

Chapter Outline

7.1 Introduction
7.2 Opening Statement
    7.2.1 Comment Lines
    7.2.2 Startup Commands
7.3 Program Body
    7.3.1 Tool Operation Opening Statement
    7.3.2 Tool Operation Program Body
    7.3.3 Tool Operation Closing Statement
7.4 Program Closing Statement

Learning Objectives

After completing this chapter, you will be able to:

- Explain the basic structure of a CNC mill program.
- List three components that make up a CNC program.
- Describe the use of parentheses to add comments to a CNC mill program.
- Explain the lines used in the opening statement of a program.
- Understand the components of the program body.
- Explain the lines used in the closing statement of a program.

Key Terms

closing statement
comment line
opening statement

program body
startup block
tool operation

zero return

7.1 Introduction

The previous chapters have covered the components that constitute a CNC mill program. These components include G-codes, M-codes, and address codes. There are specific ways to use these codes and some general rules about their placement in the program.

The next step is to build a program and organize these codes in a logical order. There are many different approaches to this process. Often, it is apparent which programmer wrote a program because it is based on a certain style or programming technique. Some companies require their programs to have certain information or a specific programming style.

The methods covered in this text are general in nature and work in almost all cases. There are other methods of programming, however. As you gain more experience, you will develop a style that you are comfortable with and suits your machinery and work environment.

Simply put, writing a program is the same as telling a story. A program has an opening statement, a program body, and a closing statement. The programming code for each tool used within the program will have the same elements. Writing programs consistently with this format will simplify the process and make it easy to identify any missing information. See **Figure 7-1**.

CNC programs can be written in most text formats on a computer and then uploaded to the machine's memory. This allows you to use all of the tools of a text editor or word processor on your computer, such as copy and paste functions.

Monkey Business Images/Shutterstock.com

Figure 7-1. A CNC program must have an organized structure. The program must be written so that the machine can interpret all commands and functions in the correct order.

From the Programmer

General Rules for Programming

As you learned in previous chapters, there are a number of rules that govern how commands and functions are programmed. The following is a summary of some of the most important programming rules to remember.

G-codes are preparatory commands used to preset the machine into a specific state of operation. Multiple G-code commands can be used on the same block of code, as long as they do not have conflicting functions. For example, the G00 and G01 commands cannot be used on the same block, but the G00 and G20 commands can be on the same block because they do not conflict with each other. Modal G-code commands, such as G00 and G01, remain in effect until they are canceled, or a new command places the machine in a different mode of operation. The G01 command must be accompanied by a feed rate command (F command) to designate the rate at which the cutting tool moves into the material.

M-codes are referred to as miscellaneous functions and are used to control machine functions and program functions. Miscellaneous codes act much like electrical switches to turn functions on and off. Only one M-code can be placed on a single line of code in a CNC program. As a recommended practice, M-codes that control program functions should not be on a line with movement commands. The M00, M01, M02, and M30 functions are examples of functions that should reside either on their own line or on lines with no movement.

On many machines, programming commands and functions beginning with a zero in the digit portion can be shortened to a single digit following the letter code. For example, G00 can be shortened to G0, M01 can be shortened to M1, T01 can be shortened to T1, and so on. This format is used in programs in this chapter and in remaining chapters of this textbook. On older machines, this format may not be recognized.

7.2 Opening Statement

The first few lines of a program are called the *opening statement*. The opening statement cancels any existing modal commands that may be in effect and places the machine in the desired starting condition for the new program. Each program should have a consistent opening statement based on the machine for which it is written.

The opening statement contains the program number and a *startup block* that cancels any previous cycles that may still be active. It should also contain information useful for the part setup, such as the tools used, offset locations, and any special instructions.

The following is an example of an opening statement in a program.

```
%
O3283 (PART NUMBER 1353283)
(PRINT REVISION L)
(DATE: 4-25-20)
(PROGRAMMED BY: R CALVERLEY)
(PART WCS ZERO IS TOP LEFT CORNER, TOP OF MATERIAL)
(USE FIXTURE #13256)
(TOOL #1 – .500 END MILL)
(TOOL #2 – .375 END MILL – DIA OFFSET #2)
G0 G20 G17 G40 G80
G91 G28 Z0.
```

The first line of this program is the percent sign (%). This symbol is only needed if using an outside source, such as a laptop computer or network server, to upload the program to or download the program from the machine controller. The percent sign signifies to the machine that a program is starting. The percent sign is also placed on its own block at the end of the program to indicate the program entry is complete. If you are entering a program manually at the machine's controller, this symbol is not necessary.

Every program must start with the program number. In this example, O3283 is the program number. The program number must start with the letter O. This line will be the only instance of using the letter O in a CNC program. The program can carry any number, but it is best practice to use a number that correlates with the part number in some way. Some controllers can only use up to four digits in the program name, but many newer controllers can accept up to five digits.

Following the program number is a comment. A comment, also referred to as a note, is enclosed in parentheses ( ). Notes can be entered anywhere in a CNC program if enclosed by parentheses. The program does not *read* any words in parentheses and no action will result from using them. Notes are very useful in operation and setup. In this case, the note following the program number gives the full part number of the part being machined by the program.

7.2.1 Comment Lines

A *comment line* is a block of code containing text enclosed in parentheses. Comment lines are important for safe operation and allow for information exchange from the programmer to the setup person and operator. See **Figure 7-2**. The following lines are comments from the program shown earlier.

```
(PRINT REVISION L)
```

This comment line identifies the print revision. The part number is important, but make sure the print revision is also correct for the corresponding program.

```
(DATE: 4-25-20)
```

Figure 7-2. Comment lines in programming provide important information for the operator and assist in safe operation of the machine.

This comment line identifies the date of program creation. Providing a date in the program ensures that the current program is being utilized.

(PROGRAMMED BY: R CALVERLEY)

The name of the programmer might be helpful if any changes or questions arise.

(PART WCS ZERO IS TOP LEFT CORNER, TOP OF MATERIAL)

This comment line identifies the origin of the work coordinate system, referred to as part WCS zero, part zero, or program zero. Often, the work origin is defined as a work offset for use with programming commands. The work origin position is vital for part setup and tool length offset settings.

(USE FIXTURE #13256)

If the program includes any special fixture or workholding requirements, note them accordingly.

(TOOL #1 – .500 END MILL)
(TOOL #2 – .375 END MILL – DIA OFFSET #2)

All tools used in the program should be noted in the opening statement. This is critical for machine setup. Diameter offsets and unusual height offsets should also be listed in the opening statement.

Communicating information in the program and keeping record of the programmer and any setup instructions can be a fail-safe method in manufacturing quality parts. Making sure the proper part revision is being machined and the correct tools are set up can save substantial time and money in the manufacturing process.

7.2.2 Startup Commands

The **G0 G20 G17 G40 G80** block in the previous example is a startup block, also referred to as a default block or safety block. This block cancels commands that may have been activated in a previous program and establishes a new starting condition for the current program. The contents of this block can vary widely based on the methodology of the programmer, but this example is a safe starting point. The following explains the commands in this block.

- The G0 command places the machine in full rapid mode. It is not a necessary command in this block, but most movements in the program start with rapid motion. If not added here, it would need to be placed on a subsequent line to activate rapid mode.
- The G20 command places the machine in inch mode. This mode specifies US Customary units, sometimes referred to as Imperial units. The G21 command is used to set the unit format to metric units.
- The G17 command designates the XY plane as the work plane. This plane is specified to generate all arcs in the XY plane. When more than one work plane is needed in arc programming, it is considered best practice to start with all arcs in one plane (typically the XY plane), and then alter to a three-dimensional plane if necessary.
- The G40 command cancels cutter compensation. This ensures that any cutter compensation used in previous programs is canceled.
- The G80 command cancels canned cycles that may still be active from previous programs.

The **G91 G28 Z0.** block in the previous example designates a safe *zero return* move along the Z axis. This is a safety move to prevent any initial movement of the spindle and any unexpected engagement of the tool or spindle with the part, fixture, or clamping. This block will return the spindle to its Z axis home position. Each entry in this block serves a specific purpose. The G91 command sets the current positioning mode to incremental mode. The G28 command is used to return the machine to the home position along the specified axis or axes. In this example, the Z0. coordinate specifies movement of the Z axis. Because only the Z axis is specified, movement to machine home will occur along that axis only. It is also possible to return to machine home along the other axes by specifying an X or Y coordinate on the same block as the G28 command. However, this can create an angular move that may cause an unexpected collision with the material or workholding. For safety reasons, the **G91 G28 Z0.** block is often used to send the machine to its Z axis home position without other axis movement. The following explains the commands in this block.

- The G91 command places the machine in incremental mode. The G91 command is always used with the G28 command.

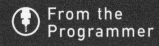

Startup Block

The commands included in a startup block can vary quite a bit. Some programmers do not use startup blocks because they are sure their programs are turning modes on and off at each use, thus making the startup block unnecessary. However, including a startup block is best practice to ensure safety. The main objective is to be safe first, and consistent second. Consistency helps to identify mistakes and oversights. Including a startup block is an easy way to make programs safe and consistent.

- The G28 command returns the machine to the home position (the machine zero position) along the specified axis in full rapid mode.
- The Z0. coordinate entry defines the axis position to which the machine will return. In this instance, the G28 command will return the Z axis to its Z0. machine home position.

7.3 Program Body

The *program body* is the heart of the program, defining all of the tools and movements to create a part according to the print. Inside the program body are all the actions of the machine, including the coordinate movements of individual tools and the entirety of their paths. The program body can be tens of thousands of lines long.

The program body contains the programming for each *tool operation*. Each operation has the same organization as the larger program. Inside of each tool operation, there is also an opening statement, body, and closing statement. From the time a tool is called into the spindle until it is put away, there is another set of programming standards that can be applied. As with the overall program build, there can be some differences in programming styles, but a repetitious consistency can also be applied to all tool operations in a program body.

7.3.1 Tool Operation Opening Statement

The following example shows the opening statement for a tool operation using a 3/8″ end mill identified as tool #1.

| T1 M6 (TOOL 1 – 3/8 END MILL – H1) |
| G0 G90 G54 X___ Y___ M3 S5000 |
| G43 H1 Z2. M8 |

The **T1 M6** block calls tool #1 and a tool change. When T1 is processed, the machine will bring tool #1 into position. For some tool changers, this will put the current tool away and then rotate to position #1. For other tool changers, this will rotate to tool #1 and then swap the spindle tool with tool #1. A machine with a side-mounted carousel allows the tool to be prepositioned for a faster tool change. On this type of machine, the T address can be programmed ahead of the tool change function. For a side-mounted carousel, the correct lines could be written as follows.

| T1 |
| ... |
| ... |
| ... |
| M6 |

The M6 function is the code signifying a tool change. On older controllers, it was necessary to send the machine home and orient the spindle before a tool change. On newer machines, the M6 function is a macro that contains all of the information to complete a tool change. The M6 function can also be altered for specific XY coordinate locations defining

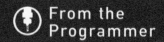

From the Programmer

G53: Machine Coordinate System

The G53 command designates that coordinate measurement is in relation to the machine zero position (machine home). This command can also be used for homing the machine and is preferred by some programmers. For example, a block written as **G53 G90 Z0.** returns the machine to the home position along the Z axis in absolute positioning mode. This is a direct move to machine home. Whether using the G53 or G28 command, the important factor is that the machine is in a safe starting position *before* the program attempts to start machining.

tool change positions. The key thing to know is that calling an M6 function will stop the spindle, send the machine to the tool change position in Z, turn off coolant, orient the spindle, and perform any tool change arm motion required.

The **G0 G90 G54 X___ Y___ M3 S5000** block executes a rapid move in absolute positioning mode to the specified XY coordinate location and turns on the spindle clockwise to 5000 rpm. This block performs several operations in the program. First it puts the machine in rapid positioning mode (G0) and absolute positioning mode (G90). It designates the work coordinate system as a work offset (G54). The work coordinate system can be set to any available coordinate location. This block also moves the machine to a specified position along the X and Y axes. At this point, the machine is still home in the Z axis. This is a safe move, but it positions the machine to the first location of the cut. The last action in this line is turning on the spindle (M3). It is a good practice to turn the spindle on while the machine is still away from the material.

The **G43 H1 Z2. M8** block activates the tool length offset for tool #1, positions the tool at Z2., and turns on coolant. As the machine is being moved down the Z axis for the first time, the specified tool length offset is referenced with the G43 command. The H address and the number that follows designates the stored tool length offset value. The offset number used is typically the same as the tool number. In this case, H1 references the offset for tool #1. In initial setup, it is advised to keep the tool away from the part a distance large enough to ensure safe operation. See **Figure 7-3**. For this example, the tool is positioned 2″ away from the final Z0. work position. The coolant is then initiated. The coolant can be turned on before or after this line, depending on preference. Initiating coolant on this line allows the coolant to stay off until the tool approaches the material, but this is early enough for the coolant to be spraying on the cutter before it engages material.

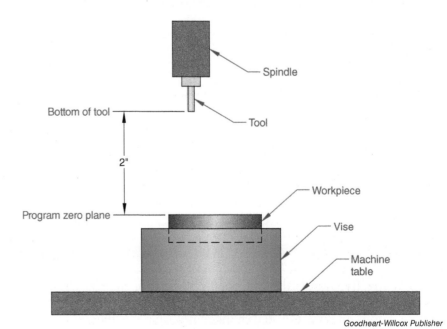

Goodheart-Willcox Publisher

Figure 7-3. When the tool moves down the Z axis for the first time, the machine references the stored tool length offset to position the tool correctly. The initial Z axis move is to the coordinate Z2. This is far enough from the workpiece to ensure safe operation.

7.3.2 Tool Operation Program Body

The next section of the program is the body of content that programs the tool's motion. It contains all of the tool movements in the X, Y, and Z axes to complete the machining operation. The following shows the basic programming format used in this section.

The **G01 Z–.5 F30.** block completes a feed rate move into the material. It moves the tool to Z–.5 at 30 inches per minute (ipm).

The **X___** block moves the tool to a specified X position in feed rate mode.

The **Y___** block moves the tool to a specified Y position in feed rate mode.

7.3.3 Tool Operation Closing Statement

At the end of a tool operation, there is also a *closing statement*. This section returns the machine to its proper modes before the next tool is brought out for further operations.

```
G80 M9
M5
G91 G28 Z0.
G28 Y0.
```

In the first block, the G80 command turns off canned cycles. Although canned cycles may not have been executed during this tool operation, this is a repetitive action intended to cancel any canned cycle modal commands that may be active.

The M9 function turns off coolant. It is not always necessary to turn off the coolant. This can make it easier for the operator to see the tool and confirm it is clear of obstructions as it moves to a safe position. See **Figure 7-4**. This too is a preference, but since the coolant will be restored in the program after each tool change, it is acceptable to shut off the coolant at this location.

In the second block, the M5 function turns off the spindle. The spindle rotation will be ceased by this function for the next tool change. If this is the final tool operation, the M30 function will also stop the spindle. Stopping spindle movement is necessary at the end of each tool operation.

The **G91 G28 Z0.** block returns the machine back safely to the Z axis home position. In this block, the G91 command places the machine in incremental mode. The Z0. coordinate specifies no incremental movement along the Z axis before the return to machine home in Z. Lifting the tool directly *up* in the Z axis is the safest move at this location. Any multiple axis movement could cause an unexpected tool and material collision.

The **G28 Y0.** block sets incremental mode and returns the machine to the Y0 home position. The Y0. coordinate specifies no incremental movement of the Y axis and straight movement to the Y axis home position. This block can be removed if another tool is about to be placed in the spindle. Often, the operator likes to have the material come out to the front of the

Figure 7-4. Turning off coolant at the end of a tool operation allows the operator to monitor the position of the tool in relation to the work.

machine in the Y axis to inspect work after each tool operation. This can add cycle time, so this block can be removed if it is not needed.

Any additional tool operations should follow this repetitive programming format for each tool's movement. Programming in a consistent manner will allow you to review your work, determine any missing information, and build successful programs.

7.4 Program Closing Statement

The closing statement contains the last lines of the program and prepares the machine to run the next part in a safe manner. The closing statement should also define the optimum position to remove the part and load new material. The following is an example of a closing statement.

| M9 |
| G91 G28 Z0. M5 |
| G28 Y0. |
| M30 |
| % |

The M9 command in the first block turns off coolant. It is preferred to turn off the coolant as early as possible to allow time for the coolant flow to stop before the machine is fully stopped.

The **G91 G28 Z0. M5** block moves the tool to the Z axis machine home position and turns the spindle off. Some programmers prefer to use an alternate homing method by utilizing the G53 command, as previously discussed. This is a good alternative, depending on machine type and normal programming practice. The most important consideration is safety. It is vitally important to return the machine to a safe and clear position well above the part surface in the Z axis. The M5 function is used to stop the spindle rotation during the tool retract.

The **G28 Y0.** block moves the tool to the Y axis machine home position. In most three-axis vertical CNC mills, sending the Y axis to the home position will bring the machined part toward the front of the machine. In this position, it is easier for the machine operator to remove and replace the material from the machine. Depending on specific machine types, this line can be altered to make changing material more accessible.

The M30 function rewinds the program. This function stops the program at the end and then resets it back to the beginning. The M30 function not only rewinds the program back to the first program line, it also shuts off all other machine functions. It stops the spindle, shuts off coolant, and ends any machine movement. The machine is now in a safe state to open the doors and inspect the machined piece.

The percent sign (%) after the M30 function is only used when programming *off line* with a computer. When sending the program to a machine controller, the percent sign indicates to the machine that all the data is complete. When sending a program out of a machine controller to a secondary source, such as a computer or network server, the percent sign appears in the file received. If programming directly at the machine's controller, the percent sign is not used.

The story is now complete. This is the format this text will use going forward to create all CNC mill programs. As you become more comfortable programming specific machines and families of parts, you may want to alter the opening and closing statements when additional information is needed. However, this format is simple and fits most applications.

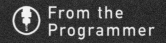

From the Programmer

Program Building and Planning
Writing a CNC program is as simple as telling a story one tool at a time. Your most important assets are the creativity and planning you bring to the project before making even one keystroke.

Chapter 7 Review

Summary

- A CNC mill program must have an organized structure that allows the machine to interpret all commands and functions in the correct order. A properly written program prepares the machine for safe operation and successful completion of a machined part.
- A CNC program consists of an opening statement, the program body, and a closing statement.
- Parentheses are used in CNC mill programs to add comments, also called notes. Any text inside of parentheses is not read by the machine, but provides information to assist in machine setup or operation.
- The opening statement of a CNC mill program cancels any previous commands left active. It contains the program number and a startup block to cancel commands or restore the proper run state. There should also be a line sending the machine to the Z axis home position to ensure the machine's first commanded move is a safe distance from the workholding and material.
- The program body contains all the tool changes and tool operations. It is the bulk of the program and may consist of thousands of lines of code.
- Within each tool operation, there is a separate opening statement, body, and closing statement to perform the programmed tasks.
- The closing statement of the CNC mill program ends the program. It returns the machine to a safe operating position and ends the program appropriately.
- Opening and closing statements can vary depending on programmer preference. However, it is important to have a consistent beginning and ending to the program to ensure optimal efficiency and safety.

Review Questions

Answer the following questions using the information provided in this chapter.

Know and Understand

1. The first lines of a CNC program are referred to as the _____.
   A. body statement
   B. body
   C. opening statement
   D. comments

2. *True or False?* The first tool change in a CNC program is located in the opening statement.

3. Comments are noted in a CNC program with the use of _____.
   A. commas
   B. parentheses
   C. quotation marks
   D. brackets

4. *True or False?* Comments are used in a program, primarily, to help the setup person and operator.

5. The body of the program is usually started with the _____ function.
   A. coolant on
   B. machine home
   C. tool change
   D. program number

6. The body of the CNC program defines all the _____.
   A. tool changes
   B. machine movements
   C. tool operations
   D. All of the above.

7. The M6 function is found in which section of the program?
   A. program opening statement
   B. program body
   C. program closing statement
   D. startup block

8. *True or False?* The # symbol is used to designate the program start and end.

9. The G91 _____ Z0. block sends the spindle to the Z axis home position.
   A. M28
   B. M30
   C. G28
   D. G30

10. *True or False?* It is mandatory to use the G28 Y0. block in the closing statement of a CNC program.

11. Which of the following codes is used to end the program in a closing statement?
    A. M30
    B. M4
    C. M7
    D. M8

12. Which of the following occurs when the M30 function is read by the machine?
    A. Coolant shuts off.
    B. Spindle rotation stops.
    C. The program resets to the beginning.
    D. All of the above.

Apply and Analyze

1. Using the blanks provided, fill in the correct codes in the opening statement to set rapid mode, set inch mode, cancel cutter compensation, activate incremental mode, and return to the Z axis home position.

   O3283 (PART NUMBER 1353283)
   _____ _____ G17 _____ G80
   _____ _____

2. Using the blanks provided, fill in the correct codes in the tool operation opening statement to activate the tool length offset, position the tool at Z2., and turn on coolant.

   T1 M6 (TOOL 1 – 3/8 END MILL – H1)
   G0 G90 G54 X–5. Y0. M3 S5000
   _____ _____ _____

3. Give three examples of comments that are typically added to the opening statement of a program to provide part information and clarify setup and operation.

4. Using the blanks provided, fill in the correct codes in the closing statement to shut off coolant, stop spindle rotation, return to machine home in the Y axis, and end the program.

   _____ _____
   G91 G28 Z0. _____
   _____ _____
   _____

Critical Thinking

1. As discussed in this chapter, writing a program is similar to telling a story. Explain some of the reasons why a CNC program needs to follow a logical, concise format that includes an opening statement, a body, and a closing statement.

2. The use of comments in a program provides details to both the setup person and operator. These notes can provide vital information to prevent catastrophic machine failures and costly mistakes. What are some comments you might need to add to a program other than the ones listed in this chapter?

3. Consider different possible formats for a CNC program. Are there different formats that might be used for special circumstances? Be creative and explain why there might be different approaches based on different machinery or more complex products.

8 Contouring

Chapter Outline

8.1 Introduction
8.2 Point-to-Point Programming
8.3 Cutter Compensation
8.4 Calculating Angular Moves
8.5 Calculating Radial Moves
    8.5.1 I, J, and K Method
    8.5.2 Radius Method
8.6 Chamfering

Learning Objectives

- Create point-to-point contour milling programs.
- Explain how to effectively use cutter compensation.
- Differentiate between cutter compensation left and right.
- Explain when and how to initiate cutter compensation.
- Apply cutter compensation to create rough and finish passes.
- Create angular moves in two axes.
- Apply the I, J, and K method to create arc programming moves.
- Create arc moves of less than 180° using the radius method.
- Explain the purpose of chamfering operations.

Key Terms

burr
chamfer
climb milling
contouring
conventional milling
cutter compensation
I, J, and K method
point-to-point programming
radius method

8.1 Introduction

Contouring is a machining operation that follows a joined path or single piece of geometry to cut material. Contouring is typically performed with an end mill. A contour is perhaps the most common type of toolpath. Simply think of contouring as cutting straight along a line, around a circle, or around a shape. Contours can be cut on the outside of a shape or on an internal feature, such as a pocket or slot. The shape can be much more complex, but the steps to creating the program will be the same.

Creating a contouring program is much like giving directions. First, the tool is positioned in a safe starting location. Then, step-by-step instructions are given to move the tool to create the desired shape. Contouring is often used to create finishing passes; however, with creative use of cutter compensation or multiple tools, roughing paths can also be created.

8.2 Point-to-Point Programming

Look at the rectangular part in **Figure 8-1**. This part is used in creating a complete contouring program. The part is .500″ thick. The part origin is located at the top-left corner, indicating that the part is oriented in quadrant 4, or the X+, Y− quadrant. This origin location is appropriate for this example.

For any part, the programmer will use the print to determine the X0. Y0. origin for the work coordinate system (WCS). The decision on where to locate the work origin is based on two very important factors—the print and the part workholding. Often, the print is dimensioned specifically, or primarily, from one feature. This feature can be the center of the material, a top edge of the part, or as in this case, a specific corner. Determining the originating location for dimensions on the print and using the same design methodology in programming and part setup will yield the most accurate dimensions in the final product. Part workholding can also dictate the work origin. The *way* a piece of material is being held in place can influence the final part dimensions. In many cases, a certain workholding device or fixture will dictate the origin position used in programming.

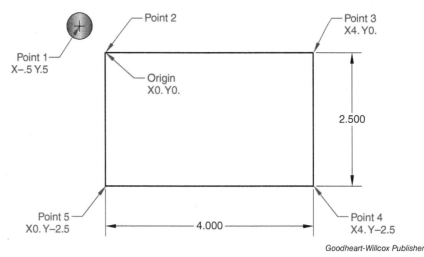

Goodheart-Willcox Publisher

Figure 8-1. A part with a rectangular contour. The top-left corner of the part is selected as the part origin. This orients the part in the fourth quadrant, or the X+,Y− quadrant.

With the top-left corner designated as the origin, the next step is to determine the path the cutter will travel. The part in **Figure 8-1** has a series of points that make it easy to determine the path for this program. Point 1, the starting point, is off the part far enough to allow the cutter to travel down to the required depth in the Z axis and not engage any material. The distance between this point and the part should be greater than the radius of the cutting tool. If using a .500″ end mill, the starting position must be more than .250″ away from the origin. This point should also not be an extended distance away, such as 2″ or 3″. That would create an extended nonproductive move and add unnecessary cycle time to machining. A good rule of thumb is to position the tool off the part the distance of the tool diameter. In this example, a 1/2″ diameter cutter is used. Point 1 is positioned at X−.5 Y.5. Point 2 is the work origin, or X0. Y0. The cutting tool in this example is positioned to cut material on the left side of the part. This is the normal orientation in most contour milling operations. Therefore, the tool proceeds to Points 3, 4, and 5 before returning to Point 1 to complete the operation. The part program is as follows.

O1234	
(PART #1234)	
(MATERIAL - ALUMINUM INCH - 6061)	
(T1 | 1/2 FLAT END MILL | H1)	
G20	Inch mode
G0 G17 G40 G49 G80 G90	Startup block
T1 M6	Tool change
G0 G90 G54 X−.5 Y.5 S2292 M3	Point 1 position
G43 H1 Z2. M8	
Z.2	
G1 Z−.5 F45.84	
X0. Y0.	Point 2 position
X4.	Point 3 position
Y−2.5	Point 4 position
X0.	Point 5 position
Y0.	Return to Point 2 position
Y.5	Safe move away from part
G0 Z2.	Rapid move above part
M5	
G91 G28 Z0. M9	
G28 X0. Y0.	
M30	

This simple form of programming is referred to as ***point-to-point programming***. Point-to-point programming is driving the machine spindle through a set of connected coordinates, regardless of the part shape.

There is an issue, however, with the program as written. The program states that a 1/2″ diameter end mill is used for the contouring operation. **Figure 8-2** shows the result of the program. The movement of the tool from Point 1 to Point 2 is shown in **Figure 8-2A**. The orange color highlighting shows the actual end mill path. The result of moving the tool to Point 3 is shown in **Figure 8-2B**.

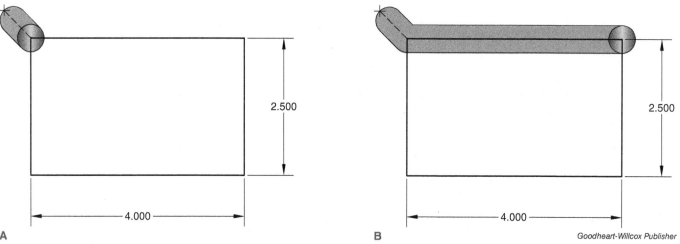

Figure 8-2. Point-to-point programming. The orange outline shows the actual end mill path. A—Movement from Point 1 to Point 2. B—Movement from Point 2 to Point 3. The center of the tool is moving through each point on the part, resulting in a part cut smaller than the dimensions specified.

The issue is that the center of the cutting tool is moving *through* the points, as programmed. The program is not compensating for the diameter of the cutting tool used to machine the part. This means that the tool is not positioned correctly and the part is cut smaller than the dimensions specified.

> **From the Programmer**
>
> **Climb Milling**
>
> There are two primary cutting methods used in milling operations: ***climb milling*** and ***conventional milling***. See **Figure 8-3**. In most CNC contour milling operations, the cutting tool is positioned for climb milling. In climb milling, the rotation of the cutting tool is in the same direction as the feed direction of the material. At the beginning of the cut, the cutting edge makes full engagement with the material and the chip width is at maximum size. The width of the chip decreases through the cut. This produces a smoother surface finish and helps increase tool life. In conventional milling, the cutting tool rotates against the feed direction of the material. At the beginning of the cut, the cutting edge starts contacting the material and the chip width is at minimum size. The width of the chip increases through the cut.
>
> Climb milling is the preferred method for most CNC milling operations. In climb milling, the cutting tool is most often positioned to the left of the material. This is the technique used in illustrations and programs in this chapter.

8.3 Cutter Compensation

CNC machine controllers have intelligent and powerful features. There are two options to solve for the improper cutting path in the previous example. The first option is to mathematically calculate the tool spindle center line for each move. Point 2 would not be X0. Y0., but rather X0. Y.25, thus accounting for the tool radius and positioning the center of the tool away

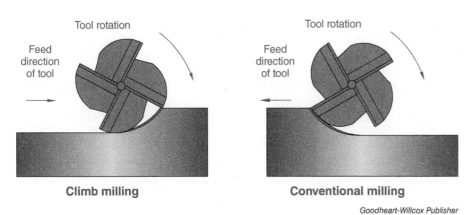

Figure 8-3. Climb milling is the preferred method for most CNC milling operations. In this method, the cutting tool is positioned to the left of the material and rotates in the same direction as the feed direction of the material. In conventional milling, the cutting tool rotates against the feed direction of the material.

from the finished contour of the part. This would require coordinates to be calculated for every move and every resulting position. Even with a simple part, this method will become complicated and programming errors can occur. The second option is to program a G-code to turn on the cutter compensation function, tell the controller what size tool is cutting, and let the controller do all the math.

Cutter compensation is a programmed offset from the center line of the cutting tool to the tool's edge along a cutting path. Cutter compensation simplifies calculations in programming and enables different offsets to be specified for tools and cutting operations. Cutter compensation is activated with the G41 or G42 command. Since these are modal commands, they will stay active until they are turned off in the program. Use the G40 command to cancel the cutter compensation mode when the operation is complete.

There are some rules and methodology to follow when using the cutter compensation feature. The first decision is whether to use G41 or G42. The G41 command activates *cutter compensation left* and the G42 command activates *cutter compensation right*. If the tool is cutting on the left-hand side of the material, the G41 command is used. If the tool is cutting on the right-hand side of the material, the G42 command is used. It is important to understand the difference. The offset direction is determined by viewing the tool from behind as it moves in the cutting direction. See **Figure 8-4**. In most milling operations, the cutting tool is positioned to cut material on the left side of the part. **Figure 8-5** illustrates the direction of cut for the sample part. Traveling in this direction, the cutting tool will be placed on the left-hand side of the part and left-hand cutter compensation is used.

The next decisions to make are the starting position of the tool and *when* the G41 or G42 command is initiated. These are important factors to consider when using cutter compensation. The following two rules apply when programming cutter compensation:

1. The initial tool position must be located away from the first cut at a distance that is more than the radius of the tool.
2. Cutter compensation must be initiated during a linear movement command.

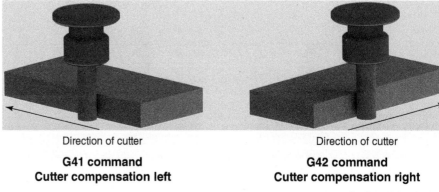

Direction of cutter | Direction of cutter
G41 command | G42 command
Cutter compensation left | Cutter compensation right

Goodheart-Willcox Publisher

Figure 8-4. The G41 command offsets the tool to the left of the programmed cutting path. The G42 command offsets the tool to the right of the programmed cutting path. The offset direction corresponds to the side of the part machined when viewing the tool from behind in the direction of the cutting path.

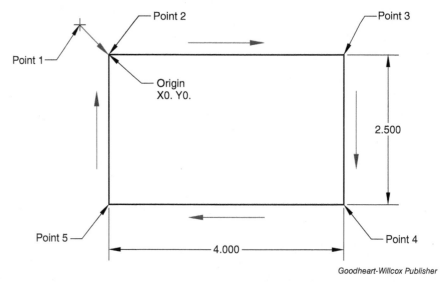

Goodheart-Willcox Publisher

Figure 8-5. The cutting tool is to be positioned on the left-hand side of the part. Shown is the resulting cutting direction. The G41 command is used in the program to establish left-hand cutter compensation.

Cutter compensation requires an initial activation move to position the tool correctly and establish the offset position relative to the part material. The distance of this move must be greater than the radius of the tool. Cutter compensation cannot be turned on while making an arc move with a circular interpolation command. The initial move must be a linear interpolation move. This allows the control to *compensate* for the cutter as it is being positioned. In **Figure 8-6**, Point 1 is the initial tool location. The distance from the part to this position needs to be more than the tool radius. In this case, the .500″ diameter end mill must start more than .250″ away from Point 2. Logically, if this distance was less than the radius, the tool would violate the finished part edge, and would also require the controller to back up the tool in the opposite direction from Point 2. The controller will not allow this to happen, and will generate a programming alarm. A good rule of thumb for programming the distance of the starting position is to use the full diameter of the tool, or .500″ in this case. This allows enough room without using an extended move that has no purpose.

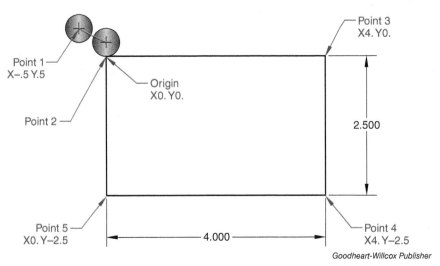

Figure 8-6. Cutter compensation must be initiated during a linear interpolation move. In this example, left-hand cutter compensation is initiated during the move from Point 1 to Point 2. The tool is offset in the correct position to follow the actual path desired. The initial move must be more than the radius of the tool.

The second rule to follow is *when* to initiate cutter compensation. In the program, the cutter compensation command must be accompanied by a linear move. This move specifies how to establish the offset position for the tool. In the following program, this move is specified by the coordinate entry X0. Y0. with linear interpolation mode active. Left-hand cutter compensation is designated on the same block with the G41 command. **Figure 8-6** shows the result of this move. Notice that the tool is offset in the correct position for cutting. The D address is used in the program to designate the diameter offset the controller will use to calculate the correct position. In this example, D1 references the tool #1 offset value stored in the controller. The following program uses the same cutting path as the previous example, but it is adjusted by using cutter compensation to follow along the correct path.

O1234	
(PART #1234)	
(MATERIAL - ALUMINUM INCH - 6061)	
(T1 | 1/2 FLAT END MILL | H1)	
G20	
G0 G17 G40 G49 G80 G90	
T1 M6	
G0 G90 G54 X-.5 Y.5 S2292 M3	Point 1 position
G43 H1 Z2. M8	
Z.2	
G1 Z-.5 F45.84	
G41 D1 X0. Y0.	Move to Point 2 using cutter compensation
X4.	Point 3 position
Y-2.5	Point 4 position
X0.	Point 5 position
Y.5	Safe move to a clearance position away from part

Continued

G40 Y1.	Safe move to exit and cancel cutter compensation
G0 Z2.	
M5	
G91 G28 Z0. M9	
G28 X0. Y0.	
M30	

Cutter compensation mode must be canceled with the G40 command after completing the cutting operation. This is done with the tool away from the part after cutting is completed.

The D address and the designated offset number must accompany the G41 (or G42) command. The specified D number tells the controller where the cutter diameter information is stored on the controller's tool offset page. **Figure 8-7** shows a tool offset page on a typical milling machine. The tool offset page shows the offset settings for the tool length offset and the tool diameter offset.

During setup, the diameter of the cutting tool is entered on the tool offset page. In this case, the tool diameter is entered under Tool #1 on the tool offset page to create the desired path. The controller references this setting when cutter compensation is activated and D1 is called in the program.

Different tool diameter offsets can be used when it is necessary to make multiple cutting passes, such as a roughing pass followed by a finish pass. Often in machining, making one cutting pass around a workpiece will leave a slightly rough surface finish or irregular dimensions. On rough cuts, the cutting tool may be cutting different amounts of material during a single pass. To compensate for this, make two passes using the same programmed path, but use two different diameter offsets. In the previous program, when the D1 value is read by the controller, the setting for Tool #1 on the tool offset page is used. Tool #1 is a .500″ diameter cutter. But if this is to be used as a roughing pass, the D1 offset setting can be changed to .510″. This will leave an additional .005″ around each edge of the finished part. Then make a finish pass using the exact same cutter path in the program, except change the D1 code to D2 (or any D number desired). On the tool offset page, enter .500″ for the D2 setting. This brings the part to the finished size. This will not require any additional programming or math. It allows the controller to calculate the tool position based on the tool diameter offset. The revised program is as follows. See **Figure 8-8**.

From the Programmer

D Offset Number

Most programmers consider it best practice to use the tool number as the same number for the D offset designation. For example, tool #1 uses D1 for the tool diameter offset, which is easy to remember. This reduces confusion and makes it easy to track offsets for tools. But you are not required to use the same number. Any D number can be used, as long as it is compatible with the program. However, use caution. If you use the wrong offset for a tool, a crash is bound to happen.

<< PROBING		TOOL OFFSET		TOOL INFO >>	
TOOL 1	COOLANT	H(LENGTH)		D(DIA)	
OFFSET	POSITION	GEOMETRY	WEAR	GEOMETRY	WEAR
1 SPINDLE	10	4.5680	0.	0.	← 0.
2	0	0.	0.	0.	0.
3	0	0.	0.	0.	0.
4	0	0.	0.	0.	0.
5	0	0.	0.	0.	0.
6	0	0.	0.	0.	0.
7	0	0.	0.	0.	0.
8	0	0.	0.	0.	0.
9	0	0.	0.	0.	0.

Goodheart-Willcox Publisher

Figure 8-7. The tool diameter is entered on the tool offset page during setup.

Chapter 8 Contouring 133

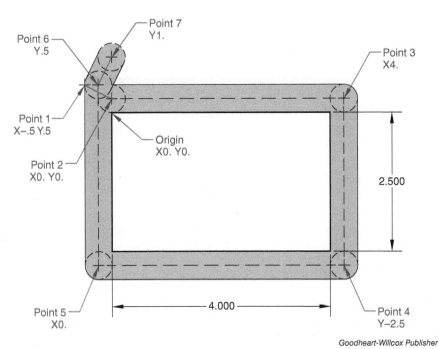

Goodheart-Willcox Publisher

Figure 8-8. The orange outline represents the end mill path programmed with cutter compensation. Each point is labeled with the corresponding coordinate entry in the program.

O1234	
(PART #1234)	
(MATERIAL - ALUMINUM INCH - 6061)	
(T1 | 1/2 FLAT END MILL | H1)	
(ROUGHING PASS)	Comment line indicating a roughing pass
G20	
G0 G17 G40 G49 G80 G90	
T1 M6	
G0 G90 G54 X-.5 Y.5 S2292 M3	Point 1 position
G43 H1 Z2. M8	
Z.2	
G1 Z-.5 F45.84	
G41 D1 X0. Y0.	Point 2 position
X4.	Point 3 position
Y-2.5	Point 4 position
X0.	Point 5 position
Y.5	Point 6 clearance position away from part
G40 Y1.	Point 7 position to exit cutter compensation
G0 Z2.	
(FINISH PASS)	Comment line indicating a finish pass
X-.5 Y.5	Point 1 position
Z.2	
G1 Z-.5 F45.84	

Continued

G41 D2 X0. Y0.	Point 2 position
X4.	Point 3 position
Y–2.5	Point 4 position
X0.	Point 5 position
Y.5	Point 6 clearance position away from part
G40 Y1.	Point 7 position to exit cutter compensation
G0 Z2.	
M5	
G91 G28 Z0. M9	
G28 X0. Y0.	
M30	

Figure 8-8 shows the correct cutter path using cutter compensation. The second cutting pass in the program follows the same path and requires no additional calculations from the programmer.

In **Figure 8-8**, the angular moves shown with red lines represent where cutter compensation is turned on and turned off. During these moves, the controller repositions the cutter to its correct position. Each point identified in **Figure 8-8** represents the center position of the tool and is labeled with the coordinate entries in the program. The programmed cutting path is represented with dashed lines.

Another advantage to using cutter compensation is tolerance control. By altering the D offset setting, the finish size of a machined part can be held to a fairly accurate dimension. A standard carbide end mill will be slightly undersized. A 1/2″ end mill, for example, is manufactured at .498″–.499″ outside diameter. In the previous example, this means that the 2.500″ and 4.000″ dimensions specified for the part will be oversize by some small amount after the machining is complete. Other factors, such as tool stickout length, machine backlash, and workholding can also alter the final results. To compensate for this, best practice is to enter a larger offset on the tool offset page for the first part, measure the part, and then alter the offset for a final machined part. Remember, the smaller the offset amount, the smaller the final part size.

Tool wear must also be considered in machining operations. Many machines have a wear compensation setting on the tool offset page to adjust the tool diameter setting. Refer to the tool offset page shown in **Figure 8-7**. This setting is used to make fine adjustments in order to keep part dimensions within tolerance.

8.4 Calculating Angular Moves

Angular moves are accomplished in a CNC mill program by using the G01 linear movement command and positioning two axes simultaneously. Angular moves are most commonly programmed along the X and Y axes, but it is also possible to program along the X and Z axes or along the Y and Z axes. The part shown in **Figure 8-9** has an angled surface that requires machining a simple angle. This feature is defined with a 45° angular dimension and a .500″ linear dimension on one side. The X and Y axis locations need to be calculated and entry and exit locations need to be determined.

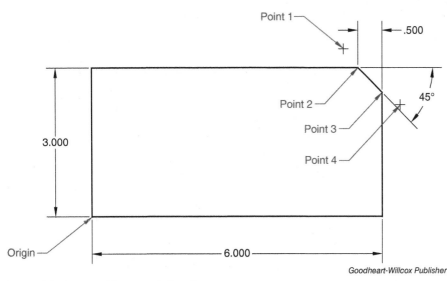

Figure 8-9. A part with an angled surface.

Points 1 and 4 will be the entry and exit coordinates. Points 2 and 3 are calculated from the dimensions given on the print. The lower-left corner of the part has been designated as the work origin, or X0. Y0. Because the angle is dimensioned as 45° and a .500″ linear dimension is given, both sides are .500″ from the corner. See **Figure 8-10.** Based on the dimensions given in **Figure 8-9**, Point 2 is located at the coordinates X5.500 Y3.000 and Point 3 is located at the coordinates X6.000 Y2.500.

In this operation, a 3/8″ diameter end mill is used to machine the angle only. The part is .500″ thick. The program is as follows.

O1022	
(T2 | 3/8 FLAT END MILL | H2 | D2 | TOOL DIA. .375)	
G20	
G0 G17 G40 G49 G80 G90	
T2 M6	
G0 G90 G54 X5.2 Y3.375 S5093 M3	Point 1 position
G43 H2 Z2. M8	
Z.2	
G1 Z-.5 F22.6	
G41 D2 X5.5 Y3.	Turn on cutter compensation left, move to Point 2
X6. Y2.5	Move to Point 3
G40 X6.375 Y2.25	Turn off cutter compensation, move to Point 4
G0 Z2.	
M5	
G91 G28 Z0. M9	
G28 X0. Y0.	
M30	

A part with multiple angles is shown in **Figure 8-11.** The work origin is at the top-left corner of the part, but the starting location of the tool can be anywhere the programmer decides. In this example, the starting

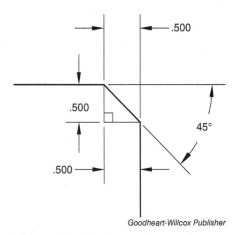

Figure 8-10. A 45° angle has equal length sides. In this case, each side is .500″.

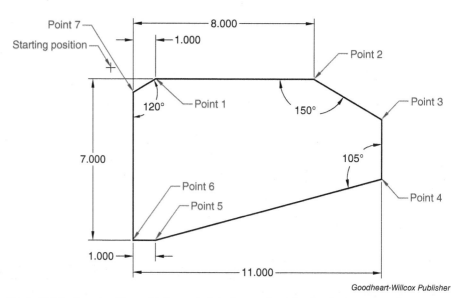

Figure 8-11. A part with multiple angled surfaces. Coordinates for programming are determined by making calculations with the trigonometric functions on a calculator.

From the Programmer

Coordinate Calculations

It is a good idea to calculate all of the points needed for a program before you begin. Coordinates can be calculated using the trigonometric functions on a calculator, as discussed in Chapter 1. Use the appropriate function based on the angular and linear dimension values you know. Some programmers like to take a pencil and write out the X and Y coordinates for every point on a photocopy of the print. If there is not enough room because of dimensions or if many coordinates are required, a separate sheet, called a coordinate sheet, can be used to record the positional coordinates. After the coordinates are calculated, the program is written by driving the tool to each of the points, step by step.

position is located in the X–, Y+ quadrant. The tool will move to Point 1 to activate cutter compensation, and then proceed through the points in a clockwise direction ending at Point 1 before moving off the part to cancel cutter compensation.

The following is a program to complete the contouring operation for the part in **Figure 8-11**. The cutting path is completed in the same manner as any other contouring operation. Start in a safe location off the part, initiate cutter compensation, and then drive the tool through all the required points. Once the contouring operation is complete, move the tool off the part safely and cancel cutter compensation.

O0425	
(T1 | 1/2 FLAT END MILL | H1 | D1 | TOOL DIA. .5)	
G20	
G0 G17 G40 G49 G80 G90	
T1 M6	
G0 G90 G54 X-1. Y.5 S2292 M3	Safe position off the part
G43 H1 Z2. M8	
Z.2	
G1 Z-.5 F45.84	
G41 D1 X1. Y0.	Move to Point 1, turn on cutter compensation
X8.	Move to Point 2
X11. Y-1.7321	Move to Point 3
Y-4.3205	Move to Point 4
X1. Y-7.	Move to Point 5
X0.	Move to Point 6
Y-.5774	Move to Point 7
X1. Y0.	Move to Point 1
G40 X1.433 Y.25	Safely exit from part and cancel cutter compensation

Continued

```
G0 Z2.
M5
G91 G28 Z0. M9
G28 X0. Y0.
M30
```

As previously discussed, angular moves can also be programmed in the Z axis direction. However, this is not a common function in contouring. It is not common because there is an inherent issue with this movement. **Figure 8-12** shows a solid model part with an angled surface representing an angle in the X and Z axis directions.

It is possible to calculate the X position of the cutting tool and then program an XZ angular move. However, consider the fact that the center of the cutting tool is moving through the coordinates. In the configuration shown, the top surface of the workpiece is the XY plane. When looking at the front view of the part, an XZ angular move positions the edge of the tool to remove material along the X and Z axes. **Figure 8-13** shows the position of the tool and the part as it is being machined.

With the tool in this position, it is clear that the tool body will violate the part boundary and not produce the desired final shape. This would result in a scrap part. The same violation will occur with a ball end mill, as shown in **Figure 8-14**. The proper way to machine this feature is to complete a contouring operation with X and Y axis movement using an angled cutting tool, as discussed later in this chapter.

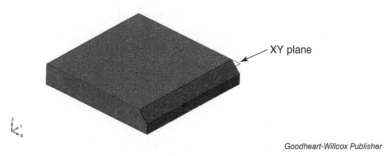

Goodheart-Willcox Publisher

Figure 8-12. A part with an angled surface. The XY work plane is aligned with the top surface of the part.

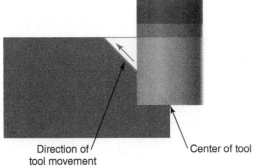

Goodheart-Willcox Publisher

Figure 8-13. If an XZ move is specified for the tool as positioned, the tool removes material along the edge, but the body of the cutting tool extends past the part boundary. This would result in a scrap part.

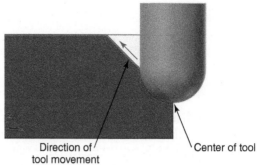

Figure 8-14. Positioning a ball end mill in the location shown violates the part boundary. An angular XZ move would result in a scrap part and cannot be used.

8.5 Calculating Radial Moves

There are three distinct types of movements in contouring: the single axis move, the angular or multiple axis move, and the circular move. Circular moves are referred to as circular interpolation or arc moves. Circular interpolation can be performed to create radial features on parts. Circular interpolation commands are used to move a tool along a circular arc to the designated end position. There are two distinct ways to accomplish this and five pieces of information needed to create the arc move.

The two methods to create an arc move are referred to as the *I, J, and K method* and the *radius method*. The radius method, also known as the R method, requires less information to perform, but it is limited to an arc of less than 360°. In comparison, the I, J, and K method requires more information to complete and does allow for full 360° circles. Older machine controllers required the I, J, and K method for arc programming, but all models since the early 2000s have incorporated the radius method. Both of these methods are described in detail in this section.

There are five pieces of information required to complete a circular interpolation command:

1. **Plane selection.** The plane on which radial moves are to occur must be specified. The G17 command designates an arc in the XY plane. The G18 command designates an arc in the XZ plane. The G19 command designates an arc in the YZ plane.

2. **Arc start position.** The start position designates the XYZ coordinates locating the start point of the arc.

3. **Arc direction.** The G02 command designates an arc in the clockwise direction. The G03 command designates an arc in the counterclockwise direction.

4. **Arc end position.** The end position designates the XYZ coordinates locating the end point of the arc.

5. **Arc center or radius.** If the I, J, and K method is used, the distance from the arc start point to the arc center point along the corresponding axes must be specified. If the radius method is used, the radius of the arc must be specified.

8.5.1 I, J, and K Method

The I, J, and K address values in the program are the incremental distances from the tool's starting point on the arc to the arc center. I, J, and K are alternate address letters for the X, Y, and Z axes.

- **I** = The incremental distance from the arc start point to the arc center point along the X axis.
- **J** = The incremental distance from the arc start point to the arc center point along the Y axis.
- **K** = The incremental distance from the arc start point to the arc center point along the Z axis.

The I and J addresses are used when programming moves on the XY plane. Examples of using I and J values with circular interpolation moves are shown in **Figure 8-15**. In **Figure 8-15A**, the I address value represents incremental movement along the X axis. In **Figure 8-15B**, the J address value represents incremental movement along the Y axis. In the I, J, and K method, the tool is positioned at the start of the arc, the arc direction is commanded, the arc end point is designated, and the incremental distance to the arc center point along each axis is designated. If the incremental distance along any axis is zero, it does not need to be entered. However, entering an incremental distance of zero for one of the axes will not have any negative impact. In **Figure 8-15A**, for example, the block could also be written as **G03 X0. Y2. I–2. J0.**

The following examples illustrate the I, J, and K method programmed with and without cutter compensation. **Figure 8-16** is the first example and is based on programming without cutter compensation.

As shown in **Figure 8-16**, the top-left corner of the part is designated as the work origin. The first coordinate position is X0. Y.1875. The tool is then programmed to move to the start of the arc at X2.5. The next line in the program contains the information to make the arc move. This line of code specifies the direction of arc movement (clockwise), the arc end point, and the incremental distance from the arc start point to the arc center. The

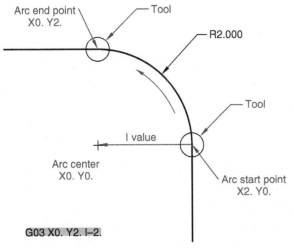
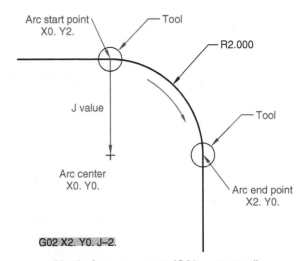

Figure 8-15. Using the I, J, and K method to program arc moves. The block of code in each example represents the corresponding program entry. A—Using the G03 circular interpolation command to create counterclockwise arc movement. B—Using the G02 circular interpolation command to create clockwise arc movement.

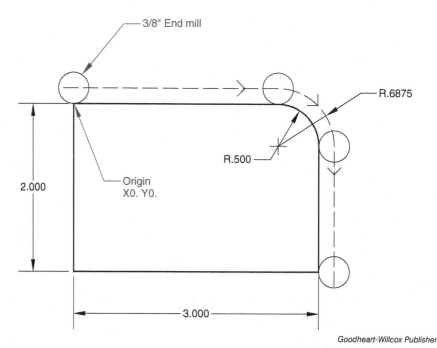

Figure 8-16. The I, J, and K method is used to program an arc move to machine the rounded end of the part. In this example, the program is written without cutter compensation.

line should read: **G2 X3.1875 Y–.5 J–.6875**. No I address value is required because the arc start point and arc center have the same X axis position. The G02 and G03 commands are modal commands. Therefore, the G01 linear interpolation command is used on the next line to specify linear movement to the Y–2. coordinate.

G1 X0. Y.1875	
X2.5	
G2 X3.1875 Y–.5 J–.6875	The J value is the incremental distance from the arc start point to the arc center; no I value is required (value is 0)
G1 Y–2.	

Without cutter compensation activated in the program, the programmer must consider where the center point of the tool is located and the additional radial measurement that must be calculated. With cutter compensation activated, the tool diameter is accounted for by the controller and the part dimensions can be input. The following example is for the same part, but uses left-hand cutter compensation.

G1 G41 D1 X0. Y0.
X2.5
G2 X3. Y–.5 J–.5
G1 Y–2.
G40 Y–.375

If the start point, end point, and I, J, and K values are not correctly specified, a programming alarm will occur and the machine will not attempt to complete the commanded line.

The advantage of the I, J, and K method is the ability to machine full 360° circles. When using the I, J, and K method to create movement in a full circle, the start point and end point will have the same X and Y coordinates. The I or J distance value determines the location of the center point about which the arc rotates. See **Figure 8-17**. In the following program, a full 360° circle move is created with cutter compensation activated.

G1 X–1.75 Y0. F10.
G41 D1 X–1.
G2 X–1. Y0. I1.
G1 Y.5
G40 Y1.

8.5.2 Radius Method

The radius method, also called the R method, is the most commonly used arc programming method on newer machines. This method is considered simpler than the I, J, and K method because it requires fewer calculations and less programming input. The information required is the arc start position, the arc direction (G02 or G03 command), the arc end position, and the R value, or radius required. For arcs less than 180°, the R value is positive. For arcs larger than 180°, the R value is negative. Remember, the radius method cannot be used for full 360° circles.

Figure 8-18 shows a 90° round corner blend. In the following example, the radius method is used to create the arc movement.

G1 X1. Y0.
G2 X1.25 Y–.25 R.25
G1 Y–1.

If the radius feature is an inside corner radius, the tool is positioned as shown in **Figure 8-19** and the G02 command is replaced with the G03 command for a counterclockwise arc move. This is a similar cutting path,

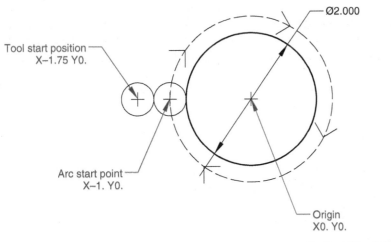

Goodheart-Willcox Publisher

Figure 8-17. The I, J, and K method must be used to machine a full 360° circle. The start point and end point have the same X and Y coordinates in the program.

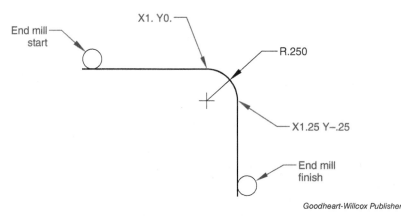

Figure 8-18. When using the radius method to create arc movement, the arc start point, arc direction, arc end point, and R value are specified in the program.

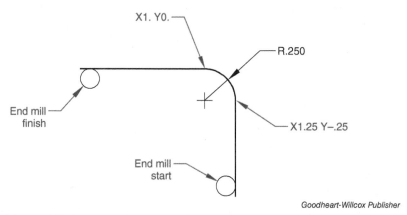

Figure 8-19. The G03 command is used to create counterclockwise arc movement to machine the 90° internal radius feature.

except the tool is positioned to machine an internal feature. The cutting direction and use of the G02 or G03 command is based on positioning the tool for climb milling. If cutter compensation is not used, the radius of the cutter is subtracted from the corner radius dimension for the R value entry.

```
G1 X1.25 Y-.25
G3 X1. Y0. R.25
G1 X0.
```

When the arc is greater than 180°, a negative R value is used. This commands the control to move the cutter past 180° to the designated arc end point. **Figure 8-20** shows a part where the arc movement is greater than 180°. The following program uses the radius method with a negative R value to indicate the extent of movement. The coordinates for the start point of the arc are determined from the dimensions on the print. The coordinates for the arc endpoint are determined using trigonometric calculations with a calculator. Calculating coordinate positions on an arc and calculation formulas for programming are covered in Chapter 1 of this text.

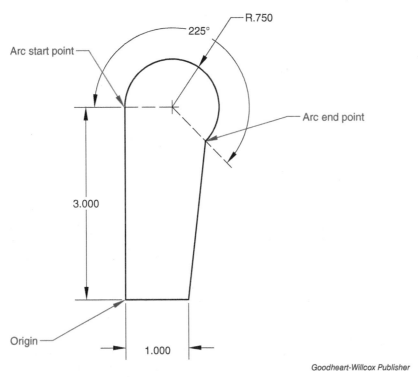

Figure 8-20. A part with a .750″ radius feature requiring arc movement greater than 180°. The coordinates for the arc end point are determined using trigonometric calculations with a calculator.

| G41 D1 X0. Y0. |
| G1 Y3. |
| G2 X1.2803 Y2.4697 R-.75 |
| G1 X1. Y0. |
| X0. Y0. |
| G40 X-.5 |

The part shown in **Figure 8-21** is .500″ thick and has a contour with three radius features. The next two programs show how the I, J, and K method or radius method can be used in arc programming. The first program uses the I, J, and K method and cutter compensation. A 1/2″ end mill is used.

| O1020 |
| (T2 | 3/8 FLAT END MILL | H2 | D2) |
| G20 |
| G0 G17 G40 G49 G80 G90 |
| T2 M6 |
| G0 G90 G54 X-.375 Y.375 S5093 M3 |
| G43 H2 Z2. M8 |
| Z.2 |
| G1 Z-.5 F61. |
| G41 D2 X0. Y0. |
| X3.5 |
| G2 X4. Y-.5 I0. J-.5 |

Continued

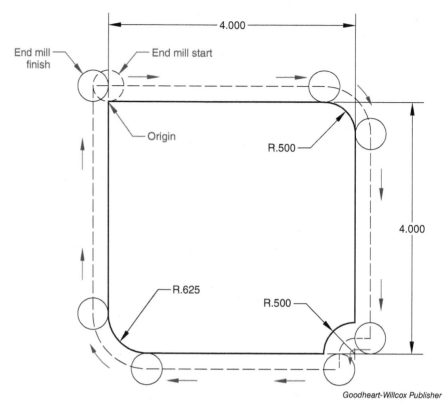

Figure 8-21. A part with three radius features requiring arc movement commands in programming.

```
G1 Y-3.5
G3 X3.5 Y-4. I0. J-.5
G1 X.625
G2 X0. Y-3.375 I0. J.625
G1 Y0.
G40 Y.375
G0 Z2.
M5
G91 G28 Z0. M9
G28 X0. Y0.
M30
```

The following program uses the radius method and cutter compensation to create the movement for each arc.

```
O1020
(T2 | 3/8 FLAT END MILL | H2 | D2)
G20
G0 G17 G40 G49 G80 G90
T2 M6
G0 G90 G54 X-.375 Y.375 S5093 M3
G43 H2 Z2. M8
Z.2
G1 Z-.5 F61.
```

Continued

```
G41 D2 X0. Y0.
X3.5
G2 X4. Y-.5 R.5
G1 Y-3.5
G3 X3.5 Y-4. R.5
G1 X.625
G2 X0. Y-3.375 R.625
G1 Y0.
G40 Y.375
G0 Z2.
M5
G91 G28 Z0. M9
G28 X0. Y0.
M30
```

8.6 Chamfering

There are a number of qualities that define a successful CNC part program. Most important are the accuracy and efficiency of the program and the quality of the part created. The aesthetics of the final part should also be a key consideration in programming. Provisions should be made for producing required surface finishes and preventing tool marks or scratches. A completed part should look good to your supervisor and the customer who will be purchasing machined products. Two conditions that are universally unacceptable are extremely sharp edges and raised material where two surfaces meet, often called a *burr*. To relieve sharp edges, a good CNC programmer makes sure to create small edge breaks, or *chamfers*, wherever possible.

Many prints designate the note *Break Sharp Edges .015 Max.* to allow for a small edge break at any location where two surfaces meet. On some prints, specific chamfer sizes and locations are dimensioned. In any case, a simple contouring path with an angled tool can be used to break sharp edges or create the required chamfers. **Figure 8-22** shows a machined part with a specified .050″ chamfer on the top edge.

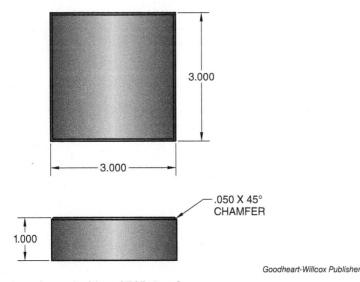

Goodheart-Willcox Publisher

Figure 8-22. A rectangular part with a .050″ chamfer.

To complete this operation, program a 1/4″ diameter cutter with a 90° tip. The 90° angle refers to the included angle on the tool end and will produce a 45° chamfer. The depth of the programmed cut and the cutter compensation amount will determine the final width of the chamfer. As a starting point, establish a .100″ diameter offset and use it for the D value in the program when cutter compensation is activated. Program the Z depth of cut to −.100″. This positions the tool so that cutting takes place further up the edge of the cutter. **Figure 8-23** shows a solid model of the part and the cutting tool.

As shown in this example, different results in machining can be obtained by altering tool settings and the cutting depth. Changing the programmed Z depth, altering the tool length offset, and changing the D value used with cutter compensation will all affect the finish size of the chamfer.

A preprogrammed contouring operation can be copied and pasted for use with a chamfer routine, with some minor alterations. For example, a 45° chamfer can be machined on the part shown in **Figure 8-21** to create a small edge break. The following program contains the contouring operation for the part in **Figure 8-21** and the additional operation to define the cutting path for a small edge break.

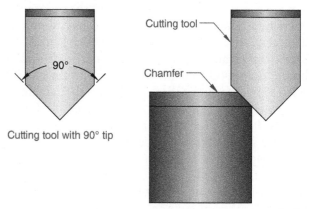

Goodheart-Willcox Publisher

Figure 8-23. A cutting tool with a 90° tip is used to create a 45° chamfer. The tool is positioned away from the part at a diameter offset of .100″. The Z depth of cut is programmed at −.100″ to cut a .050″ chamfer.

O1020	
(T2 | 3/8 FLAT END MILL | H2 | D2)	
(T3 | 1/4 X 90 DEG CHAMFER MILL | H3 | D3)	
G20	
G0 G17 G40 G49 G80 G90	
T2 M6	
G0 G90 G54 X–.375 Y.375 S5093 M3	
G43 H2 Z2. M8	
Z.2	
G1 Z–.5 F61.	
G41 D2 X0. Y0.	
X3.5	
G2 X4. Y–.5 R.5	
G1 Y–3.5	
G3 X3.5 Y–4. R.5	
G1 X.625	
G2 X0. Y–3.375 R.625	
G1 Y0.	
G40 Y.375	
G0 Z2.	
M5	
G91 G28 Z0. M9	
T3 M6 (CHAMFER MILL)	Tool change and start of chamfer operation
G0 G90 G54 X–.375 Y.375 S5000 M3	
G43 H3 Z2. M8	Altered tool length offset
Z.2	
G1 Z–.10 F30.	Altered Z depth
G41 D3 X0. Y0.	Altered cutter compensation offset
X3.5	
G2 X4. Y–.5 R.5	
G1 Y–3.5	
G3 X3.5 Y–4. R.5	
G1 X.625	
G2 X0. Y–3.375 R.625	
G1 Y0.	
G40 Y.375	
G0 Z2.	
M5	
G91 G28 Z0. M9	
G28 X0. Y0.	
M30	

Chapter 8 Review

Summary

- A contouring operation follows a joined path or single piece of geometry to cut material. Contouring is typically performed with an end mill.
- In point-to-point programming, the machine spindle is driven through a set of connected coordinates.
- The two primary cutting methods used in milling operations are climb milling and conventional milling. In climb milling, the rotation of the cutting tool is in the same direction as the feed direction of the material. In conventional milling, the cutting tool rotates against the feed direction of the material.
- Cutter compensation is a programmed offset from the center line of the cutting tool to the tool's edge along a cutting path. The G41 command activates cutter compensation left and the G42 command activates cutter compensation right.
- When cutter compensation is activated, the D address is used to designate the diameter offset the controller will use to calculate the correct position.
- Different tool diameter offsets can be used when it is necessary to make multiple cutting passes.
- Angular moves are accomplished in a CNC mill program by using the G01 linear movement command and positioning two axes simultaneously.
- Arc moves can be created using the I, J, and K method or the radius method.
- To use the I, J, and K method, the arc start position, arc direction, arc end position, and incremental distances to the arc center point are specified.
- To use the radius method, the arc start position, arc direction, arc end position, and R value (required radius) are specified.
- Chamfers are small edge breaks used to relieve sharp edges on a part.

Review Questions

Answer the following questions using the information provided in this chapter.

Know and Understand

1. *True or False?* Contouring operations are typically performed with an end mill.

2. What is one of the main factors a programmer uses to determine the work origin?
    A. Personal preference
    B. The print
    C. Available machine
    D. Finished part size

3. The initial starting position of the tool off the part edge should be more than the tool _____.
    A. diameter
    B. length
    C. radius
    D. angle

4. A tool starting position that is too far away from the part will create a(n) _____.
    A. unnecessary cycle time
    B. scrap part
    C. angled move
    D. arc move

5. Cutter compensation is used to _____.
    A. compensate for tool length
    B. offset the tool's radius away from the part
    C. compensate for tool height
    D. locate the part origin

6. *True or False?* When climb milling, the cutting tool is most often positioned to the right of the material.

7. Which command is used to activate cutter compensation left?
   A. G40
   B. G41
   C. G42
   D. G43

8. Which command is used to cancel cutter compensation?
   A. G40
   B. G41
   C. G42
   D. G43

9. After cutter compensation is initiated, which address letter designates the offset number?
   A. M
   B. J
   C. R
   D. D

10. An angular move is accomplished by moving at least two axes simultaneously with the _____ command.
    A. G01
    B. G03
    C. G41
    D. G02

11. The _____ command is programmed to create a clockwise arc.
    A. G40
    B. G42
    C. G02
    D. G03

12. *True or False?* The radius method of arc creation uses the I and J addresses to specify incremental distances to the arc center.

13. Most arc moves are made in the XY plane. Which command designates this plane for arc creation?
    A. G16
    B. G17
    C. G18
    D. G19

14. If an arc move requires that the center of the arc is located .500″ along the positive Y axis from the arc start point, what is the specified value in the program?
    A. I.500
    B. I–.500
    C. J.500
    D. J–.500

15. In the radius method, what does the *R* represent?
    A. Arc radius
    B. Cutter radius
    C. Start arc position
    D. End arc position

16. What are two factors that affect the finish size of a chamfer when creating edge breaks?
    A. Part workholding and work offset designation
    B. Tool length offset and diameter offset
    C. I and J values
    D. All of the above.

Apply and Analyze

1. Referring to the blanks provided, give the correct codes to complete the program for the part shown.

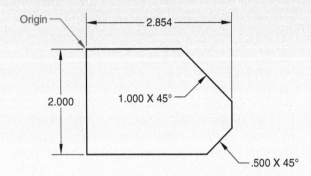

2. Referring to the blanks provided, give the correct codes to complete the program for the part shown. Use the I, J, and K method to create the part radii.

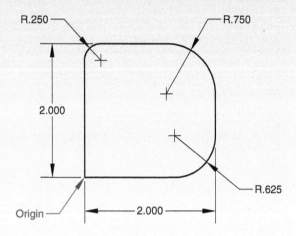

O0921
(T1 | 1/2 FLAT END MILL | H1 | D1)
G20
G0 G17 G40 G49 G80 G90
T1 M6
G0 G90 G54 X-.5 Y.5 S2292 M3
G43 H1 Z2. M8
Z.2
G1 Z-.5 F40.
____ ____ X0. Y0. F45.84
____
____ ____
____
____ ____
____
____ ____
G0 Z2.
M5
G91 G28 Z0. M9
G28 X0. Y0.
M30

O0824
(T5 | 3/8 FLAT END MILL | H5 | D5)
G20
G0 G17 G40 G49 G80 G90
T5 M6
G0 G90 G54 X-.375 Y-.375 S5093 M3
G43 H5 Z2. M8
Z.2
G1 Z-.375 F20.
____ ____ X0. Y0. F61.
____
____ ____ ____ ____
____ ____
____ ____ ____
____ ____
____ ____ ____
G1 X0.
G40 X-.375
G0 Z2.
M5
G91 G28 Z0. M9
G28 X0. Y0.
M30

3. Referring to the blanks provided, give the correct codes to complete the program for the part shown. Use the radius method to create the part radii.

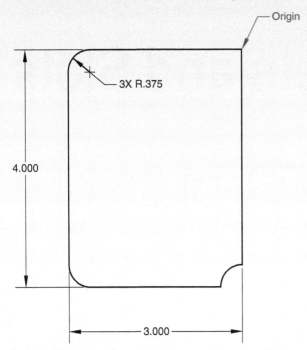

Critical Thinking

1. By using cutter compensation, can any diameter tool size be used with a program? Why or why not?
2. Explain some of the ways in which part workholding influences programming. How might a fixture alter the work coordinate system to provide access to features or machining paths?
3. As discussed in this chapter, small chamfers are machined on parts to remove sharp edges and improve appearance. What might happen if chamfers were machined to a different size than specified on the print?

```
O1995
(T4 | 7/16 FLAT END MILL | H4 | D4)
G20
G0 G17 G40 G49 G80 G90
T4 M6
G0 G90 G54 X.4375 Y.4375 S3056 M3
____ ____ Z2. M8
Z.2
G1 Z-.375 F15.
____ ____ X0. Y0. F36.
____
____ ____ ____
____ ____
____ ____ ____
G1 Y-.375
____ ____ ____
G1 X0.
____ ____
G0 Z2.
M5
G91 G28 Z0. M9
G28 X0. Y0.
M30
```

9 Pockets and Slots

Chapter Outline

9.1 Introduction
9.2 Pocket Milling
   9.2.1 Entry Strategy
   9.2.2 Pocket Roughing
   9.2.3 Depth of Cut and Stepover Amount
9.3 Pocket Finishing
9.4 Slots
   9.4.1 Open Slots
   9.4.2 Closed Slots
   9.4.3 T-Slots
   9.4.4 Dovetail Slots

Learning Objectives

After completing this chapter, you will be able to:

- Identify the correct tooling to use in pocket roughing and finishing.
- Explain entry strategies for pocket milling.
- Specify the most efficient roughing direction and pattern for pocket roughing.
- Determine depth of cut and stepover amounts in roughing.
- Determine a machining strategy for finish pocket milling.
- Determine an entry strategy for slot milling.
- Select tools for roughing and finishing a slot.
- Explain common strategies for machining T-slots and dovetail slots.

Key Terms

closed pocket	helix	slot
depth of cut	open pocket	stepover amount
dovetail slot	pocket	T-slot
helical entry	ramp entry	

Chapter opening photo credit: CNC Software, Inc.

9.1 Introduction

In Chapter 8, you learned programming and machining techniques for producing external boundary shapes. This chapter focuses on creating programs for machining two common types of internal features: pockets and slots. You will learn programming strategies used in pocket milling and slot milling and cutting techniques used to produce the desired feature accurately and efficiently.

As you learned in Chapter 8, contouring refers to cutting along a straight line or circular path to produce a desired shape. Pocket milling and slot milling are classified as contouring operations for internal shapes. Usually, pocketing and slotting operations involve a roughing pass followed by finish machining. Pockets and slots can be closed or open features. Each type of feature has unique challenges and requires distinctly different cutting strategies. Closed pockets and slots require special consideration because material removal involves an initial entry move into the material before cutting the desired contour.

9.2 Pocket Milling

A *pocket* is an internal cavity with a flat bottom. See **Figure 9-1**. A *closed pocket* is encased on all sides by the part. An *open pocket* is enclosed on three sides, but open on the fourth side.

Both types of pockets have similar characteristics and similar machining strategies for creation. The first consideration is the radii in the corners of the pocket. The size of radii will assist in determining the size of the end mill that can be used in machining.

The pocket shown in **Figure 9-2** is 1.5″ × 3″ with a 1/4″ (.250″) radius in the corners. The length and width of the pocket would allow for an end mill that is 1″ diameter or even larger. However, if a 1″ diameter end mill is used to cut the pocket, a large amount of uncut material in the corners of the pocket will remain and require additional machining. See **Figure 9-3**.

Normally, a finish pass with a finishing tool will be performed after making an initial roughing cut with a roughing tool. Roughing will leave a small amount of material in each corner of the pocket for the finish cut.

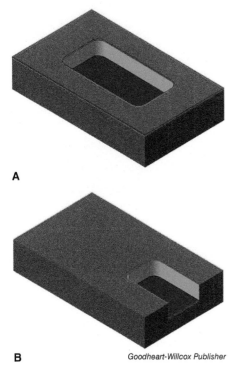

A

B

Goodheart-Willcox Publisher

Figure 9-1. Types of pockets. A—Closed pocket. B—Open pocket.

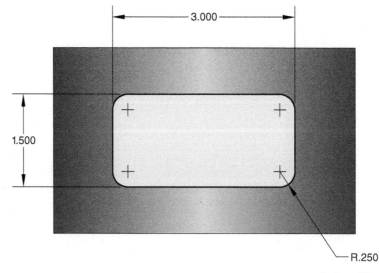

Goodheart-Willcox Publisher

Figure 9-2. A closed pocket dimensioned as 1.500″ × 3.000″ with .250″ corner radii.

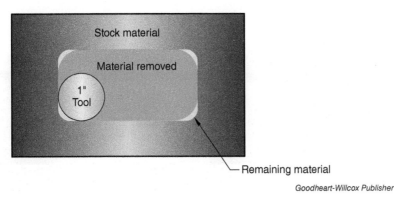

Remaining material

Goodheart-Willcox Publisher

Figure 9-3. Using a 1″ end mill to cut the pocket will leave excess material to remove in the corner areas.

A large amount of excess material, however, will result in increased tool deflection, rough surface finish, gouging, or even tool failure in the corners. This needs to be minimized in the roughing operation. A general rule of thumb is the roughing tool diameter should be no larger than 2 1/2 times the size of the corner radius. In this case, the corner radius is .250″. Therefore, the largest roughing tool used would be .625″.

.250″ (Corner Radius) × 2.5 (Constant) = .625″ (Roughing Tool Diameter)

This is not an absolute rule. For example, a larger end mill can be used when machining a large pocket with relatively small corner radii. **Figure 9-4** shows an example. The pocket in this part is 1.5″ × 3″ with 1/8″ (.125″) radii in the corners. If using the general guideline of multiplying 2.5 by the corner radius, the tool would be a 5/16″ (.3125″) end mill. This would require an excessive amount of passes to machine the entire pocket, considerably impacting total cycle time and efficiency. In this case, there are two possible alternatives.

The first option is to drill holes into the corners *before* roughing any material out of the pocket. By drilling holes, the excess material in the corners is removed, even if a larger end mill is used to rough out the inside of the pocket. See **Figure 9-5**. **Figure 9-5A** shows the result of the drilling operation. **Figure 9-5B** shows the final material left from roughing the pocket. Roughing is followed by a finish pass to complete the part.

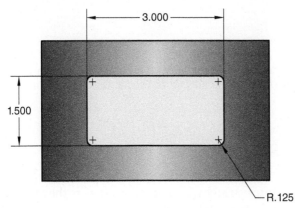

Goodheart-Willcox Publisher

Figure 9-4. A closed pocket dimensioned as 1.500″ × 3.000″ with .125″ corner radii.

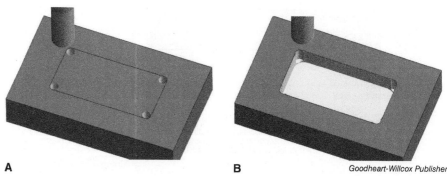

Figure 9-5. A pocket milling operation in which the corners are drilled first before roughing out the inside of the pocket. A—Corner drilling allows excess material in the corners to be easily removed. B—The part after roughing the pocket.

There are two items to consider in this method. The first is the drill size. The pocket has .125″ radius corners. Therefore, a .250″ diameter drill would fit. However, leaving some material is better, so a more appropriate size drill would be a .238″ diameter drill (B size drill), or something slightly smaller. The second consideration is drill depth. In the drilling operation, material is left for the final finishing pass. The pocket is .500″ deep, so the drill is only programmed to .490″ deep, leaving some material for the finishing pass.

The second option for machining the small radii in the corners is to rough the pocket with a larger end mill and then *semifinish* the corners. In this process, a larger roughing end mill is used to make a roughing pass first. Then a small end mill will make short movements at the corners to remove some extra stock. See **Figure 9-6**. This method takes a little more programming skill, but it is more commonly used and is the better alternative.

In both of these cases, the objectives are to optimize the cutter diameter used in roughing without sacrificing cycle time and minimize excess material to be cut. These decisions are made *prior* to programming, when making the tool list for part creation.

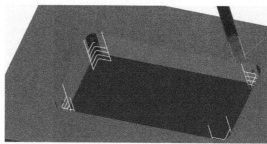

Figure 9-6. A pocket milling operation in which roughing with a large roughing end mill is completed first. A small end mill is then used at the corners to remove extra stock.

From the Programmer

Climb Milling

As discussed in Chapter 8, climb milling is the preferred method for most CNC milling operations. When roughing out pockets, one side of the tool is climb milling and one side is conventional milling. The finish pass should always be climb milling. In climb milling, the cutting tool is most often positioned to the left of the material. This is the technique used in illustrations and programs in this chapter.

9.2.1 Entry Strategy

When creating the pocket, the most difficult and important part of the program is the entry strategy. The entry strategy is the technique or process used to enter the cutter into the material. At the point where the pocket is started, the material is solid. Sending the end mill straight down into solid material is never desired. Although most end mills have the capability to cut with the bottom of the tool, the chips have nowhere to escape and can cause catastrophic tool failure. The tool needs to enter the material in a way that allows smooth entry to the desired depth and an escape path for the removed material. See **Figure 9-7**. **Figure 9-7A** shows the solid material as the tool enters. **Figure 9-7B** shows the tool plunged into the material. Imagine the additional forces required at the bottom of the tool to evacuate chips and make this entry possible. While it is possible to cut this way, this should be avoided except when no other entry is possible.

Ramp Entry

To reduce the tool load and prevent damage to the material or the cutting tool, an alternate entry strategy is needed. A common strategy is a *ramp entry*. In this type of entry, the cutting tool enters the material on an angle. A ramped entry is programmed as a simultaneous X and Z axis move (or Y and Z axis move) to enter with a descending cut. The cutting path is oriented to use the longest side of the pocket. The angled path into the material minimizes tool pressure and assists with chip evacuation.

Figure 9-8 shows a part with a pocket to be machined and a ramped entry move to enter the material. **Figure 9-8A** shows the dimensions locating the pocket. The work coordinate system origin is positioned at the top-left corner of the part. A 5/8" (.625") diameter end mill is used to make the entry. The ramped entry begins on the right side of the part at X3.5 Y–1.5.

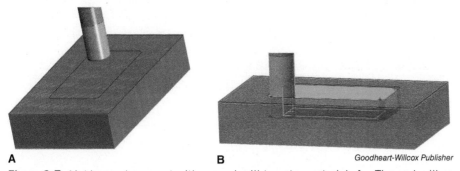

A   B   *Goodheart-Willcox Publisher*

Figure 9-7. Making a plunge cut with an end mill to enter material. A—The end mill as it enters the workpiece. B—Considerable forces are required at the bottom of the tool to make a plunge cut possible.

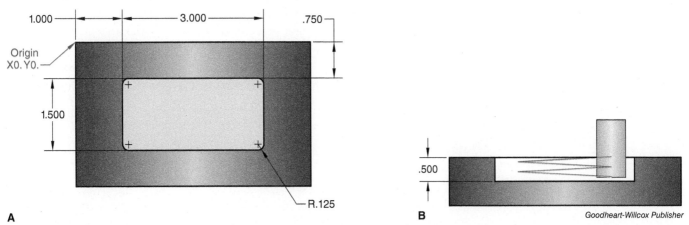

A   B   *Goodheart-Willcox Publisher*

Figure 9-8. Using a ramp entry to enter material for pocket milling. A—A part with a closed pocket dimensioned as 1.500" × 3.000" with .125" corner radii. B—A front view of the part with the programmed cutting path shown.

The next line of code will move the tool in the X axis negative direction to X1.5, while simultaneously moving it down in the Z axis negative direction. This can be repeated by making back-and-forth movement to the X3.5 and X1.5 locations with additional Z axis moves until the desired depth is reached. **Figure 9-8B** shows a front view of the workpiece and the resulting cutting path.

The tool is being *zig-zagged* back and forth to allow for easier entry to achieve the desired depth. This raises the question: How deep are the Z axis moves? The ramp entry angle can vary depending on the type and size of tool, but a safe guideline is 3°. In this example, the tool is moving 2″ in the X axis at 3°, making each Z axis move approximately .105″.

The distance of the Z axis move is calculated using the tangent trigonometric function. Use a calculator with the following formula:

$$z = x \times \tan\theta$$

In this formula, z is the depth of the Z axis move, x is the length of the horizontal move, and θ is the angle of descent of the tool. See **Figure 9-9**. For this example, use the following calculation:

$$z = x \times \tan\theta$$

$$= 2'' \times \tan(3°)$$

$$= 2'' \times .0524$$

$$= .105''$$

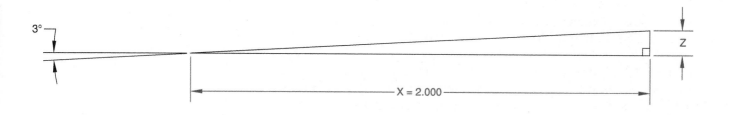

$$\text{Tangent } \theta = \frac{\text{Opposite}}{\text{Adjacent}}$$

$$\text{Tangent } \theta = \frac{Z}{X}$$

$$\text{Tangent }(3°) \times X = Z$$

Goodheart-Willcox Publisher

Figure 9-9. Use the tangent trigonometric function to calculate the depth of the Z axis move for the ramp entry. The calculation is based on an entry angle of 3°.

The following example shows how the tool entry is programmed.

T1 M6	Tool change to correct tool
G0 G90 G54 X3.5 Y1.5 S2200 M3	Move to X and Y axis position
G43 H1 Z2.	Turn on tool length offset and move to Z2.
M8	Coolant on
Z.2	Rapid move to Z.2
G1 Z0. F40.	Feed to Z0. at 40 ipm
X1.5 Z-.105	First entry move
X3.5 Z-.210	
X1.5 Z-.315	
X3.5 Z-.420	

After the tool is at final depth, the remaining pocket roughing can be completed.

Helical Entry

A *helix* is a three-dimensional spiral. Helical movement is similar to arc movement. A helical move is created with a circular interpolation command (G02 or G03) and includes a Z axis move. In a helical move, simultaneous movement occurs along three axes. This allows the tool to move gradually into the material in a circular motion. The *helical entry* method is preferred for a pocket, but it can be more complicated to program. A helical milling operation will create more space for chip evacuation and reduce the tool load significantly when compared to a plunging entry move. A helical entry can be faster and more efficient than a ramp entry.

When programming a helical entry, pick a safe starting position to use as an arc center and program small arcs to perform the entry. Ensure that the arcs used for entry do not leave any standing material on the pocket floor that will require additional material removal. Cutter compensation can be used, or it can remain off until the roughing passes are programmed. Once programmed, the helical entry should follow a path similar to that shown in **Figure 9-10**.

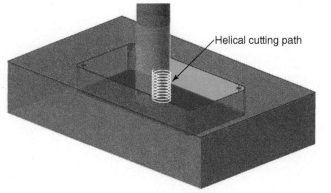

Goodheart-Willcox Publisher

Figure 9-10. Using helical movement to enter the cutter into the material. A helical move is created with a circular interpolation command by specifying arc movement and linear movement along the Z axis simultaneously.

The intention of the entry move is not to rough away large amounts of material, but simply to gain access to the inside of the pocket where more efficient roughing paths can be established. With that in mind, the helical arcs should be small and performed at high feed rates. For the part shown in **Figure 9-10**, the work coordinate system origin is positioned at the bottom-left corner of the part. The tool is initially positioned at the center of the pocket (X2.5 Y1.5). The helical moves are programmed to create .050" arcs, making sure the arcs do not violate the outer pocket boundary.

The following example shows how the helical entry is programmed to enter the tool into the pocket. Notice that the last Z axis position is at .495" deep, and not at .500" (the bottom of the pocket). Always leave material for the final pass to ensure a good, clean finished pocket.

```
T1 M6
G0 G90 G54 X2.5 Y1.5 S3056 M3
G43 H1 Z2.
M8
Z.2
G1 Z0.01 F50.
X2.55
G3 X2.55 Y1.5 Z-.05 I-.05 J0. F36.67     First helical move
X2.55 Y1.5 Z-.1 I-.05 J0.
X2.55 Y1.5 Z-.15 I-.05 J0.
X2.55 Y1.5 Z-.2 I-.05 J0.
X2.55 Y1.5 Z-.25 I-.05 J0.
X2.55 Y1.5 Z-.3 I-.05 J0.
X2.55 Y1.5 Z-.35 I-.05 J0.
X2.55 Y1.5 Z-.4 I-.05 J0.
X2.55 Y1.5 Z-.45 I-.05 J0.
X2.55 Y1.5 Z-.495 I-.05 J0.              Last Z axis position
G1 Z.1 F100.
```

After the helical entry, the material will resemble **Figure 9-11**. This allows an open position to re-enter the tool in the center of the hole and start the roughing operation.

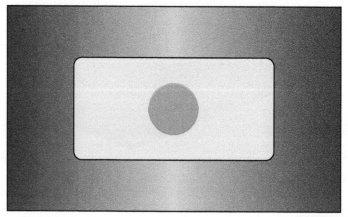

Goodheart-Willcox Publisher

Figure 9-11. Helical movement of the tool into the material creates a circular opening where the roughing operation can be started.

9.2.2 Pocket Roughing

After a successful entry is made into the pocket, the next step is to rough out the excess material. There are many ways to do this, depending on the available tools and the size, shape, and depth of the pocket. This is where experience will help, but several different techniques are available to maximize output and minimize cycle time. Roughing is no different from other types of machining in that it must be evaluated for speed and accuracy.

The shape and axis direction of the pocket can help determine the most efficient strategy for roughing. For example, **Figure 9-12** shows a part with a rectangular pocket oriented in the X axis direction. In this case, long X axis cuts are more efficient. **Figure 9-13** also shows a part with a rectangular pocket, but it is primarily in the Y axis direction. In this case, the primary cutting direction is along the Y axis.

Another common pocket shape is a circular pocket. A circular pocket can be roughed out with arc movements or circular movements. See **Figure 9-14**. This requires a little more thought, but programming a series of circular paths is normally the fastest and most efficient process for roughing circular pockets.

A circular roughing path starts in the area that was cleared away in the entry move. This is normally the center of the pocket, but it does not have to be at that location. A programmed circular roughing path can follow a circular pattern like the one shown in **Figure 9-15**.

This is a roughing operation, so remember that cutting will not produce the final dimension during this stage. The pocket diameter is 6.500″. A small amount of material—.010″ to .020″—will remain on the wall for a finish cut. This operation uses a .750″ diameter end mill to make circular cuts with a distance of .25″ between cuts. The program for the roughing operation is as follows.

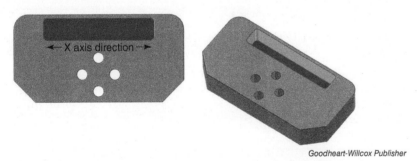

Figure 9-12. A part with a closed pocket oriented in the X axis direction.

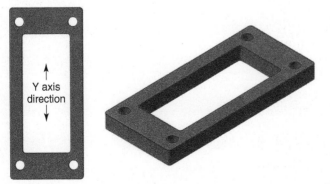

Figure 9-13. A part with a closed pocket oriented in the Y axis direction.

Figure 9-14. A machined part with a circular pocket.

Figure 9-15. A circular roughing path. The white circles represent the circular paths of the tool. The paths start on the inside and move outward toward the pocket wall at designated step sizes to rough out the existing stock. The straight white lines represent the step sizes.

```
O0425
(MATERIAL - ALUMINUM INCH - 6061)
(T1|3/4 FLAT END MILL|H1)
G20
G0 G17 G40 G49 G80 G90
T1 M6
G0 G90 G54 X-.135 Y0. S2546 M3
G43 H1 Z2.
M8
Z.2
G1 Z-1. F20.
X.615 F50.9
G3 I-.615                           First 360° circular cut
G1 X-.135
G0 Z.25
X.115
Z.2
G1 Z-1. F20.
X.865 F50.9
G3 I-.865
G1 X.115
G0 Z.25
X.365
Z.2
G1 Z-1. F20.
X1.115 F50.9
G3 I-1.115
G1 X.365
G0 Z.25
X.615
Z.2
G1 Z-1. F20.
X1.365 F50.9
G3 I-1.365
G1 X.615
G0 Z.25
X.865
Z.2
G1 Z-1. F20.
X1.615 F50.9
G3 I-1.615
G1 X.865
G0 Z.25
X1.115
Z.2
G1 Z-1. F20.
X1.865 F50.9
```

Continued

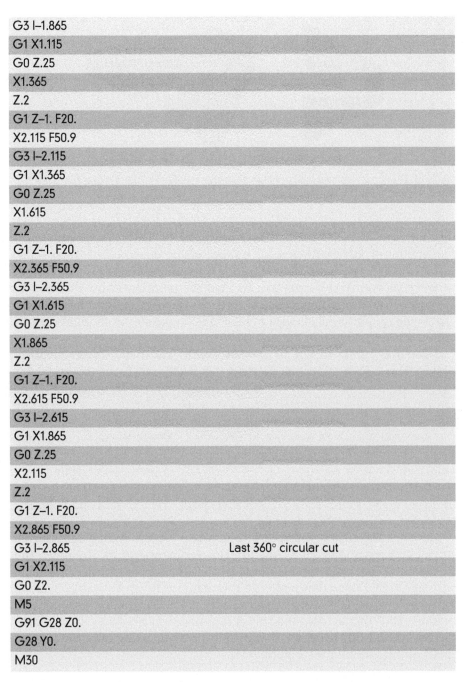

Last 360° circular cut

Open pockets do not require an entry move, because the tool can enter safely from outside the stock boundary. See **Figure 9-16**. As with other pocket types, the shape of the open pocket should determine the roughing strategy. A common approach is entering in the center of the pocket and working outward toward the pocket boundaries. However, the pocket depth and shape could dictate a different strategy.

9.2.3 Depth of Cut and Stepover Amount

As previously discussed, entry strategy and cutting direction must be determined to establish how the tool enters the material and the axis along which cutting occurs. There are two additional factors that require careful consideration in pocket cutting: the depth of cut and stepover amount. The ***depth of cut*** refers to how *deep*, in the Z axis, the tool is engaged in

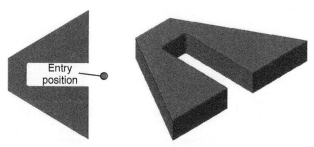

Figure 9-16. A part with an open pocket. The tool can enter from outside the stock along the center of the opening.

the material. This is also called *axial engagement*. The **stepover amount** is the amount of material being cut by the side of the end mill. This is also called *radial engagement*. In a pocketing operation, stepover is the distance between successive tool passes. This distance represents how far the tool advances from a previous position to make the next cut and how far the tool is engaged into the material. There are two schools of thought on how to determine the depth of cut and stepover amount.

One strategy is to cut to smaller depths and make larger stepover passes. This common cutting strategy has been used for many years due to the tools that were available and the maximum spindle speeds and feed rates of CNC mills. **Figure 9-17** shows the cutting path for a pocket cutting operation and the stepover amount between each pass of the tool.

The following is an example of using this strategy to cut a pocket in a similar operation. The pocket to be machined is 1.5″ × 3″ × 1″ deep with .250″ radii in the corners. Using a .500″ end mill, the program takes depth cuts of .100″ and stepover cuts of .250″, or 50% of the cutter diameter. The program uses a ramp entry and leaves .010″ finishing stock on the walls and floor. The toolpath for this operation is similar to the one shown in **Figure 9-17**.

Using a three-flute end mill in aluminum with a cutting rate of 500 surface feet per minute (sfm), the total cycle time would be 2 minutes and 51 seconds. That cycle time is not ideal for a pocket of this size, but there is an additional issue that makes this strategy even less desirable. Making repetitive cuts at .100″ deep means that the cut is being made with only the bottom .100″ of the tool, and this is repeated over and over. In this case, the pocket is 1.00″ deep, so 10 cuts will be made using only the bottom .100″ of the end mill. See **Figure 9-18**.

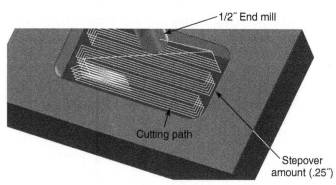

Figure 9-17. A pocket cutting operation in which the stepover amount is 50% of the cutter diameter.

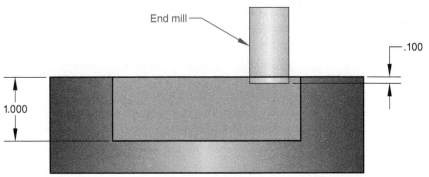

Figure 9-18. A pocket cutting operation in which the depth of cut is .100″.

This will cause early tool failure on the end of the tool. This type of strategy is required with older machines with lower spindle speeds and feed rates. Many machinists who started on manual machines use this technique because it was required on manual machines. This technique will work, but it will result in higher tool usage and slower cycle times.

The second strategy is to use the entire flute length of the end mill, or a very deep axial cut, and make very small radial cuts. There are several advantages to this style of machining, often referred to as high-speed machining. These advantages include the following:

- Improved chip evacuation. Chips will not become lodged in the flutes.
- Chip thinning. Light cuts result in thinner chips, allowing cutting feed rates to be increased.
- Shorter engagement time for the tool. This allows for higher feed rates.
- Better distribution of cutting pressure over the length of the flute. This produces less tool deflection and chatter.
- Less dulling of the cutting edge. Lower engagement improves conditions for each cutting edge.
- Longer tool life. The tool wear spreads over an extended length of the flute.

A typical toolpath for this type of machining will consist of an entry into the pocket, followed by small zig-zag moves across the pocket. See **Figure 9-19.** The stepover amount should be between 5% and 15% of the tool diameter. A 1/2″ end mill should step over between .025″ and .075″.

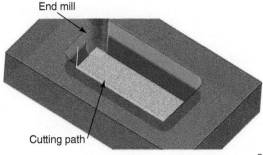

Figure 9-19. A pocket cutting operation in which the tool is deep in the material and making very small radial cuts.

From the Programmer

CAM and High-Speed Machining

Most computer-aided manufacturing (CAM) software has high-speed machining (HSM) functionality. However, high-speed machining requires CNC machines capable of rotating the spindle at high speeds (8000 rpm to 15,000 rpm and higher). In addition, there are end mills designed specifically for this type of machining. They have specific tool geometry that evacuates chips rapidly and coatings that help reduce heat and friction.

When the stepover amount is light and the entire tool is engaged, the spindle speeds and feed rates can be greatly increased. In the pocket cutting example, the spindle speed is set to 8000 rpm and the feed rate to 240 ipm. A six-flute end mill is used, which reduces the chip load (thickness of the chip) to only .005″. The cycle time was reduced to 40 seconds. With these parameters, tool life is increased. The resulting high-speed path is four times faster than the original toolpath, but it does require a machine and tooling that can run at higher rpm speeds and use higher feed rates.

9.3 Pocket Finishing

There are three parts to successfully machining a pocket: the entry, the roughing, and the finishing. In the finishing stage, one or more finishing passes are used to create the final dimensions and appearance of the pocket. It is best practice to use a separate tool to accomplish the finishing, as the roughing tool is subject to extended use and material removal.

There are two areas of consideration when finishing the pocket: the floor and the walls. In the roughing operation examples previously discussed, .010″ of material was left on the walls and .005″ of material was left on the floor. After the finishing tool is initiated, the floor can be machined to final depth by repeating the roughing toolpath machine code at the correct depth, still leaving .010″ on the walls. The walls can then be finished to the final depth with a contouring toolpath. The end mill used to finish the walls must have a radius smaller than the corner radii of the pocket. For this example, the corner radii are .250″. In addition, using cutter compensation will allow for small adjustments to the pocket size and precise control of tolerances specified on the print.

The entry move into the wall of the pocket for the finishing pass should also be considered. A direct perpendicular move into the wall will leave a mark (an imperfection in the wall). When the CNC mill abruptly changes direction, there is a momentary pause where overcutting happens. The overcutting will be very minor, but still visible.

When using cutter compensation, remember that it cannot be turned on while making an arc move. A straight linear move is needed to initiate cutter compensation, but avoid machining into the wall with this move. A more acceptable entry is to make the straight move to initiate cutter compensation, and then enter the wall with a small radial move. See **Figure 9-20.**

For this example, the work coordinate system origin is positioned at the bottom-left corner of the part. The program for the final finishing path is as follows.

```
T2 M6
G0 G90 G54 X2.55 Y1.825 S5093 M3
G43 H2 Z2.
M8
Z.2
G1 Z-.5 F20.
G41 D2 Y2.2 F102.
G3 X2.5 Y2.25 R.05
G1 X1.25
```

Continued

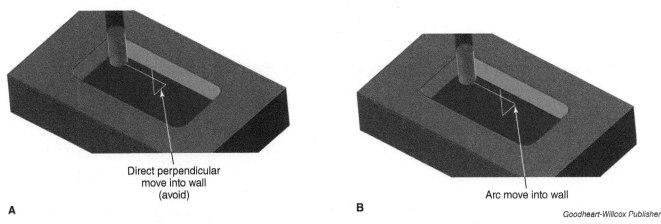

Figure 9-20. A comparison of tool entry moves in pocket finishing. A—Entering the pocket wall with a perpendicular move will leave an imperfection and should be avoided. B—Entering the pocket wall with an arc move is the proper method of entry. When using cutter compensation, the arc move must be preceded by a linear move to activate cutter compensation.

```
G3 X1. Y2. R.25
G1 Y1.
G3 X1.25 Y.75 R.25
G1 X3.75
G3 X4. Y1. R.25
G1 Y2.
G3 X3.75 Y2.25 R.25
G1 X2.5
G3 X2.45 Y2.2 R.05
G1 G40 Y1.825
G0 Z2.
```

This is the final machining for the pocket. The roughing and finishing pocket cutting strategies discussed can be used to machine either an open or closed pocket. These strategies can be used whether the pocket extends through the material or has a flat bottom.

9.4 Slots

A *slot* is an internal cavity with straight sides and open or closed ends. Toolpaths for slots often resemble pocketing toolpaths, and slots can be machined with a similar cutting strategy. Slots are categorized as three-sided machining features when they continue off one or two edges of the part boundary, but they can be fully closed, similar to a closed pocket. Slots can also vary in shape. Slots with different profile shapes include T-slots, dovetail slots, and keyway slots. In addition, slots are not always machined on the top side of the material. They can also be machined on the side walls of the material.

9.4.1 Open Slots

The traditional slotting operation is for a three-sided slot. This is a common slot with a fairly easy machining strategy. Three-sided slots require machining on three faces—two sides and the bottom. **Figure 9-21** shows

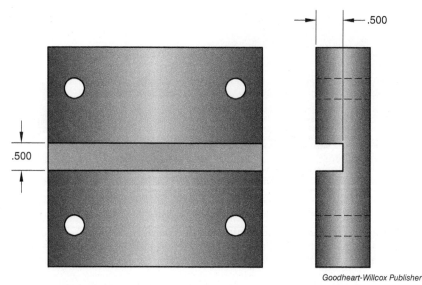

Figure 9-21. A three-sided slot.

a .500″ wide slot that is .500″ deep, and proceeds completely through the length of the material.

At first look, it appears that this slot could be machined using a .500″ end mill and machining straight along the length. As shown in **Figure 9-22**, this is not the best approach. As previously discussed, climb milling is the preferred method for most CNC milling operations. Whenever the tool is fully engaged in the material, one side of the tool is climb milling and one side is conventional milling. Another problem with using a .500″ end mill is that there is no chip evacuation. As the end mill enters the cut, the chip is pulled completely through the cut and carried to the opposite side before it can exit the tool's cutting edge. This will leave a rough surface finish on the wall. This requires a second operation.

A better strategy for machining this slot is to use an end mill smaller than .500″, make a cut through the center of the slot, and then finish cut by climb milling the walls of the slot. In this example, a 3/8″ (.375″) diameter end mill is used to center cut the slot in a roughing operation and then climb mill both sides to finish. The center cut is at a depth of .495″ and the two climb cuts are completed at .500″, creating a finish cut on the floor as well as the sides. See **Figure 9-23**.

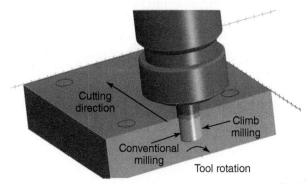

Figure 9-22. In slot cutting, the tool is fully engaged in the material during roughing and the tool is conventional milling on one side of the tool. A second operation is required to finish the slot walls.

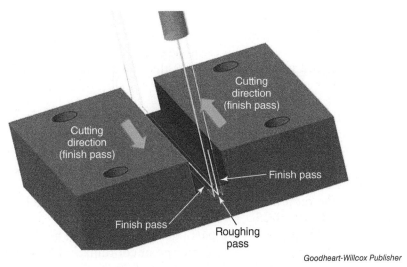

Figure 9-23. To cut a three-sided slot, a roughing operation is completed before a finishing operation. This allows the tool to climb mill on both sides of the slot during the finish pass.

Depending on the material type or depth of the slot, more than one depth cut may be required for the roughing operation. A slot dimension with a close tolerance may also require a separate finishing tool. The tolerances applied to dimensions will help make these decisions.

9.4.2 Closed Slots

Closed slots are similar to closed pockets. Normally, a closed slot will have a full radius end feature instead of a rectangular shape with four corner radii. See **Figure 9-24**.

As is the case with an open slot, it appears that the closed slot in **Figure 9-24** can be machined using a simple 1/2" (.500") end mill

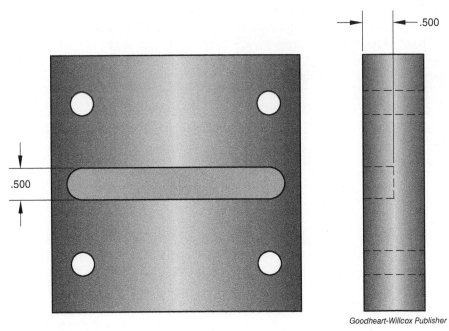

Figure 9-24. A part with a closed slot.

programmed to cut down the center of the slot. By now, it should be apparent that this is not a good strategy. As with an open slot, driving the tool straight down the center will create very poor surface finish, lead to early tool failure, and produce dimensions that are less than optimal for tolerance requirements. In addition, in the closed slot, there is the added issue of entry into the material.

As is the case with closed pocket machining, one strategy is to use a drilling operation to create a starting position for the end mill. Another entry strategy is to use a ramping motion to gain access. In a closed slot, one or both of these strategies can be used to create a starting entry position in the slot. To begin, it is imperative to know the center point coordinates of the radii in the slot. These coordinates will be the locations for drilling. See **Figure 9-25**.

The top-left corner of the part has been designated as the work coordinate system origin. Thus, the center point coordinates of the radius locations are X.500 Y–2.000 and X3.500 Y–2.000. Initially drill two holes at these locations. The diameter of the finish dimension is .500", so the drill size must be smaller than that dimension. To maintain some material on the walls for finishing, a 7/16" (.438") diameter drill or a slightly smaller drill is appropriate. The following program contains a hole drilling cycle initiated with the G81 command. Hole machining operations are covered in Chapter 10.

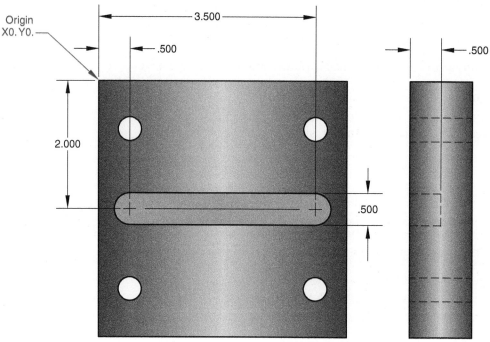

Goodheart-Willcox Publisher

Figure 9-25. Dimensions for the center point coordinates of the slot radii.

G0 G17 G40 G49 G80 G90	
T3 M6	
G0 G90 G54 X.5 Y-2. S800 M3	
G43 H3 Z2.	
M8	
G98 G81 Z-.49 R.1 F10.	Begin canned drilling cycle and drill first hole to Z depth to leave .01" material on floor
X3.5	Drill second hole
G80	Cancel canned cycle
M5	
G91 G28 Z0. M9	

The next tool will be an end mill to ramp into the slot and remove the majority of material. In this case, a 3/8" (.375") diameter end mill or even a 7/16" (.438") end mill could be used to rough out the slot. As previously discussed, a typical ramp entry angle for tool entry is 3°. The tool will ramp at this angle, back and forth between the two radius centers, until a depth of .490" is reached. The .490" depth still allows for a finish pass on the bottom of the floor of the slot. The Z depth of each ramp pass is determined using a trigonometric calculation. The program is as follows.

T2 M6 (3/8 END MILL)	
G0 G90 G54 X3.5 Y-2. S5093 M3	
G43 H2 Z2.	
M8	
Z.2	
G1 Z0. F20.	
X.5 Z-.15	Calculated Z depth from 3° ramp angle and movement along X axis
X3.5 Z-.3	
X.5 Z-.45	
X3.5 Z-.49	
X.5	
G0 Z2.	

Notice how the tool goes back and forth between X.5 and X3.5, while traveling down along the Z axis at .150" per pass. At the last depth, at .490", it moves back to X.5 to ensure the bottom of the slot is now flat. Also notice that cutter compensation is not used. The program uses the two radius centers as programmable points and the tool center moves to those points.

The last step in the process is a simple contour path to finish the slot. Be careful in designating the starting position and using cutter compensation, as the slot has very little room for a tool to maneuver. Do not violate the slot walls with the cutter compensation initiating move or cutter compensation cancel move on the exit.

9.4.3 T-Slots

A *T-slot* is a T-shaped slot used to mount work in machining. Standards for T-slots, mating nuts, and T-bolts are maintained by the American Society of Mechanical Engineers (ASME). A T-slot is designed specifically for use with a T-slot nut and a threaded clamp to position and secure workpieces on a workbench or on a machine. The T-slot nut slides along a T-slot track machined on the workbench or machine table. This is a common feature on most CNC machining centers. See **Figure 9-26**.

T-slot tracks are machined into the table or part to accept a T-slot nut. A stud is threaded into the T-slot nut and receives a hex nut to secure a fixture, machine vise, or the material itself. See **Figure 9-27**.

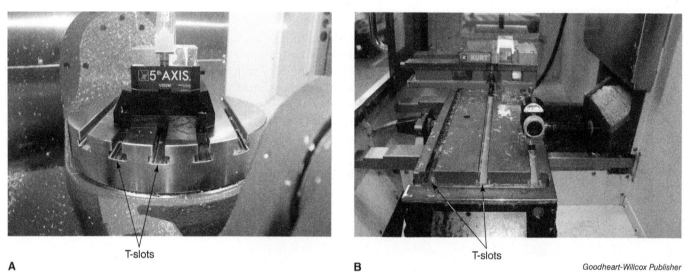

Goodheart-Willcox Publisher

Figure 9-26. T-slots are open slots used to secure a workpiece to a machine table. A—A T-slot rotary table. B—A T-slot mill table.

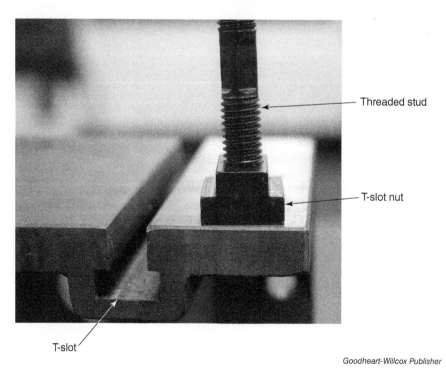

Goodheart-Willcox Publisher

Figure 9-27. A T-slot nut assembled with a threaded stud.

Special tooling is needed to machine a T-slot. The cutting tool used is called a T-slot cutter, **Figure 9-28**. This tool cuts the lower shape at the bottom of the T-slot. This type of cutter will cut the wider dimension of the T-slot, but the shank will not machine, so the upper shape must be machined as a slot prior to cutting the bottom portion of the T-slot.

Figure 9-29 shows an isometric view and front view of a part with machined T-slots. The top opening is .562″ and the T-slot portion at the bottom is 1.250″ across. The cutting strategy will be to use a 1/2″ (.500″) end mill to machine the .562″ slot at the top. This slot will be machined down in the Z axis direction to a depth of .757″. This depth is calculated from the .288″ dimension at the top and the .469″ dimension at the bottom of the T-slot.

$$.288″ + .469″ = .757″$$

After the 1/2″ end mill machines the .562″ slot, the T-slot cutter is used to machine the T-slot. The following program is used to finish the first slot (the .562″ slot) all the way down to the Z–.757 depth. See **Figure 9-30**. The work origin for this part is located at the top-right corner, indicating the

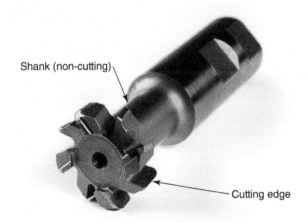

Nikola Bilic/Shutterstock.com

Figure 9-28. A T-slot cutter is used to cut the lower profile shape of a T-slot.

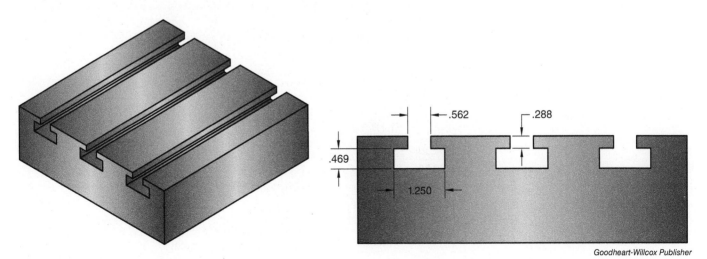

Goodheart-Willcox Publisher

Figure 9-29. A machined part with T-slots and the dimensions required for the T-slot machining operation.

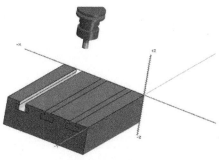

Figure 9-30. The highlighted slot is a .562″ slot cut to a depth of Z–.757 with a 1/2″ end mill. This is the first slot milled to machine the T-slot.

part is oriented in the X–, Y– quadrant. The part is 8″ wide and 10″ long. Cutter compensation is used in the program to position the tool at the appropriate coordinates. The first two cuts are roughing cuts made in the negative Y axis direction to depths of Z–.742 and Z–.757 on one side of the slot. Then, a finish cut is made on the same side of the slot to the final depth. Two roughing cuts and a finish cut are then made on the other side of the slot by driving the tool in the positive Y axis direction.

O1020 (T SLOT)	
(T1|1/2 FLAT END MILL|H1|D1)	
G20	
G0 G17 G40 G49 G80 G90	
T1 M6	
G0 G90 G54 X–6.766 Y.5 S3820 M3	
G43 H1 Z2.	
M8	
Z.2	
G1 Z–.742 F76.4	Move to depth of first roughing cut
G41 D1 Y0.1	
Y–10.	First roughing cut
G40 Y–10.5	
G0 Z.25	
Y.5	
Z.2	
G1 Z–.757	Move to depth of second roughing cut
G41 D1 Y0.1	
Y–10.	Second roughing cut
G40 Y–10.5	
G0 Z.25	
X–6.781 Y.5	
Z.2	
G1 Z–.757	
G41 D1 Y0.1	
Y–10.	First finish cut
G40 Y–10.5	
G0 Z2.	
X–6.234	
Z.2	
G1 Z–.742	
G41 D1 Y–10.1	
Y0.	Roughing cut
G40 Y.5	
G0 Z.25	
Y–10.5	
Z.2	
G1 Z–.757	
G41 D1 Y–10.1	
Y0.	Roughing cut

Continued

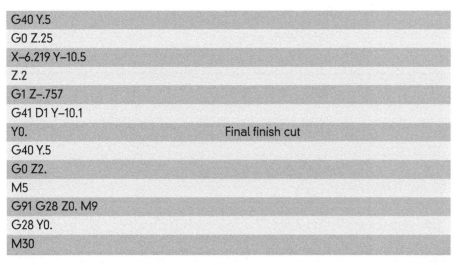

```
G40 Y.5
G0 Z.25
X–6.219 Y–10.5
Z.2
G1 Z–.757
G41 D1 Y–10.1
Y0.                              Final finish cut
G40 Y.5
G0 Z2.
M5
G91 G28 Z0. M9
G28 Y0.
M30
```

The selection of the T-slot cutter is important to the programming method. For this application, a standard 1.250″ diameter cutter that is .250″ thick is used. See **Figure 9-31**. This tool also has a .500″ shank, which has clearance for the .562″ slot that was previously machined. The cutter diameter is the same size as the final T-slot in this example. Another option is to use an undersized cutter to allow for finish passes. However, be aware that a smaller diameter cutter will have to move outward to finish the T-slot. This might cause the shank to collide with the .562″ slot and damage the part.

The cutter thickness is .250″ and the final cutting depth of the T-slot is .757″. As shown in **Figure 9-32**, the height dimension of the T-slot portion at the bottom is .469″. This means the cutter must make multiple passes at different depths to completely machine the T-slot. In calculating for these depths, be aware that the tool length offset is in reference to the bottom of the tool, and not the cutting edge on top of the tool. If the top of the cutter is positioned at the Z–.288 edge, the bottom of the cutter is at Z–.538. This is calculated from the .288″ dimension and the .250″ cutter thickness (.288″ + .250″ = .538″).

In this case, the optimal cutting strategy is to make the first pass at the middle of the T-slot depth (Z–.650) to rough, then machine to the bottom of the slot depth (Z–.757), and then make a final pass at the top

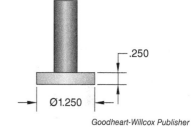

Goodheart-Willcox Publisher

Figure 9-31. A T-slot cutting tool.

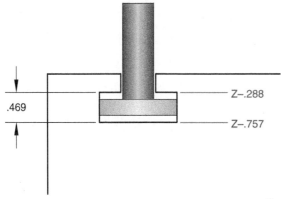

Goodheart-Willcox Publisher

Figure 9-32. The cutting tool is .250″ thick and requires multiple passes to machine the .469″ T-slot portion at the bottom. Calculations are made to position the bottom of the tool at the appropriate Z depth for each pass.

of the slot (Z–.538). These depths represent the position of the bottom of the cutting tool. In the following program, the cutting tool travels straight through the middle of the slot without using cutter compensation. If necessary, depending on surface finish requirements, cuts can be made by using a smaller diameter cutter for roughing followed by a different tool for finish machining.

```
O0715 (T SLOT BOTTOM)
(T2|1.25 SLOTTING TOOL|H2)
G20
G0 G17 G40 G49 G80 G90
T2 M6
G0 G90 G54 X–6.5 Y.75 S1600 M3
G43 H2 Z2.
M8
Z.2
G1 Z–.65 F25.              Roughing depth
Y0. F8.
Y–10.
Y–10.75
G0 Z2.
Y.75
Z.2
G1 Z–.757 F25.             Bottom of slot depth for second cut
Y0. F8.
Y–10.
Y–10.75
G0 Z2.
Y.75
Z.2
G1 Z–.538 F25.             Slot depth for final cut
Y0. F8.
Y–10.
Y–10.75
G0 Z2.
M5
G91 G28 Z0. M9
G28 Y0.
M30
```

Similar cutting strategies are used to machine side slots. See **Figure 9-33**. Be aware of the clearance of the shank and the Z axis height locations in programming. In side slots and T-slots, it is best practice to use an undersized cutter, and traverse the tool in the Z axis to make finish cutting passes. Extra caution should be used when machining. A tremendous amount of cutting force is required in machining these types of slots. If work is not mounted securely or improper feed rates are used, poor surface finishes or tool failure can occur.

Goodheart-Willcox Publisher

Figure 9-33. Machining side slots on a part.

9.4.4 Dovetail Slots

A *dovetail slot* has two angled sides and a flat bottom. Dovetail slots are commonly used in woodworking joints. See **Figure 9-34**. The woodworking industry has been heavily involved in CNC machining for many years. With the exception of custom furniture making, all production furniture making is done on machines such as CNC routers and CNC gantry mills. Dovetail slots are also seen in metalworking, as they form very strong joints and accurate slides. Many CNC machine tables slide through dovetail joints.

Dovetail slots are machined using special cutting tools. Dovetail slots can be cut at any angle, but most dovetail cuts are made at 45° or 60°, depending on the application. The machining strategy for dovetail slotting is similar to the strategy for T-slot machining. An end mill is used to rough out material from the slot, and then a dovetail cutter is used to finish. Because of the angled shape of the slot, an undersized cutter must be used if it is necessary to make multiple finishing passes. **Figure 9-35** shows a 3″ × 3″ part with a dovetail slot.

robcocquyt/Shutterstock.com

Figure 9-34. Dovetail joints are commonly used in wood furniture products.

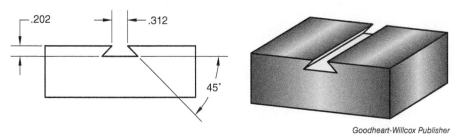

Goodheart-Willcox Publisher

Figure 9-35. A dovetail slot has two angled sides and a flat bottom. In machining, an end mill is used to rough out material and a dovetail cutter is used to finish.

The opening at the top of the slot is .312″ wide and the dovetail angle is 45° on each side. A 5/16″ (.312″) end mill can be used to machine the top opening, or a 1/4″ (.250″) end mill can be used to rough and finish. Machining for the .312″ slot will be programmed down to the full Z−.202 depth, leaving no material at the bottom for the dovetail cutter. A .715″ dovetail cutter is used for the dovetail cutting operation. See **Figure 9-36**. The work origin for this part is located at the bottom-left corner, indicating the part is oriented in the X+, Y+ quadrant. The program is as follows.

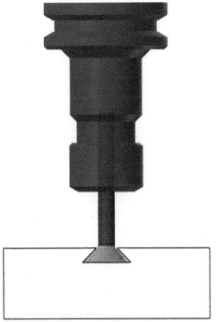

Figure 9-36. The dovetail cutter is used to make the cut for the dovetail slot.

```
O0824 (DOVETAIL SLOT)
(T2|5/16 FLAT END MILL|H2)
(T3|0.715 DOVETAIL CUTTER|H3)
G20
G0 G17 G40 G49 G80 G90
T2 M6
G0 G90 G54 X1.500 Y3.3125 S6112 M3
G43 H2 Z2.
M8
Z.2
G1 Z-.202 F6.4
Y3. F122.2
Y0.
Y-.3125
G0 Z2.
M5
G91 G28 Z0. M9
M01
T3 M6
G0 G90 G54 X1.500 Y3.715 S1049 M3
G43 H3 Z2.
M8
Z.2
G1 Z-.202 F25.
Y3. F5.2
Y0.
Y-.715
Z.2 F50.
G0 Z2.
M5
G91 G28 Z0. M9
G28 Y0.
M30
```

Chapter 9 Review

Summary

- A pocket is an internal cavity with a flat bottom. A closed pocket has four sides and is encased on all sides by the part. An open pocket is enclosed on three sides, but open on the fourth side.
- The size of the corner radii in a pocket will assist in determining the size of the end mill that can be used in machining.
- When considering tooling to machine a pocket, both the roughing and finishing operations should be considered. The roughing tool should be large enough to remove maximum amounts of material without leaving excessive material in the corners. The finishing tool should have a smaller radius than the pocket radii to allow for clean finishing.
- The most difficult part of roughing a pocket is the entry into the material. A ramp or helical entry will reduce the tool load and prevent damage to the material or cutting tool.
- When roughing a pocket, the direction of the pocket should determine the direction of cutting. Circular pockets should use circular toolpaths to remove excess material efficiently.
- The depth of cut refers to the amount of material removed in the Z axis direction, or axial engagement. The stepover amount refers to the amount of tool engagement on the side of the tool, or radial engagement.
- High-speed machining refers to the removal of material using high spindle speeds, high feed rates, and small radial or stepover cuts.
- Pocket roughing should leave a small amount of material (010″ to .020″) for the finish cut. In pocket finishing, one or more finishing passes are used to create the final dimensions of the pocket.
- Open slots have an entrance and an exit off the material. A closed slot is completely enclosed in the material. Many closed slots have full radius ends and are narrow in shape.
- Making an entry into a closed slot is best accomplished with a ramp entry move. In some cases, a drill can be used to clear material at the end of the slot and establish a plunge entry point for the end mill.
- Often, a slot can be roughed and finished with the same tool. It is not recommended to use a tool with the same size as the slot width, as it will cause poor surface finish.
- A T-slot requires two machining operations. A straight end mill is used to clear the top portion and machine down to the T-slot bottom. The second operation uses a T-slot cutter to form the larger feature on the bottom of the T-slot. The T-slot cutter can be used in multiple passes to clear excess material, taking caution of the shank diameter and cutter width.
- Machining a dovetail slot is similar to machining a T-slot. An end mill is used in the first operation to clear material away. The second operation uses an angled dovetail cutter with the correct angle and cutting diameter.

Review Questions

Answer the following questions using the information provided in this chapter.

Know and Understand

1. *True or False?* A closed pocket is enclosed on three sides, but open on the fourth side.
2. *True or False?* Pockets, by definition, have an angled floor.
3. *True or False?* In climb milling, the cutting tool is most often positioned to the left of the material.
4. *True or False?* In pocket milling, the roughing tool can be larger than the finishing tool.
5. What is an acceptable entry method for a pocket?
    A. Helical entry
    B. Ramp entry
    C. Drilled clearance hole
    D. All of the above.
6. When finishing a pocket, a _____ toolpath is recommended.
    A. contouring
    B. ramping
    C. drilling
    D. roughing
7. *True or False?* The tool used on the first cut of an open slot should be the same size as the slot width.
8. *True or False?* It is not critical to complete finish passes on an open slot.

9. In a closed slot, what is the most acceptable entry strategy?
   A. Helical entry
   B. Tapping
   C. Ramp entry
   D. Contouring

10. *True or False?* It is best practice to leave some material on the slot floor for finishing when roughing.

11. When performing a ramp entry, an angle of _____ is recommended for the tool entry.
   A. 0°
   B. 3°
   C. 30°
   D. 45°

12. T-slots are machined into a table to accept a(n) _____.
   A. machine screw
   B. T-slot nut
   C. hex nut
   D. hex bolt

13. Because of the tool length offset setting, which dimension will need special attention in programming a T-slot operation?
   A. The height dimension of the T-slot portion
   B. The angle of the T-slot portion
   C. The diameter of the T-slot cutter
   D. The width of the T-slot portion

14. The first tool used in machining a T-slot is a(n) _____.
   A. drill
   B. T-slot cutter
   C. end mill
   D. dovetail cutter

15. *True or False?* An end mill is the first tool used in a side slot milling operation.

16. A dovetail slot is made up of _____.
   A. two straight sides and a flat bottom
   B. two straight sides and an angled bottom
   C. two angled sides and an angled bottom
   D. two angled sides and a flat bottom

17. *True or False?* A dovetail cutter is the first tool used in a dovetail slot milling operation.

Apply and Analyze

Use the drawing below to answer Questions 1–4.

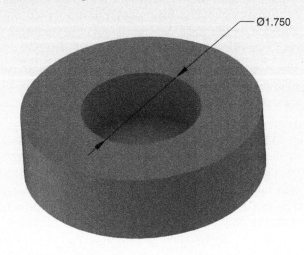

1. What type of entry motion should be used to enter the tool into the material?
2. What type of roughing path pattern should be used?
3. Approximately how much material should be left for the finish cut?
4. What type of toolpath should be used for the finishing operation?

Use the drawing below to answer Questions 5–8.

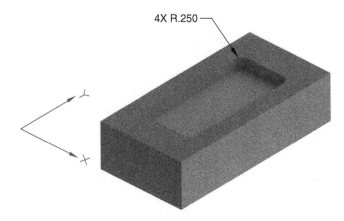

5. What type of entry motion should be used to enter the tool into the material?
6. Along which axis should the roughing pattern travel?
7. When selecting a tool for finishing the walls, how should the size of the tool be determined?

8. What type of toolpath should be used for the finishing operation?

Use the drawing below to answer Questions 9–11.

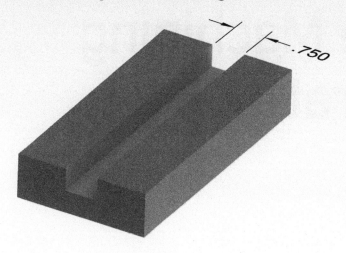

Critical Thinking

1. Briefly describe a typical part that contains a pocket. Explain the characteristics of the feature, such as the shape and whether it is open or closed, and describe the best strategy for machining.
2. Identify a part that contains a closed slot with full radius ends. Explain which entry strategy would be best for machining the feature.
3. Explain how determining a toolpath strategy and the tooling required for a machining operation can improve efficiency. Explain how these decisions impact cycle time and identify other factors that can be considered to make machining more efficient.

9. What type of slot is this?
10. When selecting a tool for machining the slot, how should the size of the tool be determined?
11. Briefly describe the toolpath strategy for machining the slot.

Use the drawing below to answer Questions 12–14.

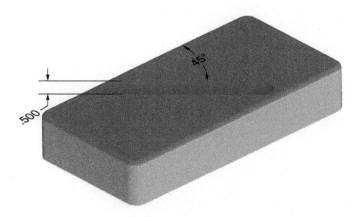

12. What type of slot is this?
13. When selecting a tool for machining the slot, how should the size of the tool be determined?
14. Briefly describe the toolpath strategy for machining the slot.

10 Hole Machining Operations

Chapter Outline

10.1 Introduction
10.2 Hole Shapes and Types
10.3 Machining Holes
10.4 Canned Cycles
- 10.4.1 G98 and G99: Initial Plane and Return Plane
- 10.4.2 G73: High-Speed Peck Drilling Cycle
- 10.4.3 G74: Left-Hand Tapping Cycle
- 10.4.4 G76: Precision Boring Cycle
- 10.4.5 G80: Cancel Canned Cycles
- 10.4.6 G81: Standard Drilling Cycle
- 10.4.7 G82: Spot Drilling Cycle
- 10.4.8 G83: Peck Drilling Cycle
- 10.4.9 G84: Right-Hand Tapping Cycle
- 10.4.10 G85: Boring Cycle—Feed In and Feed Out
- 10.4.11 G86: Boring Cycle—Feed In and Spindle Stop
- 10.4.12 G89: Boring Cycle with Dwell

10.5 Hole Patterns
- 10.5.1 Straight Line Hole Pattern
- 10.5.2 Angled Hole Pattern
- 10.5.3 Bolt Circles

10.6 Helical Milling
10.7 Thread Milling

Learning Objectives

After completing this chapter, you will be able to:

- Describe different types of holes and related machining operations.
- Identify a basic type of hole defined on a print.
- Explain how to create edge breaks using a 90° spot drill.
- Identify canned cycles for machining holes.
- Identify and program hole patterns.
- Explain the purpose and correct application of helical milling.
- Explain the purpose of thread milling and describe how a thread milling operation is programmed.

Key Terms

bell mouth
blind hole
bolt circle
boring
canned cycle

counterbore
countersink
peck drilling
point-to-point programming

spot drill
spotface
spring pass
tapped hole
through hole

Chapter opening photo credit: Aumm graphixphoto/Shutterstock.com

10.1 Introduction

Hole machining is the most common operation performed on a CNC mill or machining center. For many years, holes have served as critical features to facilitate assemblies, from the earliest hand-bored holes in wooden assemblies to today's precision-machined holes in aerospace parts. There are many types of holes and many different types of operations to produce them.

Hole making involves much more than simply drilling a hole. Other operations, such as spot drilling, tapping, boring, spotfacing, counterboring, and reaming, may be needed to produce the required hole. This chapter covers the interpretation of holes on prints, different types of hole machining operations, and determining the most efficient way to program a hole to be machined. Special machining cycles, called canned cycles, are used to simplify programming and are also discussed in this chapter.

10.2 Hole Shapes and Types

The requirements for a hole are specified on the print of the part. The features of the hole determine the machining operations that will be needed. When reviewing the print, identify the type of hole, evaluate the dimensions and tolerances, examine all part views, and consider any notes on the print.

A *blind hole* is a hole that does not extend through the part completely. In **Figure 10-1**, the hole is noted as a .625″ (5/8″) diameter hole that is .625″ deep. In this example, no tolerances are specified on the hole diameter and depth dimensions. If a tolerance is not specified with a dimension on the print, the general tolerances located in the tolerance block apply. See **Figure 10-2**. Also, be sure to check any notes on the print for information related to holes. In this example, the dimensions for the diameter and

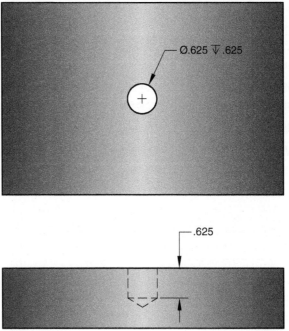

Figure 10-1. A rectangular part with a blind hole.

UNSPECIFIED TOLERANCES
.XXX = ±.005 .XX = ±.01 .X = ±.050 ANGLES = ±.1 MACHINE FINISH 125 MAX.
REMOVE ALL BURRS BREAK SHARP CORNERS
THIRD ANGLE PROJECTION

 Tolerance block

NOTES:
1. ±.005 ON FRACTIONAL DIMENSIONS
2. BREAK SHARP EDGES EXCEPT AS NOTED .015 MAX
3. REMOVE ALL BURRS
4. DELIVER WITH RUSTPROOFING LPS #1 OR EQUIV.

Tolerances in title block   **Tolerances in notes**

Goodheart-Willcox Publisher

Figure 10-2. The tolerance block in the title block on the print specifies general tolerances for dimensions. These tolerances apply to dimensions that do not have a specified tolerance on the print. General notes provide additional information that may include applicable tolerances.

depth are both three-place decimals (.XXX), thus making the tolerance for each dimension ±.005″.

For blind holes, the depth indicated on the print is the depth of the full-diameter portion of the hole. Due to the angled shape at the bottom of the drill, the tip extends beyond this depth, as shown in **Figure 10-1**. The CNC programming code must account for the additional depth of the drill tip to achieve the .625″ depth.

From the Programmer

Drill Point Calculation

Most standard drills have tips that form an angle of 118°. A simple formula is used to calculate the depth to use in programming for a drill with a 118° tip. In this formula, the drill diameter is multiplied by the constant value of .3:

Programmed depth = hole depth + (.3 × drill diameter)

To calculate for a .500″ drill with a 118° tip to travel .750″ deep, use the following calculation:

Programmed depth = .750 + (.3 × .500)
= .750 + .150
= .900 deep

For a drill with a 135° tip, the constant value of .21 is used. To calculate for a .500″ drill with a 135° tip to travel .750″ deep, use the following calculation:

.750 + (.21 × .500)
.750 + .105 = .855 deep

> **From the Programmer**
>
> **Drilling Pilot Holes**
>
> In manual machining, a center drill is commonly used to machine a pilot hole before drilling, but on CNC machines, a spot drill should be used exclusively for this purpose. A spot drill produces a better starting surface for drilling. In addition, a spot drill with a 90° tip is suitable for creating a 45° chamfer on the edge of the hole when an edge break is required. See **Figure 10-4**.

A *through hole* is a hole that completely extends through the material. On a print, a hole that does not include a depth specification is assumed to be a through hole. See **Figure 10-3**. A drill operation can be used to produce the through hole, but it is important to consider that a through hole may require several different operations. In addition, when drilling a through hole, the drill tip length must still be accounted for. Both the blind hole and the through hole shown in **Figure 10-3** will require a spot drill operation first.

Another common type of hole in manufacturing is a tapped hole. A *tapped hole* is a hole with internal threads. Tapped holes can be blind holes or through holes. A tapping operation requires a spot drill, followed by a drill, and lastly the tap. The top view of the hole on the print will show a solid circle and a hidden, or broken, circle. The solid circle represents the drill diameter and the hidden circle represents the tap diameter. See **Figure 10-5**.

When drilling a tapped hole, the drill must be programmed to drill deep enough to allow full threads to extend to the designated depth. In **Figure 10-5**, the full tap depth is .750″. The additional depth required to be drilled will depend on the type of tap used. A taper tap, plug tap, and bottom tap all require different amounts of clearance at the bottom of the hole. These taps have different numbers of chamfered threads at the end to help start the thread. At this point, the important thing to know is that the full drill diameter has to be programmed deeper than the tap depth.

A *spot drill* creates a starting position for the drill. A drill is thin in the middle and presents difficulty in starting perpendicular to the surface being drilled. For example, if you attempt to drill into a steel workpiece, the drill *walks* (moves from side to side) and almost never enters straight or perpendicular to the surface. The solution to this is to use a short, rigid

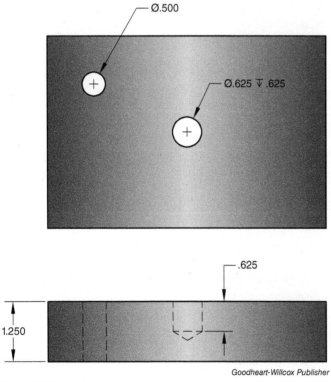

Figure 10-3. A rectangular part with a blind hole and a through hole.

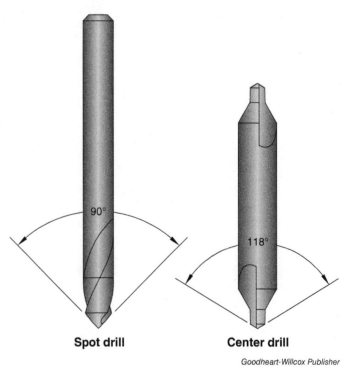

Figure 10-4. Types of drills used to drill pilot holes. A spot drill is used in CNC machining operations.

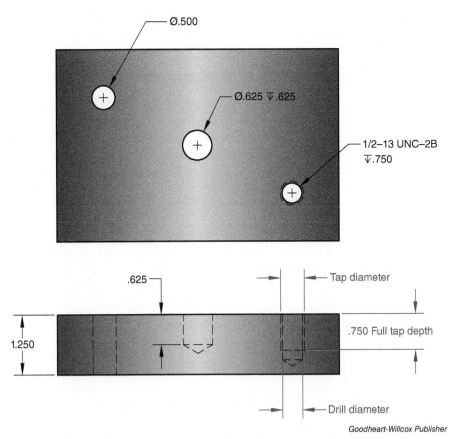

Figure 10-5. A tapped hole is represented in the top view with a solid circle and a hidden-line circle. The solid circle represents the drill diameter and the hidden-line circle represents the tap diameter. As shown in the front view, the drilled hole must be machined deep enough to allow for full threads to be cut to the specified depth.

tool to create a starting surface for the drill. See **Figure 10-6**. The spot drill shown has a 90° tip. The spot drill produces a shallow 90° point that serves as a guide for the drill to start and remain perpendicular to the surface being machined.

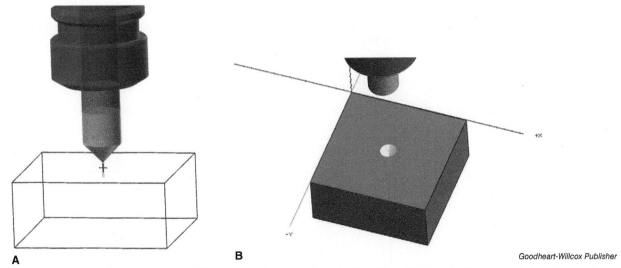

Figure 10-6. Spot drills are used to start holes in CNC applications. A—A spot drill with a 90° tip. B—The hole produced by a spot drill provides a starting position for the drilling operation that follows.

Careful determination of the spot drill depth can prevent sharp edges and burrs at the edge of the hole. In a spot drilling operation, only the angular portion at the end of the tool is used in cutting. If the tool tip is 90°, the depth drilled will be equal to the radius of the resulting hole. See **Figure 10-7**. In addition, a 90° spot drill can be used to produce a small 45° chamfer at the edge of the drilled hole. By spot drilling to a depth equal to half the diameter of the hole size, the spot drilling operation produces a chamfer of the correct size at the edge of the hole. In this application, a spot drill larger than the size of the hole to be drilled is used in order to form the chamfer. The following formula is used for calculating spot drill depth:

Spot drilling depth = .5 × hole diameter

A spot drilling example is shown in **Figure 10-8**. This example requires a .375″ diameter hole with a .020″ chamfer. The total diameter of the chamfered hole is the hole diameter plus twice the chamfer size, or .415″ (.375 + .020 + .020). The programmed depth of the spot drill operation is calculated as follows:

Spot drilling depth = .5 × hole diameter

= .5 × .415

= .2075

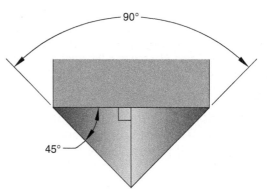

Goodheart-Willcox Publisher

Figure 10-7. The end portion of a 90° spot drill. In addition to producing a starting position for the drill to be used in drilling, the spot drill forms a 45° chamfer at the edge of the hole.

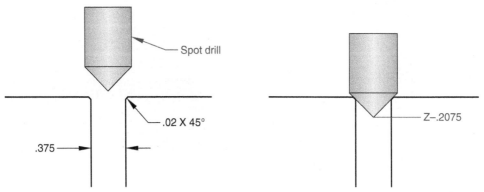

Goodheart-Willcox Publisher

Figure 10-8. A spot drilling operation used for a .375″ diameter hole with a .020″ chamfer.

Another type of hole produced in manufacturing is a countersunk hole. A *countersink* is an enlarged conical feature at the top of a hole used to create clearance and allow the top of a fastener to sit flush with the part face. Countersinks are commonly used with fasteners such as flat head screws and rivets. Depending on the type of fastener to be used, the countersink angle can be 60°, 82°, 100°, or 110°. An angled countersink tool is most often used to machine the countersink after a hole is drilled. On a print, a countersink is specified by its diameter and included angle. The countersink symbol precedes the countersink diameter in the dimension. The hole diameter is also dimensioned. See **Figure 10-9**.

Another common type of hole is a counterbored hole. A *counterbore* is an enlarged cylindrical feature at the top of a hole that provides a flat surface for a fastener to sit flush below the part face. See **Figure 10-10**. A counterbore is dimensioned by specifying the hole diameter, the counterbore diameter, and the counterbore depth. The counterbore symbol precedes the counterbore diameter in the dimension. Counterbores are often used with socket head cap screws. There are specific tools used to machine counterbores to industry standards, but much of the counterboring in CNC machining is done with flat end mills and helical toolpaths.

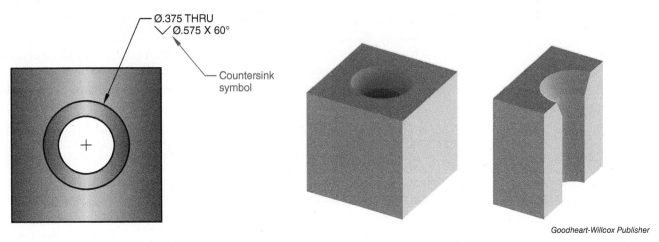

Goodheart-Willcox Publisher

Figure 10-9. A part with a countersunk hole. Countersinks are dimensioned by specifying the diameter of the hole, the countersink diameter, and the included angle of the countersink.

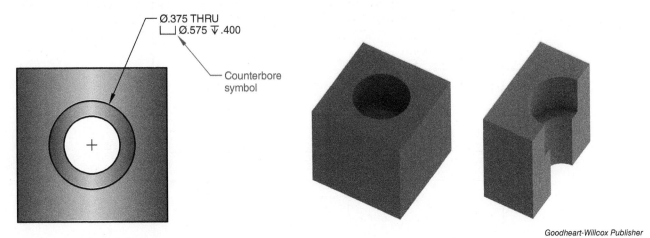

Goodheart-Willcox Publisher

Figure 10-10. A part with a counterbored hole. Counterbores are dimensioned by specifying the diameter of the hole, the counterbore diameter, and the depth of the counterbore.

A spotface is similar to a counterbore, but not as deep. A *spotface* is a shallow bearing surface normally used to provide a smooth flat spot on a casting or forging. The diameter of a spotface is dimensioned on the print. The spotface depth may be specified, but often the depth is designated as 100% clean up or a maximum depth allowed.

10.3 Machining Holes

Machining a hole is not a very complicated operation. The only motion it requires is to position the X and Y axis locations of the tool, rapid down to a safe place above the part, feed down the Z axis to the desired depth, and then rapid out to a safe position. The coordinates in the program define the XY position of the hole and the Z axis positions of the tool. This type of programming is referred to as *point-to-point programming*. This is similar to contouring or pocketing, where the written code designates every position and movement of the tool. **Figure 10-11** shows a 1″ thick part with four .375″ (3/8″) diameter holes drilled through the part. These holes can be machined with a simple drilling program.

As shown in **Figure 10-11**, the part origin is designated at the top-left corner and the holes are numbered for easy reference. The first consideration is the tooling to be used. A 1/2″ diameter spot drill and a 3/8″ diameter drill will be the only tooling needed. Spot drilling establishes a starting location for each hole drilling operation and allows an edge break to be created for each hole. The next step is to calculate the drill point locations in the X and Y axes. The coordinate locations for the four holes are listed in **Figure 10-11**.

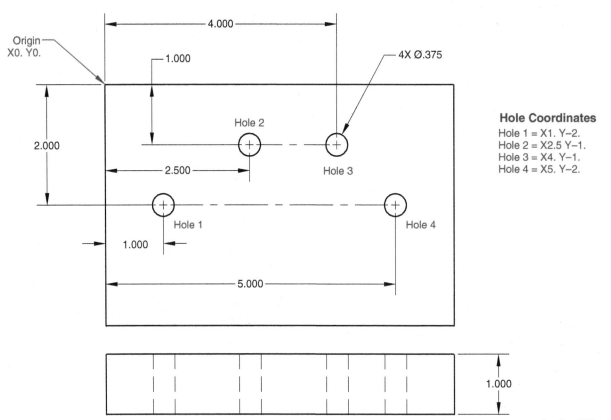

Goodheart-Willcox Publisher

Figure 10-11. A rectangular part with four drilled holes.

The last calculations to make are the programmed Z axis depths in drilling. The Z0. point is the top of the workpiece. The spot drill depth is calculated as previously discussed. The Z−.208 depth in the program creates a small .020″ chamfer on each hole. The through hole drilling depth is determined by calculating the thickness of the part and the length of the drill tip. The drilling depth is calculated as follows:

Part thickness (1″) + drill tip length (.3 × .375″) = 1.1125″

The program for machining the holes is as follows.

O1020	
(T1|1/2 SPOT DRILL|H1)	
(T2|3/8 DRILL|H2)	
G20	
G0 G17 G40 G49 G80 G90	
T1 M6	1/2″ Spot drill
S4584 M3	Spindle on
G0 G90 G54 X1. Y−2.	Hole 1 location
G43 H1 Z2.	Turn on tool length offset
M8	Coolant on
G0 Z.1	Rapid to safe position
G1 Z−.208 F45.8	Feed to spot drill depth
G0 Z.1	Rapid to safe position
X2.5 Y−1.	Hole 2 location
G1 Z−.208 F45.8	
G0 Z.1	
X4.	Hole 3 location
G1 Z−.208 F45.8	
G0 Z.1	
X5. Y−2.	Hole 4 location
G1 Z−.208 F45.8	
G0 Z.1	
Z2.	Safe retract
M5	Spindle off
G91 G28 Z0. M9	Machine home
M01	
T2 M6	3/8″ Drill
S5093 M3	Spindle on
G0 G90 G54 X1. Y−2.	Hole 1 location
G43 H2 Z2.	Turn on tool length offset (Tool 2)
M8	Coolant on
G0 Z.1	Rapid to safe position
G1 Z−1.1125 F30.2	Feed down through part
G0 Z.1	
X2.5 Y−1.	Hole 2 location
G1 Z−1.1125 F30.2	
G0 Z.1	

Continued

X4.	Hole 3 location
G1 Z–1.1125 F30.2	
G0 Z.1	
X5. Y–2.	Hole 4 location
G1 Z–1.1125 F30.2	
G0 Z.1	
Z2.	Safe retract
M5	
G91 G28 Z0. M9	Machine home
G28 Y0.	
M30	

This program is approximately 50 lines of machine code to spot drill and drill four simple holes. Imagine if the drilling operation is more complicated and requires 50 holes, or if a peck drilling operation is required. *Peck drilling* is a technique used for drilling deep holes and hard materials. In peck drilling, the drill feeds down to a partial depth and then retracts to remove chips and allow coolant into the hole, then re-enters the material to continue cutting. See **Figure 10-12**.

As you can imagine, peck drilling 50 holes would require a large number of programmed points, significantly increasing the programming time and the opportunity for error. Fortunately, CNC controllers provide canned cycle commands to simplify these repetitive operations.

10.4 Canned Cycles

A *canned cycle* is an abbreviated machining cycle that automates complex repetitive operations, such as hole machining operations. Canned cycles are regularly used in CNC milling to make programs efficient. They also provide easy editing for multiple holes without requiring extensive program rewrites. The canned cycle commands discussed in this chapter are intended for use in Fanuc and Haas controllers. There are additional commands beyond those covered. Each CNC machining center controller has its own custom codes and these can vary by controller type, so it is important to check your machine programming manual before attempting any of these cycles.

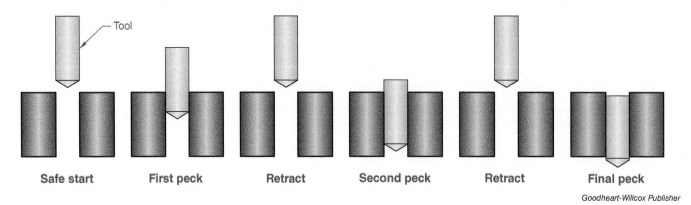

Figure 10-12. In peck drilling, the drill feeds down into the material to a partial depth and then retracts before it re-enters to continue cutting.

Some of the available canned cycles for hole machining are shown in **Figure 10-13**. In the following sections, these commands are explained further. The G80 command must be used after these cycles to cancel them.

10.4.1 G98 and G99: Initial Plane and Return Plane

The G98 and G99 commands are modal commands used in canned cycles to control the retract position of the tool between cycle operations. These commands reference Z axis positions defined in the program. The G98 and G99 commands establish the position, along the Z axis, for the tool to begin the cycle and retract after completing the cycle. In a canned cycle, there are times when the tool will need to move up in the Z axis to clear steps in the part, clamps, or other workholding. In a hole machining cycle, the G98 and G99 commands allow the programmer to precisely specify the retract position so that the tool can safely clear obstructions as it moves to each hole position. See **Figure 10-14**.

Before a canned cycle begins, a couple of things will happen. First, the tool will be positioned in a rapid move to the correct X and Y axis locations for the first hole location. Second, the tool will rapid down to a safe Z axis height above the part. This Z axis position is known as the initial Z position. In selecting this position, the programmer needs to consider any clearance issues the tool might encounter when completing the entire canned cycle. The initial Z position is the retract position that will be referenced by the G98 command when it is programmed. Inside the canned cycle, to speed up the cycle and make it more efficient, the tool retract position can be set to a different height so that the tool does not have to return to the initial Z position after each operation. In the activation block for the canned cycle, the R address is used to set the retract plane, also called the rapid plane or R plane, above the material. The height of this plane is usually set closer to the material where the tool can retract safely and clear of any obstructions above the part. This amount of retraction is used at

Hole Machining Canned Cycles

Command	Description	Application
G73	High-speed peck drilling cycle	Deep hole drilling with minor retract
G74	Left-hand tapping cycle	Tapping cycle for left-hand taps
G76	Precision boring cycle	Precision boring with shift amount
G80	Canned cycle cancel	Cancels any active cycle; used at the end of all canned cycles
G81	Standard drilling cycle	Shallow drilling
G82	Spot drilling cycle	Spotface drilling, counterboring, countersinking, and spot drilling
G83	Deep hole drilling cycle	Full retract peck drilling
G84	Right-hand tapping cycle	Tapping cycle for right-hand taps
G85	Boring and reaming cycle	Bore cycle—feed in and feed out
G86	Boring cycle	Bore cycle—bore in, spindle stop, rapid retract
G89	Boring cycle	Blind hole boring cycle with dwell at depth

Goodheart-Willcox Publisher

Figure 10-13. Canned cycle commands for hole machining.

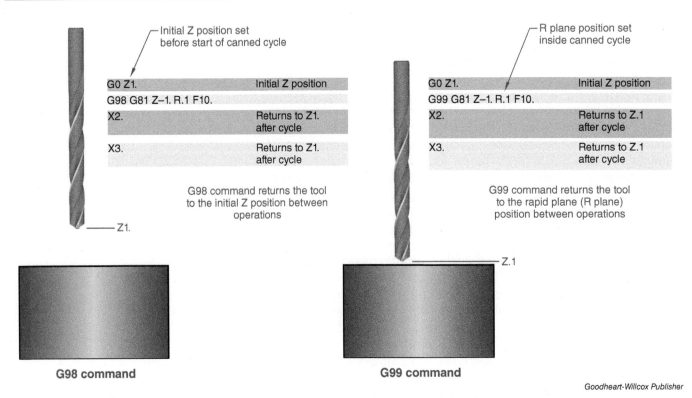

Figure 10-14. The G98 and G99 commands are used to control the retract position of the tool between cycle operations.

hole locations where the tool does not need to return to the initial Z position. The G99 command references the retract plane position when it is programmed. Using the G99 command allows the programmer to shorten cycle time as the tool travels between holes.

Figure 10-15 shows the difference between using the G98 and G99 commands in a canned cycle for drilling holes. The part is clamped between the second and third holes. After the second hole is drilled, the tool needs to return to the initial Z position to clear the workholding. The use of the G98 or G99 command will determine if, in between each hole, the tool returns to the initial Z position or to the retract plane position defined by the R address. The G98 command returns the tool to the initial Z position and the G99 command returns the tool to the R plane position.

10.4.2 G73: High-Speed Peck Drilling Cycle

As previously discussed, peck drilling is used for machining deep holes. The G73 canned cycle is designed to simplify the task of drilling medium-depth holes. The cycle will feed down in incremental steps and make a short retraction or pull-off between steps without retracting completely out of the part. The retraction allows coolant to reach the cutting edge and prevents any long, stringy chips from forming. The peck amount is an incremental value programmed with the Q address. The retraction distance is a stored setting in the machine controller. The tool retracts by this amount after each peck. After the cycle is complete, the tool retracts to the Z axis position specified by the G98 or G99 command. The following shows the programming format used with the G73 canned cycle, commonly referred to as a chipbreaker cycle.

G73 G98 Z__ R__ Q__ F__

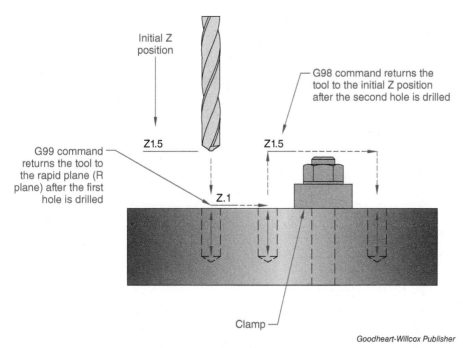

Figure 10-15. The G98 command returns the tool to the initial Z position programmed before the start of the canned cycle. The G99 command returns the tool to the R plane position defined with the R address in the canned cycle. The G98 command is used when it is necessary to clear obstructions such as clamps. The G99 command is used when the tool can move between hole positions without obstructions.

The entries are read as follows:

- **G73.** Chipbreaker canned cycle.
- **G98.** Return to the initial Z position. The G99 command can be used here to return to the R plane.
- **Z.** Final drilling depth.
- **R.** Z position to rapid to before the drill cycle begins and after the drill cycle is complete. The tool will retract to this position in rapid mode after the drill cycle is complete if the G99 command is programmed.
- **Q.** Incremental amount of motion along the Z axis for each peck.
- **F.** Feed rate in inches per minute.

The following program is an example of using the G73 canned cycle for drilling.

G0 G90 G54 X1. Y–2.	Position at first hole location
G43 H2.	Turn on tool length offset
Z1.	Initial Z position for use with the G98 command
G73 G98 Z–1.25 R.1 Q.2 F10.	Complete canned cycle at first location
X2. Y–3.	Complete same canned cycle at second location
X3. Y–4.	Complete same canned cycle at third location
G80	Cancel canned cycle

10.4.3 G74: Left-Hand Tapping Cycle

The G74 canned cycle is designed to perform tapping cycles with left-hand taps only. This cycle synchronizes the spindle speed and feed rate and automatically reverses the spindle at the final Z depth to allow the tap to exit properly. See **Figure 10-16**. Using a tapping cycle requires a calculation of spindle speeds and feed rates. The feed rate calculation is based on the threads per inch (TPI) of the tap. The following shows the programming format used with the G74 canned cycle.

G74 G98 Z__ R__ F__

The entries are read as follows:

- **G74.** Left-hand tapping cycle.
- **G98.** Return to the initial Z position. The G99 command can be used here to return to the R plane.
- **Z.** Final tapping depth.
- **R.** Z position to rapid to before the tapping cycle begins and retract after the tapping cycle is complete. The tool will retract to this position if the G99 command is programmed.
- **F.** Feed rate in inches per minute. Calculated as spindle speed (rpm) ÷ threads per inch (TPI).

The following program is an example of using the G74 canned cycle for tapping holes.

M4 S1000 (1/2–13 Tap)	Spindle on to 1000 rpm
G0 G90 G54 X1. Y–2.	Position at first hole location
G43 H2.	Turn on tool length offset
Z1.	Initial Z position for use with the G98 command
G74 G98 Z–.75 R.1 F76.92	Complete canned cycle at first location
X2. Y–3.	Complete same canned cycle at second location
X3. Y–4.	Complete same canned cycle at third location
G80	Cancel canned cycle

Mr.1/Shutterstock.com

Figure 10-16. A tapping canned cycle simplifies repetitive hole tapping operations.

A tap designation of 1/2–13 indicates a 1/2″ diameter tap with 13 threads per inch. The feed rate is calculated by dividing the spindle speed (in rpm) by the number of threads per inch. In this case, the spindle speed is 1000 rpm:

$$\text{Feed rate} = \text{spindle speed} \div \text{threads per inch}$$
$$= 1000 \div 13 = 76.92 \text{ ipm}$$

10.4.4 G76: Precision Boring Cycle

Boring is a machining operation used to enlarge an existing hole to a finish size. The G76 canned cycle is designed for precision boring with a single-point boring bar. See **Figure 10-17**. This cycle will feed down in the Z axis to the programmed depth and then make a short dwell. The spindle will stop and orient before making a small shift in the X and Y axes to pull away from the machined surface. The tool then rapids up in the Z axis to the programmed retract position. The shift at the bottom allows the tool to pull out and not make any marks on the machined surface. The amount of shift is an incremental value programmed with the Q address. The P address is used to specify the dwell time in milliseconds. The following shows the programming format used with the G76 canned cycle.

`G76 G98 Z__ R__ Q__ P__ F__`

The entries are read as follows:

- **G76.** Precision boring cycle.
- **G98.** Return to the initial Z position. The G99 command can be used here to return to the R plane.
- **Z.** Final boring depth.
- **R.** Z position to rapid to before the boring cycle begins and after the boring cycle is complete. The tool will retract to this position if the G99 command is programmed.
- **Q.** Incremental shift amount in the X and Y axes at the bottom of the hole.
- **P.** Dwell time at the bottom of the hole in milliseconds. A number entered without a decimal point represents milliseconds. A number entered with a decimal point represents seconds.
- **F.** Feed rate in inches per minute.

The following program is an example of using the G76 canned cycle for a boring operation.

G0 G90 G54 X1. Y-2.	Position at first hole location
G43 H2.	Turn on tool length offset
Z1.	Initial Z position for use with the G98 command
G76 G98 Z-.75 R.1 Q.005 P10 F10.	Complete canned cycle at first location
X2. Y-3.	Complete same canned cycle at second location
X3. Y-4.	Complete same canned cycle at third location
G80	Cancel canned cycle

Goodheart-Willcox Publisher

Figure 10-17. A boring bar with an insert. Boring produces a fine surface finish without scratches or marks.

Using the G76 Canned Cycle

The G76 canned cycle takes practice to program correctly. The shift at the bottom requires some practice and the wrong shift can have very negative effects. The G85 canned cycle will feed in and feed out, leaving a nice finish, but it is considerably slower. If you use the G76 canned cycle, practice it outside of the part and make sure the tool is moving in the direction you anticipate.

10.4.5 G80: Cancel Canned Cycles

The G80 command is often used in the startup block at the beginning of a program to ensure that no canned cycle is still active from the previous program. Canned cycle commands are modal and will remain active until they are shut off. The G80 command should always be used directly after the last hole is machined in a canned cycle.

10.4.6 G81: Standard Drilling Cycle

The G81 canned cycle is designed to simplify the task of drilling shallow holes or spot drilling prior to drilling. The cycle will feed down to the programmed Z depth and then rapid retract before starting the next drilling cycle. This cycle is used when the depth of the hole is less than four times the diameter of the drill. This cycle should also be used for solid carbide drills, as carbide will chip or shatter if peck drilling cycles are used. The following shows the programming format used with the G81 canned cycle.

G81 G98 Z___ R___ F___

The entries are read as follows:

- **G81.** Drilling cycle.
- **G98.** Return to the initial Z position. The G99 command can be used here to return to the R plane.
- **Z.** Final drilling depth.
- **R.** Z position to rapid to before the drilling cycle begins and after the drilling cycle is complete. The tool will retract to this position if the G99 command is programmed.
- **F.** Feed rate in inches per minute.

The following program is an example of using the G81 canned cycle for a drilling operation.

G0 G90 G54 X1. Y–2.	Position at first hole location
G43 H2.	Turn on tool length offset
Z1.	Initial Z position for use with the G98 command
G81 G98 Z–1.25 R.1 F10.	Complete canned cycle at first location
X2. Y–3.	Complete same canned cycle at second location
X3. Y–4.	Complete same canned cycle at third location
G80	Cancel canned cycle

10.4.7 G82: Spot Drilling Cycle

The G82 canned cycle is specifically designed for programming a short dwell at the hole bottom. This cycle is particularly useful when performing a spotfacing, counterboring, or countersinking routine. It is also used for spot drilling operations that require a short dwell. The dwell time allows the tool to make at least one additional revolution and leave no unwanted tool marks on the surface finish. Usually, this cycle is programmed when a slow spindle speed is utilized. The following shows the programming format used with the G82 canned cycle.

```
G82 G98 Z__ R__ P__ F__
```

The entries are read as follows:

- **G82.** Spot drilling cycle.
- **G98.** Return to the initial Z position. The G99 command can be used here to return to the R plane.
- **Z.** Final hole depth.
- **R.** Z position to rapid to before the drilling cycle begins and after the drilling cycle is complete. The tool will retract to this position if the G99 command is programmed.
- **P.** Dwell time at the bottom of the hole in milliseconds. A number entered without a decimal point represents milliseconds. A number entered with a decimal point represents seconds.
- **F.** Feed rate in inches per minute.

The following program is an example of using the G82 canned cycle for a drilling operation.

G0 G90 G54 X1. Y–2.	Position at first hole location
G43 H2.	Turn on tool length offset
Z1.	Initial Z position for use with the G98 command
G82 G98 Z–1.25 R.1 P20 F10.	Complete canned cycle at first location
X2. Y–3.	Complete same canned cycle at second location
X3. Y–4.	Complete same canned cycle at third location
G80	Cancel canned cycle

10.4.8 G83: Peck Drilling Cycle

The G83 canned cycle is a peck drilling cycle used for deep hole drilling. Peck drilling is used when the depth of the hole is four times greater or more than the diameter of the drilling tool. See **Figure 10-18**. For example, a .500″ diameter drill that has to drill a hole more than 2″ deep needs a pecking cycle to remove chips and allow coolant down to the cutting surface. This cycle can be used at shallower depths, but it is not intended for that purpose. The cycle will feed down in incremental steps and fully retract between steps out of the hole. When the tool retracts, it returns to the R plane or the initial Z position. Avoid using any peck drilling cycle with solid carbide drills. Carbide is brittle by nature and the re-entry into the cut after a pecking cycle can cause tool failure. The following shows the programming format used with the G83 canned cycle.

```
G83 G98 Z__ R__ Q__ F__
```

The entries are read as follows:

- **G83.** Full retract peck drilling cycle.

guruXOX/Shutterstock.com

Figure 10-18. Peck drilling is used for drilling deep holes to allow for chip removal and coolant delivery to the cutting edge.

- **G98.** Return to the initial Z position. The G99 command can be used here to return to the R plane.
- **Z.** Final drilling depth.
- **R.** Z position to rapid to before the drilling cycle begins and after the drilling cycle is complete. The tool will retract to this position if the G99 command is programmed.
- **Q.** Incremental amount of motion along the Z axis for each peck.
- **F.** Feed rate in inches per minute.

The following program is an example of using the G83 canned cycle for drilling.

G0 G90 G54 X1. Y–2.	Position at first hole location
G43 H2.	Turn on tool length offset
Z1.	Initial Z position for use with the G98 command
G83 G98 Z–1.25 R.1 Q.3 F10.	Complete canned cycle at first location
X2. Y–3.	Complete same canned cycle at second location
X3. Y–4.	Complete same canned cycle at third location
G80	Cancel canned cycle

10.4.9 G84: Right-Hand Tapping Cycle

The G84 canned cycle is designed to perform tapping cycles with right-hand taps only. This cycle synchronizes the spindle speed and feed rate and automatically reverses the spindle at the final Z depth to allow the tap to exit properly. See **Figure 10-19**. The feed rate calculation is based on the threads per inch (TPI) of the tap. The following shows the programming format used with the G84 canned cycle.

G84 G98 Z___ R___ F___

Aumm graphixphoto/Shutterstock.com

Figure 10-19. Machining operation used to tap multiple holes along the perimeter of a part.

The entries are read as follows:

- **G84.** Right-hand tapping cycle.
- **G98.** Return to the initial Z position. The G99 command can be used here to return to the R plane.
- **Z.** Final tapping depth.
- **R.** Z position to rapid to before the tapping cycle begins and retract after the tapping cycle is complete. The tool will retract to this position if the G99 command is programmed.
- **F.** Feed rate in inches per minute. Calculated as spindle speed (rpm) ÷ threads per inch (TPI).

The following program is an example of using the G84 canned cycle for tapping holes.

M3 S1000 (3/8–16 Tap)	Spindle on to 1000 rpm
G0 G90 G54 X1. Y–2.	Position at first hole location
G43 H2.	Turn on tool length offset
Z1.	Initial Z position for use with the G98 command
G84 G98 Z–.75 R.1 F62.5	Complete canned cycle at first location
X2. Y–3.	Complete same canned cycle at second location
X3. Y–4.	Complete same canned cycle at third location
G80	Cancel canned cycle

A tap designation of 3/8–16 indicates a 3/8″ diameter tap with 16 threads per inch. The feed rate is calculated by dividing the spindle speed (in rpm) by the number of threads per inch.

$$\text{Feed rate} = \text{spindle speed} \div \text{threads per inch}$$
$$= 1000 \div 16 = 62.5 \text{ ipm}$$

> **From the Programmer**
>
> **G74 and G84 Canned Cycles**
>
> The M3 and M4 functions are not required with the G74 and G84 canned cycles. Although the functions are often programmed out of habit or for consistency, the G74 command will automatically turn the spindle on counterclockwise and the G84 command will automatically turn the spindle on clockwise. These two cycles will also automatically reverse the spindle direction at the final Z depth and synchronize the spindle speed and feed rate to safely remove the tap from the hole without damaging the threads. If not for this synchronization, the tool would break and the threads would be damaged.

10.4.10 G85: Boring Cycle—Feed In and Feed Out

The G85 canned cycle is the preferred boring cycle. This cycle will feed down to the final Z depth and then feed back to the retract plane using the programmed feed rate. This feeding in both directions creates a fine surface finish and allows for a spring pass while the tool travels back to the retract plane. A *spring pass* is a final pass that removes any material that remains as a result of tool deflection on the initial cutting pass. The final cutting pass will not leave any scratches or tool marks on the surface finish. The benefit of boring cycles is the high-quality surface finish, excellent hole location, and the roundness (circularity) or cylindricity of the hole being machined. The following shows the programming format used with the G85 canned cycle.

G85 G98 Z___ R___ F___

The entries are read as follows:

- **G85.** Bore in and bore out boring cycle.
- **G98.** Return to the initial Z position. The G99 command can be used here to return to the R plane.

- **Z.** Final boring depth.
- **R.** Z position to rapid to before the boring cycle begins and retract after the boring cycle is complete. The tool will retract to this position if the G99 command is programmed.
- **F.** Feed rate in inches per minute.

The following program is an example of using the G85 canned cycle for a boring operation.

G0 G90 G54 X1. Y–2.	Position at first hole location
G43 H2.	Turn on tool length offset
Z1.	Initial Z position for use with the G98 command
G85 G98 Z–.75 R.1 F5.	Complete canned cycle at first location
X2. Y–3.	Complete same canned cycle at second location
X3. Y–4.	Complete same canned cycle at third location
G80	Cancel canned cycle

10.4.11 G86: Boring Cycle—Feed In and Spindle Stop

The G86 canned cycle is similar to the G85 canned cycle, except the tool rapids out of the hole. At the final Z depth, the spindle stops and the boring bar rapids back to the retract plane. This cycle can leave a small scratch on the bore diameter, but it is faster than the G85 canned cycle. The following shows the programming format used with the G86 canned cycle.

G86 G98 Z___ R___ F___

The entries are read as follows:

- **G86.** Bore in and rapid out boring cycle.
- **G98.** Return to the initial Z position. The G99 command can be used here to return to the R plane.
- **Z.** Final boring depth.
- **R.** Z position to rapid to before the boring cycle begins and after the boring cycle is complete. The tool will retract to this position if the G99 command is programmed.
- **F.** Feed rate in inches per minute.

The following program is an example of using the G86 canned cycle for a boring operation.

G0 G90 G54 X1. Y–2.	Position at first hole location
G43 H2.	Turn on tool length offset
Z1.	Initial Z position for use with the G98 command
G86 G98 Z–.75 R.1 F5.	Complete canned cycle at first location
X2. Y–3.	Complete same canned cycle at second location

Continued

X3. Y-4.	Complete same canned cycle at third location
G80	Cancel canned cycle

10.4.12 G89: Boring Cycle with Dwell

The G89 canned cycle is also similar to the G85 canned cycle, except the tool dwells at the final Z depth. This cycle is primarily used for blind holes or counterbored holes that require a boring operation. The tool will feed out after the dwell time is complete. The following shows the programming format used with the G89 canned cycle.

G89 G98 Z__ R__ P__ F__

The entries are read as follows:

- **G89.** Bore in, dwell, and bore out boring cycle.
- **G98.** Return to the initial Z position. The G99 command can be used here to return to the R plane.
- **Z.** Final hole depth.
- **R.** Z position to rapid to before the boring cycle begins and retract after the boring cycle is complete. The tool will retract to this position if the G99 command is programmed.
- **P.** Dwell time at the bottom of the hole in milliseconds. A number entered without a decimal point represents milliseconds. A number entered with a decimal point represents seconds.
- **F.** Feed rate in inches per minute.

The following program is an example of using the G89 canned cycle for a boring operation.

G0 G90 G54 X1. Y-2.	Position at first hole location
G43 H2.	Turn on tool length offset
Z1.	Initial Z position for use with the G98 command
G89 G98 Z-.75 R.1 P10 F5.	Complete canned cycle at first location
X2. Y-3.	Complete same canned cycle at second location
X3. Y-4.	Complete same canned cycle at third location
G80	Cancel canned cycle

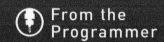

From the Programmer

Canned Cycle Programming
The previously discussed canned cycles are the most often used canned cycles. Each of the examples used three different hole positions to demonstrate the simplicity of machining the same size hole at multiple locations. But, it is just as effective to program one hole with a canned cycle as it is to program 100 holes. In the previous examples, each hole in the canned cycle was machined to the same final Z depth. It is possible, however, to repeat the same canned cycle with different depths by adding a Z coordinate on the same line as the X and Y coordinate position. This is useful when spot drilling many holes with different diameters. Become familiar with these canned cycles, because they are regularly used in CNC milling.

10.5 Hole Patterns

It is common in a machining program for holes to be positioned in a regular pattern. A hole pattern can consist of a straight line of evenly spaced holes, a set of holes aligned at an angle, or a circular pattern of holes on a common radius. There are some programming shortcuts that can assist with these types of patterns. All of the canned cycle examples discussed in the previous section were accomplished with the G90 command, or absolute positioning. However, using the G91 command, or incremental positioning, with a looping cycle, can create some nice shortcuts for hole patterns.

10.5.1 Straight Line Hole Pattern

Equally spaced holes aligned in a straight line can be easily programmed with just a couple of lines of machine code. In the example shown in **Figure 10-20**, the first hole is dimensioned off the bottom-left corner of the part, so that will be the work coordinate system origin. The part has 10 holes. The hole size is .438″ (7/16″) diameter, and the holes are spaced .750″ apart. These holes will require a spot drilling operation using a 90° spot drill and then a drilling operation using a .438″ drill to drill 1″ deep. The G81 canned cycle will be used to drill the holes. This will be accomplished by looping the canned cycle in incremental positioning mode. In incremental positioning, coordinate values are measured from the current point to the next point.

The K or L address is used to create loops (repetitive cycles) in a CNC program. By setting incremental mode current and using the K or L address with the number of times to repeat, the drilling cycle is repeated at equal distances apart. The program for machining the holes is as follows.

T1 M6 (1/2 Spot drill)	
G0 G90 G54 X.75 Y1. M3 S1500	Position at first hole location
G43 H1 Z2. M8	
Z1.	
G81 G99 Z-.225 R.1 F10.	Complete canned cycle at first location
G91 X.75 K9 (L9)	Incremental mode, incremental shift of X.75, repeat cycle nine times
G80	Cancel canned cycle
G91 G28 Z0.	
T2 M6 (7/16 Drill)	
G0 G90 G54 X.75 Y1. M3 S2200	
G43 H2 Z2. M8	
Z1.	
G81 G99 Z-1. R.1 F15.	
G91 X.75 K9 (L9)	Incremental mode, incremental shift of X.75, repeat cycle nine times
G80	Cancel canned cycle
G28 Z0.	
M30	

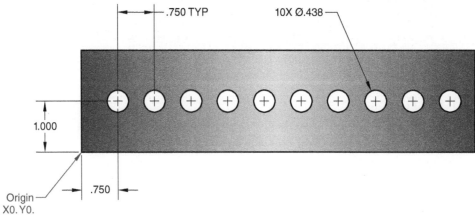

Goodheart-Willcox Publisher

Figure 10-20. A rectangular part with a pattern of 10 holes aligned in a straight line.

By using canned cycles and incremental positioning in this example, 20 drilling operations are completed using a relatively small number of lines of code. Note the use of the G91 command to move the tool incrementally in the appropriate direction (X+.75), and the use of K9 or L9 for the loop count. There are 10 holes to drill, but the first one is completed when the canned cycle is initiated on the line with the G81 command. The cycle only needs to repeat nine more times to produce 10 holes.

10.5.2 Angled Hole Pattern

Holes can also be aligned in a straight line along an angle. An angled hole pattern can be programmed in a few different ways, depending on the controller type. The first way is similar to the previous example using the G91 command for incremental positioning and looped cycles. An angled hole pattern requires a calculation of the distance between each hole along the X and Y axes. Trigonometric formula calculations can be used to quickly determine the coordinate shift amount between each hole.

In the example shown in **Figure 10-21**, the first hole is dimensioned off the bottom-left corner of the part, so that will be the work coordinate system origin. The information needed to complete the calculations for the incremental shift are the angle and center-to-center distance between holes. **Figure 10-22** shows how a right triangle can be used to calculate the X and Y shift amount.

The hypotenuse of the right triangle represents the 1.000″ center-to-center distance between the holes, and the angle used in trigonometric calculations is 15°. To calculate the X and Y linear distances, or the adjacent and opposite sides of the triangle, use the following formulas:

$$x = H \times \cos\theta$$
$$y = H \times \sin\theta$$

Debugging the Code

Canned Cycle Loop Count

The difference between the K and L address is related to the controller type. Traditionally, Fanuc machine controllers use the K address to perform loops, while Haas machine controllers use the L address. Check the operator's manual to verify the correct code before you attempt to run a looped cycle.

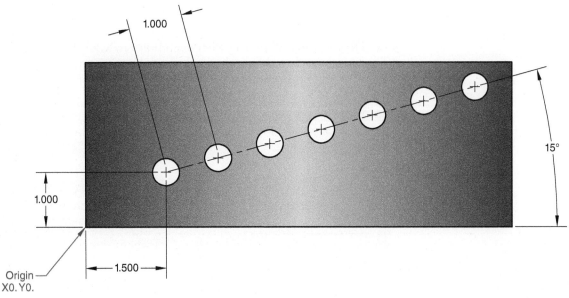

Figure 10-21. A rectangular part with a pattern of seven holes aligned along an angle.

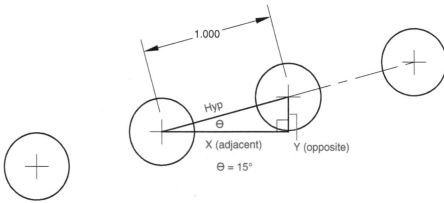

Goodheart-Willcox Publisher

Figure 10-22. The distances along the X and Y axes between holes can be calculated using trigonometric formulas based on the given angle (15°) and hypotenuse length (1.000″) of a right triangle. The hypotenuse represents the 1.000″ center-to-center distance between holes.

In these formulas, *H* represents the hypotenuse. The X distance formula uses the cosine function with the given angle and given hypotenuse value. For the X distance, use a calculator with the following formula:

$$\begin{aligned} x &= H \times \cos\theta \\ &= 1'' \times \cos(15°) \\ &= 1'' \times .9659 \\ &= .9659'' \end{aligned}$$

The Y distance formula uses the sine function with the given angle and given hypotenuse value. For the Y distance, use a calculator with the following formula:

$$\begin{aligned} y &= H \times \sin\theta \\ &= 1'' \times \sin(15°) \\ &= 1'' \times .2588 \\ &= .2588'' \end{aligned}$$

The incremental distance between holes is X.9659 Y.2588. These coordinates are used to define the incremental coordinate shift between holes. The following program is used for spot drilling the holes.

T1 M6 (1/2 Spot drill)	
G0 G90 G54 X1.5 Y1. M3 S1500	Position at first hole location
G43 H1 Z2. M8	
Z1.	
G81 G99 Z−.225 R.1 F10.	Complete canned cycle at first location
G91 X.9659 Y.2588 K6 (L6)	Incremental mode, incremental shift of X.9659 Y.2588, repeat cycle six times
G80	Cancel canned cycle
G91 G28 Z0.	

The Haas machine controller provides an easier way to accomplish this task. The G72 canned cycle is used to calculate and machine evenly spaced holes along an angle. The following shows the programming format used with the G72 canned cycle.

```
G72 I__ J__ L__
```

The entries are read as follows:

- **G72.** Hole pattern along an angle canned cycle.
- **I.** Center-to-center distance between holes along an angle.
- **J.** Angle of holes measured from the 3 o'clock position to 360° counterclockwise.
- **L.** Number of evenly spaced holes along an angle.

A spot drilling operation for the hole pattern in **Figure 10-21** can be programmed as follows.

T1 M6 (1/2 Spot drill)	
G0 G90 G54 X1.5 Y1. S1451 M3	Start position of holes along an angle
G43 H03 Z1. M8	
G81 G99 Z−.225 R.1 G72 I1.0 J15. L7 F8.	
G80	
G0 Z1.	

Notice that the G81 canned cycle and G72 canned cycle commands are on the same line. The G72 canned cycle requires a hole machining canned cycle to be active in order to function.

10.5.3 Bolt Circles

A *bolt circle* is a theoretical circle on which the center points of holes lie in a circular pattern of holes. The holes in the pattern are equally spaced, with equal angles between holes. See **Figure 10-23**.

Figure 10-24 shows a circular pattern of eight holes with equal spacing between holes. Each hole is positioned at the same distance from the theoretical arc center. This is an eight-hole circular pattern with 45° spacing on a 6″ diameter bolt circle. The 3″ dimension is the radius, or half the diameter.

In CNC programming and computer-aided manufacturing (CAM), angles are measured from the 3 o'clock position (the 0° angle position). See **Figure 10-25**. Angles with positive values are measured counterclockwise from the 0° angle position. The 12 o'clock position is 90°, the 9 o'clock position is 180°, and the 6 o'clock position is 270°. This method of angular measurement is vitally important to hole pattern calculations.

To determine the angle between equally spaced holes on a bolt circle, divide 360° by the number of holes in the pattern. For example, in an eight-hole pattern with equally spaced holes, the holes are 45° apart:

$$360° \div 8 = 45°$$

This is the spacing for the pattern in **Figure 10-24**. The 3″ radius and the 45° spacing are used in calculating the position of each hole. However, notice that the first hole in the pattern, or the one closest to 0°, is at 10° relative to the 0° angle position. This must be accounted for in programming.

It is possible to calculate the hole center positions with trigonometric functions, but there are commands on Haas and Fanuc machine controllers

Goodheart-Willcox Publisher

Figure 10-23. A bearing housing with two circular patterns of holes. The hole pattern on the bottom surface consists of eight counterbored holes spaced 45° apart. The hole pattern on the top surface consists of six blind holes spaced 60° apart. In each pattern, the hole centers lie on a theoretical bolt circle.

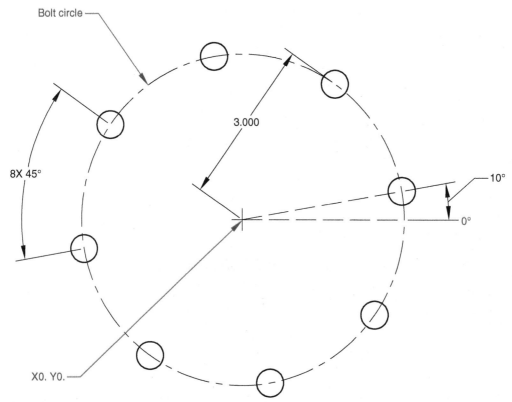

Figure 10-24. A circular pattern of eight equally spaced holes. The holes are spaced 45° apart on a 6″ diameter bolt circle. The holes can be programmed using polar coordinates or a canned cycle designed for bolt hole circle programming.

that make this easier. On a Fanuc controller, the G16 command activates the polar coordinate system to program the hole locations. In this system, coordinates are located by specifying a radius value and an angular position relative to a center point. When the G16 command is initiated, the X axis designation represents the radius value and the Y axis designation represents the polar angle measured from the zero angle position. The following program uses the G16 command to specify the position for each hole. The starting location is positioned at X0. Y0. This establishes the center point from which polar coordinates are measured. The location of the first hole is X3. Y10. This represents a 3″ radius value and a polar angle of 10° measured from the zero angle position. The remaining holes use the same X coordinate and can be located by simply specifying the Y angular value. Each successive Y value is 45° counterclockwise. The G15 command is used after the operation to cancel polar coordinate system entry.

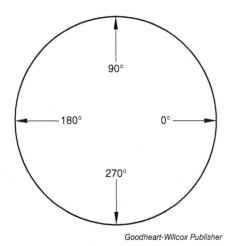

Figure 10-25. Angle positions on a circle with 0° oriented at the 3 o'clock position.

O0425 (G16 BOLT HOLE CIRCLE–FANUC)	
T1 M6	
G0 G90 G54 X0. Y0. M3 S1500	Position at center of bolt hole circle
G43 H1 Z1. M8	Turn on tool length offset
G16 X3. Y10.	Polar coordinates on, move to X3. and 10° angle
G98 G81 Z–.255 R.1 F10.	Spot drill cycle
Y55.	Perform drill cycle at 55°

Continued

Y100.	Perform drill cycle at 100°
Y145.	Perform drill cycle at 145°
Y190.	Perform drill cycle at 190°
Y235.	Perform drill cycle at 235°
Y280.	Perform drill cycle at 280°
Y325.	Perform drill cycle at 325°
G80	Cancel drill cycle
G15	Polar coordinates off
G91 G28 Z0.	
M30	

Using the polar coordinate system saves considerable time in making calculations and reduces opportunities for operator or programmer error.

On Haas controllers, the G70 bolt hole circle canned cycle is used to program holes in a circular hole pattern. It is quite a bit different from the Fanuc cycle. The following shows the programming format used with the G70 canned cycle.

G70 I___ J___ L___

The entries are read as follows:

- G70. Bolt hole circle canned cycle.
- I. Radius of the bolt hole circle.
- J. Starting angle for the pattern measured from the 3 o'clock position to 360° counterclockwise. Note: This entry is optional only if the starting angle is 0°. The best practice is to always use the J address to prevent unexpected results.
- L. Number of evenly spaced holes around the bolt hole circle.

In the program, the tool must be positioned at the center point of the circle from which coordinates are measured. This coordinate location is programmed on a block before the G70 command or on the same block. In the following program, the tool is initially positioned at X0. Y0. to machine the holes shown in **Figure 10-24**.

O0931 (G70 BOLT HOLE CIRCLE)	
T1 M06	
G0 G90 G54 X0. Y0. S1000 M03	Position at center of bolt hole circle
G43 H1 Z0.1 M8	Turn on tool length offset
G81 G98 Z–1. R0.1 F15. L0	Begin G81 canned cycle; L0 skips drilling at X0. Y0. position
G70 I3. J10. L8	Begin G70 cycle at 10°
G80	Cancel canned cycles
G91 G28 Z0.	

In this program, notice that the L address is used twice. In the first instance, L0 is specified on the block containing the G81 canned cycle command. Because the value is zero, this instructs the machine to skip the drill cycle at the starting location (the X0. Y0. location). On the next block, L8 is specified. This designates the number of evenly spaced holes to machine.

10.6 Helical Milling

Helical milling is another viable machining strategy for creating holes. Like most machining operations, it is important to know its strengths and limitations. Helical milling is best suited for shallow, flat-bottom holes. Some programmers and machinists attempt to use this technique to create very close tolerance holes or even deep holes. The physics of using an end mill for these types of holes will cause some inconsistencies in machining.

As discussed in Chapter 9, a helical move is created with a circular interpolation command (G02 or G03) and includes a Z axis move. During the move, simultaneous movement occurs along three axes.

Figure 10-26 shows a helical milling operation using a 5/8″ end mill to produce a 3/4″ diameter hole that is 2.000″ long. The end mill is held in place at the tool holder, where it has the most rigidity and least runout. Runout refers to surface variation relative to a center axis.

As the end mill extends away from the tool holder, the variation in runout becomes increasingly higher, as does the amount of deflection or tool pressure. Depending on the hole depth, material, and size of the end mill, this condition can result in a bell mouth shape. A *bell mouth* is a tapered hole shape that is larger at the top and smaller toward the bottom. See **Figure 10-27**.

There are many variables that affect how much of a bell mouth will be created, and the part might still meet the print specifications. The bell mouth shape is a disadvantage of using helical milling.

The advantage to helical milling is that it works great for short counterbore or spotface holes. By nature, helical milling will create hole diameters that are perpendicular to the face being machined. By using cutter compensation in programming, the hole diameters can be easily controlled.

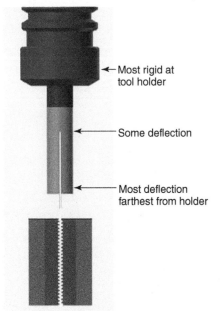

Goodheart-Willcox Publisher

Figure 10-26. In helical milling, an end mill makes helical movement as it cuts material. The amount of tool deflection is higher toward the bottom of the tool.

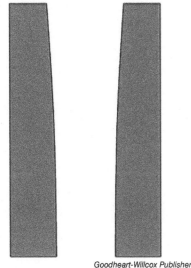

Goodheart-Willcox Publisher

Figure 10-27. A bell mouth is a hole shape that tapers from a larger dimension at the top of the feature to a smaller dimension at the bottom.

10.7 Thread Milling

One of the most common functions in machining is making threads. As previously discussed, the G84 and G74 tapping cycles are used for creating threaded holes with taps. Tapping a hole is a very efficient method and for most operations, it is the preferred method. However, as with most machining operations, there are limitations when using taps and instances where a tapping cycle is not the best choice. For example, breaking a tap while it is in the drilled hole can cause the machined part to be scrapped.

Imagine spending 30 minutes machining contours, pockets, and slots, followed by drilling all of the holes requiring threads. The last operation is tapping the drilled holes. Then, the tap breaks in a hole and cannot be removed, or damages the threads beyond repair. This is an even greater concern when tapping hard materials such as Inconel®, Waspaloy®, titanium, or even stainless steel. These materials are very tough and the chances of breaking a tap are relatively high.

There is another option to machine threads besides tapping with the G84 and G74 canned cycles. Threaded holes can be thread milled. Thread milling uses a special tool designed for thread cutting, **Figure 10-28**, and a helical path to machine the threads. A thread mill can be a single-point cutter or a multi-point cutter. A single-point cutter has a single cutting edge and can cut almost any standard thread size, but much more programming is required and it can be difficult to achieve the desired thread size. A multi-point cutter has multiple cutting edges along its length. This allows

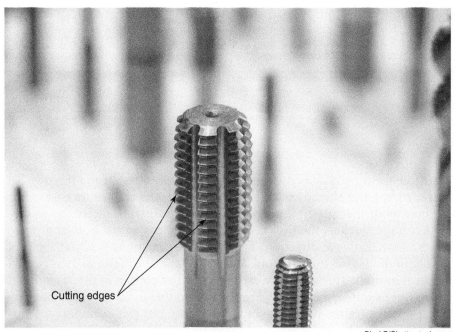
Pixel B/Shutterstock.com

Figure 10-28. A multi-point thread mill has multiple cutting edges along the body of the tool.

the full length of the thread to be cut in a 360° helical pass. A multi-point cutter must have the same thread pitch as the desired thread. For example, a 3/8–16 thread must be machined with a 16 pitch thread mill. Thread mills have a number of advantages in comparison to taps. Thread milling is an excellent machining method when threading hard materials or deep holes. The same tool can be used to cut internal or external threads and is capable of cutting right-hand or left-hand threads. Thread milling uses lower cutting forces than tapping and produces high-quality threads with a fine surface finish. Multiple passes can be made to control how the final thread is produced. In addition, thread mills are available in solid carbide. Solid carbide thread mills are well-suited for machining hard materials.

As is the case with tapping threads, thread milling requires the hole to be drilled first. For example, to create a 5/16–18 threaded hole 1/2" deep, the hole is spot drilled and drilled first. The spot drill should be deep enough to extend beyond the .312" (5/16") thread diameter. The drill to be used in this example will be a .257" diameter drill. The size of the tap drill to be used can be found on a tap drill chart. Tap drill sizes are given in the Reference Section of this textbook. In this example, the drilling depth is .625" to allow for clearance at the bottom of the hole for the thread mill. The following is the programming code for the two drilling operations.

```
O2501
T1 M6 (1/2 Spot drill)
G0 G90 G54 X1. Y1. M3 S1250
M8
G43 H1 Z.1
G98 G81 Z-.160 R.1 F10.
G80
```

Continued

```
G0 G91 G28 Z0.
T2 M6 (.257 Drill)
G0 G90 G54 X1. Y1. M3 S2500
M8
G43 H2 Z1.
G98 G81 Z-.625 R.1 F12.
G80
G0 G91 G28 Z0.
```

After the hole is drilled, the thread mill is used to cut the threads. To create right-hand threads with the thread mill, the tool must first go to the bottom of the hole and then move up in a positive Z direction. To create left-hand threads, the tool must start at the top of the hole and move down in a negative Z direction. These cutting directions position the tool for climb milling so that the tool is to the left of the material as it cuts in a helical motion. This is the preferred method of cutting in thread milling.

In this example, the threads are right-hand threads. There are some different options for the helical thread mill moves, but this example uses cutter compensation and incremental programming with the G91 command. The thread mill is .232″ diameter with a thread pitch of 18. The diameter and pitch of the thread mill are laser engraved on the mill body. The thread mill used in this example is a multi-point cutter. The cutting edges along the length allow the tool to cut the entire length of the thread in a single 360° helical pass. The following is the machine code for the thread milling operation. The program is broken down into portions with explanations of the corresponding blocks.

```
T3 M6 (Thread mill)
G0 G90 G54 X1. Y1. M3 S2500
M8
G43 H3 Z1.
G0 Z.1
```

To this point, the standard tool callout, positioning, and tool length offset are programmed. The next line will feed the tool down to the final Z depth of Z−.500. Remember, the hole was drilled .625″ deep for clearance for the bottom of the tool and the tool is to feed up from the bottom of the hole. The feed rate for the move to the bottom can be high, because there is no cutting happening at this point.

```
G1 Z-.50 F50.
```

After the tool is positioned at the bottom of the hole, cutter compensation is activated and a switch to incremental programming is initiated with the G91 command.

```
G41 D3 G91 X.04 F20.
```

As cutter compensation is turned on, the tool is programmed to make a X.04 move in incremental mode. This position is calculated from the .312″ finish thread diameter and the .232″ thread mill diameter. This is a linear move to one edge of the thread equal to the radial amount calculated from the difference in diameters. The movement is calculated as follows:

.312 (Thread diameter) − .232 (Tool diameter) = .080 diameter or .040 radius

```
G3 I−.04 Z.055
```

This is the helical move to cut the thread. It is not necessary to specify the X and Y axis position because the cutter is programmed to make a full revolution and ends in the same XY position. Note that incremental mode is still active. This is a 360° helical move with a positive Z axis incremental linear move of .055. The Z axis move is equal to the thread pitch. The thread pitch is the length from one tooth to the next tooth on the cutter and is calculated as 1″ divided by the number of threads per inch:

1/18 = .055

```
G1 X−.04
```

This block moves the cutter back to the center of the hole diameter.

```
G90
G0 Z.1
G0 G91 G28 Z0.
```

The first line in this block sets the machine back into absolute mode before any more moves are made. Most programmers prefer to make multiple passes with a thread mill to make sure the thread is straight and very clean. Repeat the above programming lines twice if needed. The following is the thread mill operation in its entirety.

```
T3 M6 (Thread mill)
G0 G90 G54 X1. Y1. M3 S2500
M8
G43 H3 Z1.
G0 Z.1
G1 Z−.50 F50.
G41 D3 G91 X.04 F20.
G3 I−.04 Z.055
G1 X−.04
G90
G0 Z.1
G0 G91 G28 Z0.
```

The multi-point cutting tool used in this program provides an efficient way to machine the threads by reducing the number of helical moves required. A single-point cutting tool has a single cutting edge, which would require additional helical moves for the cutter to make each 360° helical pass as it cuts the entire length of the thread. This would require more calculations and add to the length of the program.

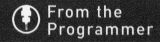

From the Programmer

Hole Programming

Hole making is an essential skill for CNC programmers. The ability to select canned cycles and program cutting paths for a specific type of hole is a skill that is learned through testing and experience. Be creative and use all of the tools available in creating the holes that the part requires.

Chapter 10 Review

Summary

- There are many different types of hole machining operations. Knowing how to program different types of holes is critical in machining.
- Holes on a print are defined by their diameter and depth. Holes can be described as through holes or blind holes. A through hole is a hole that completely extends through the part. A blind hole is a hole that does not extend through the part completely.
- A spot drill produces a better starting surface for drilling than a center drill. The use of a 90° spot drill also facilitates hole chamfering, reducing sharp edges or burrs.
- A number of canned cycles are available for hole machining operations. Correct application of these cycles will significantly reduce program length and programming errors. It is important to understand how each cycle operates and the programming format required for each cycle.
- The G98 and G99 commands are used to control the Z height retract plane between hole machining canned cycles. The G98 command will retract the tool to the last Z height programmed before the cycle began, and the G99 command will return the tool to the R plane (retract plane) programmed inside the canned cycle. The G99 command will save cycle time in programming as long as there is no interference between programmed holes.
- Holes are often located in patterns. Using looped cycles or control-specific cycles to machine hole patterns can save calculation time and programming errors.
- Helical milling operations are best used for shallow, flat-bottom holes. Attempting to use helical milling to machine deep holes can produce undesirable results. Helical milling can be useful if applied correctly.
- Thread milling is an excellent machining method when threading hard materials or deep holes. Thread milling uses a special thread mill cutter with a helical path to create threads.

Review Questions

Answer the following questions using the information provided in this chapter.

Know and Understand

1. Which of the following is considered to be a hole machining operation?
   A. Drilling
   B. Contouring
   C. Slot cutting
   D. Pocket finishing

2. *True or False?* A blind hole extends completely through a piece of material.

3. What is the difference between a counterbore and a spotface?
   A. A counterbore is very shallow.
   B. A spotface is very shallow.
   C. A counterbore has a larger diameter.
   D. A spotface has a larger diameter.

4. *True or False?* The depth of a drilled hole on a print refers to the point where the drill tip ends.

5. To perform high-speed peck drilling with minimal tool retract, what drilling canned cycle is used?
   A. G73
   B. G81
   C. G84
   D. G89

6. In the G81 canned cycle, what does the R address represent?
   A. Peck amount
   B. Feed rate
   C. Stepover
   D. Retract plane

7. To perform left-hand tapping, which canned cycle is used?
   A. G85
   B. G74
   C. G86
   D. G73

8. To perform a spot drilling operation with a 90° spot drill and create a .350 diameter chamfer, which line of code is used?

   A. G99 G81 Z−.175 R.1 F10.
   B. G99 G81 Z−.350 R.1 F10.
   C. G99 G81 Z.1 R−.175 F10.
   D. G99 G81 Z.1 R−.350 F10.

9. To correctly program a 3/8–16 tap for right-hand thread, 1/2″ deep, at 1000 rpm, the correct line of code should read ____.

   A. G74 G98 Q.1 Z−.5 F62.5
   B. G84 G98 R.1 Z−.5 F.0625
   C. G84 G98 R.1 Z−.5 F62.5
   D. G74 G98 R.1 Z−.5 F.0625

10. *True or False?* It is best practice to spot drill prior to a drilling operation.

11. The correct boring canned cycle for a hole that does not go completely through the material is ____.

    A. G81
    B. G85
    C. G86
    D. G89

12. In a canned cycle, which letter normally represents an incremental pecking amount?

    A. P
    B. Q
    C. R
    D. L

13. In a canned cycle, which letter normally represents a dwell time amount?

    A. P
    B. Q
    C. R
    D. L

14. On a Fanuc machine controller, the letter ____ creates a loop cycle.

    A. P
    B. K
    C. Q
    D. L

15. *True or False?* Helical milling is best suited for shallow, flat-bottom holes.

16. When creating right-hand hole threads with a thread mill, in which direction should the tool be programmed to make linear movement in a helical path?

    A. +Z
    B. −Z
    C. +X
    D. −X

Apply and Analyze

1. Referring to the blanks provided, give the correct codes to complete the program for the part shown.

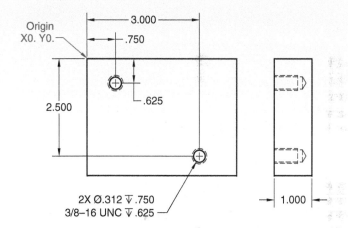

| O2020 |
(T1	1/2 SPOT DRILL	H1	D1	TOOL DIA. .5)
(T2	5/16 DRILL	H2	D2	TOOL DIA. .3125)
(T3	3/8–16 TAP RH	H3	D3	TOOL DIA. .375)
G20
G0 G17 G40 G49 G80 G90
T1 M6
G0 G90 G54 X.75 Y−.625 S1500 M3
G43 H1 Z2.
M8
G98 ____ Z−.188 ____ F10.
____ ____
G80
M5
G91 G28 Z0. M9
M01
T2 M6
G0 G90 G54 ____ ____ S1800 M3

Continued

```
G43 H2 Z2.
M8
____ G73 Z-.844 ____ ____ F10.
____ ____
G80
M5
G91 G28 Z0. M9
M01
T3 M6
G0 G90 G54 ____ ____ S800 M3
G43 H3 Z2.
M8
____ ____ ____ R.1 ____
X3. Y-2.5
G80
M5
G91 G28 Z0. M9
G28 Y0.
M30
```

2. Referring to the blanks provided, give the correct codes to complete the program for the part shown. In the program, a 5/8" (.625") diameter hole is drilled before a feed in and feed out boring cycle is used to produce the .6500±.0005 diameter. Note: Because the hole will be bored, no spot drilling is required.

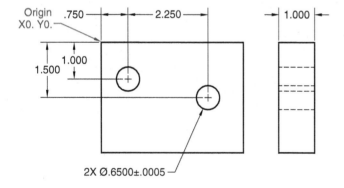

```
O0931
(T1|5/8 DRILL|H1|TOOL DIA. .625)
(T2|.65 FINE BORING TOOL|H2)
G20
G0 G17 G40 G49 G80 G90
T1 M6
G0 G90 G54 ____ ____ S1800 M3
G43 H1 Z2.
M8
G98 ____ ____ R.1 F18.
____ ____
____
M5
G91 G28 Z0. M9
M01
T2 M6
G0 G90 G54 X.75 Y-1. S1538 M3
G43 ____ Z2.
M8
G98 ____ Z-1.005 R.1 F7.7
____ ____
____
M5
G91 G28 Z0. M9
G28 Y0.
M30
```

3. Referring to the blanks provided, give the correct codes to complete the program for the part shown. This part requires a 3/8" (.375") diameter through hole, a drill and tap for a 1/2–13 threaded hole, and a .650±.001 diameter blind hole bore.

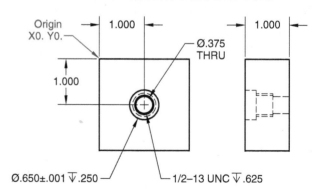

```
O0718
(T1|1/2 SPOT DRILL|H1|D2|TOOL DIA. .5)
(T2|3/8 DRILL|H2|D2|TOOL DIA. .375)
(T3|27/64 DRILL|H3|D3|TOOL DIA. .421875)
(T4|1/2-13 TAP RH|H4|D4|TOOL DIA. .5)
(T5|.65 FINE BORING TOOL|H5|D5|TOOL DIA. .65)
G20
G0 G17 G40 G49 G80 G90
T1 M6
G0 G90 G54 ____ ____ S4584 M3
G43 ____ Z2.
M8
G98 ____ Z-.25 R.1 F45.8
G80
M5
G91 G28 Z0. M9
M01
____ M6
G0 G90 G54 ____ ____ S3056 M3
G43 ____ Z2.
M8
G98 G83 ____ R.1 ____ F18.
G80
M5
G91 G28 Z0. M9
M01
T3 M6
G0 G90 G54 X1. Y-1. S2716 M3
____ H3 Z2.
M8
____ G81 Z-.738 ____ F18.
____
M5
G91 G28 Z0. M9
M01
T4 M6
G0 G90 G54 X1. Y-1. S500 M3
G43 H4 Z2.
M8
____ ____ Z-.625 ____ ____
G80
M5
G91 G28 Z0. M9
M01
T5 M6
G0 G90 G54 ____ ____ S1538 M3
G43 H5 Z2.
M8
G98 ____ ____ ____ P20 F7.7
____
M5
G91 G28 Z0. M9
G28 Y0.
M30
```

Critical Thinking

1. Based on what you have learned in this chapter, explain the tooling and programming that would be required to produce the following.

    A. Two .500" diameter blind holes drilled .750" deep, each with a .0625" × 45° chamfer.
    B. Three .375" through holes to be drilled in a 45° pattern in a 1" thick part.
    C. The six .125" blind holes in **Figure 10-23**. The depth of each hole is .25".

2. Briefly explain what might happen if the G99 command is programmed instead of the G98 command by mistake when moving between hole locations in a hole drilling canned cycle.

3. Thread milling has a number of advantages in comparison to tapping threads, such as the tooling used and control of the final thread. Based on production needs, explain a machining requirement in which tapping would be more efficient than thread milling.

11 Facing and Island Machining

Chapter Outline

11.1 Introduction to Facing
11.2 Facing Tools
11.3 Facing Program Strategy
11.4 Creating the Facing Program
11.5 Island Machining
    11.5.1 Island Machining Strategy
    11.5.2 Creating an Island Machining Program

Learning Objectives

After completing this chapter, you will be able to:

- Explain the purpose of facing operations in CNC milling.
- Select the correct tool for a facing operation.
- Determine the appropriate cutting direction and programming strategy for face milling.
- Identify an island feature on a part.
- Determine the programming strategy for an island feature.

Key Terms

boss
face mill
face milling
facing
fly cutter
island

11.1 Introduction to Facing

The machining term *facing* is used in both CNC mill and CNC lathe operations. In lathe operations, facing refers to cutting the end of the part material furthest from the spindle. In milling operations, facing refers to machining any surface perpendicular to the Z axis. Facing is commonly called *face milling* when cutting the top surface of a part to a finish size. Often in milling, the first operation is a face milling operation to cut a piece of stock to a final part thickness. For example, if the overall part thickness has a finish dimension of 1.950″, but the stock material is 2.000″, a face milling operation is required to remove the extra .050″. When necessary, a facing pass is made on each side of the stock to remove excess material in equal amounts. In this case, if .050″ of material is to be removed, .025″ is removed from each side and face milling will create two flat parallel surfaces.

11.2 Facing Tools

Before programming techniques for facing are covered, a discussion is needed to address the types of tools used for facing. The selection of the tool used depends on the size of the workpiece being machined and the amount of material to be removed. There are specific types of tools made for facing operations. Although an end mill can be used for facing, it is not the best option. An end mill will take too many passes to cover the top of the stock, and each pass between transitions will leave a ridged surface that negatively affects the surface finish. A face mill can machine the same face in fewer passes, making it considerably more efficient. See **Figure 11-1**.

A *face mill* is a cutter specifically designed for cutting the top face of a part, as opposed to creating a pocket or cutting a contour. The cutting edges of a face mill are located along the outer edges on the cutting face. Cuts are made in a horizontal direction at a given depth and width, and entry of the tool must come from outside the stock. Face mills come in a variety of sizes and normally have cutting inserts.

In manual machining, it was common to use a single-point cutting tool called a *fly cutter* for face milling. See **Figure 11-2**. With the development of modern insert technology and greater CNC machining capabilities, however, fly cutters are not as versatile and efficient as face mills. Face mills with multiple inserts allow for cutting at higher spindle speeds and feed rates. The benefit of a fly cutter is that it can cover a large surface area, and the single cutting edge will leave a very fine finish. The downside is that a fly cutter is limited to very small depths of cuts, less than .015″, and it requires slow feed rates, as low as 5 inches per minute (ipm).

A shell mill is another type of tool used in face milling. Sometimes, the term *shell mill* is used to refer to a face mill. However, a shell mill is not the same as a face mill. A shell mill is a hollow steel cutting tool with cutting edges along the entire length on the sides. A shell mill is a solid body tool similar to a solid end mill and is not designed to hold cutting inserts.

Modern face mills have multiple cutting inserts and range in size from 2″ to 8″ and larger in diameter. See **Figure 11-3**. With multiple cutting edges, deeper cuts can be machined while still leaving exceptional finishes. Face mills can be run at high spindle speeds and feed rates. The programmed speeds and feeds will vary considerably depending on material and insert type. Different inserts can be used to allow for light machining

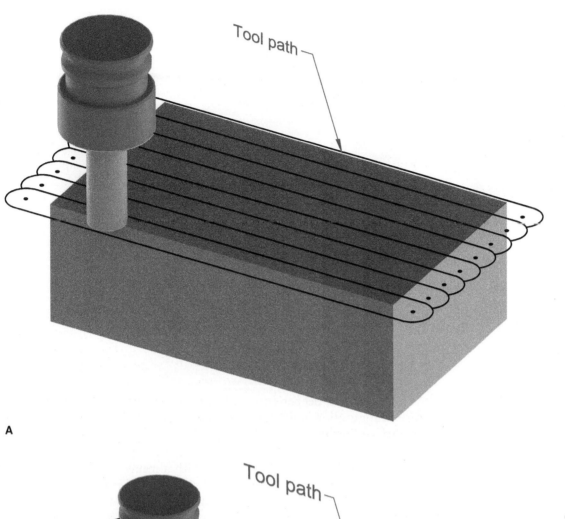

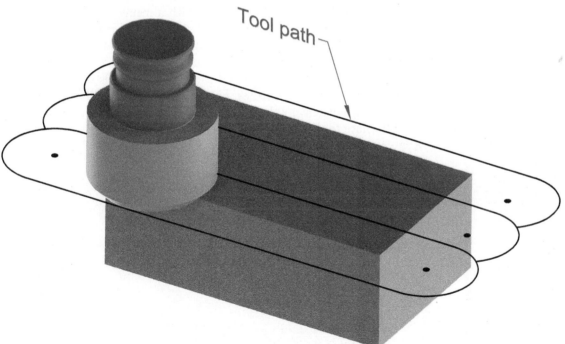

Goodheart-Willcox Publisher

Figure 11-1. The tool selected for a facing operation should be appropriate for the workpiece and the amount of material to be removed. A—Using an end mill in a facing operation requires a large number of passes to cover the entire workpiece. B—A face mill can machine the same face in fewer passes and is the preferred tool.

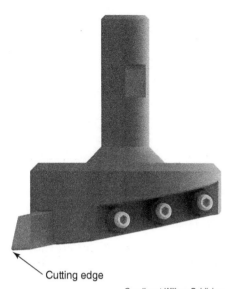

Figure 11-2. A fly cutter is a single-point cutting tool. Fly cutters are designed for shallow cuts and are used on smaller machines with lower metal removal rates.

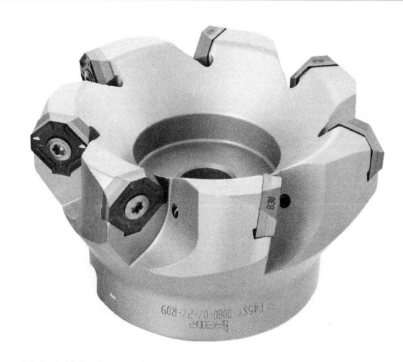

Figure 11-3. A 4″ face mill with multiple cutting inserts.

with a fine finish or deeper cuts with a rougher surface finish. The type of insert configuration used depends on material type, depth of cut, and desired surface finish.

A round insert will not handle a deeper cut, but it provides superior surface finish. See **Figure 11-4**. An insert with a more traditional triangular or rectangular shape provides deeper cuts and generally good surface finishes, **Figure 11-5**. Triangular and rectangular inserts are the most common styles for facing off material.

There are also face mills designed for extreme depths of cut, but they produce very poor surface finishes, **Figure 11-6**. A face mill with serrated inserts is used when heavy volumes of material need removal during roughing and alternate tools will be used to complete finish passes. The serrated

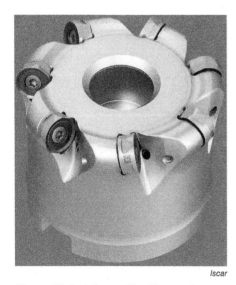

Figure 11-4. A face mill with round inserts.

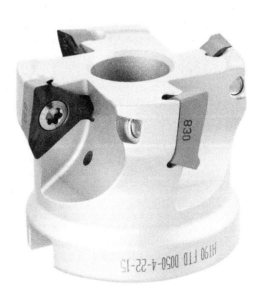

Figure 11-5. A face mill with triangular inserts.

edges allow for heavy chip removal and clearance in deep cuts. An example of a part that might require this type of face mill is shown in **Figure 11-7**. In this example, there is a large step feature that is 4″ wide and 3″ deep. A traditional end mill path would take considerable time and cause significant tool wear with each part machined. A large roughing face mill can remove this volume of material in just a few passes at high feed rates.

11.3 Facing Program Strategy

After determining the cutting tool to be used, the next step is to analyze the cutting strategy. Just as in other types of machining, there are two main concerns. The first is the quality of the part, and the second is the machining time, or cycle time, to complete the machining operation. The most important concern is the quality of the part. Consider that every part has to meet tolerance requirements, and this is the standard that the finished part will be measured against. A tolerance establishes a permissible amount of variation from a given dimension. The specified amount of variation defines a permissible range and allows for parts to be machined efficiently. For example, if a certain feature has a 1.000″ ± .010 dimension, there is no need to make multiple extra finish passes to achieve an exact dimension of 1.000″. A general rule to adhere to is to make sure the machining path is adequate and repetitive to within half of the tolerance. For this example, make sure the programmed toolpath can repetitively machine the 1.000″ feature to be within a tolerance of ±.005. Making the feature better than that will add unnecessary cycle time. The same holds true for the facing operation. Many facing operations will require a maximum surface finish specified in microinches, such as 125 microinches or 63 microinches. The lower the number, the *smoother* the finish required. Other common requirements for facing operations can be size tolerances and controls for flatness and parallelism. All of these factors need to be considered before the programming strategy is determined.

Iscar

Figure 11-6. A face mill with serrated cutting inserts.

Goodheart-Willcox Publisher

Figure 11-7. A face milling operation used for heavy roughing. The blue lines represent the path of the tool.

> **From the Programmer**
>
> **Efficiency in Machining**
> Mr. Calverley was a classic, apprenticeship-trained tool and die maker. He was an expert in the field of manual machining. One day, he noticed a young machinist milling some parts. He asked what was taking so long. The young machinist told him that he was "trying to make these parts perfect." Mr. Calverley responded and said, "There's no such thing as a perfect part." He offered this valuable advice: "Focus on being efficient, make sure your toolpath strategies are fast, and leave every part inside the print tolerances. Taking way too long to make a great part is just as bad as making a scrap part fast."

When analyzing a part for a facing operation, decisions need to be made about cutting direction, stepover amounts, and transitions from cut to cut. Study the part shown in **Figure 11-8**. Before writing a program for the part, it is necessary to determine how it should be machined.

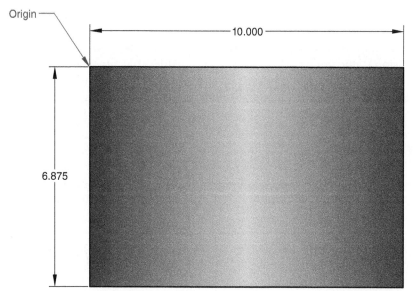

Figure 11-8. A part requiring a facing operation.

The first determination to make is what *direction* the machining pass will take. The cutting direction can be in the X axis direction or the Y axis direction. For this example, a 2 1/2″ diameter face mill is used. If cutting passes are made in the X axis direction, four cuts will be required. See **Figure 11-9**.

If the cutting direction is changed to the Y axis direction, six passes will be required to cut the material. See **Figure 11-10**. For this part, it is more efficient to align the milling direction along the X axis.

In this example, a 2 1/2″ diameter face mill was chosen. A larger face mill could be utilized in order to reduce the amount of passes. The diameter of the tool is used in determining the width of the cut, or stepover amount. A rule of thumb for stepover amounts is 75% of the cutter diameter. This will normally result in a good surface finish and flatness on the stock.

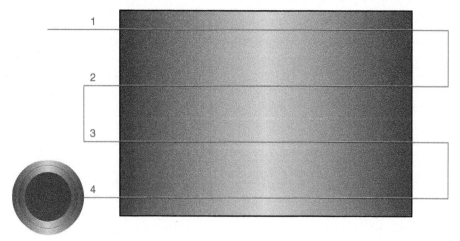

Figure 11-9. Positioning the face mill to cut in the X axis direction will require four cuts.

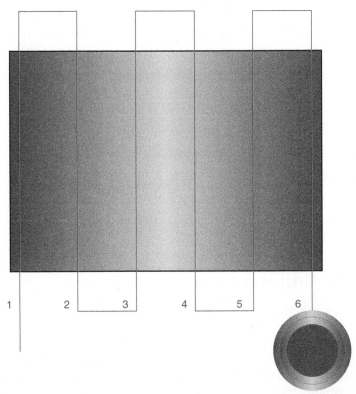

Figure 11-10. Positioning the face mill to cut in the Y axis direction will require six cuts.

In certain cases, where a very close flatness or parallelism tolerance must be met, a smaller stepover amount (perhaps 50% of the cutter diameter) might yield better results. For this example, a 2 1/2″ diameter cutter is used and the stepover amount will be 1.875″.

$$2.5 \times .75 = 1.875$$

When removing a larger amount of material stock, a roughing Z axis pass and a finishing pass may be required. Generally, if a roughing and finishing pass are required, it is recommended to leave about .010″–.015″ for the finish pass. Leaving too much material can result in poor surface finish. Leaving too little material does not give the cutter an opportunity to get underneath the stock and machine efficiently.

The last consideration in machining is the transition movement between passes. The cutter needs to fully exit the stock before moving over in the Y axis to complete the next machining pass. This requires a small amount of clearance past the edge of the part between passes. But remember to be efficient—exiting too far off the part is wasted machine movement and cycle time. For this example, a sufficient amount of clearance past the part is .100″. The transition movement between passes can be a straight-line move or an arc move. In manual programming, it is easier to program straight-line moves between passes. When using computer-aided manufacturing (CAM) software, tools in the software simplify the process of programming radial arcs for transition movement. Looped transitions save cycle time by engaging the tool more rapidly. See **Figure 11-11**.

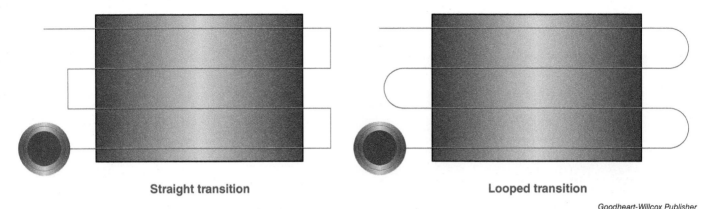

Figure 11-11. Transition moves between passes of the cutting tool can be programmed as straight-line moves or looped moves with radial arcs. Looped transitions increase the efficiency of the toolpath.

11.4 Creating the Facing Program

Once all of the programming decisions are made, the machine code can be generated. Because the facing operation is only cutting the top of the part, cutter compensation is not used. The tool positions are defined using point-to-point programming. The work origin is designated as the top-left corner of the part, so that location will be used in defining the tool starting position. However, this could be at any corner. Remember that the tool spindle center is being programmed, so consider the cutting tool radius (1.25″). The starting location for the tool will be at a position that is more than 1.25″ in the negative X axis direction from the origin. The Y axis starting position will be −.625″ from the origin. The Y axis starting position is calculated as the distance added to the cutter radius (1.25″) to produce the stepover amount (1.875″). This distance is calculated by subtracting the cutter radius from the stepover amount (1.875 − 1.25 = .625). **Figure 11-12** shows the part and the path of the cutting tool. Each position along the path is numbered. These numbered positions correlate to the specified coordinates in the following program.

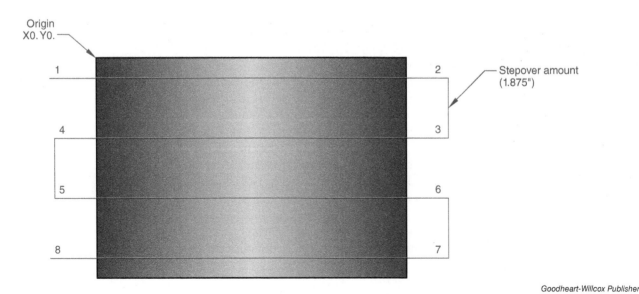

Figure 11-12. The cutting path for the face milling operation. The numbered positions correspond to the programmed coordinates.

```
O0804
(1000 SFM – .004 CHIP LOAD)
(T1|2-1/2 FACE MILL|H1)
G20
G0 G17 G40 G49 G80 G90
T1 M6
G0 G90 G54 X–1.5 Y–.625 S1528 M3      Point 1 (start position)
G43 H1 Z2.
M8
Z.100
G1 Z0. F24.4
X11.35                                 Point 2
Y–2.500                                Point 3
X–1.35                                 Point 4
Y–4.375                                Point 5
X11.35                                 Point 6
Y–6.25                                 Point 7
X–1.5                                  Point 8 (exit position)
G0 Z2.
M5
G91 G28 Z0. M9
G28 X0. Y0.
M30
```

Although the facing operation can be a simple program, it is vital in creating an accurate top surface on the finished part. Trying some different tool geometry and using a different number of inserts can greatly impact the results of the facing operation. The goal of the program is to create a flat, smooth starting face for the required part.

11.5 Island Machining

Island is a term in machining that can be used in different contexts. Sometimes the term *island* is used to describe a raised feature that stands alone from the part surface, as shown in **Figure 11-13**. However, the correct nomenclature for this type of feature is a *boss*. The term *boss* is used in most CAM systems to describe a protruding feature that is part of an existing solid part.

The more accurate description of an island is a protruding standalone feature that is located inside of a pocket. An island will usually be fully encapsulated by the walls of a pocket. There may be a single island or multiple islands. Their shapes can be round, rectangular, or irregular, and the pocket can be closed or open. Different types of islands encountered in machining are shown in **Figure 11-14**.

11.5.1 Island Machining Strategy

Island machining is a hybrid process that involves pocket machining and contour machining. The program will rough out open areas of the pocket while leaving enough stock for the island. Once the pocket area is

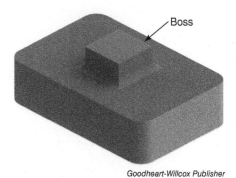

Goodheart-Willcox Publisher

Figure 11-13. The term *island* is sometimes used in machining to describe a raised feature above a part surface. However, the correct term for this feature is a *boss*.

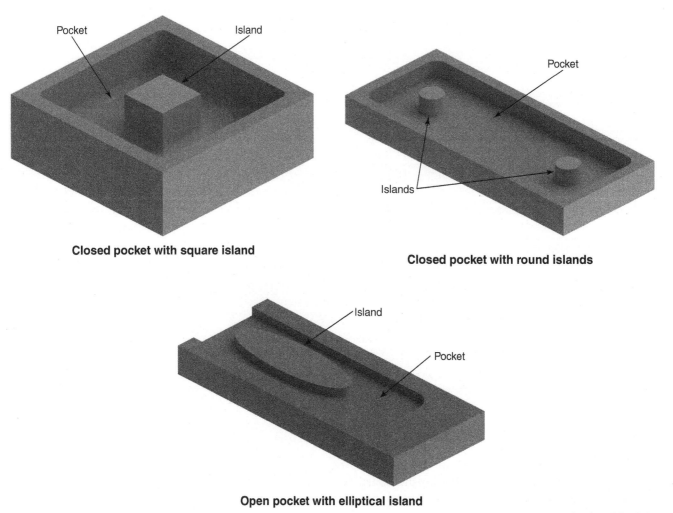

Figure 11-14. An island is a protruding feature that is located inside the walls of a pocket.

machined, finish contour paths can be used to complete the pocket profile and island profile. This type of programming is difficult if done manually, but it can be accomplished.

An example of a part with an island requiring pocket and contour machining is shown in **Figure 11-15**. Notice that the wall thickness around the part is dimensioned as .500″. This dimension is noted as *typical*, meaning it applies on all sides. There are also .250″ radii in each corner and a 1.000″ square island in the middle of the pocket. The pocket is .500″ deep and the island is .500″ in height. For this example, the top-left corner of the part is the work origin. A solid model of the part is shown in **Figure 11-16**.

The machining strategy for this part is fairly straightforward. The pocket will be roughed using a 1/2″ end mill, leaving .010″ on the outside walls and .010″ around all four sides of the island for finishing. The first step is to map out the cutting path. Then, the X and Y axis positions are calculated for each move. **Figure 11-17** shows a map of the cutting path with numbered positions and a table with coordinate points correlated to the positions. The position coordinates shown in the table are the X and Y axis positions used in the program. The cutting path uses .250″ stepovers, where possible, to reduce program length. This stepover amount is calculated as 50% of the cutter diameter. Where needed, the stepover amount is

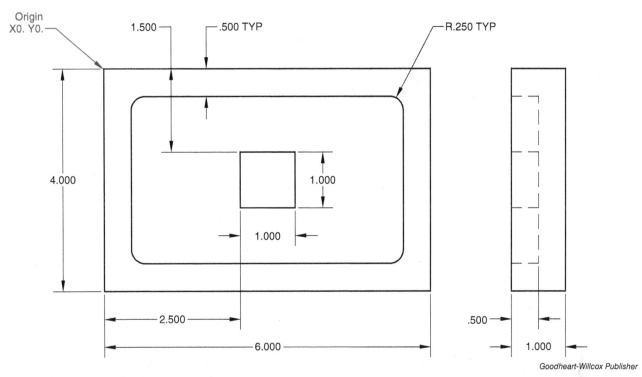

Figure 11-15. A part with a closed pocket and a square island located in the middle of the pocket.

Figure 11-16. A solid model of the part requiring island machining.

adjusted in the program to provide clearance between the cutting tool and the island feature or side wall. This clearance must be accounted for when calculating position coordinates.

Once the cutting path is developed and coordinate points are calculated, an entry strategy for roughing can be determined. For this operation, a 1/2″ end mill was selected. The end mill cuts at 500 surface feet per minute with a chip load of .005″. For this example, the tool descends to the .500″ final depth on the entry and then proceeds to each X and Y axis

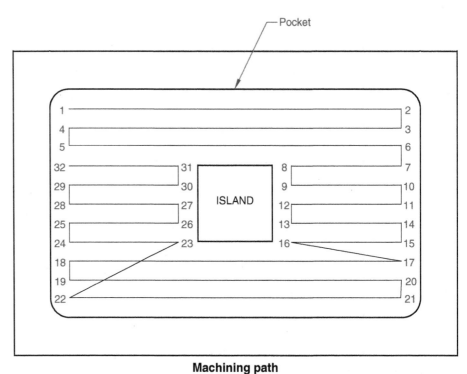

Programming Coordinates

Point	X Position	Y Position	Point	X Position	Y Position
1	0.760	−0.760	17	5.240	−2.760
2	5.240	−0.760	18	0.760	−2.760
3	5.240	−1.010	19	0.760	−3.010
4	0.760	−1.010	20	5.240	−3.010
5	0.760	−1.240	21	5.230	−3.240
6	5.240	−1.240	22	0.760	−3.240
7	5.240	−1.510	23	2.240	−2.510
8	3.760	−1.510	24	0.760	−2.510
9	3.760	−1.760	25	0.760	−2.260
10	5.240	−1.760	26	2.240	−2.260
11	5.240	−2.010	27	2.240	−2.010
12	3.760	−2.010	28	0.760	−2.010
13	3.760	−2.260	29	0.760	−1.760
14	5.240	−2.260	30	2.240	−1.760
15	5.240	−2.510	31	2.240	−1.510
16	3.760	−2.510	32	0.760	−1.510

Calculated coordinates

B

Goodheart-Willcox Publisher

Figure 11-17. To plan the roughing operation, a path for machining is developed. Then, coordinate points for programming are determined. A—The cutting path for roughing the pocket is laid out with numbered positions. B—Coordinates are calculated for each position based on the dimensions given on the print. The coordinates establish the X and Y axis positions for each tool move and are recorded in a table.

location. The entry is programmed as a series of helical moves into the upper-left corner of the pocket. This allows the tool to enter the material gradually in a spiral motion. Helical moves are programmed as circular interpolation moves with a Z axis move so that simultaneous movement occurs along the three axes. Cutter compensation is not used in this operation because it may cause the tool to "over travel" the desired cutting path and collide with a side wall or the island feature.

11.5.2 Creating an Island Machining Program

Once all of the programming decisions are made, the machine code can be generated. As previously discussed, island machining involves a roughing operation for the pocket and a finish operation for the pocket walls and island. The program for the rough machining is as follows.

O0615	
(T1|1/2 FLAT END MILL|H1)	
G20	
G0 G17 G40 G49 G80 G90	
T1 M6	
G0 G90 G54 X1.4626 Y−.8769 S3820 M3	
G43 H1 Z2.	
M8	
Z.2	
G1 Z.1 F30.	
G2 Z−.0235 I.0374 J−.3731 F61.1	Helical entry move
Z−.147 I.0374 J−.3731	Helical entry move
Z−.2704 I.0374 J−.3731	Helical entry move
Z−.3939 I.0374 J−.3731	Helical entry move
X1.5374 Y−1.6231 Z−.4557 I.0374 J−.3731	Helical entry move
X1.1873 Y−1.043 Z−.5 I−.0374 J.3731	Helical entry move to final Z depth
G1 X.760 Y−.760	Point 1
X5.240 Y−.760	Point 2
X5.240 Y−1.010	Point 3
X.760 Y−1.010	Point 4
X.760 Y−1.240	Point 5
X5.240 Y−1.240	Point 6
X5.240 Y−1.510	Point 7
X3.760 Y−1.510	Point 8
X3.760 Y−1.760	Point 9
X5.240 Y−1.760	Point 10
X5.240 Y−2.010	Point 11
X3.760 Y−2.010	Point 12
X3.760 Y−2.260	Point 13
X5.240 Y−2.260	Point 14
X5.240 Y−2.510	Point 15
X3.760 Y−2.510	Point 16
X5.240 Y−2.760	Point 17
X.760 Y−2.760	Point 18

Continued

X.760 Y−3.010	Point 19
X5.240 Y−3.010	Point 20
X5.230 Y−3.240	Point 21
X.760 Y−3.240	Point 22
X2.240 Y−2.510	Point 23
X.760 Y−2.510	Point 24
X.760 Y−2.260	Point 25
X2.240 Y−2.260	Point 26
X2.240 Y−2.010	Point 27
X.760 Y−2.010	Point 28
X.760 Y−1.760	Point 29
X2.240 Y−1.760	Point 30
X2.240 Y−1.510	Point 31
X.760 Y−1.510	Point 32
G0 Z.25	
M5	
G91 G28 Z0.	
G28 Y0.	

This part of the program will create all of the roughing moves for the pocket and the island feature. Notice that the M30 function is not used at the end of the program. This is because there is one more operation to finish the pocket walls and island walls. This operation will be accomplished with a 3/8″ end mill. The 1/4″ corner radii require a tool with a smaller radius to ensure that the correct arc move is programmed in each corner. The program will use left-hand cutter compensation, initiated with the G41 command, to complete the finish pass for the pocket walls and island walls. This positions the tool so that it is climb cutting. The program for the finish operation is as follows.

```
T2 M6
G0 G90 G54 X5.25 Y−.875 S5000 M3
G43 H2 Z.25
M8
Z.2
G1 Z−.5 F50.
G41 D2 Y−.5 F100.
X.75
G3 X.5 Y−.75 R.25
G1 Y−3.25
G3 X.75 Y−3.5 R.25
G1 X5.25
G3 X5.5 Y−3.25 R.25
G1 Y−.75
G3 X5.25 Y−.5 R.25
G1 G40 Y−.875
Z.2 F6.3
G0 Z.25
```

Continued

```
X2.5 Y-1.125
Z.2
G1 Z-.5 F50.
G41 D2 Y-1.5 F100.
X3.5
Y-2.5
X2.5
Y-1.5
G40 X2.125
Z.2 F6.3
G0 Z.25
M5
G91 G28 Z0. M9
G28 X0. Y0.
M30
```

Bear in mind that a machined part with a pocket and internal island can have various pocket shapes. A single part can also have multiple islands within a pocket. However, the strategy for programming and machining remains the same. First, rough out the pocket, making sure the island boundaries are not violated. Then, use a contour toolpath to finish the pocket walls and island. For a better finish, material can be left on the pocket floor and a finish pass programmed for the final depth. Depending on the material type and print requirements, more material can be left on the walls. This example is designed to show the general machining strategy without regard to actual material or dimensional requirements.

Chapter 11 Review

Summary

- The term *facing* in CNC milling refers to machining any surface perpendicular to the Z axis.
- Face milling is a specific facing operation generally used to create a flat top face on a part.
- Facing tools come in a variety of types and sizes. For tools with inserts, the number of inserts can be varied to allow for light machining with a fine finish, or deeper cuts with a rougher surface finish. The selection of the cutting tool used depends on the size of the workpiece being machined and the amount of material to be removed.
- A face mill with triangular or rectangular inserts produces deep cuts and good surface finishes. A face mill with round inserts will not handle a deeper cut, but it provides superior surface finish. A face mill with serrated inserts is used when heavy volumes of material need removal during roughing.
- Determining the appropriate direction of cut and proper stepover amounts will make the facing operation efficient.
- A basic face milling program will consist of straight-line moves that cover the entire face to be machined. Entry and exit amounts should be kept at a minimum to reduce cycle time.
- An island feature is best defined as a protruding standalone feature inside of a pocket. The island shape can be round, rectangular, or irregular, and the pocket can be closed or open.
- The programming strategy for an island is to rough out as much material as possible while not violating the island feature.
- An island machining operation consists of pocket roughing and one or more finish contour passes.

Review Questions

Answer the following questions using the information provided in this chapter.

Know and Understand

1. Face milling refers to machining a surface perpendicular to the ____ axis.
   A. X
   B. Y
   C. Z
   D. None of the above.

2. The face mill shown below is best suited for ____.

Iscar

   A. light finish passes
   B. closed pockets with islands
   C. island contours
   D. heavy face roughing

3. What are some of the factors considered when selecting a face mill for machining?
   A. Surface finish
   B. Surface flatness
   C. Number of inserts
   D. All of the above.

4. *True or False?* Since the amount of material to cut is the same, it does not make a difference which direction the face mill travels.

5. What is a common characteristic of an *island* in machining?

   A. It is a recessed cavity.
   B. It extends from the top surface of a part.
   C. It is the same as a boss.
   D. It is located in a pocket.

6. Island machining is a hybrid process that involves which two types of machining?

   A. Pocket cutting and facing
   B. Pocket cutting and contouring
   C. Facing and contouring
   D. Drilling and facing

7. *True or False?* An island feature can be round, rectangular, or irregular.

8. *True or False?* The pocket around an island must be closed.

9. When roughing the pocket for an island, it is important not to _____.

   A. take deep cuts
   B. use climb milling
   C. collide with the island feature
   D. use a helical entry

10. The finish toolpath for an island is a _____ toolpath.

    A. roughing
    B. drilling
    C. facing
    D. contour

Apply and Analyze

1. If face milling is used to machine the part shown, along which axis should the tool travel to make programming efficient?

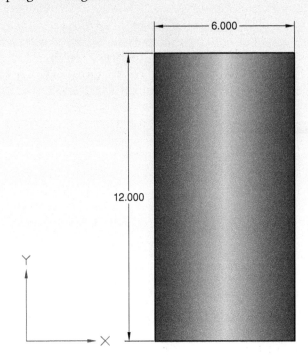

   A. X axis
   B. Y axis
   C. Z axis
   D. Either the X or Z axis.

2. For a face milling operation, which of the following tools should be selected to remove .010" of material and produce a fine surface finish?

   A.

   B.

   Iscar

   C.

   Iscar

3. Which of the following parts contains an island feature?

   A.

   B.

   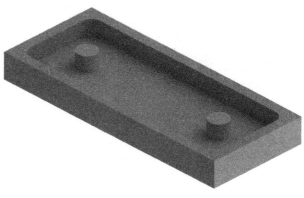

   C.

4. Study the part drawing shown. What is the most appropriate roughing tool and finishing tool for machining?

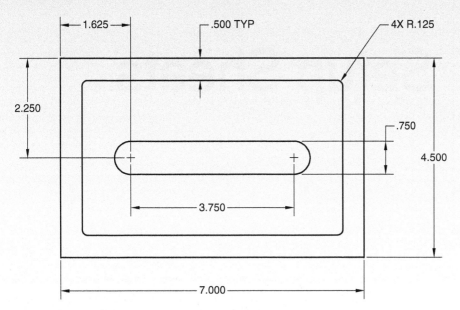

   A. 1/2" roughing tool and 3/8" finishing tool
   B. 5/8" roughing tool and 3/16" finishing tool
   C. 1/2" roughing tool and 1/4" finishing tool
   D. 3/8" roughing tool and 1/4" finishing tool

Critical Thinking

1. The intention of a facing operation is to create a flat surface on top of the part. If done correctly, it should provide a flat surface with a smooth finish. Why is it important to start with a flat surface on the top of the part? What are some probable issues that can result if the top surface is not flat?

2. In a pocket roughing operation, what are some of the advantages of using helical entry motion over other types of entry?

3. The island machining example in this chapter was for a closed pocket with a square island. How would the programming change for the same type of pocket with a round island?

12 Setup Sheets

Chapter Outline

12.1 Introduction

12.2 Exchange of Information
    12.2.1 Efficiency of Setup and Operation

12.3 Setup Sheets
    12.3.1 Header
    12.3.2 Rough Stock
    12.3.3 Tool Data
    12.3.4 Workholding and Fixturing
    12.3.5 Special Instructions

12.4 Setup Sheet Formats

12.5 Efficiency in Production

Learning Objectives

After completing this chapter, you will be able to:

- Explain the importance of the exchange of information between a programmer and setup person.
- Explain the importance of efficiency and how it applies to machine setup and operation.
- List the components that are included on a setup sheet.
- Describe different documentation formats used for setup sheets.

Key Terms

edge finder	setup	setup sheet
mill stop	setup person	work offset

Chapter opening photo credit: Travelpixs/Shutterstock.com

12.1 Introduction

One of the most important aspects of CNC programming is the transfer of information from the programmer to the individual in charge of job setup and operation. In many machine shops, the programmer is not responsible for setup work or machine operation. Except for the smallest of shops, job setup and operation is normally handed off to a CNC machinist who has the skills needed to complete the setup and run the machine. *Setup* is the phase of CNC machining that involves setting up the workholding for the part, setting up the tooling, and making a test run of the program before the first part is machined.

12.2 Exchange of Information

In the pre-programming stages, there are usually several people who are involved in the planning of a production run of machined parts. Often, one of these individuals is a setup person. The exchange of information between the programmer and the setup person is critical. The *setup person* is a technician who has first-hand knowledge of the machinery, the part workholding equipment, and the available tooling. During the programming phase, the programmer reviews the print and considers the machine to be used, the workholding method, and the available tooling. Once these requirements are determined and the program is created, the programmer makes a list of operations, tooling, workholding devices, and work coordinates. This makes the setup process easy to understand for the setup person.

Many machine shops have developed organized processes to communicate the required information for machining. Most computer-aided manufacturing (CAM) packages have functions for generating documentation for the program. Depending on the software and capabilities of the machine, program documentation generated in the software can be displayed on the control screen. Some of the newer machine controllers have the ability to display PDF files or other types of image files with setup and operational instructions. Regardless of the documentation method used, the way that information is transferred from the programmer to the machinist should be based on a systematic approach. What is most important is to not be the shop that relies on the setup person to "figure it out."

12.2.1 Efficiency of Setup and Operation

The word used most often to describe a good machining process is *efficient*. Just saying that a process is *fast* does not indicate how effective the process is based on what it is intended to achieve. Being efficient means the machining process is taking into account the speed that a tool can travel, but it also means that additional factors are considered in improving productivity. An efficient machining process limits unnecessary moves, factors in part changeover times, and analyzes the total time needed to process a part. In machining, part cycle time refers to the amount of time required to produce a part to completion. The part cycle time is used in calculating the total number of parts that can be produced in a given work shift.

For example, **Figure 12-1** shows a simple part to be machined. This part has a 1" thick base and a .25" boss on the top surface with four machined holes. One way to set up this part is to place the material in a

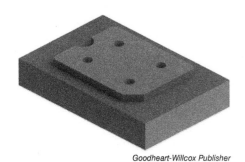

Goodheart-Willcox Publisher

Figure 12-1. A part with a raised boss feature and four machined holes.

vise and use an edge finder to locate a side of the part in relationship to the machine. As discussed in Chapter 13, an *edge finder* is a cylindrical alignment tool that is brought into contact with a side of the part to establish coordinates for the work coordinate system. Based on this type of setup, it could take two minutes to machine the part and two minutes to change out the part each time. At a total of four minutes per part, it is possible to produce 15 parts per hour or 120 parts per eight-hour shift.

To be more efficient, a mill stop could be used to reduce the part changeover time by eliminating the edge finding operation. A *mill stop* is a positioning tool with an end stop for locating the part when the part is mounted in a vise. See **Figure 12-2**. This allows identical parts to be aligned quickly and precisely in the same position in a production run. Although this does not affect the two-minute cutting time, it does reduce the part changeover time to 15 seconds. Now, each part takes only two minutes and 15 seconds to produce. This yields more than 26 parts per hour, or approximately 213 parts per eight-hour shift. This increases the efficiency of production by more than 75%, while keeping the machining time constant. Depending on the shape and size of this part, it might be possible to run many parts in one operation and reduce overall production time even further.

Setup time required for a production run can have a significant impact on total run time. The most important objective in a machine shop is to maximize *spindle turn time*. If the spindle is not running, it means the machine is not making parts, and the shop is not making money. Although customers can be charged for setup time and programming, the customer will evaluate the total price when considering quotes for machining. Therefore, efficiency in setup is just as important as efficiency in operation.

12.3 Setup Sheets

Setup sheets provide the most efficient way to exchange information between the programmer and the setup person. A *setup sheet* is a document that shows detailed information for setting up the part on the machine. A setup sheet lists items such as general information about the part, the type of stock

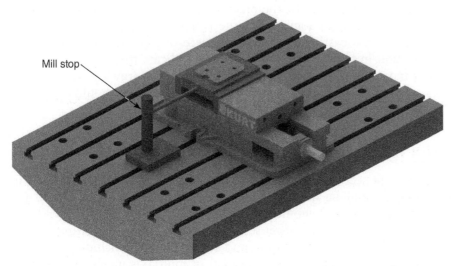

Goodheart-Willcox Publisher

Figure 12-2. A mill stop is added to the setup to reduce changeover time, thus increasing efficiency in production.

to be machined, the tooling required, and the type of workholding used. Often, a setup sheet contains a diagram of the part showing the part orientation and how the work setup should be completed for a specific machining operation. When necessary, the program documentation is made up of a series of setup sheets with information about each operation. Setup sheets vary in format and there can be a variety of opinions about what information needs to be included. The most important consideration is that all of the information required to make the setup easy and seamless is provided.

A setup sheet should provide a complete description of the required workholding, tooling, and origin of the work coordinate system. Most programs are written with a designated work offset. A *work offset* is a coordinate setting that establishes the location of the work origin relative to the machine home position. When a work offset is programmed, the designated work offset and physical location of the work origin are normally illustrated on the setup sheet. As an example, the G54 command is used in programming to activate work coordinate offset 1 as the work coordinate system. On the setup sheet, the G54 work offset is identified on the part diagram to show the location of the X0. Y0. work origin.

Setup sheets provide specific information about the work setup and tooling. Different machine shops will have normal operating procedures in place defining *how* parts are machined. The purpose of the setup sheet is to answer all of the unknown questions. The following sections discuss the general components that make up setup sheets and the standard information that most setup sheets should contain.

12.3.1 Header

The area at the top of a setup sheet is called the header. The header contains information such as the part name, part number, part revision, program number, program revision, customer name, and date. The date should indicate the date of completion for the program, and that date should also be included at the beginning of the written program.

12.3.2 Rough Stock

The material to be used for the machining operation should be included on the setup sheet. In some cases, the rough stock will be a previously machined part, a casting, or a saw cut piece of raw material. The rough stock section in the documentation should indicate the size and material type of the stock used for the operation. Usually, a drawing of the rough stock is included in the rough stock section. The drawing should include an illustrated representation of the work coordinate system.

12.3.3 Tool Data

The tool data section describes the tooling to be used. Each tool used in machining should be listed and detailed with information required for setup. The information should specify the amount the tool protrudes from the holder, the tool length offset setting, the diameter offset setting if used, and the toolholder used to mount and secure the tool. Including the toolholding information is especially critical in multiaxis machining because the toolholder can interfere with the part workholding, work table, or workpiece material.

12.3.4 Workholding and Fixturing

The method used to secure the part material must be clearly defined. When necessary, a detailed illustration of the method and device used to mount the work should be included in the workholding section. If using a specific fixture or soft jaws, define the required device by number or type. If the material is simply being held in a vise, define the minimum amount of grip length to prevent interference with the vise while machining. If work parallels or any other devices are being used in conjunction with the vise or fixture, specify them in the workholding section.

12.3.5 Special Instructions

Any important information that is not covered elsewhere on the setup sheet should also be included. Special instructions might be needed for specific dimensions that need inspection, specific places where the program may pause, and anything else that does not fall under normal operations and requires special attention.

12.4 Setup Sheet Formats

Setup sheets vary in the amount of information provided. As previously discussed, the amount of detail included on a setup sheet depends on the information needed to completely describe the work setup and tooling requirements. This section presents different examples of setup sheets to illustrate common formats and the options available to the programmer.

Setup sheets can be created using the tools of a CAM program, or they can be created using standard word processing or spreadsheet software. **Figure 12-3** shows an example of a setup sheet created as a spreadsheet document in Microsoft Excel®. This setup sheet has a simple format and includes basic information about the rough stock, the tooling used, and the part orientation and work coordinate system.

As shown in **Figure 12-3**, the stock size dimensions of the part are 6″ × 4″ × 1.25″. The part is mounted in a vise and four different tools are used in machining. Each tool is identified. The first tool is an end mill used to rough and finish the boss feature on the top surface. This is detailed in the special instructions below the tooling data. Then, a spot drill is used to spot drill the four threaded holes. A letter O drill (.316″ diameter drill) and 3/8–16 tap are used to produce the four holes. The tooling data includes the tool length and the tool offset settings. At the bottom of the setup sheet, the designated work offset and part orientation are illustrated. The work origin is located at the top-left corner of the part boundary, with the top surface of the boss feature positioned on the Z0. plane.

Depending on the type of job being processed, a list of tooling may be the only program documentation. **Figure 12-4** shows a tool list with a breakdown of the four tools required. The tool list provides detailed information about each tool, including an illustration of the specified toolholder with dimensions. Although this type of documentation is often used, it is the least desirable format. Much of the vital information needed for the setup person is not included. The benefit of this format is that it is short and simple, with good detail of the tooling and toolholders being used.

Another common setup sheet format is shown in **Figure 12-5**. This is perhaps the most compact but complete example of a setup sheet. This

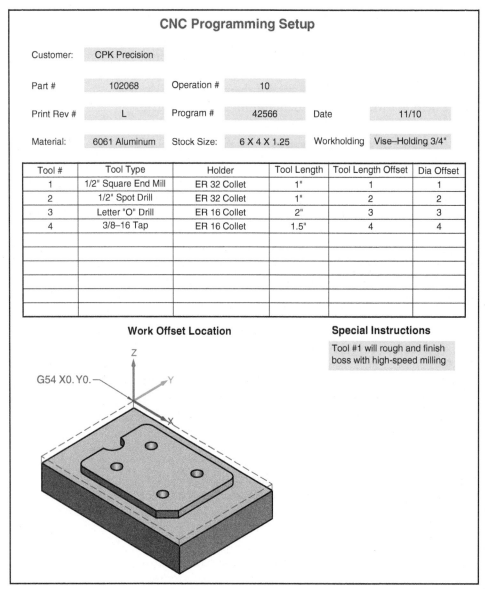

Goodheart-Willcox Publisher

Figure 12-3. A setup sheet generated in spreadsheet software.

format provides general information about the program, an illustration of the part orientation with the work coordinate system origin, an illustration of the part workholding, and related comments. Also included are the tooling data with toolholder information and a breakdown of cycle times by operation. Although more information and illustrations can be added, this format is generally considered to be fairly comprehensive.

Sometimes too much information is the right amount of information. For some jobs, the program documentation can be many pages long and can have an extensive amount of data, including explanations of each operation, tool lists, toolpath representations, and cycle times. See **Figure 12-6**. This is a comprehensive setup sheet report made up of four pages. Most CAM systems are capable of generating multiple views of parts, workholding, and toolpaths. The operation lists in **Figure 12-6** show toolpaths corresponding to each operation. The lists include information on speeds, feed rates, and final machining depths. This helps ensure that the setup person and operator have enough information for every step of the process. Do not be hesitant to add photos or in-depth explanations of each operation.

As previously discussed, setup sheets can be developed with CAM systems or other types of software. If these resources are not available, a simple layout sheet with photographs and tooling information will be helpful in setup. See **Figure 12-7**.

12.5 Efficiency in Production

By providing a means of communication between the programmer and machinist, setup sheets serve a key purpose in production. Sometimes, programmers assume that the setup person or operator can "figure out" where the program is going or what it is doing. Obviously, this can be a bad assumption. Making sure that the program is executed exactly as it was designed, and programmed, is critical to the success of the operation. The proper transfer of information from the programmer's desk to the machine shop floor can ensure an efficient setup and limited down time for the machine.

A more detailed setup sheet can improve efficiency by allowing the setup person to build the tool and holder assemblies before the machine has completed its current job. In addition, any workholding or fixturing devices can be prepared in advance. Working ahead keeps the machine making product for as much of the available time as possible. The period of transition from finishing one part, or operation, to the next part is a great opportunity to minimize machine down time. Depending on the work being machined, this can save several minutes or hours of total operation time.

Many new machine controllers have functions for displaying external files on the control screen. This allows the machinist to view setup instructions and tool lists directly at the machine during setup. This also enables the setup sheet to be stored with the program, eliminating the need for additional paper copies. This is a great advancement in technology and will become the standard in the future. No matter what method is in place for managing the program documentation, the importance of providing detailed setup information from the programming desk to the shop floor cannot be emphasized enough.

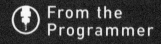

From the Programmer

Teamwork in Machining

The setup person is the programmer's best friend. The same is true of the machine operator. No matter how good the program might be, if it is not set up the way you intended or the operator has to change the program, the fault goes to the programmer. Work with your setup person and operator and make them part of the process. Machining involves teamwork, so work with your teammates to be as successful as you can. The more information the setup person and operator have and the more they are part of the planning process, the better the program will be and the more successful your company will become.

CNC Programming Setup Sheet

Tool Report

GENERIC HAAS 3 - AXIS VMC

TOOL LIST   Sorted: NO

TOOL INFO   1/2 FLAT END MILL

TYPE:	End mill 1 flat
NUMBER:	1
DIAMETER:	0.5
CORNER RADIUS:	0.0
LENGTH OFFSET:	1
DIAMETER OFFSET:	1
MATERIAL:	Carbide
NUMBER OF FLUTES:	4
FPT: 0.0094	SFM: 1047.1204
MFG CODE:	
ASSEMBLY:	
HOLDER:	B2C3-0020
TIME:	00:04:29

USED BY OPERATION: # 1   1 - 2D High Speed (2D Dynamic Mill)
USED BY OPERATION: # 2   2 - 2D High Speed (2D Dynamic Contour Mill)

TOOL INFO   1/2 SPOT DRILL

TYPE:	Spot drill
NUMBER:	2
DIAMETER:	0.5
CORNER RADIUS:	0.0
LENGTH OFFSET:	2
DIAMETER OFFSET:	2
MATERIAL:	HSS
NUMBER OF FLUTES:	2
FPT: 0.005	SFM: 600.0
MFG CODE:	
ASSEMBLY:	
HOLDER:	B2C4-0020
TIME:	00:00:03

USED BY OPERATION: # 3   3 - Drill/Counterbore

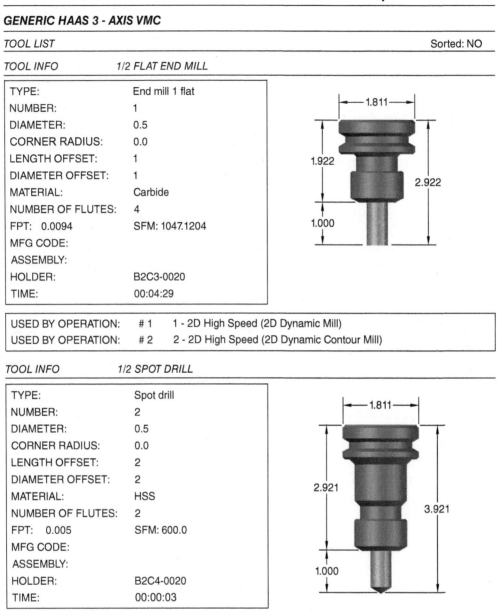

Continued

Figure 12-4. A setup sheet consisting of a tool list showing the details for each tool required.

TOOL INFO	LTR. O DRILL
TYPE:	Drill
NUMBER:	3
DIAMETER:	0.316
CORNER RADIUS:	0.0
LENGTH OFFSET:	3
DIAMETER OFFSET:	3
MATERIAL:	HSS
NUMBER OF FLUTES:	2
FPT: 0.0025	SFM: 499.9749
MFG CODE:	
ASSEMBLY:	
HOLDER:	B2C4-0020
TIME:	00:00:09

USED BY OPERATION:	# 4	4 - Drill/Counterbore

TOOL INFO	3/8-16 TAPRH
TYPE:	Tap RH
NUMBER:	4
DIAMETER:	0.375
CORNER RADIUS:	0.0
LENGTH OFFSET:	4
DIAMETER OFFSET:	4
MATERIAL:	Carbide
NUMBER OF FLUTES:	1
FPT: 0.0625	SFM: 49.0838
MFG CODE:	
ASSEMBLY:	
HOLDER:	B2C3-0016
TIME:	00:00:27

USED BY OPERATION:	# 5	5 - Tap

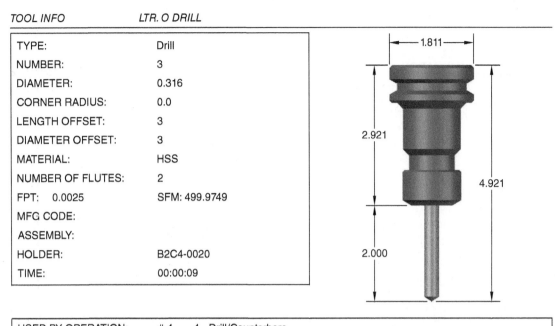

Goodheart-Willcox Publisher

Figure 12-4. *(Continued)*

GENERIC HAAS 3-AXIS VMC

CNC Programming Setup Sheet

GENERAL INFORMATION

PROJECT NAME:	CAM Sample Part	
CUSTOMER NAME:	CPK Precision Aerospace	
PROGRAMMER:	R. Calverley	
DRAWING:	102068	REVISION: L
DATE:	Wednesday, January 06	
TIME:	1:05 PM	

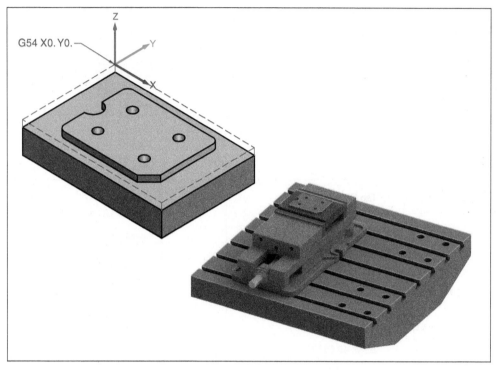

COMMENTS

Place raw stock in vise - hold 3/4" or less of material.

Continued

Figure 12-5. A setup sheet with an illustration of the part to be machined, the part workholding, and the tooling data.

OPERATION LIST

OP #	OPERATION NAME	TOOL #	MIN-Z	MAX-Z	CYCLE TIME
1	1 - 2D High Speed (2D Dynamic Mill)	1	–0.25	0.375	00:04:14
2	2 - 2D High Speed (2D Dynamic Contour Mill)	1	–0.25	0.25	00:00:15
3	3 - Drill/Counterbore	2	–0.2	2.0	00:00:03
4	4 - Drill/Counterbore	3	–0.845	2.0	00:00:09
5	5 - Tap	4	–0.7	2.0	00:00:27

TOOL LIST SORTED: ASCENDING

TYPE:	End mill 1 flat	DIAMETER:	0.5
MFG CODE:		CORNER RADIUS:	0.0
HOLDER:	B2C3-0020	TIP ANGLE:	NA
NUMBER:	1	FLUTE LENGTH:	0.5
LENGTH OFFSET:	1	OVERALL LENGTH:	1.0
DIAMETER OFFSET:	1	# OF FLUTES:	4

#1 - 0.5000 END MILL 1 FLAT - 1/2 FLAT END MILL

TYPE:	Spot drill	DIAMETER:	0.5
MFG CODE:		CORNER RADIUS:	0.0
HOLDER:	B2C4-0020	TIP ANGLE:	118.0
NUMBER:	2	FLUTE LENGTH:	0.5
LENGTH OFFSET:	2	OVERALL LENGTH:	1.0
DIAMETER OFFSET:	2	# OF FLUTES:	2

#2 - 0.5000 SPOT DRILL - 1/2 SPOT DRILL

TYPE:	Drill	DIAMETER:	0.316
MFG CODE:		CORNER RADIUS:	0.0
HOLDER:	B2C4-0020	TIP ANGLE:	118.0
NUMBER:	3	FLUTE LENGTH:	1.5
LENGTH OFFSET:	3	OVERALL LENGTH:	2.0
DIAMETER OFFSET:	3	# OF FLUTES:	2

#3 - 0.3160 DRILL - LTR. O DRILL

TYPE:	Tap RH	DIAMETER:	0.375
MFG CODE:		CORNER RADIUS:	0.0
HOLDER:	B2C3-0016	TIP ANGLE:	NA
NUMBER:	4	FLUTE LENGTH:	1.25
LENGTH OFFSET:	4	OVERALL LENGTH:	1.5
DIAMETER OFFSET:	4	# OF FLUTES:	1

#4 - 0.3750 X 16.00 TAP RH - 3/8–16 TAP RH

Goodheart-Willcox Publisher

Figure 12-5. *(Continued)*

GENERIC HAAS 3-AXIS VMC

GENERAL INFORMATION

PROJECT NAME:	CAM Sample Part
CUSTOMER NAME:	CPK Precision Aerospace
PROGRAMMER:	R. Calverley
DRAWING:	102068 REVISION: L
DATE:	Wednesday, January 06
TIME:	1:05 PM

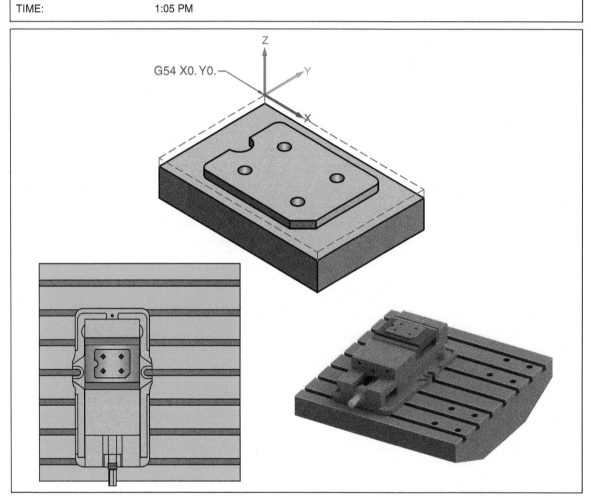

COMMENTS

Place raw stock in vise - hold 3/4" or less of material.

Continued

Figure 12-6. Setup sheet reports can be several pages long. The sheets in this report show toolpaths and cycle times for each operation.

OPERATION LIST

OPERATION INFO 1 - 2D High Speed (2D Dynamic Mill)

SPINDLE SPEED:	8000 RPM
FEED RATE:	300.0 inch/min
CLEARANCE PLANE:	0.0
RETRACT PLANE :	0.25
FEED PLANE:	0.1
DEPTH:	0.0
STOCK TO LEAVE:	0.01
COMP TO TIP:	YES
WORK OFFSET:	54

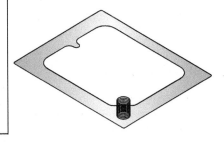

TOOL INFO 1/2 FLAT END MILL

TYPE:	End mill 1 flat
NUMBER:	1
DIAMETER:	0.5
LENGTH OFFSET:	1
DIAMETER OFFSET:	1
MATERIAL:	Carbide
NUMBER OF FLUTES:	4
FPT: 0.0094	SFM: 1047.1204
HOLDER:	B2C3-0020
TIME:	00:04:14

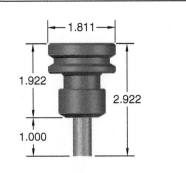

OPERATION INFO 2 - 2D High Speed (2D Dynamic Contour Mill)

SPINDLE SPEED:	3820 RPM
FEED RATE:	76.4 inch/min
CLEARANCE PLANE:	0.25
RETRACT PLANE :	0.1
FEED PLANE:	0.1
DEPTH:	0.0
STOCK TO LEAVE:	0.0
COMP TO TIP:	YES
WORK OFFSET:	54

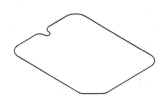

TOOL INFO 1/2 FLAT END MILL

TYPE:	End mill 1 flat
NUMBER:	1
DIAMETER:	0.5
LENGTH OFFSET:	1
DIAMETER OFFSET:	1
MATERIAL:	Carbide
NUMBER OF FLUTES:	4
FPT: 0.005	SFM: 500.0
HOLDER:	B2C3-0020
TIME:	00:00:15

Continued

OPERATION INFO 3 - Drill/Counterbore

SPINDLE SPEED:	4584 RPM
FEED RATE:	45.84 inch/min
CLEARANCE PLANE:	2.0
RETRACT PLANE	0.1
FEED PLANE:	0.1
DEPTH:	−0.2
STOCK TO LEAVE:	0.0
COMP TO TIP:	NO
WORK OFFSET:	54

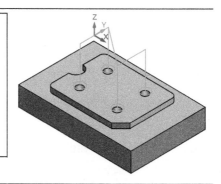

TOOL INFO   1/2 SPOT DRILL

TYPE:	Spot drill
NUMBER:	2
DIAMETER:	0.5
LENGTH OFFSET:	2
DIAMETER OFFSET:	2
MATERIAL:	HSS
NUMBER OF FLUTES:	2
FPT: 0.005	SFM: 600.0
HOLDER:	B2C4-0020
TIME:	00:00:03

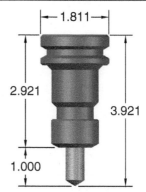

OPERATION INFO   4 - Drill/Counterbore

SPINDLE SPEED:	6044 RPM
FEED RATE:	30.0 inch/min
CLEARANCE PLANE:	2.0
RETRACT PLANE	0.1
FEED PLANE:	0.1
DEPTH:	−0.845
STOCK TO LEAVE:	0.0
COMP TO TIP:	NO
WORK OFFSET:	54

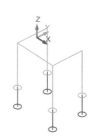

TOOL INFO   LTR. O DRILL

TYPE:	Drill
NUMBER:	3
DIAMETER:	0.316
LENGTH OFFSET:	3
DIAMETER OFFSET:	3
MATERIAL:	HSS
NUMBER OF FLUTES:	2
FPT: 0.0025	SFM: 499.9749
HOLDER:	B2C4-0020
TIME:	00:00:09

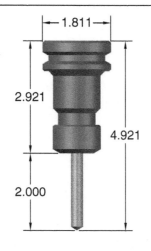

Continued

OPERATION INFO   5 - Tap

SPINDLE SPEED:	500 RPM
FEED RATE:	31.25 inch/min
CLEARANCE PLANE:	2.0
RETRACT PLANE :	0.1
FEED PLANE:	0.1
DEPTH:	−0.7
STOCK TO LEAVE:	0.0
COMP TO TIP:	NO
WORK OFFSET:	54

TOOL INFO   3/8–16 TAP RH

TYPE:	Tap RH
NUMBER:	4
DIAMETER:	0.375
LENGTH OFFSET:	4
DIAMETER OFFSET:	4
MATERIAL:	Carbide
NUMBER OF FLUTES:	1
FPT: 0.0625	SFM: 49.0838
HOLDER:	B2C3-0016
TIME:	00:00:27

Goodheart-Willcox Publisher

Figure 12-6. *(Continued)*

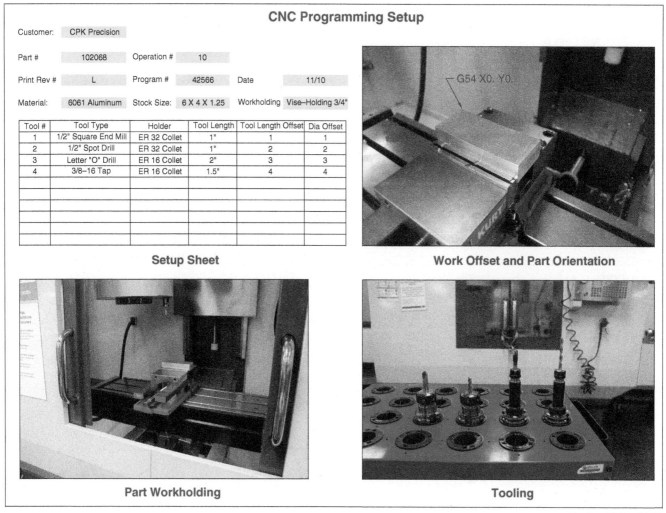

Figure 12-7. A setup sheet layout with tooling data and photographs illustrating the work offset location, part workholding, and tools.

Chapter 12 Review

Summary

- It is critical for the programmer to relay all pertinent information for a job to the setup person and machine operator.
- A setup sheet provides a complete description of the stock to be machined, the tooling required, and the type of workholding used.
- In machining, efficiency refers to the ability to successfully produce parts as rapidly as possible. Reductions in setup times, changeover times, and cycle times all have a positive impact on the total amount of time it takes to produce parts.
- Setup sheets vary in format. A typical setup sheet contains a header and sections detailing the rough stock, tooling, and workholding. The complexity and amount of detail on a setup sheet can vary, but it is critical to provide a complete explanation of the required processes to the setup person.
- Setup sheets can be created using a CAM program or standard word processing or spreadsheet software. The format of the setup sheet is not as important as the information contained on the sheet. Use a format that works for the machine shop's specific needs and add information as needed.
- Modern CNC machine controllers have the ability to display external files, including setup sheets. This greatly improves efficiency and allows the machinist to view setup instructions and tool lists directly at the machine during setup.

Review Questions

Answer the following questions using the information provided in this chapter.

Know and Understand

1. After the programming is completed, which professional is responsible for the next step in the manufacturing process?
   A. Operator
   B. Quality control inspector
   C. Setup person
   D. Engineer

2. The exchange of information between the setup person and the programmer should be accomplished by means of the _____ sheet.
   A. revision
   B. setup
   C. production
   D. data

3. In manufacturing, which of the following defines efficiency?
   A. Producing parts in as little time as possible.
   B. Minimizing unnecessary moves.
   C. Producing as many good parts as possible.
   D. All of the above.

4. Which of the following components is typically included in the header of the setup sheet?
   A. Special instructions
   B. Part revision number
   C. Toolholder dimensions
   D. Toolpath layout

5. Which of the following items is included in the tool data section of the setup sheet?
   A. Tool length offset
   B. Fixture identification
   C. Rough stock material
   D. Work offset position

6. *True or False?* Most computer-aided manufacturing (CAM) programs have the ability to create setup sheets.

7. What is a possible documentation format for a setup sheet?
   A. Spreadsheet document
   B. Word processing document
   C. CAM setup sheet
   D. All of the above.

Apply and Analyze

1. The following setup sheet is partially completed. Use the CNC mill program that follows to complete the setup sheet with the information that is missing.

CNC Programming Setup

Customer:			
Part #		Operation #	10
Print Rev #		Program #	Date: 11/10
Material:		Stock Size: 6 X 4 X 1.25	Workholding: Vise–Holding 3/4"

Tool #	Tool Type	Holder	Tool Length	Tool Length Offset	Dia Offset
1		ER 32 Collet	1"		
2		ER 32 Collet	2"		
3		ER 16 Collet	2"		

```
O2050 (PART #061599)
(PART REV – S)
(CUSTOMER: SEC PRECISION POG'S)
(MATERIAL 303 SST)
(T1|1/2 SPOT DRILL|H1|TOOL DIA. .5)
(T2|5/16 DRILL|H2|TOOL DIA. .3125)
(T3|3/8-16 TAP RH|H3|TOOL DIA. .375)
G20
G0 G17 G40 G49 G80 G90
T1 M6
G0 G90 G54 X.75 Y-.625 S1500 M3
G43 H1 Z2.
M8
```

```
G98 G81 Z-.188 R.1 F10.
X3. Y-2.5
G80
M5
G91 G28 Z0. M9
M01
T2 M6
G0 G90 G54 X.75 Y-.625 S1800 M3
G43 H2 Z2.
M8
G98 G73 Z-.844 R.1 Q.1 F10.
X3. Y-2.5
```

Continued

```
G80
M5
G91 G28 Z0. M9
M01
T3 M6
G0 G90 G54 X.75 Y-.625 S800 M3
G43 H3 Z2.
M8
G98 G84 Z-.625 R.1 F50.
X3. Y-2.5
G80
M5
G91 G28 Z0. M9
G28 Y0.
M30
```

2. The following setup sheet is partially completed. Use the drawing shown and the CNC mill program that follows to complete the setup sheet with the information that is missing.

CNC Programming Setup

Customer:

Part #   Operation #   10

Print Rev #   Program #   Date   11/10

Material: Stock Size: Workholding Vise–Holding 3/4"

Tool #	Tool Type	Holder	Tool Length	Tool Length Offset	Dia Offset
1		ER 32 Collet	1"		
2		ER 32 Collet	2"		
3		ER 16 Collet	2"		
4		ER 32 Collet	1.5"		
5		Boring Head	1.25"		

Copyright Goodheart-Willcox Co., Inc.

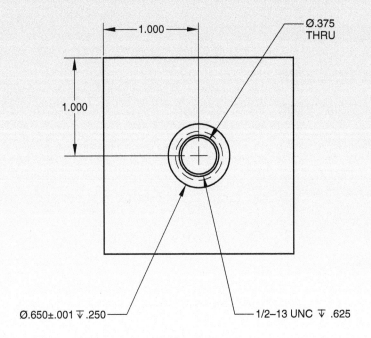
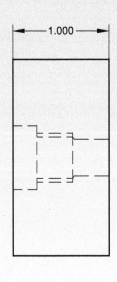

Ø.650±.001 ▽ .250   1/2–13 UNC ▽ .625

O0718 (PART #42566)	M5
(REV C PRINT)	G91 G28 Z0. M9
(CUSTOMER: CPK PRECISION)	M01
(MATERIAL 2 X 2 X 1 BRASS)	T3 M6
(T1|1/2 SPOT DRILL|H1)	G0 G90 G54 X1. Y–1. S2716 M3
(T2|3/8 DRILL|H2)	G43 H3 Z2.
(T3|27/64 DRILL|H3)	M8
(T4|1/2–13 TAP RH|H4)	G98 G81 Z–.738 R.1 F18.
(T5|.65 FINE BORING TOOL|H5)	G80
G20	M5
G0 G17 G40 G49 G80 G90	G91 G28 Z0. M9
T1 M6	M01
G0 G90 G54 X1. Y–1. S4584 M3	T4 M6
G43 H1 Z2.	G0 G90 G54 X1. Y–1. S500 M3
M8	G43 H4 Z2.
G98 G81 Z–.25 R.1 F45.8	M8
G80	G98 G84 Z–.625 R.1 F38.5
M5	G80
G91 G28 Z0. M9	M5
M01	G91 G28 Z0. M9
T2 M6	M01
G0 G90 G54 X1. Y–1. S3056 M3	T5 M6
G43 H2 Z2.	G0 G90 G54 X1. Y–1. S1538 M3
M8	G43 H5 Z2.
G98 G83 Z–1.113 R.1 Q.1 F18.	M8
G80	

Continued

```
G98 G89 Z-.25 R.1 P20 F7.7
G80
M5
G91 G28 Z0. M9
G28 Y0.
M30
```

Critical Thinking

1. Setup sheets are a key tool in the exchange of information between the programmer and machining personnel. By including important information such as the tooling required and the material type and size, setup sheets eliminate questions that may arise during setup. It is critical that the setup person and operator have all of the information needed to complete setup and run the machine. Explain what can happen if the setup sheet does not include a reference to the most current print revision for the part to be machined.

2. As discussed in this chapter, the format used for setup sheets can vary considerably. Setup sheets can range from a single page with notes to reports made up of multiple pages. If you currently work in a machine shop, is the process for developing setup sheets adequate, or are there often unanswered questions? What are some improvements that could be made to the documentation process? If you have not started your machining career yet, what type of setup sheet do you think is most adequate for the field? Explain your answer.

3. The period of time between machine operations when the machine is running presents opportunities to make processes more efficient. Other than preparing tool assemblies and workholding devices, what are some additional ways to improve efficiency and limit down time for the machine?

13 | Machine Setup

Chapter Outline

13.1 Introduction
13.2 Setting the Workholding
13.3 Establishing Work Coordinates
  13.3.1 Locating Work Coordinates on the Material
  13.3.2 Setting Work Coordinates on the Machine Control
13.4 Establishing Tool Length Offsets
13.5 Tool Diameter Offsets
13.6 Running the First Piece

Learning Objectives

After completing this chapter, you will be able to:

- List and explain the processes used in the machine setup operation.
- Describe different ways to establish coordinates for work offsets.
- Explain how to enter work offset coordinates on the machine controller.
- Describe the relationship between the tool length offset and the machine home position.
- Describe different ways the tool length offset can be established.
- Explain the effects of tool diameter offsets on the cutting path of the tool.
- Identify the correct tool diameter offset and input the setting into the machine controller.
- Explain safe methods used to initiate the first piece run after the setup is complete.

Key Terms

crash
dial indicator
edge finder
machine home

machine setup
tool length offset
work coordinate system (WCS)

work offset
zero return

13.1 Introduction

After planning and programming have been completed, and the setup sheet has been turned over to the setup person, the program moves to the shop floor and the physical setup of the machine begins. At this point in the process, the tools, workholding, and machining toolpaths have been determined. *Machine setup* refers to mounting the work in the correct position, setting the work coordinates for the part, and establishing tool length and diameter offsets.

A CNC machine, on its own, is only a machine, and is not capable of running and producing parts without input. A CNC machine has no idea what kinds of tools are being used, how long the tools protrude from the spindle, or where the material is placed in the machine. All of this information needs to be entered into the machine controller during the setup process. As you can imagine, this is critical information that must be verified when setting up and machining the first part. If tool offset settings are a little bit off, the first part might not meet print specifications and the settings will have to be adjusted. An additional part would then have to be made to bring the part dimensions into the required specifications. If the tool offsets are a lot off, the machine can have a catastrophic failure, or crash. A *crash* is an accident that occurs when the tool or spindle collides with the machine, material, or workholding. A machine crash can damage the material or workholding, or in the worst-case scenario, can disable the machine. A machine crash can also injure the operator or other personnel. Crashing a machine can cost thousands of dollars and can result in a week or more of downtime in production. Proper setup of work on the machine is critical to ensure that the machine operates safely and that parts are made accurately.

Efficiency is another key consideration in machine setup. There are a number of ways to improve the setup process to make it more efficient. By reducing setup times, the overall production time is reduced and more parts can be processed in a production run.

13.2 Setting the Workholding

Before work coordinates or tooling can be set, the part must be secured to the workholding on the CNC machine's work table. It is very rare in machining to clamp a piece of material directly to the table. It is possible, but normally a vise or fixture is attached to the work table and used to mount the workpiece. The workholding device must be aligned with the machine axes. See **Figure 13-1**. A CNC program is written assuming the part material has a specific orientation. Normally, the part orientation is aligned with the X and Y axes. In simpler terms, the material is *square* on the machine table.

The program for the part in **Figure 13-1** was written so that the part material is aligned with the X and Y axes, and the vise was placed in alignment with the X and Y axes of the machine. In the example shown, the toolpath is performing exactly as expected. But what would be the result if the CNC program was written with alignment to the X and Y axes, but the vise was not aligned correctly?

In the setup shown in **Figure 13-2**, the vise is misaligned and the part is not square to the X and Y axes of the machine. This poses an obvious problem with the toolpath. It is easy to see that the path of the tool (shown

Goodheart-Willcox Publisher

Figure 13-1. In machine setup, the part is mounted to the workholding. The part orientation is normally aligned with the X and Y axes of the machine. In this example, the workholding is secured correctly to the work table and the part is square on the table.

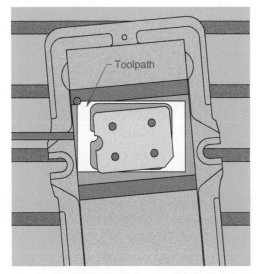

Goodheart-Willcox Publisher

Figure 13-2. A misaligned vise where the part is not square with the X and Y axes of the machine.

in white) is not following the limits of the part as desired. The vise in this example was rotated 5° to make the alignment issue obvious, but even a minor error in axis alignment will directly impact the dimensions of the finished workpiece.

Some machining fixtures are built with alignment holes where pins are placed through the fixture and into previously machined holes in the table or a subplate fixture. This ensures quick and accurate alignment. Using this type of fixture is a good example of how to make setup efficient.

>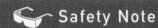
>
> **Safety Note**
>
> As discussed in Chapter 12, the setup sheet provides critical information about the part workholding, tooling, and stock to be used. The setup person must ensure that all requirements specified on the setup sheet are met.

Many fixtures will require the setup person to manually align the fixture. When a vise is being used, a *dial indicator* can be used to *sweep* across the fixed jaw of the vise to check for alignment. See **Figure 13-3**. This process includes touching the indicator to one edge of a jaw and traversing along the jaw. Any variation in indicator reading from one side to the opposite side constitutes misalignment.

More complicated fixtures might require different techniques to align. When fixtures are designed and manufactured, the ability to align the fixture to an axis is a critical consideration. More complicated fixtures might require some additional features to establish orientation. See **Figure 13-4**. The fixture shown has a circular base, offset pins to machine both sides of the part, and a top plate. This setup provides full access to the top surface of the part material for machining.

13.3 Establishing Work Coordinates

Now that the part material is aligned and secured in place, the work coordinates for the part need to be established. The machine does not know where the part is located on the table until it is defined by the work coordinate system. The *work coordinate system (WCS)* defines where the zero point of the part physically sits within the work envelope of the machine. Most programs are written with a work offset designation to establish the work coordinate system. A *work offset* is a coordinate setting that establishes the location of the work origin relative to the *machine home* position. The appropriate work offset command is used in the program to designate the work offset. For example, the G54 command is used to activate work coordinate offset 1 as the work offset. During setup, coordinates for the work offset are entered in the corresponding work offset setting in

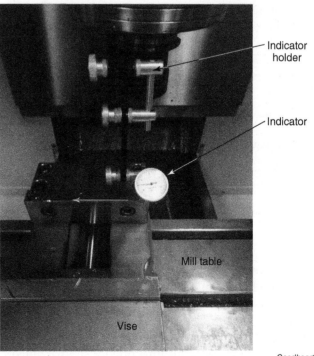

Goodheart-Willcox Publisher

Figure 13-3. A dial indicator is used to check for precise alignment of a vise by sweeping the tool across the fixed jaw of the vise.

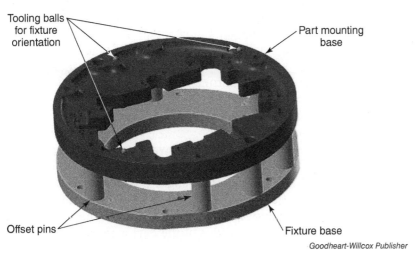

Figure 13-4. A fixture used to secure a part and provide full machining access to the top surface of the part.

the machine. Although there are more work offsets available, most programs are written with one of the G54–G59 work offset designations.

In the program, a specific point on the part is considered to be at the X0. Y0. origin location, also called the work origin, part zero, or program zero. The programmer must identify this location on the setup sheet, and the setup person must establish this position to tell the machine where the zero point is, physically, in the work envelope of the machine. See **Figure 13-5**. As discussed in Chapter 12, when a work offset is programmed, the designated work offset and physical location of the work origin are normally illustrated on the setup sheet.

When a machine is initially started up and returned to machine zero, it travels to its machine home position. This return is called a *zero return*. The machine zero position is defined as the machine coordinate X0. Y0. Z0. Now, the setup person must physically move the machine to locate program zero (part zero) on the part material as defined in the program. This zero point can be a corner on the part, the center of the part boundary, or the center of an existing feature such as a hole or boss.

13.3.1 Locating Work Coordinates on the Material

There are various ways to locate the work coordinate system on the part material. A common method is to bring an alignment tool into contact

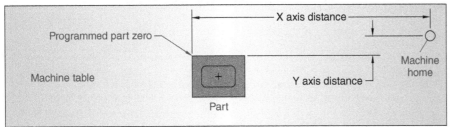

Figure 13-5. Part zero is the programmed location on the part defining the work origin relative to machine home. The distance from machine home along the X and Y axes establishes the part zero coordinate location.

> **From the Programmer**
>
> **Repeatable Workholding**
>
> In the program planning stages, ensure that the part workholding is repeatable. If the material can easily be placed back in the same position on the machine with each subsequent part, the work coordinate system only has to be set once.

with the sides of the part to establish the X and Y axis positions. This can be done manually with an edge finder, a dial indicator, or a solid post. The X and Y axis positions are found individually when using one of these methods. There is also special equipment available for locating work coordinates on the part material automatically. What is critical to understand is that the exact location of the machine will be recorded when the part edges are precisely located and the coordinates are entered for the work offset in the machine control.

Using an *edge finder* is a common way to locate a part edge. An edge finder is a cylindrical alignment tool with a shank and a magnetic tip. The tip can be moved off center with a slight push. With the spindle rotating at approximately 500 revolutions per minute (rpm), the tip wobbles as the material is slowly moved into position against the tip. When the tip is realigned with the shank, the edge of the part material is lined up with the edge of the tip. See **Figure 13-6**.

Figure 13-6A shows the edge finder locating the Y axis and **Figure 13-6B** shows the edge finder locating the X axis. It is important to note that the tip of the edge finder is .200″ in diameter. This means the centerline of the spindle is still .100″ from the edge of the part. Depending on where the edge finder is located in relation to the part, the coordinate location must be adjusted by .100″ or –.100″. This adjustment can also be made by raising the edge finder up to a safe location along the Z axis and then moving the machine an additional .100″ in the appropriate axis direction before recording the location. This brings the spindle centerline into alignment with the X or Y axis zero position on the part, depending on which axis is being located. This process is repeated to establish the other axis zero location.

If a probing system is available, the work coordinates can be set by contacting the probing tool against the sides of the part. This is a fast and

A B *Goodheart-Willcox Publisher*

Figure 13-6. An edge finder is used to locate the X and Y axis positions on a part. A—The tip of the edge finder is aligned with the long edge of the part. B—The tip of the edge finder is aligned with the adjacent edge of the part.

very accurate method for establishing the work coordinate system. Probing cycles are discussed in depth later in this textbook. Probing cycles reduce setup times significantly and make the setup process much more efficient. If the machine has a probing system, it is important to learn how to use it. See **Figure 13-7**.

13.3.2 Setting Work Coordinates on the Machine Control

The coordinates for the X and Y zero axis positions are entered into the machine control after they are established. Once the machine's spindle centerline is aligned directly over the programmed zero axis location, this location is recorded on the machine's *Work Offset* page. This process establishes the X and Y zero axis positions of the work offset relative to machine home. Depending on the machine controller, the name and location of the *Work Offset* page may vary. On a Haas machine controller, the *Work Zero Offset* page is used to enter coordinates for work offsets. See **Figure 13-8**.

The process of setting the work offset coordinates is similar on different machines. To set the X axis coordinate, use the cursor to highlight the X axis entry in the correct G54–G59 work offset row, and press the *Part Zero Set* button. See **Figure 13-9**. This records the machine position in the corresponding work offset setting. Repeat this process for the Y axis entry.

In **Figure 13-9**, notice there is also a work offset setting for the Z axis. For this text, the Z axis work offset setting will be left as 0. Placing a coordinate entry in the Z axis column will change the Z0. plane on which all of the tools are established. There are instances when tools can be set to alternate Z planes, and this is discussed briefly in the following section on tool length offsets. In those instances, a Z axis work offset setting will be used. Notice also in **Figure 13-9** that there are additional work offset settings other than the G54 setting. If necessary, these settings can be used for multiple parts requiring additional work coordinate systems. See **Figure 13-10**. This allows for manufacturing multiple parts with the same program by simply altering the designated work offset in the program.

Goodheart-Willcox Publisher

Figure 13-7. Using a probing system is an efficient way to set work coordinates.

<< WORK PROBE		WORK ZERO OFFSET		WORK PROBE >>
G CODE	X AXIS	Y AXIS	Z AXIS	
G52	0.	0.	0.	
G54	0.	0.	0.	
G55	0.	0.	0.	
G56	0.	0.	0.	
G57	0.	0.	0.	
G58	0.	0.	0.	
G59	0.	0.	0.	
G154 P1	0.	0.	0.	
G154 P2	0.	0.	0.	
G154 P3	0.	0.	0.	

Goodheart-Willcox Publisher

Figure 13-8. A work coordinate offset page on a Haas controller. Coordinates are entered in the X and Y axis columns next to the programmed work offset to establish the part zero location.

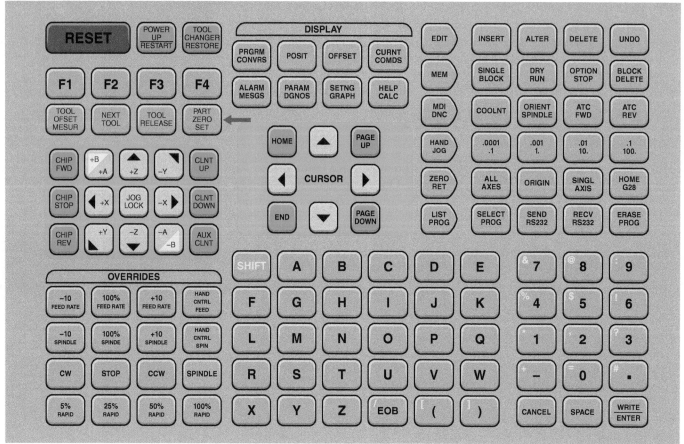

Figure 13-9. Pressing the *Part Zero Set* button records the machine position in a work offset setting.

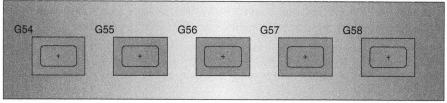

Figure 13-10. Multiple parts can be manufactured with the same program by using different work offsets in the program.

13.4 Establishing Tool Length Offsets

Tool length offsets, sometimes referred to as tool height offsets, are measurements used to compensate for different lengths of tools used in machining. The *tool length offset* represents the distance the tool travels from the machine home Z axis position to the work coordinate system Z0. position. Because each tool has a different length, the machine controller must be able to reference a measurement for the Z axis position of the tool relative to the machine home position. In machine setup, after the work offset coordinates are set, the tool length offsets are established and recorded. In the program, the G43 command is used with an H address to specify the tool length offset. Remember that a program is written with reference to a Z0. plane. When the tool is commanded to travel down to a Z–.500

location, that position is referencing the Z0. plane. Normally this plane is at the top of the stock, but it does not have to be at that location. Programmers can write programs where the bottom of the part is designated as the Z0. plane and all Z axis moves will be in the positive Z direction.

In **Figure 13-11**, the machine spindle is at the machine home location, also referred to as machine zero. As previously discussed, the machine has no idea how long the tool is, where the top plane of the material is located, or where the programmer designated the Z0. location. If the top of the part is the Z0. plane, the distance from the bottom of the tool to the top of the stock must be measured and recorded in the machine controller. See **Figure 13-12**.

There are a few different ways to obtain this measurement. Just as with the work offset measurement, once the machine is positioned at the correct location, the tool length offset measurement is recorded. During this measurement process, it is important to not engage the tool directly with the workpiece. Most tools are made from solid carbide, so engaging the tool with the material can cause the carbide to fracture.

A common way to measure the tool length offset is to *touch off* the tool while holding a piece of paper on the part material. However, this is the least accurate method. A piece of paper is about .003″ thick. If the paper is placed on the part material and the tool is carefully moved down along the Z axis until it traps the paper between the tool and material, without damaging the tool, then the tool is within .001″ to .002″ of the material. See **Figure 13-13**.

Once the tool is in this Z axis position, the tool length offset can be recorded on the machine's *Tool Offset* page. Depending on the machine controller, the name and location of this page may vary. Navigate to the offset page and highlight the tool number that corresponds to the tool length offset being used in the program. Then, press the *Tool Offset Measure* button

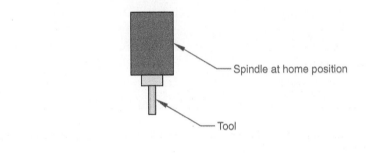

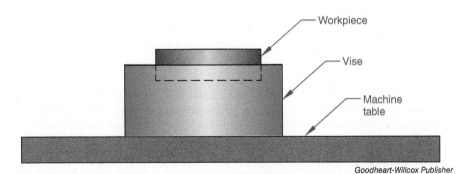

Goodheart-Willcox Publisher

Figure 13-11. During machine setup, the tool length offset must be established so that the machine controller can reference the Z axis position of the tool relative to machine home. In the current configuration, the spindle is at the machine home location.

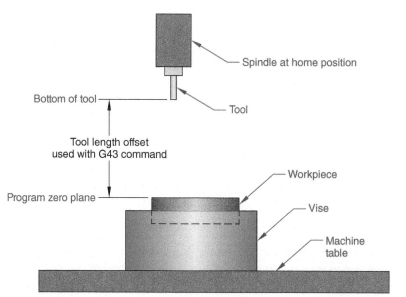

Figure 13-12. When the spindle is at machine home, the distance from the bottom of the tool to the program zero plane represents the tool length offset value. This distance must be measured during machine setup.

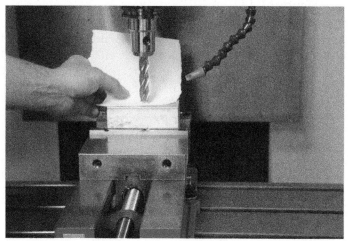

Figure 13-13. Touching off the tool with a piece of paper to establish the tool length offset.

(or the appropriate button) on the controller to record the offset setting, **Figure 13-14**. The resulting setting is shown in **Figure 13-15**. On a Haas machine controller, this setting is displayed in the Geometry column of the *Tool Offset* page. This is the distance from the bottom of the tool at machine home to the top of the part material. This is the setting used when the G43 command is programmed with the H1 tool offset designation to reference the offset for tool #1. In most cases, programmers will use the tool number as the same number for the H offset designation. The number of each tool and offset setting used in the program must be identified by the programmer on the setup sheet.

On the *Tool Offset* page, the Wear column is located to the right of the Geometry column. In this column, small adjustments to the tool length offset can be made. For example, a completed part is measured and it is

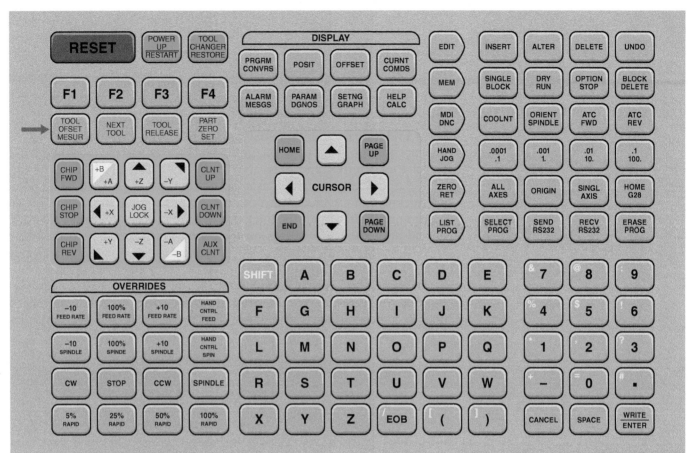

Figure 13-14. Pressing the *Tool Offset Measure* button records the tool length offset setting for the highlighted tool number on the tool offset page.

<< PROBING		TOOL OFFSET		TOOL INFO >>	
TOOL 1	COOLANT	H(LENGTH)		D(DIA)	
OFFSET	POSITION	GEOMETRY	WEAR	GEOMETRY	WEAR
1 SPINDLE	10	−4.5680	0.	0.	0.
2	0	0.	0.	0.	0.
3	0	0.	0.	0.	0.
4	0	0.	0.	0.	0.
5	0	0.	0.	0.	0.
6	0	0.	0.	0.	0.
7	0	0.	0.	0.	0.
8	0	0.	0.	0.	0.
9	0	0.	0.	0.	0.

Tool length offset setting

Figure 13-15. On a Haas machine controller, the *Tool Offset* page is where the tool length offset setting is stored.

determined that a feature was not cut deep enough. In inspecting the part, it is determined that the position of tool #1 is too high by .002″. In the Wear column, −.002 can be entered to adjust the offset by .002″. The Wear column allows small adjustments to the tool length offset without changing the total offset value in the Geometry column.

> **Safety Note**
>
> An improperly recorded tool length offset value can cause a severe machine crash. Setting an incorrect tool length offset is probably the most common way to cause a machine crash. If the offset is incorrect, the tool can rapidly move down into the material and damage the tool, spindle, and material. Take the time needed to ensure that the offsets are correct.

Another common way to measure the tool length, without using paper, is to touch off the tool using a reference block with a known thickness. This can be a gage block, a 1-2-3 block, or a block of a known size that is specifically designed for tool length measurements. The block is placed on top of the material and the tool is moved down slowly to just above the block. Then, the tool is jogged down in .001″ increments and the block is slid back and forth until there is a slight drag felt between the tool and material. See **Figure 13-16**. Be careful not to damage the tool during this process.

Once the tool is in position, go to the *Tool Offset* page, highlight the correct tool number, and press the *Tool Offset Measure* button. At this point, the tool length offset is established on top of the block being used. The thickness of the block must be subtracted from the tool length offset. To accomplish this, input the thickness of the block as a negative (–) number and press the *Enter* button on the controller. This will shift the offset down to the top of the material. While doing this, observe the offset value before and after the adjustment to ensure no entry errors were made.

The fastest and most accurate way to set tool offsets is to use a tool setter probe. See **Figure 13-17**. This probe is an electronic device that provides a calibrated, predetermined work plane. Tool probing is performed in various

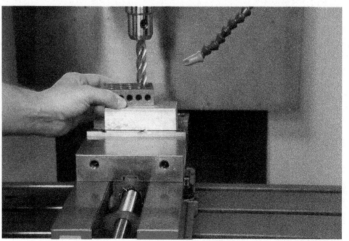

Goodheart-Willcox Publisher

Figure 13-16. Using a 1-2-3 block to measure tool length.

Goodheart-Willcox Publisher

Figure 13-17. Using a tool setter probe to establish the tool length offset setting.

ways on different machine and controller types. If your machine is equipped with a probing system and a tool setter probe, learn how it works and use it. Using probing cycles reduces setup time significantly. If you are using a tool setter probe, the top of the material must also be probed with a spindle probe. Probing functions are discussed in more detail later in this text.

The methods that have been described in this section are based on touching off tools on the part material to establish the tool length offset. Now that you have a clear understanding as to how tool offsets work, the following discussion presents another way to establish tool length offsets, as well as the positives and negatives of using this alternate approach. The practice of touching off all tools to the top surface of the stock makes sense because the top surface is a single plane where the programmed Z0. plane was established. The key to this process is that the tools are being set off on a single, flat plane inside the machine's work space. However, it is possible to select a different plane and have equally good results.

Consider the following example. All of the required tools can be touched off by using the machine table, a gage block, or any other plane. The additional step is to tell the machine the distance from the tool plane to the programmed Z0. plane, or the top of the stock. See **Figure 13-18**.

After all of the tools are touched off of the gage block, a Z height coordinate must be added to the work offset. Earlier in this chapter, the work offset was set with X and Y axis coordinates, but not a Z axis coordinate entry. The method shown in **Figure 13-18** is an example of where the Z axis work offset setting would be entered. See **Figure 13-19**. In the example shown in **Figure 13-18**, the Z axis setting would be a positive (+) value to establish a positive Z axis position. The distance value for this setting can be measured with an indicator.

What are the positives and negatives of this method? There really is only one negative—if the Z axis height of the work offset is not updated when different parts are machined, or the Z axis setting in the work offset is incorrect, a machine crash will happen. When using this method, it is essential to check and verify the Z axis work offset setting. The main positive impact of this method is that the tools do not have to be touched off again on new setups when new parts are machined. Imagine the following: What if the machine could hold 50 tools, and the tools remained in the machine carousel for every job. Instead of having to set up and touch off 20 or 30 tools for each job, the Z height from the gage block to the top of the part could be measured and the value added to the Z axis entry of the work offset. This can be a very efficient technique to use in work setup and part changeover.

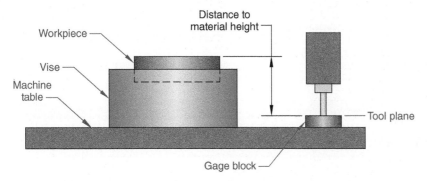

Goodheart-Willcox Publisher

Figure 13-18. Using a gage block to touch off tools to establish tool length offsets. The distance between the top of the gage block and the top of the stock must be used to adjust the Z axis setting of the work offset.

<< WORK PROBE		WORK ZERO OFFSET		WORK PROBE >>
G CODE	X AXIS	Y AXIS	Z AXIS	
G52	0.	0.	0.	
G54	0.	0.	0.	
G55	0.	0.	0.	
G56	0.	0.	0.	
G57	0.	0.	0.	
G58	0.	0.	0.	
G59	0.	0.	0.	
G154 P1	0.	0.	0.	
G154 P2	0.	0.	0.	
G154 P3	0.	0.	0.	

Z axis work offset setting G54

Goodheart-Willcox Publisher

Figure 13-19. A value for the Z axis setting is entered for the work offset when touching off tools on a plane other than the top surface of the stock.

13.5 Tool Diameter Offsets

As discussed in Chapter 8, tool diameter offset settings are referenced by the controller when cutter compensation is used in a program. Cutter compensation is activated with the G41 and G42 commands. Cutter compensation simplifies programming in contouring operations and gives the operator the ability to make small adjustments to the path that the side of the tool is cutting. During setup, the diameter of the cutting tool is entered on the *Tool Offset* page of the controller. Before this is discussed in more detail, the following three programs are presented to show the different ways in which cutter compensation is used in programming. These are contour milling programs used in machining a rectangular boss feature for a basic part.

Program O0001 does not use cutter compensation. The toolpath is programmed with point-to-point programming to allow an extra .250″ around the boss feature.

```
O0001
(No Comp)
(MATERIAL ALUMINUM 6061)
(T1|1/2 EM|H1)
G20
G0 G17 G40 G49 G80 G90
T1 M6
G0 G90 G54 X-.275 Y.275 S4584 M3
G43 H1 Z2.
M8
Z.2
G1 Z-.5 F30.
X.1 Y-.1 F55.
```

Continued

```
X3.9
Y-2.9
X.1
Y.15
Y.525
G0 Z2.
M5
G91 G28 Z0. M9
G28 X0. Y0.
M30
```

Program O0002 uses left-hand cutter compensation. The path around the boss feature is programmed from the dimensions on the print representing each point on the path. In this case, the full diameter of the cutting tool will be entered on the *Tool Offset* page of the controller.

```
O0002
(Control)
(MATERIAL ALUMINUM 6061)
(T1|1/2 FLAT EM|H1|D1)
G20
G0 G17 G40 G49 G80 G90
T1 M6
G0 G90 G54 X-.275 Y.275 S4584 M3
G43 H1 Z2.
M8
Z.2
G1 Z-.5 F30.
G41 D1 X.35 Y-.35 F55.
X3.65
Y-2.65
X.35
Y.15
G40 Y.275
G0 Z2.
M5
G91 G28 Z0. M9
G28 X0. Y0.
M30
```

Program O0003 looks almost exactly like Program O0001, except it is using cutter compensation. But in this case, while cutter compensation is used, the diameter of the cutting tool is set at zero (0). This technique is often used when creating programs with computer-aided manufacturing (CAM) software. In this case, the toolpath calculated by the software automatically compensates for the diameter of the cutting tool.

```
O0003
(Computer)
(MATERIAL ALUMINUM 6061)
(T1|1/2 FLAT EM|H1|D1)
G20
G0 G17 G40 G49 G80 G90
T1 M6
G0 G90 G54 X-.275 Y.275 S4584 M3
G43 H1 Z2.
M8
Z.2
G1 Z-.5 F30.
G41 D1 X.1 Y-.1 F55.
X3.9
Y-2.9
X.1
Y.15
G40 Y.525
G0 Z2.
M5
G91 G28 Z0. M9
G28 X0. Y0.
M30
```

After looking at all three programs, it might be surprising to learn that they all produce the same boss feature. The programmed part has a simple block shape and a .500″ raised boss on the top surface. See **Figure 13-20**. **Figure 13-20A** shows the finished part, **Figure 13-20B** shows the toolpath created in Program O0001, and **Figure 13-20C** shows the toolpath created in Program O0002. The difference in the programs relates to how cutter compensation is used.

The reason to consider all three programs is to determine *if* an offset amount needs to be entered on the *Tool Offset* page, and *what* amount needs to be entered. This is where the setup sheet should help.

To enter the tool diameter offset for use with cutter compensation, navigate to the *Tool Offset* page on the controller. The tool diameter offset is entered in the same row as the corresponding number of the tool that is used in the program. See **Figure 13-21**. This is the number used for the D address in the program when cutter compensation is activated. This number does not have to be the same number as the tool number, but it usually will correspond. In **Figure 13-21**, the diameter value .500 is entered for use with tool #1 in Program O0002. On a Haas machine controller, this setting is displayed in the Geometry column under D(DIA) on the *Tool Offset* page.

For the part in this example, Program O0001 and Program O0003 will have zero (0) entered in the Geometry diameter offset column. In Program O0002, a positive value will be entered in the Geometry diameter offset column. On older machines, the controller only allowed for tool radius amounts in this column. In this case, the offset would read .250. On most newer machine controllers, the diameter of the tool is used, with an option to change the setting to a radius value if desired. For a 1/2″ end mill, the diameter offset entry is .500. The diameter offset setting allows for a tool

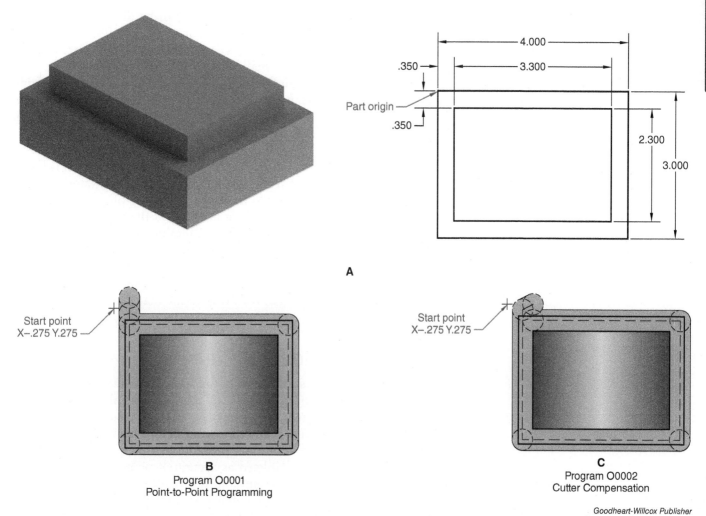

Figure 13-20. Programming a contouring operation for a boss feature. A—An isometric view and dimensioned top view of the finished part. B—The toolpath programmed without cutter compensation. C—The toolpath programmed with cutter compensation.

Goodheart-Willcox Publisher

Figure 13-21. The diameter offset value for use with cutter compensation is entered in the same row as the corresponding number of the tool that is used in the program. This setting is referenced by the corresponding D offset number used in the program.

to be measured, and then that measurement can be directly entered into the machine controller. Just as with the Geometry tool offset column, the Geometry diameter offset column will have the initial tool size, and the Wear column will be used for small incremental changes to that value.

When using cutter compensation, consider exactly how the setting might affect the final part. In the example shown in **Figure 13-22**, the tool diameter offset used is .500 and the tool is cutting on the left-hand side of the part. When the offset is made smaller, the controller thinks the tool is smaller, so it positions the tool further into the material, or cuts additional material. If the offset is made larger, the controller thinks the tool is larger and positions the tool away from the material, or cuts less material.

If the tool is traveling around more than one side of the part, adjusting the tool diameter offset will affect all sides of the part, **Figure 13-23**. It is important to consider this with programming and entering offsets at the machine. This is true with straight cuts and when machining holes. An increase in the tool diameter offset, or positive offset, will make the controller back away from the cut. A decrease in the tool diameter offset, or negative offset, will make the tool move closer to the material, **Figure 13-24**.

13.6 Running the First Piece

At this point in the process, the work setup has been completed. The setup person has secured the part, built all of the tool assemblies, established the work coordinate system, and recorded the tool length and diameter offsets. The first part is now ready to be run. Before the cycle start button is pushed, there are some additional precautions that should be taken to make sure the setup was completed as accurately as possible. In order to verify the accuracy of the program and the work setup, it is essential to test the part program safely on the machine. Making sure the program is running as it was designed is critical to proper and safe part completion.

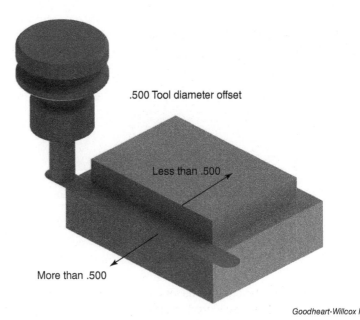

Goodheart-Willcox Publisher

Figure 13-22. Adjusting the tool diameter offset setting changes the position of the tool to cut more or less material. Using a smaller offset value results in cutting additional material. Using a larger offset value results in cutting less material.

Goodheart-Willcox Publisher

Figure 13-23. The tool diameter offset setting affects the entire cutting path in a contouring operation.

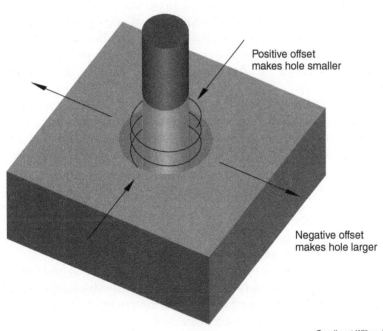

Goodheart-Willcox Publisher

Figure 13-24. Increasing the tool diameter offset setting (positive offset) results in cutting less material when machining a hole with an end mill. Decreasing the setting (negative offset) results in cutting more material.

Earlier in this chapter, the work offset was set with the Z axis offset left at zero (0). See **Figure 13-25.** The Z axis offset can be used to test drive the program *in the air* without cutting the material. This is done by manually setting the Z axis offset to a positive value so that the work plane is shifted to a safe distance above the programmed Z0. plane. In order to do this, the program needs to be analyzed. Review the program and determine the deepest move made by the cutting tool along the Z axis. The Z axis offset value to use in testing the program must be greater than the absolute value of the deepest Z axis move in the program.

<< WORK PROBE		WORK ZERO OFFSET		WORK PROBE >>	
G CODE	X AXIS	Y AXIS	Z AXIS		
G52	0.	0.	0.		
G54	0.	0.	0.		
G55	0.	0.	0.		
G56	0.	0.	0.		
G57	0.	0.	0.		
G58	0.	0.	0.		
G59	0.	0.	0.		
G154 P1	0.	0.	0.		
G154 P2	0.	0.	0.		
G154 P3	0.	0.	0.		

Z axis work offset setting G54

Goodheart-Willcox Publisher

Figure 13-25. The Z axis work offset entry can be adjusted to a positive value in order to test drive the program. The value entered must be sufficient to shift the work plane to a safe distance above the programmed Z0. plane.

For example, if the deepest Z axis move in the program is Z–2.25, a safe Z axis offset value for testing the program is +2.5. Enter this value in the Z axis offset column on the machine's *Work Offset* page. This will allow the program to run at a sufficient distance above the programmed Z0. plane without the tool making contact with the part material. After the program is verified, the Z axis offset can be reset to 0, or the original offset value.

After verifying that the machine runs as expected, the machine operator is ready to cut material. However, there is one more safety measure to take when cutting the first piece. Before machining the first part, consider reducing the rapid move rate. This will slow down the rapid moves in the program for safety while machining the first part. Every controller has the ability to reduce rapid move rates, **Figure 13-26**. Machining the first part at 25% rapid will allow the operator time to react to an unexpected machine move. If the tool is extremely close to the part, the machine can be run at 5% rapid. This will slow the machine down further and allow the machinist to ensure operation is safe.

Once the program is verified, and the first part is machined and inspected, production is ready to run. As previously discussed, repeatability of the setup is critical to efficiency. No matter how good the program may be, the setup must be both accurate and efficient. A good setup person and machine operator are always looking for opportunities to improve the total machining time in producing parts.

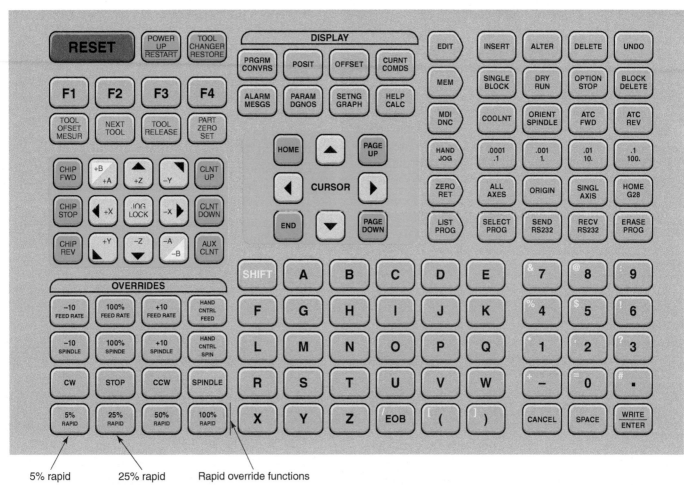

Figure 13-26. Before machining the first part, the rapid move rate can be adjusted to a slower speed by using the appropriate rapid override function on the controller.

Chapter 13 Review

Summary

- Machine setup is a vital step in the machining process. Setup involves mounting the work in the correct position, setting the work coordinates for the part, and establishing tool length and diameter offsets.
- The part must be properly oriented and secured on the machine using the required workholding. Proper alignment will allow the program to function as designed.
- The work coordinate system defines the location of the part within the work envelope of the machine.
- Most programs are written with a work offset designation to establish the work coordinate system. A work offset is a coordinate setting that establishes the location of the work origin relative to the machine home position. The work origin, also called part zero or program zero, is the X0. Y0. location on the part.
- Coordinates for the work coordinate system can be located manually with an edge finder, a dial indicator, or a solid post. If a probing system is available, work coordinates can be set by using a probing tool. This is a fast and very accurate method for establishing the work coordinate system.
- Coordinates for a work offset are entered on the appropriate offset page on the controller.
- The tool length offset represents the distance the tool travels from the machine home Z axis position to the work coordinate system Z0. position. Often, the tool plane is the top surface of the part material, but alternate tool planes can be used.
- The tool length offset can be set by manually moving the tool down to the top of the material and recording that distance for the corresponding tool number on the machine's tool offset page.
- Tool diameter offsets are used in conjunction with cutter compensation. Cutter compensation simplifies programming and gives the operator the ability to make small adjustments to the toolpath.
- The tool diameter offset is entered for the corresponding tool number on the machine's tool offset page. Using a larger offset value results in cutting less material. Using a smaller offset value results in cutting additional material.
- Establishing the correct work coordinates and tool offsets is critical in producing an acceptable part and preventing a machine crash.
- It is essential to safely test the part program on the machine before production. Setting the Z axis work offset value to a positive value sufficiently above the programmed Z0. plane will allow the program to run above the material at a safe distance.
- Reducing the rapid move speed rate and test running the program can help identify any setup or programming errors.

Review Questions

Answer the following questions using the information provided in this chapter.

Know and Understand

1. What is a task that needs to be completed during machine setup?
   A. Write the program.
   B. Set tool length offsets.
   C. Identify toolpaths.
   D. Identify tools to be used.

2. *True or False?* It is just as important to be efficient with the setup as it is with the program and operation.

3. One way to properly align a vise on a CNC mill is to _____.
   A. eyeball it
   B. make it look straight
   C. use a dial indicator
   D. use a tool setter probe

4. What is an acceptable tool to locate the work coordinate system position?
   A. Probe
   B. Edge finder
   C. Solid stop
   D. All of the above.

5. On which offset page is the work offset recorded?
   A. Work Zero Offset
   B. Tool Offset
   C. Diameter Offset
   D. Geometry Offset

6. Once the work offset coordinates are established, which button is pushed to record the work offset location on the controller?
   A. Work Coordinate button
   B. Part Zero Set button
   C. Tool Offset button
   D. Offset button

7. *True or False?* Only one work offset can be used in a program.

8. The tool length offset is the distance the tool travels from machine home to the ____.
   A. fixture
   B. workholding
   C. tool plane
   D. table

9. *True or False?* The tool plane position must be established on the programmed Z0. plane.

10. On which offset page are the tool length offsets stored?
    A. Work Zero Offset
    B. Tool Offset
    C. Diameter Offset
    D. Geometry Offset

11. When using cutter compensation, the diameter of the tool is stored in the ____ column on the corresponding offset page.
    A. Geometry tool offset
    B. Wear tool offset
    C. Geometry diameter offset
    D. Wear diameter offset

12. *True or False?* The tool diameter offset can be a diameter or radius setting depending on the machine controller settings.

13. Increasing the tool diameter offset value will signal the controller to ____.
    A. cut less material
    B. cut more material
    C. cut the same amount of material
    D. position the tool closer to the material

14. *True or False?* When machining a hole with cutter compensation activated, setting the tool diameter offset to a smaller value will make the hole smaller.

15. What is a safety measure to use before running the first part?
    A. Increase the tool diameter offset.
    B. Reduce the tool diameter offset.
    C. Increase the rapid rate.
    D. Reduce the rapid rate.

Apply and Analyze

Use the illustration below to answer Questions 1 and 2.

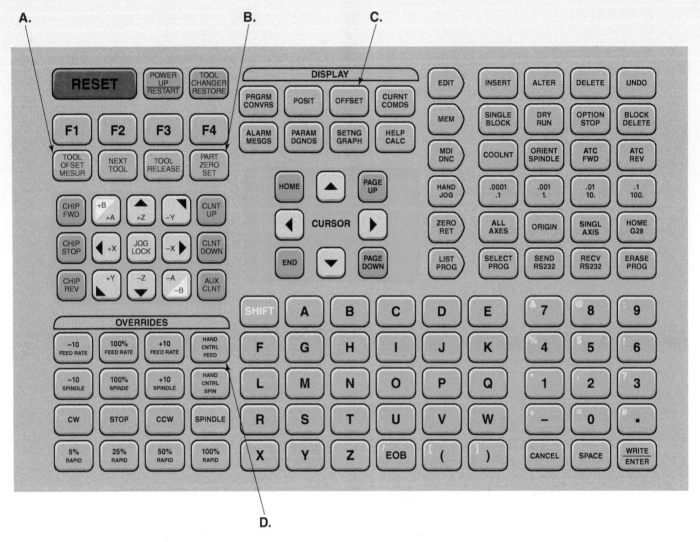

1. Which button is pushed on the controller to set the tool length offset when the tool is at the appropriate Z axis position?
   A. Tool Offset Measure button
   B. Part Zero Set button
   C. Offset button
   D. Hand Control Feed button

2. When setting coordinates for the work offset, which button is pushed on the controller?
   A. Tool Offset Measure button
   B. Part Zero Set button
   C. Offset button
   D. Hand Control Feed button

3. Identify the names of the terms labeled A, B, and C in the following work setup.

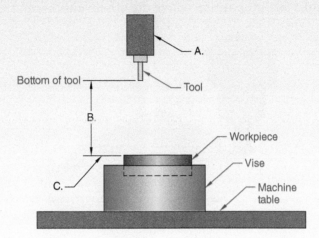

4. After the first piece is run, it is determined the tool length offset for tool #3 is .002″ too high. On the offset page shown, identify the labeled position where the offset value is adjusted.

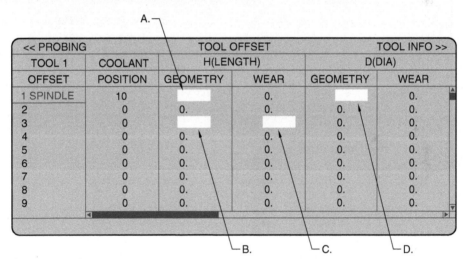

5. After machining the part contour with tool #4, it is determined that the feature is .002″ oversize. On the offset page shown, identify the labeled position where the offset value is adjusted.

Critical Thinking

1. Proper planning is an essential part of machine setup and operation. To successfully produce a part, considerable thought and planning must happen well before the process begins. Consider a project you may have at home—perhaps repairing a vehicle, assembling shelves, or building a workbench. Write out an explanation of what is needed to start the project. What supplies or tools are needed? What are some of the steps to take to make sure the project is completed safely?

2. Setting the workholding or fixturing for a part to be machined requires precision and care. The part workholding must be in alignment and secure through the entire machining process. What are some of the possible failures that can happen with the tooling, the part, or the machine if the workholding is not in alignment or secure?

3. As discussed in this chapter, it is essential to test the part program for accuracy and safety before production begins. Safety should always be the primary goal in any work environment. Running the program in the air and slowing down rapid move rates are two ways to protect the part material and the machine as it is being run. What safety measures can be taken to protect the machine operator? Make a list of personal protective equipment (PPE) that machinists can use to protect themselves.

Section 2

CNC Lathe Programming

- 14 Machining Mathematics for Turning
- 15 Cartesian Coordinate System and Machine Axes for Turning
- 16 Preparatory Commands: Lathe G-Codes
- 17 Miscellaneous Functions and Address Codes for Lathe Programming
- 18 Lathe Program Planning
- 19 Lathe Contour Programming
- 20 Hole Machining on a Lathe
- 21 Programming Grooves and Parting Off Operations
- 22 Threading
- 23 Live Tooling
- 24 Lathe Setup

Pixel B/Shutterstock.com

14 Machining Mathematics for Turning

Chapter Outline

14.1 Introduction
14.2 Converting Fractions to Decimals
    14.2.1 Communicating Precision
14.3 Geometric Shapes
    14.3.1 Circles
    14.3.2 Triangles
    14.3.3 Lines
14.4 Angles
    14.4.1 Supplementary Angles
    14.4.2 Complementary Angles
14.5 Trigonometry
14.6 Tapers
14.7 Thread Measurements
    14.7.1 Major Diameter and Minor Diameter
    14.7.2 Thread Depth and Pitch
14.8 Turning Speeds and Feeds

Objectives

After completing this chapter, you will be able to:

- Convert fractions to decimals.
- Explain how required precision is communicated in machining.
- Identify two-dimensional and three-dimensional geometric shapes encountered in machining.
- Name the parts of a right triangle.
- Use the Pythagorean theorem to solve for sides of triangles.
- Differentiate between supplementary and complementary angles.
- Use basic trigonometry to solve for sides and angles in a right triangle.
- Explain how to calculate tapers.
- Describe the basic elements of a thread and explain how to make calculations for thread pitch and thread depth.
- Calculate speeds and feeds in turning operations.

Key Terms

acute angle	isosceles triangle	round
angle	lathe	scalene triangle
arc	line	sine
chamfer	line segment	straight angle
circle	major diameter	supplementary angles
complementary angles	minor diameter	tangent
cosine	nominal size	tangent line
curved line	numerator	taper
denominator	obtuse angle	thread depth
diameter	pitch	triangle
equilateral triangle	Pythagorean theorem	trigonometry
fillet	radius	turning
geometric shape	right angle	
hypotenuse	right triangle	

14.1 Introduction

Math is essential in the machining and manufacturing trades. The math principles and applications covered in this chapter are limited to the basics needed for creating CNC lathe programs. Machinists must be able to make calculations based on information provided on prints. This chapter explains how to calculate linear distances and angles and how to determine locations of features for turning operations. *Turning* is a machining operation that removes material from a rotating workpiece. This chapter also covers trigonometric functions and making calculations for tapers, threads, and cutting speeds and feeds.

Most prints used in manufacturing in the United States are dimensioned in decimal inches or fractional inches. The inch is the basic unit of linear measurement in the US Customary system. All measurements that you will encounter in this text are based on this system. In most countries outside the United States, prints are dimensioned using the International System of Units (abbreviated SI), known as the metric system. Metric prints used in the manufacturing industry are commonly dimensioned in millimeters. Most CNC machines can read programs prepared in inches or millimeters. As is the case with mill programs, the correct mode for working units must be set when writing a lathe program, as discussed in Chapter 16.

When making unit conversions and other machining calculations, a calculator is often useful. This text assumes access to a simple scientific calculator, such as the one shown in **Figure 14-1**. However, any calculator with a fraction key and sine, cosine, and tangent functions is sufficient to perform machining calculations. Consider the calculator as a tool, just like a file or wrench, to support your work in the shop.

In this text, green buttons are used to display exactly which buttons on the calculator to press and in which order. The prompt will look like this:

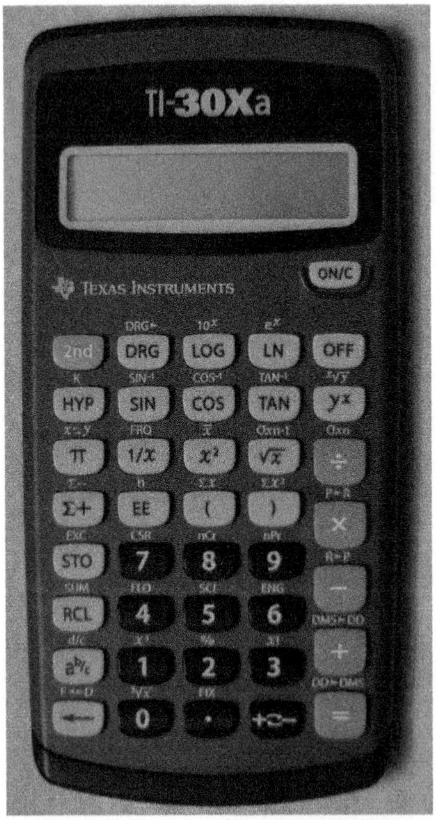

Figure 14-1. A scientific calculator is a useful tool for completing mathematical calculations.

14.2 Converting Fractions to Decimals

Most prints used in manufacturing today are dimensioned in decimal inches. However, you will also encounter prints that use fractions to specify dimensions. It is common to see fractional dimensions on older prints and drawings of simple parts. The use of fractions normally signifies a noncritical measurement or a feature that is not designed to meet a high degree of precision in manufacturing. See **Figure 14-2**.

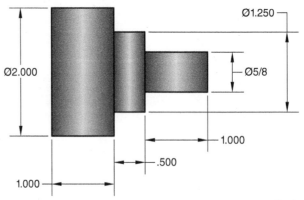

Figure 14-2. A basic part with dimensions in decimal inches and fractional inches.

Fractions cannot be used in CNC programs, however. Instead, they must be converted to decimal values. Knowing how to convert fractions is essential for machinists. Having this knowledge is also important when making measurements because many measuring tools make readings in decimal units.

Fractions have a numerator and a denominator. The *numerator* is the number on top of the fraction and indicates the number of parts in the fraction. The *denominator* is the number on the bottom of the fraction. It indicates the whole number quantity into which the parts are divided. For example, in the fraction 5/8, the numerator is 5 and the denominator is 8.

The process of converting fractions to decimals is a straightforward division problem. To convert a fraction into a decimal, divide the denominator into the numerator. The result is a decimal.

Take a look at this sample problem:

$$\frac{5}{8} = 8\overline{)5} \text{ or } 5 \div 8$$

$$\frac{5}{8} = .625$$

The process is even easier using a calculator. Use the following sequence:

$$5 \div 8 =$$

14.2.1 Communicating Precision

The answer to the problem in the previous example is .625. Mathematically, it would also be correct to express the answer as .6250. The trailing zero after the decimal point does not hold any value. Similarly, any leading zero on the leftmost side of the decimal point does not hold any value.

$$00.625 = 0.625 = .625 = 0.625000 = 0000.6250000$$

All of these numbers are equal. However, machinists have a specific way of communicating. To prevent confusion, machinists *talk* or *work* to three places or four places to the right of the decimal point, depending on the required precision. The third place to the right of the decimal point, for example, is the thousandths place:

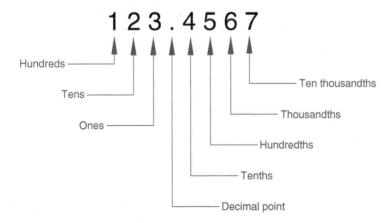

As an example of how machinists communicate, a print dimension of .750 is read out as *seven hundred fifty thousandths* (.750). If you were measuring a drill bit that was .5 in diameter, you would say *five hundred thousandths* (.500). This system provides an efficient and effective method of communication.

A similar method is used when the required precision is in ten thousandths of an inch. As shown previously, the fourth place to the right of the decimal point is the ten thousandths place. This place is referred to by machinists using the abbreviation *tenths*. For example, a print dimension of .0005 is read out as *five tenths*. In actual measurement, the referenced dimension is five ten-thousandths of an inch.

14.3 Geometric Shapes

Geometric shapes include two-dimensional objects, such as circles, arcs, and triangles; and three-dimensional objects, such as spheres, cylinders, and cones. See **Figure 14-3**. Two-dimensional geometric shapes are flat. Imagine drawing a shape on a piece of paper. You can draw left and right or up and down, but only in two dimensions. A three-dimensional shape has depth. Consider the difference between a circle and a sphere. A circle is two-dimensional, and spheres are three-dimensional. You should know how to recognize basic geometric shapes and how the geometry of a part can be evaluated to determine dimensional information.

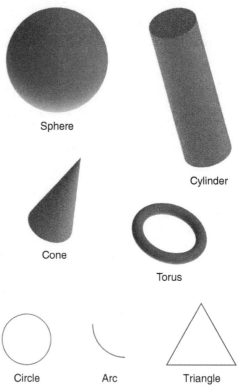

Goodheart-Willcox Publisher

Figure 14-3. Assorted three-dimensional and two-dimensional geometric shapes.

14.3.1 Circles

Circles are two-dimensional objects often encountered in machining. Circular shapes are the most common shapes on parts produced in lathe operations. For example, circular shapes are produced in drilling holes and cutting round shafts. A *circle* is defined as a closed plane curve that is an equal distance at all points from its center point. A circle can be thought of as a closed loop with a center point. If the distance from that center point to every point on the outside is the same, the closed loop is a circle, **Figure 14-4**. An *arc* is a portion of a circle. An arc has a start point, end point, center point, and defined direction (clockwise or counterclockwise). Many shapes encountered in machining are formed by arcs. For example, a 90° round corner blend represents one-quarter of a circle.

The distance from the center point to the edge of a circle or arc is defined as the *radius*. The radius is frequently used in creating CNC programs and calculating locations. The distance across a circle through the center point is the *diameter*. The radius of a circle is one-half the diameter. See **Figure 14-5**. Hole sizes and full circular profiles are often defined on a print in terms of diameter, whereas corner blends and other partial circles are defined by radius. In machining, an external radial edge is called a *round*. An internal radial edge is called a *fillet*. When you encounter a circular feature on a print, pay close attention to specifications and verify whether the dimension defines a radius or a diameter.

Circular motion is the underlying principle behind the operation of a lathe. A *lathe* is a metal cutting machine that holds and rotates the workpiece while the cutting tool removes material. The tool can be fixed in place or driven through the material. Because the workpiece is rotating, the machine produces round parts. Understanding the difference between radius and diameter and how related features are defined is critical in lathe programming, **Figure 14-6**.

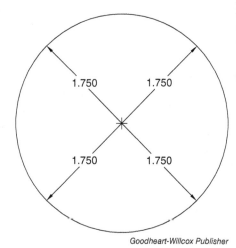

Goodheart-Willcox Publisher

Figure 14-4. A circle is a closed plane curve defined by a center point that is equidistant from all points on the curve.

14.3.2 Triangles

Triangles are particularly important shapes in machining applications. A *triangle* is made up of three sides that form three angles. There are several types of triangles that you will commonly encounter in machining work.

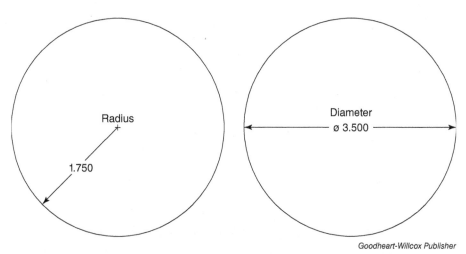

Goodheart-Willcox Publisher

Figure 14-5. The radius of a circle is the distance from the center point to the edge of the circle. The diameter is the distance from edge to edge through the center point. The radius is one-half the diameter.

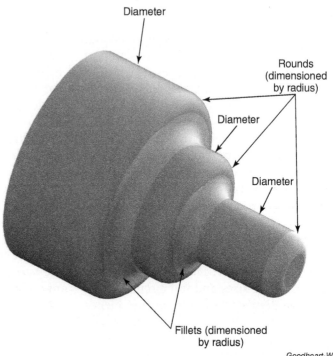

Figure 14-6. Partial circular features on a part are defined by radius. Full circular profiles are defined by diameter. In machining, external rounded edges are called rounds. Internal rounded edges are called fillets.

The following types are defined by the relationship between the sides and angles.

- A *scalene triangle*, **Figure 14-7,** has no equal sides or equal angles.
- An *isosceles triangle*, **Figure 14-8,** has two equal sides and two equal angles.
- An *equilateral triangle*, **Figure 14-9,** has all equal sides and all equal angles.
- A *right triangle*, **Figure 14-10,** has one 90° angle. A 90° angle is defined as two lines that are exactly perpendicular to each other. On a print, a right triangle is designated by a square symbol in the 90° corner of the triangle. The longest side in a right triangle, called the *hypotenuse*, is always across from the 90° angle. The right triangle has special mathematical properties that make it advantageous for use in machining.

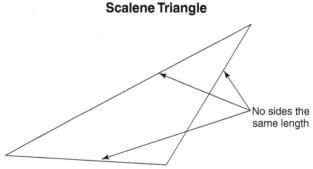

Figure 14-7. A scalene triangle has no equal sides or angles.

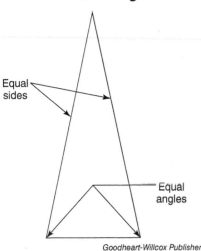

Figure 14-8. An isosceles triangle has two equal sides and two equal angles.

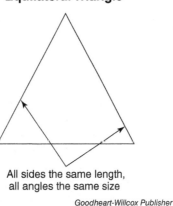

Figure 14-9. An equilateral triangle has all equal sides and all equal angles.

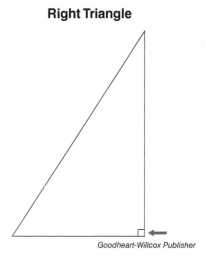

Figure 14-10. A right triangle, denoted by a right angle symbol.

All triangles share the same mathematical property: the sum of all three angles must equal 180°.

$$180° = \text{Angle 1} + \text{Angle 2} + \text{Angle 3}$$

The ability to calculate angles without all of the angles defined is often necessary in print reading. Using the previous formula, you can calculate any angle of a triangular shape if you know the other two angles. If you have a triangle where you know one angle is 45° and another angle is 55°, calculate the third angle as follows:

$$180° - \text{Angle 1} - \text{Angle 2} = \text{Angle 3}$$
$$180° - 45° - 55° = \text{Angle 3}$$
$$80° = \text{Angle 3}$$

Right triangles have a unique characteristic defined by the Pythagorean theorem. The **Pythagorean theorem** states that, in a right triangle, the square of one side plus the square of the second side is equal to the square of the hypotenuse, **Figure 14-11**:

$$a^2 + b^2 = c^2$$

Consider the following example. In the triangle shown in **Figure 14-11**, side a measures 2″ and side b measures 1.5″. What is the length of side c?

$$a^2 + b^2 = c^2$$
$$(2″)^2 + (1.5″)^2 = c^2$$
$$4″ + 2.25″ = c^2$$
$$c = \sqrt{6.25} = 2.5″$$

To solve this problem on your calculator, press the following keys:

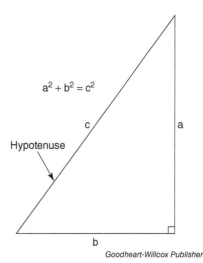

Figure 14-11. The sides of a right triangle can be calculated using the Pythagorean theorem.

The last input key takes the square root. This last step changes c^2 to the desired solution, *c* or 2.5″.

As shown in these examples, using standard algebra, you can calculate any side of a right triangle if you know the other two sides. Additional mathematical methods for working with right triangles and finding all of their sides and angles are covered later in this chapter.

14.3.3 Lines

Lines are important geometric features in machining. With the exception of commands for circular moves, most CNC machining commands are programmed along a line. A ***line*** is a continuous, straight, one-dimensional geometric element with no end. This means a line is infinite. It is impossible to machine an infinitely long part. When the term *line* is used in machining, it actually refers to a line segment. A ***line segment*** is a line with a definitive beginning and end. Line segments, like lines, do not curve or waver. A ***curved line*** is an arc, partial circle, or spline.

A line that touches an arc or circle at exactly one point is referred to as a ***tangent line.*** If a line crosses or touches an arc at more than one point, it is not tangent, but intersecting, **Figure 14-12.**

Goodheart-Willcox Publisher

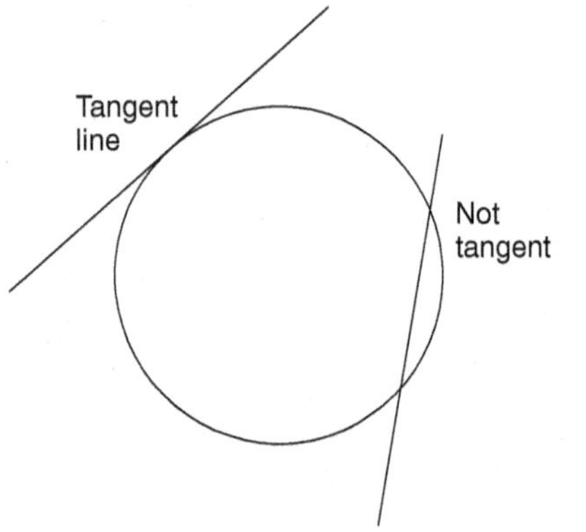

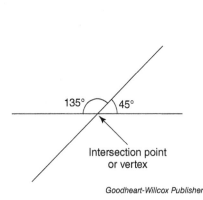

Figure 14-12. A circle displaying a tangent and nontangent (intersecting) line.

14.4 Angles

Most machining requires a good working knowledge of angles. An ***angle*** measures the rotational distance between two intersecting lines or line segments from their point of intersection (or vertex). Angles are usually given in degrees, **Figure 14-13.**

There are 360° in a circle (one degree is 1/360 of a circle). A straight line passing through the center point of a circle divides the circle in half and forms two semicircles, each representing 180°. See **Figure 14-14.**

There are four common types of angles:

- A ***right angle*** measures exactly 90° between two intersecting lines. When two lines intersect and form a right angle, **Figure 14-15,** all other angles of intersection are also 90°. Notice that the sum of the angles is 360°.

Goodheart-Willcox Publisher

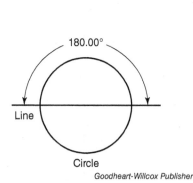

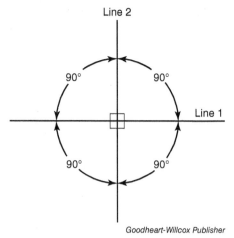

Figure 14-13. Two lines intersecting at known angles.

Goodheart-Willcox Publisher

Figure 14-14. A circle contains 360°. A line passing through the center point of the circle divides the circle in half and forms two semicircles. Each semicircle is a 180° arc.

Goodheart-Willcox Publisher

Figure 14-15. Two lines intersecting at 90° angles.

- An *acute angle* measures less than 90°. Anything less than a right angle is an acute angle, **Figure 14-16**.
- An *obtuse angle* measures more than 90°, **Figure 14-17**. Notice that the second leg of the angle intersects the first at an angle greater than a right angle.
- A *straight angle* measures exactly 180°. A straight line represents a straight angle. Refer to the horizontal line shown in **Figure 14-14**.

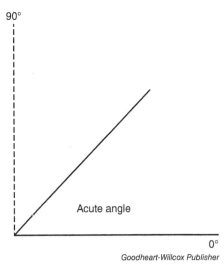

Figure 14-16. An acute angle is one that is less than 90°.

14.4.1 Supplementary Angles

Supplementary angles are two angles that add up to 180°. Recall that a straight angle is 180°. When one straight line intersects another at an angle, the angles on the opposite sides of the vertex are *supplementary*. If one angle is known, the other can be calculated. For example, if one line intersects another at 50°, calculate the second angle as follows. See **Figure 14-18**.

$$50° + x = 180°$$
$$x = 180° - 50°$$
$$x = 130°$$

The prints used in a normal machining environment do not define every feature. In fact, it is incorrect to overdimension an object on a print. Often, one angle is defined, and you must calculate the second or supplementary angle.

14.4.2 Complementary Angles

Complementary angles are two angles that add up to 90°. Problems involving complementary angles normally start with a 90° angle that has been intersected by a line. You can find the missing angle that completes the 90° angle if the other angle is known, **Figure 14-19**.

In this example, the given angle is 25°. What is the complementary angle?

$$90° = 25° + x$$
$$90° - 25° = x$$
$$x = 65°$$

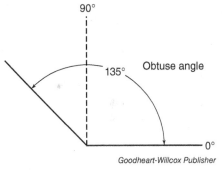

Figure 14-17. An obtuse angle is one that is more than 90°.

In **Figure 14-20**, the right angle symbol in the lower-left corner indicates a 90° angle. If the known angle is 34°, what is the complementary angle?

$$90° = 34° + x$$
$$90° - 34° = x$$
$$x = 56°$$

As you move forward with print reading, you will see prints that have right angles or right triangles. Often, the print may define only one angle. You can calculate unknown angles using the principles previously discussed as well as basic trigonometry.

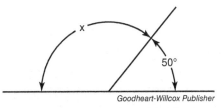

Figure 14-18. When one supplementary angle is known, the other supplementary angle can be calculated by subtracting the known angle from 180°.

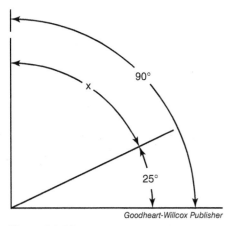

Figure 14-19. An unknown angle can be calculated by subtracting its known complementary angle from 90°.

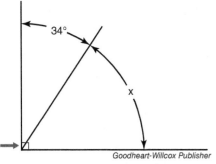

Figure 14-20. A complementary angle is easily calculated when a right angle is given (as shown by the right angle symbol) and one of the complementary angles is known.

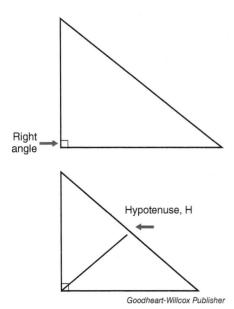

Figure 14-21. The right angle symbol denotes a right angle. The side directly across from the 90° angle in a right triangle is the hypotenuse.

14.5 Trigonometry

Trigonometry is a branch of mathematics that deals with the relationships between the sides and angles of triangles and with functions of angles. Essentially, trigonometry is the mathematics of triangles and angles. This text focuses on right triangles.

The three fundamental principles for working with right triangles are:

- Right triangles always have one angle of 90°.
- The sum of all three angles in a triangle is 180°.
- The Pythagorean theorem states that $a^2 + b^2 = c^2$.

Start by reviewing the parts of right triangles, **Figure 14-21**. Immediately, you should identify each triangle as a right triangle because it has the right angle symbol. The hypotenuse, or longest side, is directly across from the right angle and is designated by the letter H.

The other two sides are named based on their relation to a given angle or for the angle for which you are trying to solve. They are called the *opposite side* and *adjacent side*, indicated by the letters O and A, respectively, **Figure 14-22**.

The mathematical relationships between sides and angles are given by three trigonometric functions:

- The *sine* of a given angle is the ratio of the opposite side to the hypotenuse.
- The *cosine* of a given angle is the ratio of the adjacent side to the hypotenuse.
- The *tangent* of a given angle is the ratio of the opposite side to the adjacent side.

"SOHCAHTOA" is a helpful mnemonic for remembering the definitions of sine, cosine, and tangent and the mathematical calculations to solve for right triangle sides and angles.

$$S = \text{sine}$$
$$O = \text{opposite}$$
$$H = \text{hypotenuse}$$
$$C = \text{cosine}$$
$$A = \text{adjacent}$$
$$H = \text{hypotenuse}$$
$$T = \text{tangent}$$
$$O = \text{opposite}$$
$$A = \text{adjacent}$$

$$\text{Sine (sin)} = \frac{\text{Opposite (O)}}{\text{Hypotenuse (H)}}$$

$$\text{Cosine (cos)} = \frac{\text{Adjacent (A)}}{\text{Hypotenuse (H)}}$$

$$\text{Tangent (tan)} = \frac{\text{Opposite (O)}}{\text{Adjacent (A)}}$$

Trigonometric functions can be used to determine all of the sides and angles of a right triangle. Sine, cosine, and tangent calculations can be made using reference tables or a scientific calculator. Although reference tables are helpful tools, the probability of error increases with manual calculations. The quickest and most accurate calculations are made with a calculator. First, identify what sides and angles you know and what you are trying to find. Then, choose the appropriate formula and use the correct calculator inputs.

The following is an example of finding the length of a side in a right triangle when one angle and one side are known. The first step is to identify the sides. The side opposite the right angle is always the hypotenuse. The side opposite the known angle is the *opposite side*. The side touching the known angle is the *adjacent side*. In **Figure 14-23**, the length of the hypotenuse and one angle are known. To calculate the opposite side, use the sine function. The known angle (55°) is represented as *a* in the formula:

$$\sin a = \frac{\text{Opposite (O)}}{\text{Hypotenuse (H)}}$$

$$\sin 55 \times \text{Hypotenuse (H)} = \text{Opposite (O)}$$

$$.8192 \times 2.179 = O$$

$$1.785'' = O$$

To solve this problem using a calculator, press the following keys:

| 55 | SIN | × | 2.179 | = |

The adjacent side in **Figure 14-23** can be calculated using the cosine function. The known angle (55°) is represented as *a* in the formula:

$$\cos a = \frac{\text{Adjacent (A)}}{\text{Hypotenuse (H)}}$$

$$\cos 55 \times \text{Hypotenuse (H)} = \text{Adjacent (A)}$$

$$.5736 \times 2.179 = A$$

$$1.250'' = A$$

To solve this problem using a calculator, press the following keys:

| 55 | COS | × | 2.179 | = |

The following summarizes the calculator inputs used for sine, cosine, and tangent formulas when solving for one side of a right triangle. These formulas are used when one angle and one side are known. The first two

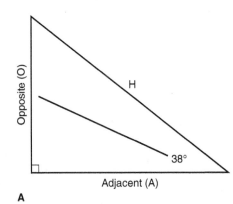

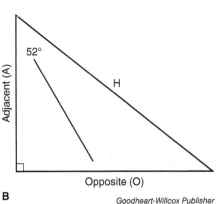

Figure 14-22. The opposite and adjacent sides of a right triangle are determined relative to the known angle. A—This right triangle has one angle defined as 38°. The side directly opposite that angle is known as the opposite side. The side of the triangle that is next to or touching the 38° angle is called the adjacent side. B—This is the same triangle, but the known angle is now 52° and the opposite and adjacent sides change positions.

From the Programmer

Trigonometric Calculations

On a scientific calculator, the sine of a given angle is calculated by entering the angle and then pressing the SIN key. The COS and TAN keys are used in the same manner to calculate cosine and tangent.

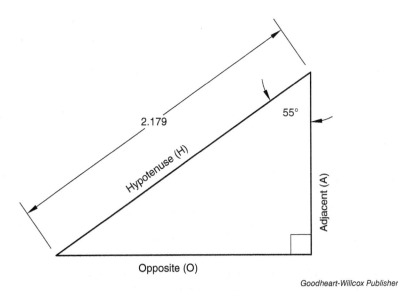

Figure 14-23. A right triangle with one known side and one known angle.

formulas were used in the previous examples. Use the appropriate input based on the side and angle you know.

1. Hyp × Angl SIN = (Opp)
2. Hyp × Angl COS = (Adj)
3. Adj × Angl TAN = (Opp)
4. Opp ÷ Angl SIN = (Hyp)
5. Adj ÷ Angl COS = (Hyp)
6. Opp ÷ Angl TAN = (Adj)

Other trigonometric functions are used when you know two sides of a right triangle and need to find an angle. The inverse sine, cosine, and tangent functions are used to find an unknown angle relative to two known sides. On a typical calculator, the inverse trigonometric functions are labeled SIN^{-1}, COS^{-1}, and TAN^{-1}. These are typically second functions located on or above the SIN, COS, and TAN keys. Second functions on a calculator are accessed by pressing the second function key before pressing the appropriate key. The second function key may be color coded and is typically labeled 2nd or Shift.

The inverse trigonometric functions are also known as arcsine, arc-cosine, and arctangent. In the following formulas, –1 denotes an inverse function. The angle to be calculated is represented as a.

$$\text{Angle } a = \sin^{-1}\left(\frac{\text{Opposite (O)}}{\text{Hypotenuse (H)}}\right)$$

$$\text{Angle } a = \cos^{-1}\left(\frac{\text{Adjacent (A)}}{\text{Hypotenuse (H)}}\right)$$

$$\text{Angle } a = \tan^{-1}\left(\frac{\text{Opposite (O)}}{\text{Adjacent (A)}}\right)$$

The first step in solving for an unknown angle in a right triangle is to identify the two known sides. In **Figure 14-24**, there are two known sides. Since the known sides are the opposite and adjacent sides, the inverse tangent function is used to solve for the unknown angle:

$$\text{Angle } a = \tan^{-1}\left(\frac{\text{Opposite (O)}}{\text{Adjacent (A)}}\right)$$

$$\text{Angle } a = \tan^{-1}\left(\frac{1.750}{2.500}\right)$$

$$\text{Angle } a = \tan^{-1}(0.7)$$

$$\text{Angle } a = 34.992°$$

To solve this problem using a calculator, press the following keys:

1.75 ÷ 2.5 = 2nd Tan

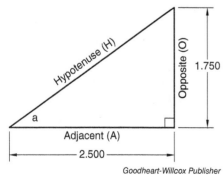

Figure 14-24. A right triangle with two known sides and an unknown angle to solve.

The following summarizes the calculator inputs used for inverse sine, cosine, and tangent formulas when solving for an unknown angle. These formulas are used when two sides are known. Use the appropriate input based on the sides you know.

1. Opp ÷ Hyp = 2ⁿᵈ SIN

2. Adj ÷ Hyp = 2ⁿᵈ COS

3. Opp ÷ Adj = 2ⁿᵈ TAN

As previously discussed in this chapter, the Pythagorean theorem can be used to calculate a side of a right triangle when you know two sides. First, identify the sides you know. Then, select one of the following formulas and calculate the unknown side. Use the appropriate calculator inputs based on the given information for the triangle.

1. Adj x^2 + Opp x^2 = $\sqrt{x}$ (Hyp)

2. Hyp x^2 − Opp x^2 = $\sqrt{x}$ (Adj)

3. Hyp x^2 − Adj x^2 = $\sqrt{x}$ (Opp)

In **Figure 14-25**, two sides are known and the hypotenuse must be calculated. When considering how to label the two known sides, it is irrelevant which side is defined as opposite or adjacent. To understand why, look at the Pythagorean theorem equation:

$$a^2 + b^2 = c^2$$

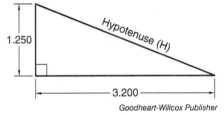

Goodheart-Willcox Publisher

Figure 14-25. A right triangle with two known sides and one unknown side to solve. The unknown side is the hypotenuse.

Side c is always the longest side, or the hypotenuse. The hypotenuse must always be identified in this manner. In comparison, sides a and b can have an interchangeable order in the equation. Just as 6 + 3 = 9 and 3 + 6 = 9, the designation of opposite and adjacent will not affect the final answer. Since the unknown side in this example is the hypotenuse, use the following calculator formula and key entry.

Calculator formula:

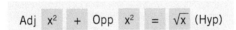

Key entry:

1.250 x^2 + 3.200 x^2 = $\sqrt{x}$

Solution:

$$1.250^2 + 3.200^2 = 3.435$$

Figure 14-26 shows a practical example of determining distances from dimensions on a print. The part view shown is a profile of a cylindrical part produced in a lathe contouring operation. On this part, there is an angled feature dimensioned as .250 × 45°. This feature is a *chamfer*. A chamfer is a beveled edge machined to relieve a sharp corner. Because the chamfer angle is 45° and a .250″ linear dimension is given, both sides are .250″ from the corner. Figure 14-27 shows an enlarged view of the feature with a right triangle established to represent the sides. Because this is a right triangle and one of the angles is 45°, the other angle is 45°. Both legs of the triangle are equal, so each side measures .250″. This example shows how a right triangle can be used to establish relationships between angles and sides.

Figure 14-28 shows an example of an angled feature on the right end of a part where the angle is not 45°. In this case, a right triangle can be established with one angle and one side known. The unknown side can be calculated using trigonometric functions. Figure 14-29 shows an enlarged view of the feature with a right triangle established to represent the sides. One side measures .260″ and one of the angles is calculated as a complementary angle from the known 60° angle. The complementary angle is 30°. With this information, the tangent function can be used to calculate the opposite side.

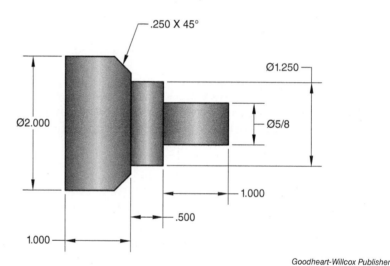

Figure 14-26. A cylindrical part with a 45° chamfer.

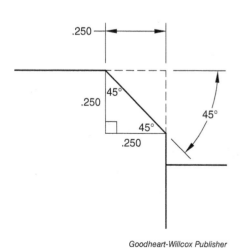

Figure 14-27. A 45° chamfer has equal length sides. In this case, each side is .250″.

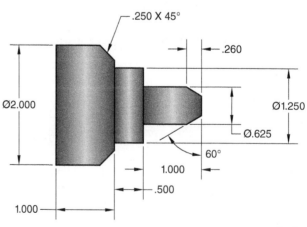

Figure 14-28. A cylindrical part with a 60° angled feature.

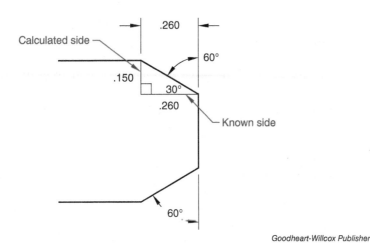

Goodheart-Willcox Publisher

Figure 14-29. A theoretical right triangle used to calculate the unknown side of an angled feature. The known side is .260″. The 30° angle is the complement of the known 60° angle.

$$\tan a = \frac{\text{Opposite (O)}}{\text{Adjacent (A)}}$$

$$\tan 30 \times \text{Adj (A)} = \text{Opp (O)}$$
$$.5774 \times .260 = \text{Opp (O)}$$
$$.150″ = \text{Opp (O)}$$

To solve this problem using a calculator, use the following formula and key entry.

Calculator formula:

Adj × Angl TAN = (Opp)

Key entry:

.260 × 30 TAN =

Solution:

.260 × .5774 = .150″

Programming methods for angular moves in lathe programs are discussed later in this textbook.

14.6 Tapers

Tapered diameters are common features on parts produced in lathe operations. A *taper* refers to a diameter that changes in size uniformly along its length. A taper has a conical shape. On a print, tapers can be dimensioned in different ways. Often, one of the taper dimensions must be calculated from the given information for programming purposes. With an understanding of right triangles and the trigonometric functions previously discussed, this can be a straightforward calculation.

Figure 14-30 shows a tapered part with the small diameter, length, and angle of the taper dimensioned. The large diameter at the end is unknown and is needed to create the lathe program. Because one end diameter and the angle are known, the large diameter is calculable. In **Figure 14-31**, a right triangle is drawn to solve for one of the side lengths. This establishes one known angle and the length of the adjacent side. The side to be solved is the opposite side. With this information, the tangent function is used to calculate the opposite side. To solve this problem using a calculator, use the following formula and key entry.

Calculator formula:

Key entry:

Solution:

$$3.000 \times .0524 = .157''$$

The opposite side of the triangle is .157″. But the calculation is not complete because this only establishes part of one side in the total calculation. To calculate the large diameter, add .157″ to both sides of the diameter and add the resulting sum to the 1.000″ small diameter, **Figure 14-32**.

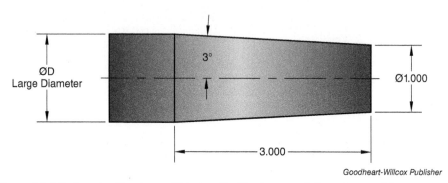

Goodheart-Willcox Publisher

Figure 14-30. A part with a taper 3″ in length. The large end diameter must be calculated for programming.

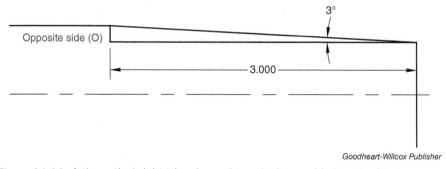

Goodheart-Willcox Publisher

Figure 14-31. A theoretical right triangle used to calculate a side length with the tangent function. The known angle is 3° and the adjacent side is 3.00″.

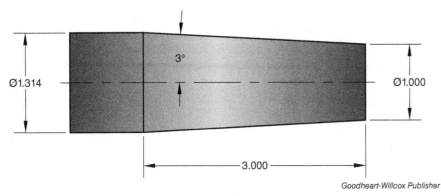

Figure 14-32. The calculated large diameter is 1.314″.

$$1.000 + .157 + .157 = 1.314$$

Another way a taper can be dimensioned is by specifying the angle and the small and large diameters. See **Figure 14-33**. On this part, the length of the taper needs to be calculated for programming.

For this example, the small diameter is subtracted from the large diameter and the result is divided in half to establish the opposite side of a right triangle. See **Figure 14-34**.

$$1.144 - .750 = .394$$
$$.394 \div 2 = .197$$

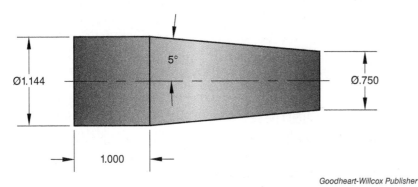

Figure 14-33. A part with a taper dimensioned by specifying the angle and the end diameters. The length of the taper must be calculated for programming.

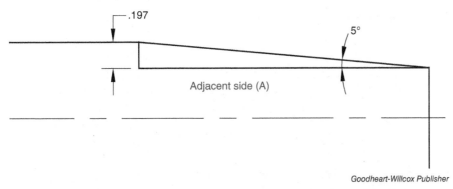

Figure 14-34. A theoretical right triangle used to calculate a side length with the tangent function. The known angle is 5° and the opposite side is .197″. The .197″ measurement is calculated by subtracting the small diameter from the large diameter and dividing the difference in half.

The angle and opposite side are known. The adjacent side is calculated using the tangent function. This establishes the length of the taper. To solve this problem using a calculator, use the following formula and key entry.

Calculator formula:

Opp ÷ Angl TAN = (Adj)

Key entry:

.197 ÷ 5 TAN =

Solution:

.197 ÷ .0875 = 2.25

The length of the taper is 2.25″.

14.7 Thread Measurements

Cutting threads is one of the most common functions in machining, **Figure 14-35**. When machining threads on a CNC lathe, precise calculations are required. Usually, there are more calculations required for thread cutting in lathe programming than in mill programming. This section covers basic thread terminology and typical calculations to be made when writing lathe programs for threading operations. It is important to be familiar with the different elements of threads and specific nomenclature used in threading applications.

14.7.1 Major Diameter and Minor Diameter

Just as the name implies, the *major diameter* is the largest diameter of a thread. Whether the thread being measured is external or internal, the major diameter will be the largest diameter. See **Figure 14-36**.

Pixel B/Shutterstock.com

Figure 14-35. A thread cutting operation for external threads on a lathe.

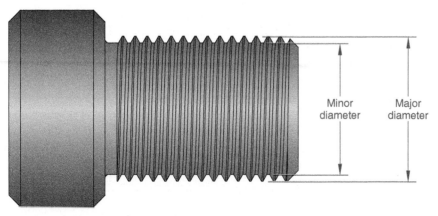

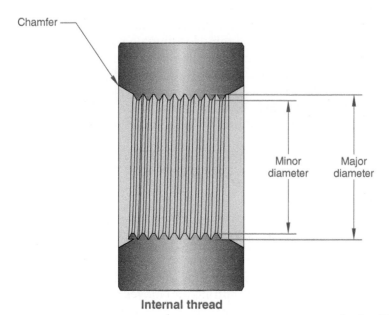

Goodheart-Willcox Publisher

Figure 14-36. Elements of external and internal threads. The major diameter is the largest diameter of the thread. The minor diameter is the smallest diameter of the thread.

In creating a lathe program, it is essential to know the required major diameter for the threading operation. This dimension must be inspected before threading.

The *minor diameter* is the smallest diameter of a thread. Refer to **Figure 14-36**. In a lathe program for external threads, the minor diameter will be established by the threading tool. In a lathe program for internal threads, the minor diameter will be established by a drill or a boring tool. In machining, it is important to verify these dimensions through inspection.

14.7.2 Thread Depth and Pitch

There are normally two calculations needed for a lathe program used for cutting threads. The first is the thread depth calculation. The ***thread depth***, also called the *thread height*, is the perpendicular distance between the major diameter and minor diameter of the thread. The thread depth calculation is used in canned cycles for threading on a lathe. The thread depth represents a radial distance. To calculate the thread depth, subtract the minor diameter from the major diameter, then divide the difference by 2. The major diameter for a given thread designation can be determined from standard diameter-pitch reference tables. In the thread designation given on a print, the major diameter is a *nominal size*. A *nominal size* is a designated value used for general identification purposes. Usually, for an external thread measurement, the measured size is smaller than the nominal size. For example, the thread designation 1/2–13UNC–2A specifies a major diameter of 1/2″. The allowable limits of size for the major diameter of this thread are .4876–.4985. For this example, the major diameter is .491″ and the minor diameter is .405″.

$$\text{Thread depth} = \frac{\text{Major diameter} - \text{Minor diameter}}{2}$$

$$.491 - .405 = .086$$
$$.086 \div 2 = .043'' \text{ thread depth}$$

The second calculation required is the distance between threads. This distance is called the *pitch*. Thread pitch is the distance from a point on one thread to another point at the same location on the next thread, measured parallel to the axis of the cylinder. In a thread designation, the number of threads per inch follows the major diameter. For example, in the thread designation 1/2–13UNC–2A, the number of threads per inch is 13. The pitch is calculated by dividing 1″ by the number of threads per inch. This calculation applies to inch-based threads. See **Figure 14-37**.

$$\text{Pitch} = \frac{1''}{\text{Number of threads per inch}}$$

$$1 \div 13 = .0769''$$

14.8 Turning Speeds and Feeds

When cutting material on a CNC lathe, the relationship between how fast the spindle turns and how fast the cutting tool moves across the material is vitally important. Machine shops try to get parts cut as quickly as possible. In addition, tooling reacts differently at different speeds and feeds. For

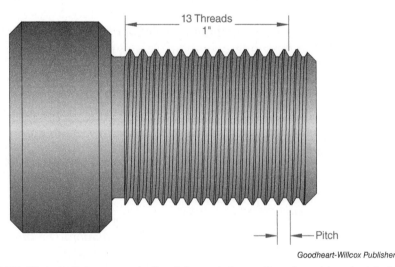

Goodheart-Willcox Publisher

Figure 14-37. Thread pitch represents the distance between threads and is calculated by dividing 1″ by the number of threads per inch.

example, if you turn a piece of material very slowly and feed the tool too rapidly, the tool may suffer a catastrophic failure and shatter. Conversely, if you run a piece of stock at a high speed but feed the tool slowly, it takes excessive time to cut the material and can dull the cutting edges of the tool.

Calculating speeds and feeds in the modern machining environment is one of the most overlooked and miscalculated aspects of machining. Fortunately, there is a relationship between the material, the cutting tool, and the machine that can be easily calculated. Every CNC programmer should be able to accurately calculate proper speeds and feeds.

The first factor in calculating speeds and feeds is the cutting tool. The machining industry today primarily uses tools made from solid carbide, so this is the material assumed in this text. When working with other cutter materials, consult manufacturer specifications. Lathe calculations also depend on the diameter of the material being machined.

The next important factor is the type of material being cut. Stainless steels do not cut as easily as aluminum, for example. Different materials have different rates of material removal and require careful consideration of cutting speeds. Cutting speed is a measure of movement in feet per minute. It is expressed in surface feet per minute (sfm) and is commonly referred to as surface footage. In simple terms, surface footage refers to the number of linear feet a point on a rotating component travels in one minute. Tooling manufacturers provide charts with recommended cutting speeds for common materials. On a chart, recommended speeds are expressed in ranges of values. These values are obtained through empirical testing. Cutting speeds can vary based on the tooling manufacturer, so it is best to get the range of recommended values directly from the manufacturer. See the Reference Section in this text for information on recommended cutting speeds for common materials.

Surface footage is used in calculating the spindle speed. The rotating speed of the spindle is measured in revolutions per minute (rpm).

The following examples are used to calculate the spindle speed for two different part sizes with two different types of materials. Consider cutting a 1 1/2″ bar of aluminum at 700 sfm and cutting a 2″ bar of 304 stainless steel at 300 sfm. The formula for calculating spindle speed is given below:

$$\text{rpm} = \frac{3.82 \times \text{sfm}}{\text{Diameter of material}}$$

Calculate the spindle speed for the 1 1/2″ (1.500″) bar of aluminum.

$$\frac{3.82 \times 700}{1.500} = 1783 \text{ rpm}$$

The speed for this operation is 1783 rpm. Now, calculate the spindle speed for the 2″ bar of stainless steel.

$$\frac{3.82 \times 300}{2} = 573 \text{ rpm}$$

The feed rate in a turning program is normally expressed in inches per revolution (ipr). This defines the amount of movement in the tool every time the spindle completes one revolution, **Figure 14-38**.

Depending on the type of tool and material, the feed rate can vary from .005″ to .030″. In some special cases, the feed rate can be programmed in inches per minute (ipm) for multiaxis machines, but in most cases the feed rate will be stated in ipr.

Additional coverage on programming methods for setting speeds and feeds is given in later chapters of this textbook.

> ### From the Programmer
>
> **Speeds and Feeds**
>
> Calculating the correct spindle speeds and cutting feeds is the best way to make your tools last as long as possible. Too slow of feed means the tool is cutting for longer amounts of time to do the same amount of work. Too fast of feed means the tool does not have enough time to cut, and can fail on the cutting edge. Often the best resource for speed and feed settings can be the tool distributor or a tooling engineer. Always calculate speeds and feeds to remain as efficient as possible.

> ### From the Programmer
>
> **Calculating Spindle Speeds**
>
> The mathematical constant 3.82 is used in the calculation for spindle speed. This value is the result of dividing the number of inches per foot by pi: 12/3.14 = 3.82.

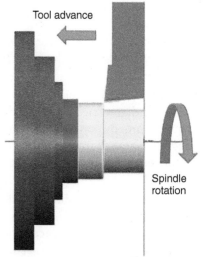

Goodheart-Willcox Publisher

Figure 14-38. Feed rate in a turning operation is a measurement of how far the tool advances in one revolution of the spindle.

Chapter 14 Review

Summary

- A good working knowledge of math principles is needed to create CNC lathe programs.
- A scientific calculator is a useful tool for making machining calculations.
- To convert a fraction to decimal format, divide the denominator into the numerator.
- To ensure a proper level of accuracy and ease of language, machinists express decimals in the thousandths and ten thousandths place.
- When evaluating a print, machinists identify basic geometric shapes and evaluate part geometry to determine dimensional information.
- Four common types of triangles encountered in machining work are scalene, isosceles, equilateral, and right triangles.
- The Pythagorean theorem states that, for a right triangle, the square of one side plus the square of the second side is equal to the square of the hypotenuse.
- Right triangles can be solved for all sides and all angles using standard formulas.
- Supplementary angles are two angles that add up to 180°. Complementary angles are two angles that add up to 90°.
- The hypotenuse is the longest side of a right triangle and is always across from the 90° angle.
- Basic trigonometry is used to solve for sides and angles in a right triangle.
- Required dimensions for tapers can be calculated by using right triangles and trigonometric functions.
- The major diameter of a thread represents the largest diameter. The minor diameter of a thread represents the smallest diameter. Thread pitch is the distance from a point on one thread to another point at the same location on the next thread.
- Cutting speed is a measure of movement in feet per minute and is expressed in surface feet per minute (sfm). Surface footage is used in calculating the spindle speed, which is measured in revolutions per minute (rpm).

Review Questions

Answer the following questions using the information provided in this chapter.

Know and Understand

1. Most prints used in manufacturing today are dimensioned in _____.
   A. fractional inches
   B. decimal inches
   C. millimeters
   D. centimeters

2. The position three places to the right of the decimal point is called the _____ place.
   A. hundredths
   B. tenths
   C. ten thousandths
   D. thousandths

3. *True or False?* A circle is a closed plane curve that is an equal distance from its center point to all points.

4. The distance across a circle through the center point is the _____.
   A. length
   B. width
   C. diameter
   D. radius

5. The _____ triangle has no equal sides or equal angles.
   A. isosceles
   B. equilateral
   C. obtuse
   D. scalene

6. A right triangle must contain one _____ angle.
   A. 30°
   B. 45°
   C. 60°
   D. 90°

7. *True or False?* The sum of all angles in any triangle is 90°.

8. The Pythagorean theorem is written as _____.
   A. $a^2 \times b^2 = c^2$
   B. $a^2 + b^2 = c^2$
   C. $a^2 - b^2 = c^2$
   D. $b^2 - a^2 = c^2$

9. An acute angle is defined as _____.
   A. greater than 180°
   B. equal to 90°
   C. greater than 0°
   D. less than 90°

10. *True or False?* Two angles are complementary if they add up to 180°.

11. _____ is defined as a branch of mathematics that deals with the relationships between the sides and angles of triangles and with functions of angles.
    A. Solid modeling
    B. Plane geometry
    C. Pythagorean theory
    D. Trigonometry

12. The longest side in a right triangle, or the side directly across from the right angle, is called the _____.
    A. apex
    B. adjacent side
    C. hypotenuse
    D. opposite side

13. *True or False?* Besides the hypotenuse, the other two sides in a right triangle are called the opposite and the adjacent.

14. Thread depth is determined by subtracting the minor diameter from the major diameter and _____.
    A. multiplying by 2
    B. adding the pitch
    C. subtracting the pitch
    D. dividing by 2

15. Thread pitch is calculated by dividing 1 by the _____.
    A. major diameter
    B. minor diameter
    C. threads per inch
    D. thread depth

16. _____ is a measure of movement in feet per minute and is expressed in surface feet per minute (sfm).
    A. Chip load
    B. Feed rate
    C. Spindle speed
    D. Surface footage

Apply and Analyze

1. Convert 3/8 to its decimal equivalent. Round to the nearest thousandths.

2. Convert 1 5/8 to its decimal equivalent. Round to the nearest thousandths.

3. Identify the following geometric shape. Assume all sides are equal.

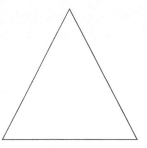

4. List and briefly describe the four types of triangles discussed in this chapter.

5. Use the Pythagorean theorem to solve for the unknown side of this right triangle.

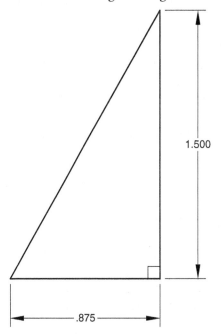

6. Use the Pythagorean theorem to solve for the unknown side of this right triangle.

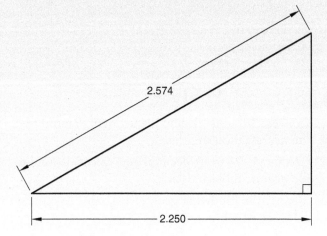

7. Identify the unnamed side of this right triangle.

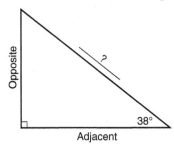

8. Label all of the sides of this right triangle.

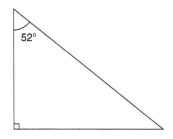

9. Use a calculator to solve for the missing angle of this right triangle.

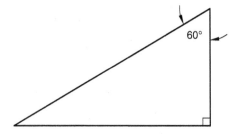

10. Use a calculator to solve for the missing side of this right triangle. Give your answer to the nearest thousandths.

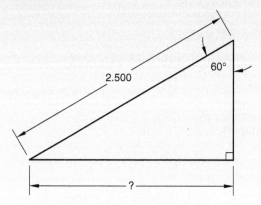

11. Use a calculator to solve for the two missing angles of this right triangle. Round answers to the nearest whole angle.

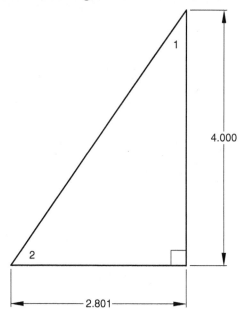

12. Calculate the small diameter for the tapered part shown. Give your answer to the nearest thousandths.

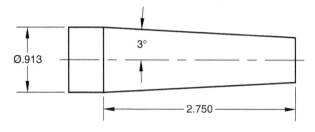

13. When making a cut on 1 5/8" diameter aluminum stock, what is the correct speed, in rpm, for cutting at 750 surface feet per minute?

Critical Thinking

1. Take a look around your home, school, or workplace and identify various geometric shapes. Can you measure and calculate any irregularities in these shapes? How do these irregularities affect the construction of these shapes?
2. Consider how right triangles are used in everyday life. Consider the construction of your home, the school building, or other built structures, such as a storage shed. Determine the approximate angle of a roof or calculate the length of a diagonal between two walls that meet at 90°.
3. The lathe was the earliest machine tool used in making functional parts. What are some of the earliest uses of turning? How do CNC lathes today compare with the machinery that was originally used in turning operations?

15 Cartesian Coordinate System and Machine Axes for Turning

Chapter Outline

15.1 Introduction
15.2 Number Line
15.3 Two-Axis Coordinate System
    15.3.1 Diameter Programming
    15.3.2 Absolute and Incremental Positioning
    15.3.3 Inside Diameter Features
15.4 Machine Home and Part Origin
15.5 C Axis Coordinate Programming

Learning Objectives

After completing this chapter, you will be able to:

- Describe how the Cartesian coordinate system is applied on a CNC lathe.
- Plot X and Z coordinate positions in a two-axis coordinate system.
- Plot coordinates locating inside diameter and outside diameter positions.
- Describe the relationship between machine home and the part origin.
- Explain how to plot coordinates about a third machine axis to establish rotational positions in lathe programming.

Key Terms

absolute positioning
boring
Cartesian coordinate system
incremental positioning
live tool
machine home
origin
work envelope
work offset

Chapter opening photo credit: Aumm graphixphoto/Shutterstock.com

15.1 Introduction

The basic principle of creating any CNC program is positioning a tool at a known location and moving that tool to a series of defined locations. The system used to plot and calculate machine positions and movement is known as the Cartesian coordinate system. The *Cartesian coordinate system* specifies each point uniquely in a plane with a pair of alphanumeric coordinates. This system is named for its developer, French scientist and mathematician René Descartes. An influential philosopher, Descartes is considered the father of analytic geometry. His developments led to the coordinate plotting system that is widely used today.

An example of the Cartesian coordinate system in use is a vending machine. See **Figure 15-1**. When you purchase a snack or beverage from a vending machine, you select the row of the item you want and then the number in that row. Specifying two points of intersection tells the machine where to locate your item.

Lissandra Melo/Shutterstock.com

Figure 15-1. The organization of rows and columns in a vending machine is a common example of the Cartesian coordinate system.

15.2 Number Line

To understand how points are located in the Cartesian coordinate system, look at the simple number line shown in **Figure 15-2**. A number line is a straight line with numbered divisions spaced equally along the line. The zero mark in the middle of the line establishes the starting point and is known as the *origin*. The numbered divisions along the line are located in relationship to the origin. The numbers can have any defined value and unit format. For example, each number can represent one inch, one foot, or one mile. In this example, the numbers represent whole number inches.

The numbers to the right of the origin are positive numbers. The numbers to the left of the origin are negative numbers. Movement in the right direction is positive and movement in the left direction is negative. The numbers define the amount of linear movement in a given direction. For example, moving five units to the right of the origin represents a movement of 5″ in the positive direction. Moving three units to the left of the origin represents a movement of 3″ in the negative direction. These distances are represented by Point 1 and Point 2 in **Figure 15-2**. Specifying distances in relation to a known origin is a basic way to describe tool movement.

The horizontal number line shown in **Figure 15-2** establishes a single axis of movement. In CNC programming, additional axes are used to define the direction of the cutting tool.

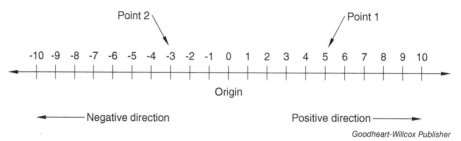

Goodheart-Willcox Publisher

Figure 15-2. A number line measures distance and direction. Point 1 represents a movement of 5″ from the origin in the positive direction. Point 2 represents a movement of 3″ from the origin in the negative direction.

15.3 Two-Axis Coordinate System

A number line represents a single axis of travel. However, all CNC machines have at least two axes of travel. Adding a second number line establishes the second coordinate axis in the Cartesian coordinate system. See **Figure 15-3**. In CNC lathe programming, these axes are designated as the X axis and the Z axis. These axis designations are different from those in mill programming, where the axes in the two-axis coordinate system are designated X and Y. The X and Z axes intersect at the origin and are perpendicular to each other (oriented at 90°). Points representing distances from the origin are specified with X and Z coordinates. The coordinate system origin is designated as X0. Z0. This two-axis coordinate system allows the programmer to define the location of the workpiece and the direction of tool movement.

As shown in **Figure 15-3**, the coordinate system is divided into four quadrants. Coordinates are specified as positive or negative based on their location from the origin. A coordinate to the right of the origin has a positive Z value and a coordinate to the left of the origin has a negative Z value. In **Figure 15-3**, notice that each quadrant has positive or negative X and Z values based on its location in relation to the origin. The quadrants are numbered 1–4 in a counterclockwise direction. Coordinates in the upper-right quadrant have a positive X value and positive Z value. Coordinates in the upper-left quadrant have a positive X value and negative Z value.

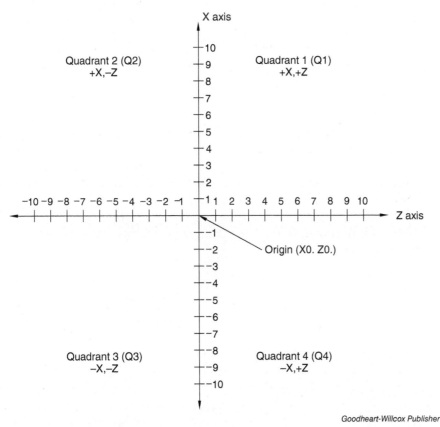

Goodheart-Willcox Publisher

Figure 15-3. A two-axis coordinate system is used to define point locations. In lathe programming, the two axes making up this system are designated as the X axis and Z axis. The axes intersect at the origin and divide the system into four quadrants. Points are located in relation to the origin and have positive or negative values based on their location from the origin.

On a two-axis CNC lathe, the X and Z axes establish a theoretical flat work plane known as the XZ plane. In a two-axis lathe configuration, the Y axis is not used. This is different from a milling machine, which uses a three-axis coordinate system with X, Y, and Z axis designations. See **Figure 15-4**. On a milling machine, the X axis represents the horizontal movement of the machine table from side to side. The Y axis represents the movement of the table from front to back. The Z axis represents the movement of the spindle. The Z axis orientation is the same on every mill, whether using a vertical or horizontal milling machine. The Z axis defines the tool approach and retract direction. A move away from the table is a movement in the positive Z direction (+) and a move toward the table is a movement in the negative Z direction (−).

Compare this configuration to a two-axis lathe, **Figure 15-5**. A two-axis lathe has the same relationship between the Z axis and the spindle. The difference is that the coordinate system has to "lay over" in the lathe to maintain that relationship. The X axis is perpendicular to the Z axis and extends outward from the spindle centerline. The X axis defines the tool approach and retract direction, **Figure 15-6**. There are multiaxis lathes that utilize the Y axis, where the tooling moves off the centerline of the spindle. The sole focus of this chapter, however, is the two-axis lathe. Multiaxis lathe functions are discussed in Chapter 23 of this textbook.

On a CNC mill, the origin established for the part can vary in location and is defined at a suitable position, such as on a corner of the part. In a lathe setup, the X axis origin is always set in the same location—at the spindle centerline. The spindle centerline represents the axis of rotation for the part. The Z axis aligns with the spindle centerline. The Z axis origin can be established at any desired location, but for ease of programming, it is usually set at the finished front face of the workpiece. This is the orientation shown in **Figure 15-6**. Study the two-axis coordinate system closely to

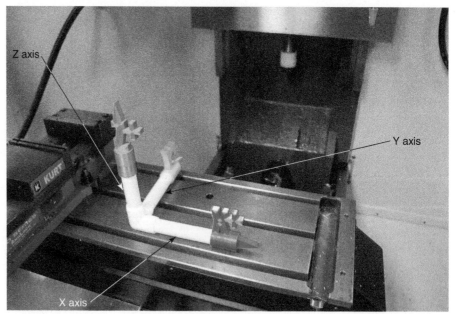

Goodheart-Willcox Publisher

Figure 15-4. A three-axis coordinate system is used to define point locations on a milling machine and consists of the X, Y, and Z axes. The X and Y axes establish a theoretical flat work plane known as the XY plane. The Z axis is aligned with the spindle and represents movement of the spindle toward or away from the workpiece.

Pixel B/Shutterstock.com

Figure 15-5. A two-axis coordinate system is used to define point locations on a two-axis lathe and consists of the X and Z axes. The Z axis is aligned with the spindle. The X axis is perpendicular to the Z axis and defines the tool approach and retract direction. The machine shown is a slant bed lathe.

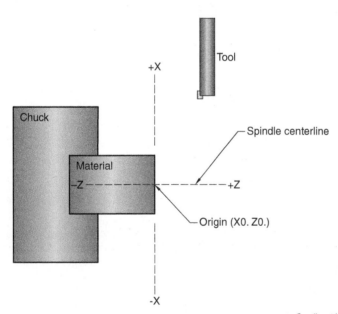

Goodheart-Willcox Publisher

Figure 15-6. On a two-axis lathe, the X and Z axes establish a theoretical flat work plane known as the XZ plane. The X and Z axes intersect at the origin and are perpendicular to each other. The work origin is typically oriented on the front face of the part and is always aligned with the spindle centerline.

become familiar with the lathe axes of travel. Moving toward the spindle represents movement in the negative Z axis direction. Moving away from the spindle represents movement in the positive Z axis direction. Moving toward the spindle centerline represents movement in the negative X axis direction. Moving away from the spindle centerline represents movement in the positive X axis direction.

15.3.1 Diameter Programming

For ease in programming, coordinates used to define distances along the X axis in lathe operations are specified as diameter measurements. This is a unique characteristic of the two-axis coordinate system used in lathe programming. Because the X axis origin is the center of rotation, or the part center, the distance from the origin to the edge of a cylindrical part represents a radial distance. On most prints, however, cylindrical part features are dimensioned with diameter dimensions. In addition, most measurements made on parts that are made from round stock will be made on the diameter of the part. For example, if a print defines a 2.000″ diameter for an external feature, the distance from the part center to the edge is 1.000″. In measuring this distance, however, an outside micrometer is used to measure the 2.000″ diameter. If the print is defining a 2.000″ diameter and the measuring tool is measuring the 2.000″ diameter, it makes sense that the program should also reference the 2.000″ diameter, **Figure 15-7**.

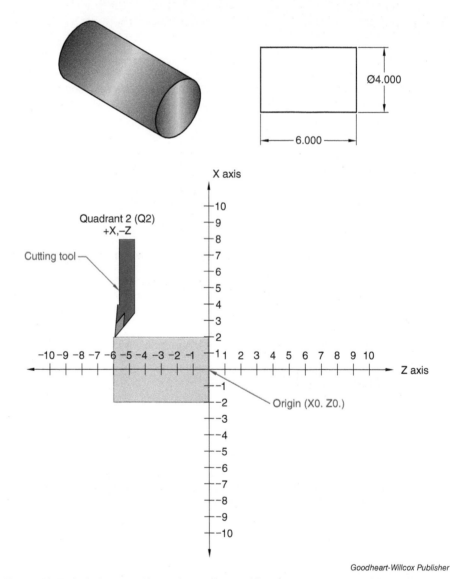

Goodheart-Willcox Publisher

Figure 15-7. In lathe operations, the cutting tool is programmed to machine at the diameter of the part. Diameter programming simplifies this process by specifying X coordinate values as diameter measurements. This allows dimensions from the print to be used directly in programming.

When CNC lathes were first built, the issue with how cylindrical parts are dimensioned and measured became apparent. There were two possible solutions to this in programming. The first solution was to program movements for lathe operations in radial distances, which required the programmer to divide in half all of the diameter values on the print. The second solution was to alter the amount of movement read by the machine by using diameter programming. This became the obvious solution because it enabled programmers to program the machine to the dimensions used on the print. In diameter programming, when a CNC lathe is programmed to 2.000″, the tool will move 1.000″ from the part center and machine at the 2.000″ diameter. When an offset of .010″ is specified, the machine adjusts the tool location .005″ radially, and then cuts .005″ off both sides of the part, or .010″ in diameter. The coordinate input in diameter programming represents twice the actual (radial) measurement. In radial programming, the coordinate input represents the radial measurement, or half the diameter.

To create movement in a CNC machine, a series of coordinate points is given in the program. These coordinates will move the tool from point to point. **Figure 15-8** shows a set of points defining the path of travel to machine a cylindrical part. The coordinates are specified using diameter programming. The X axis coordinates represent diameter measurements and the Z axis coordinates represent length measurements. The part origin is located at the front face of the part.

The following is an explanation of each coordinate position. Point 1 is 1″ from the origin in the positive X axis direction and 0″ from the Z axis origin. The 1″ measurement is a radial value. However, for programming purposes, this coordinate location defines the 2″ diameter given on the print. Thus, the coordinate is expressed in the program as X2. Z0. Point 2 is 1.5″ from the origin in the negative Z axis direction and remains at 1″ from the origin in the positive X axis direction. In the program, this coordinate is expressed as X2. Z–1.5. To create movement from Point 1 to Point 2, and then to Point 3, Point 4, and Point 5, the CNC program is written as follows. Notice that each X axis coordinate is expressed as a diameter value.

> X2. Z0. (Point 1)
>
> X2. Z–1.5 (Point 2)
>
> X4. Z–1.5 (Point 3)
>
> X4. Z–4. (Point 4)
>
> X6. Z–4. (Point 5)

 Debugging the Code

Coordinate Format

In general mathematics, Cartesian coordinates are commonly written in the format X,Y. However, in CNC programming, this format is not used. In lathe programming, entries for XZ coordinates are expressed as decimals. For example, in the program, the coordinate entry X2. Z0. represents Point 1 in **Figure 15-8**. Notice that this format does not require a trailing zero in the decimal. For example, the coordinate X2.0 is shortened to X2. Similarly, a leading zero is not required for decimal values below zero. For example, the coordinate Z–.5 does not require a leading zero. Older controllers did not accept the decimal point programming format and entries were read in ten thousandths of an inch. For example, the entry X6 represented .0006″ and the entry X60000 represented 6″. Machines today use decimal point programming. This is the format used throughout this textbook.

From the Programmer

Coordinate Entry Format

The X axis coordinate value does not need to be first in the program, but the X and Z movement to each position must be on the same line to create a direct move to the next position when movement is along both axes. Specifying the X value first, followed by the Z value, is common practice in programming. This is the format used in this textbook.

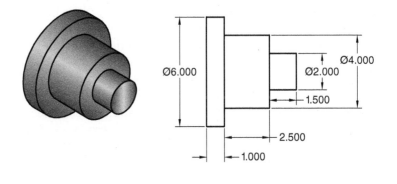

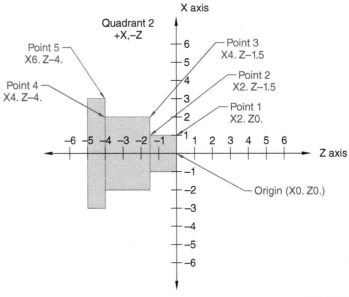

Goodheart-Willcox Publisher

Figure 15-8. Plotted coordinates representing the path of travel in a lathe program. The coordinates are specified using diameter programming. The X axis coordinates represent diameter measurements and the Z axis coordinates represent length measurements.

In the example shown in **Figure 15-8**, notice that the part is oriented in the second quadrant and all moves take place in that quadrant. In most cases, lathe programs are written with movement occurring in the second or first quadrant. In these quadrants, all X coordinates have a positive value. Although it is possible, it is not common to use the third or fourth quadrant in lathe programming, unless a special tool is used or a feature must be machined on the other side of the part.

Figure 15-9 shows a part oriented in the first quadrant with the part origin located on the finished rear face. Programming coordinates relative to this location means that all Z axis coordinates will have positive values.

Study the part dimensions shown in **Figure 15-9**. As in the previous example, diameter programming is used, so the X axis coordinates are expressed as diameter values. The programmed coordinates for each point in **Figure 15-9** are as follows.

X4. Z10. (Point 1)

X6. Z5.5 (Point 2)

X6. Z3. (Point 3)

X9. Z3. (Point 4)

X9. Z0. (Point 5)

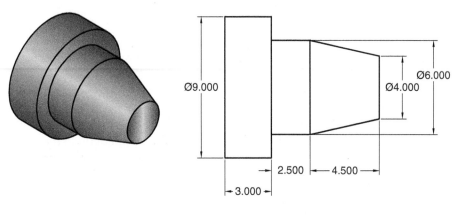

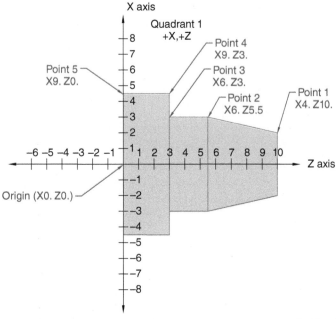

Figure 15-9. Programmed coordinates for a part oriented in the first quadrant of the two-axis coordinate system. In this orientation, all Z axis coordinates have positive values.

15.3.2 Absolute and Incremental Positioning

The programmed coordinates specified in the previous examples are based on absolute coordinate entry. Absolute coordinates are measured relative to the part origin (X0. Z0.). In a CNC program, moves specified with absolute coordinates are referred to as *absolute positioning* moves, meaning that they represent absolute measurements in reference to the origin. Absolute positioning is the appropriate method of movement for most purposes. A second way to specify movement is to use incremental positioning. *Incremental positioning* refers to measuring from the current location to a second location. Incremental coordinates represent a relative measurement from the current position of the tool, not from the origin.

Although it is possible, and sometimes preferred, to program in incremental moves, most CNC lathe programs are written using absolute positioning. An origin, or zero point, is established, and all machine or tool movements are made relative to that zero point.

> **Safety Note**
>
> Coordinates must be entered carefully in programming. The importance of defining the correct coordinates cannot be emphasized enough. A missed minus sign or decimal point can cause a catastrophic machine failure or part damage. Be careful plotting the points and entering them into the CNC program.

15.3.3 Inside Diameter Features

Many cylindrical parts produced on a lathe also have inside diameter features, **Figure 15-10**. These features are machined with drills or boring bars. *Boring* is a machining operation used to enlarge an existing hole to a finish size. To create the CNC program for inside diameter features, the coordinate locations must be calculated.

In cases where the part has both outside diameter (OD) and inside diameter (ID) features to be machined, the part orientation will remain in the same quadrant (quadrant 1 or 2) for each operation, based on the programmed origin position. **Figure 15-11** shows the part oriented on the XZ plane with the plotted points required for programming. The part origin is located on the front face of the part and moves are programmed in the second quadrant. Coordinates for machining the outside diameter and inside diameter features are shown. Study the part dimensions and the corresponding X axis coordinates used for the diameter positions in the program.

Figure 15-10. A cylindrical part with inside diameter features requiring coordinates to be defined in programming.

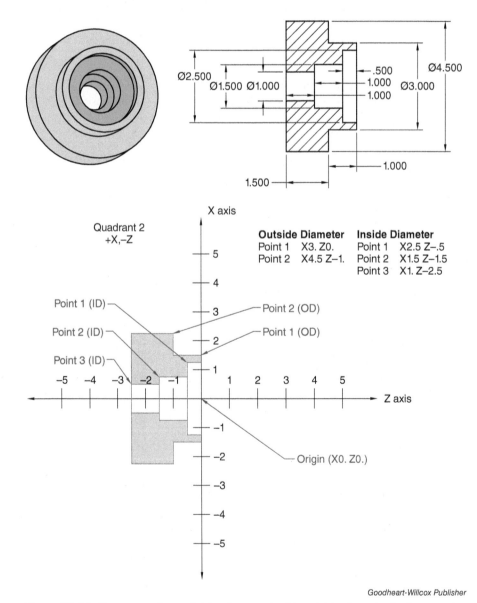

Figure 15-11. Programmed coordinates for a part with outside diameter features and inside diameter features to be machined.

From the Programmer

Part Orientation

In lathe programming, some programmers always program from the front face of the part and the spindle centerline. Some programmers prefer to program from the back face of the part or from the face of the lathe chuck. There can be advantages to both approaches. A common method used by programmers is to establish the finished part length along the Z axis and program all tool moves in the negative Z axis direction. Programming methods can vary depending on the machine shop and the part requirements. The bottom line is to be consistent. Programmers must make sure that the setup person and machine operator know where the part origin is established to prevent a serious machine crash.

15.4 Machine Home and Part Origin

In a CNC lathe operation setup, the machine controller must know where the part material is located. This is achieved by measuring the distance from the machine's home position to the part origin location. Every machine has a fixed location called *machine home*. This position varies based on the machine type and manufacturer and can be anywhere within the *work envelope* of the machine. On a standard two-axis lathe, this position is most often located at the limit of the positive X axis travel and the positive Z axis travel. **Figure 15-12** shows a typical machine home position for a two-axis lathe.

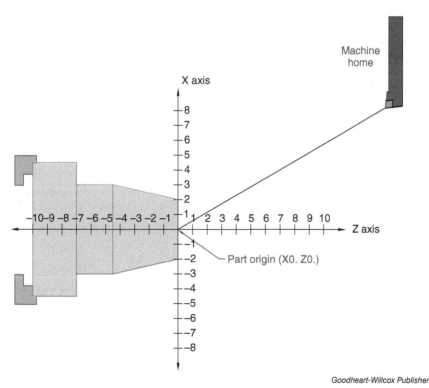

Goodheart-Willcox Publisher

Figure 15-12. Machine home on a two-axis lathe is a fixed position usually located at the limits of machine travel along the positive X axis and positive Z axis.

> **From the Programmer**
>
> **Work Offsets**
>
> Most modern CNC lathes have the ability to use a work offset, just like a CNC mill. A ***work offset*** is a coordinate setting that establishes the location of the work origin relative to the machine home position. The appropriate work offset command is used in the program to designate the work offset. During setup, coordinates for the work offset are entered in the corresponding work offset setting in the machine. Although there are more work offsets available, most programs are written with one of the G54–G59 work offset designations.

In a practical setup, the part material will be secured in the lathe chuck or collet, and the distance from machine home to the part origin will be measured. The active tool can be used to establish this distance by touching off the tool on the part face to measure the Z axis position, then touching off the tool on the outside diameter of the part to measure the X axis position. Each distance is entered into the machine controller to establish the X axis and Z axis tool offsets. The tool offsets represent the distance the tool travels from machine home to the X0. Z0. part origin. In writing the CNC program, a specific point on the part is considered to be at the X0. Z0. part origin, also called part zero or program zero. Program zero is the location on the part from which coordinates are measured in the program. Refer to **Figure 15-12**. The part origin represents an established position of the machine relative to machine home.

15.5 C Axis Coordinate Programming

A CNC lathe can be equipped with a third axis for controlling the angular orientation of the machine spindle. The third axis defines rotational motion and is used in conjunction with live tooling. A ***live tool*** is a driven tool used to perform drilling or milling operations while the workpiece remains in orientation with the main spindle of the lathe. The third axis is programmed to rotate the spindle to a specific angle for machining features on the periphery of the part. This axis is referenced as the C axis in the program. Because the third axis is a rotational axis, the programmed coordinates are expressed in degrees.

A lathe live tool can be driven radially or axially to machine features on the part's outside diameter or face. This allows features such as holes or slots to be created on a cylindrical part in the same machine setup used for turning operations. This reduces the number of setups required and enables parts to be made more efficiently. Along with the X and Z axis locations, an angular direction is programmed to orient the workpiece to the desired angle prior to the live tooling operation. As in mill programming, angles are measured from 0°. While looking directly at the face of the chuck or collet nose on the machine, the 3 o'clock position is designated as 0°. See **Figure 15-13**.

When programming rotational movement about a machine axis, an address code is used to specify the axis. On a typical machine, rotation about the C axis is defined with the C address code. The rotational movement can be specified individually or in combination with other axis movements. Address codes in programming vary based on the controller and machine type, but the following is a typical example of programming rotation about the C axis.

G0 C45.

This line of code will engage the C axis and rotate the spindle to a 45° position. Some machines require a spindle brake to hold the spindle in place once it is oriented. At this position, the live tooling can be used to drill holes or create slots. See **Figure 15-14**. In this type of machining, the tool remains in alignment with the centerline of the machine spindle.

A C axis coordinate for rotation can be programmed as a positive (+) or negative (–) value. The rotation can be programmed using absolute or incremental positioning.

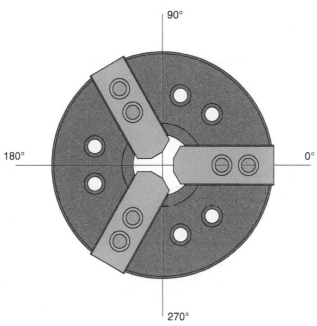

Figure 15-13. Angle positions for the C axis on a lathe with 0° oriented at the 3 o'clock position.

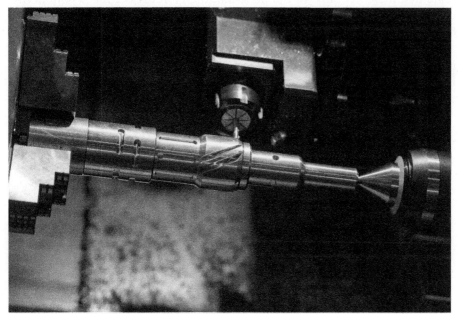

Figure 15-14. A live tooling operation used to mill a slot on a part mounted in a lathe.

Figure 15-15 shows typical angles of rotation for features in lathe programming. The angles are in 30° increments. Points 1 and 2 correspond to C axis coordinates programmed as C30. and C330. These coordinates are based on absolute positioning.

From Point 1, the machine axis can be rotated to Point 2 in several different ways. Using absolute positioning, it can be rotated to C330° by programming the coordinate C330. This movement can also be programmed as C–30. If the origin is at the 3 o'clock position, or C0., then a negative

> **From the Programmer**
>
> **Multiaxis Turning**
>
> The C axis can be used in conjunction with the X or Z axis. When drilling a cross hole, the C axis is positioned and the drill is programmed to drill in the X axis direction. For slotting, the C axis is positioned and the tool is programmed to cut in the Z axis direction. This slotting method is used for lateral slots. To create radial slots, the tool can be programmed to an X and Z axis position, and then the C axis is rotated simultaneously with Z axis movement. This operation is an example of three-axis turning. Multiaxis machining is discussed in Chapter 23.

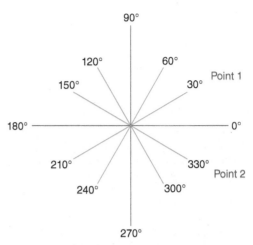

Goodheart-Willcox Publisher

Figure 15-15. Common angles of rotation used in programming C axis coordinates. Points 1 and 2 represent the angles to be programmed.

C axis coordinate will rotate the spindle counterclockwise. The difference between programming a positive C axis move and a negative C axis move is the direction of travel of the rotation. It can be more efficient to program the shortest distance. The move from Point 1 to Point 2 can also be programmed incrementally. In the program, the H address code is commonly used to specify incremental movement for the C axis. The incremental movement from Point 1 to Point 2 would be specified as H–60. or H300.

Figure 15-16 shows a typical part made from round stock and the rotation angles corresponding to the orientation of the C axis. As in **Figure 15-15**, the angles are in 30° increments.

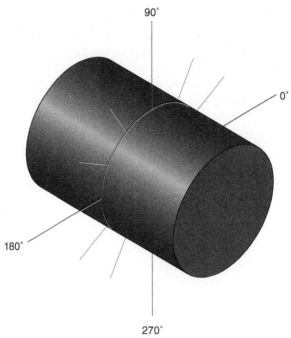

Goodheart-Willcox Publisher

Figure 15-16. A cylindrical part in a default position with angle positions based on 0° oriented at the 3 o'clock position. A feature can be machined on the outside diameter with live tooling by programming the appropriate angle to rotate the C axis.

Chapter 15 Review

Summary

- A standard two-axis CNC lathe uses a two-axis coordinate system. The axes are designated as the X axis and Z axis and the coordinate system is divided into four quadrants. The X and Z axes establish a theoretical flat work plane known as the XZ plane.
- On a two-axis lathe, the Z axis aligns with the spindle centerline. The X axis is perpendicular to the Z axis and extends outward from the spindle centerline.
- Coordinates defined in lathe programming are based on diameter programming. The X axis coordinates represent diameter measurements and the Z axis coordinates represent length measurements.
- Program zero represents the part origin and is the location on the part from which coordinates are measured in the program. On the machine, the part origin represents an established position of the machine relative to machine home. Machine home is a fixed location normally located at the limit of the positive X axis travel and the positive Z axis travel.
- Parts machined on a CNC lathe often contain both inside diameter (ID) and outside diameter (OD) features. In cases where outside diameter and inside diameter features are to be machined, the part orientation will remain in the same quadrant for each operation, based on the programmed origin position.
- A CNC lathe equipped with a third axis and live tooling has the ability to orient the machine spindle at different angular positions. The third axis is designated as the C axis.

Review Questions

Answer the following questions using the information provided in this chapter.

Know and Understand

1. The Cartesian coordinate system was developed by ____.
   A. Ray Cartesian
   B. Carte Desysteme
   C. René Descartes
   D. Remy Descarme

2. On a number line, the starting point is known as ____.
   A. the origin
   B. point zero
   C. coordinate home
   D. machine home

3. In two-axis lathe programming, the coordinate system is made up of the ____ axes.
   A. X and Z
   B. A and Z
   C. A and B
   D. X and Y

4. *True or False?* The most common way to specify movement in a CNC program is to use incremental positioning.

5. Tool offsets established during machine setup represent the distance the tool travels from machine home to the part origin.

6. Based on the programmed points shown for the given part, in which quadrant are coordinates located?

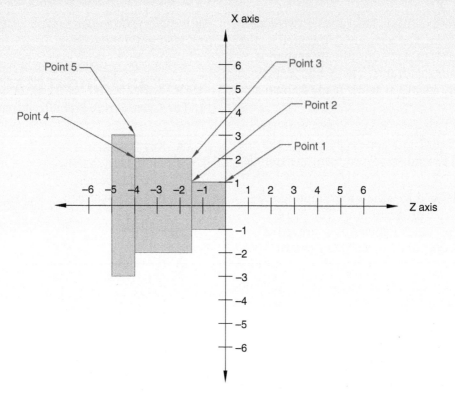

A. Quadrant 1
B. Quadrant 2
C. Quadrant 3
D. Quadrant 4

7. In a lathe program, the _____ coordinates represent diameter measurements.

A. C axis
B. Z axis
C. Y axis
D. X axis

8. What is the coordinate position for Point 1 on the part shown?

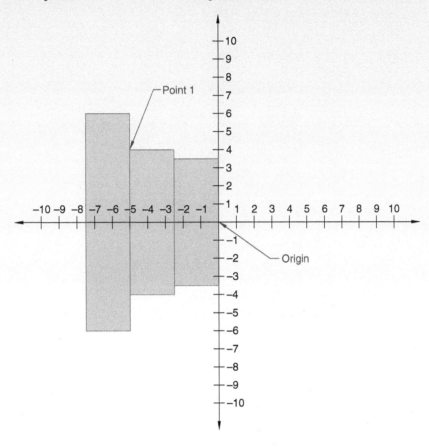

A. X8. Z–5.
B. X8. Z–10.
C. X–5. Z4.
D. X–5. Z8.

9. What is the coordinate position for Point 1 on the part shown?

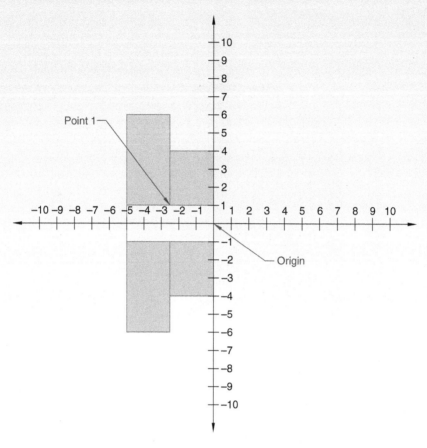

   A. X–2.5 Z2.
   B. X–2.5 Z1.
   C. X–5. Z2.
   D. X2. Z–2.5.

10. *True or False?* In a lathe program, a C axis coordinate represents an angular position.

11. The 0° position, or starting position, for the third axis spindle orientation is located at the _____ position.

   A. 3 o'clock
   B. 6 o'clock
   C. 9 o'clock
   D. 12 o'clock

12. *True or False?* The C axis spindle orientation can be programmed in a positive or negative direction.

13. What is the C axis coordinate position for the drilled hole at Point 1?

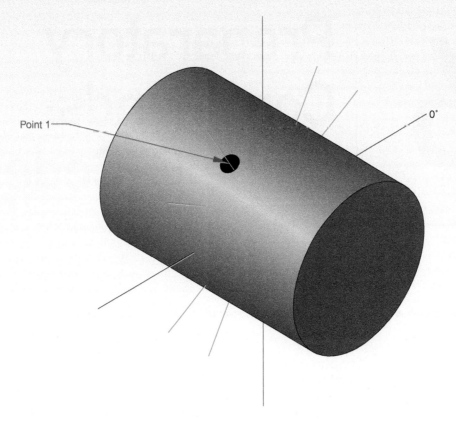

A. C30.
B. C120.
C. C–30.
D. C–120.

Apply and Analyze

1. What is the purpose of using the Cartesian coordinate system in CNC programming?
2. What are the two designated axes of movement on a two-axis CNC lathe?
3. Where is machine home normally located on a standard two-axis lathe?
4. What is the purpose of live tooling?
5. Explain the additional benefit provided by the third axis on a CNC lathe.

Critical Thinking

1. The Cartesian coordinate system has important uses in the machining industry and other industries, including mechanical design and engineering. What are some examples in everyday life where coordinate systems are used to designate distance and direction?
2. In a typical CNC lathe setup, the part origin is oriented on the front face of the part. What considerations need to be made in programming when the origin is aligned with the back face of the part or the face of the lathe chuck?
3. What are some of the benefits of purchasing a lathe equipped with a third axis? What are some examples of common parts that might benefit from three-axis machining?

16 Preparatory Commands: Lathe G-Codes

Chapter Outline

16.1 Introduction
16.2 Using G-Codes in a Program
16.3 G-Code Commands
   16.3.1 G00: Rapid Positioning
   16.3.2 G01: Linear Interpolation
   16.3.3 G02 and G03: Circular Interpolation
   16.3.4 G20 and G21: Measurement Modes
   16.3.5 G28: Machine Zero Return
   16.3.6 G40, G41, and G42: Tool Nose Radius Compensation
   16.3.7 G70, G71, G72, and G73: Multiple Repetitive Cycles
16.4 Startup Blocks

Learning Objectives

After completing this chapter, you will be able to:

- Explain the purpose of preparatory commands.
- Identify the most common G-code commands for lathe operations.
- Identify G-code groups and their purpose.
- Explain the difference between modal and nonmodal commands.
- Describe the purpose of the G00 and G01 movement commands.
- Explain the use of tool nose radius compensation commands.
- Identify G-code commands used to execute canned cycles.

Key Terms

address
block
canned cycle
G-code

linear interpolation
modal command
nonmodal command
preparatory command

startup block
word

16.1 Introduction

A CNC program is a series of letters and numbers that, when combined, create action from the machine. A letter and number grouping used to execute a command in a program is called a *word*. The letter preceding the number in a word is called an *address*. An address is a single-letter character that defines what a machine should do with the numerical data that follows. Addresses are used in programs to designate commands and machine functions. The most commonly used program address is the letter G, or *G-code*. A G-code identifies a *preparatory command*. The purpose of the preparatory command is to prepare the machine controller, or preset the machine, into a specific state of operation. This chapter covers the most commonly used preparatory commands in CNC lathe programming.

A G-code positioning command by itself will not create any motion or movement, but it places the machine in the operational mode for the program entries that follow. Conversely, the machine will not operate without the operational mode being set. For example, the line of machine code below, by itself, will not create any movement from the machine.

X6. Z5.

Although there is a designated set of coordinates to a target position, the controller has not been put into an operational mode. Is this a rapid move or a move at a feed rate? Are these coordinates in inches or in metric units? These questions cannot be answered without an accompanying G-code address.

16.2 Using G-Codes in a Program

Once the machine is preset into a specific operating mode with the appropriate G-code command, a motion or movement can occur. For example, the G01 command creates straight-line movement. The programming example shown in **Figure 16-1** moves the machine from its current location, the origin, to the X6. Z5. position. The G01 code commands a linear move in a straight line.

The X6. coordinate in **Figure 16-1** represents a diameter measurement. As discussed in Chapter 15, diameter programming is used to specify X axis coordinates in lathe programming. This allows diameter dimensions from the print to be input in the program. The coordinate input represents twice the actual (radial) measurement. Therefore, in **Figure 16-1**, the tool moves 3.000" in the positive X axis direction when programmed to move to the X6. Z5. position from the origin.

The single line of code shown in **Figure 16-1** is called a *block*. A *block* is a single word or a series of words forming a complete line of CNC code. Notice that F.010 appears at the end of the block as the last word. The address code F designates the feed rate. The programmed feed rate is used by the G01 command and must be specified. The feed rate is the rate at which the cutting tool moves into the material in inches per revolution (ipr).

Most G-code commands are *modal commands*, meaning once they are turned on, the machine stays in that condition until the mode is canceled or until a subsequent command changes the machine's condition. **Figure 16-2** shows a series of straight-line movements programmed with the G01 command. All of the movements are in a linear direction and occur in succession because the G01 command is modal and presets the mode of operation to straight-line movement.

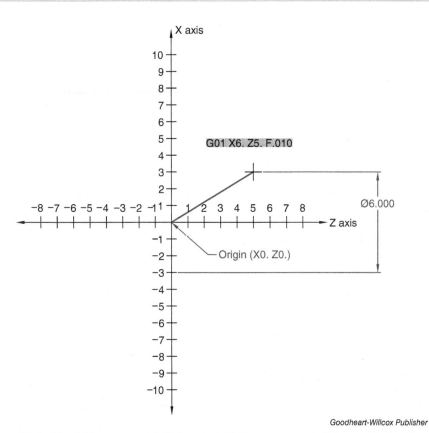

Figure 16-1. The G01 command initiates straight-line movement. It is used in the line of code shown to move the machine from its current location to the X6. Z5. location.

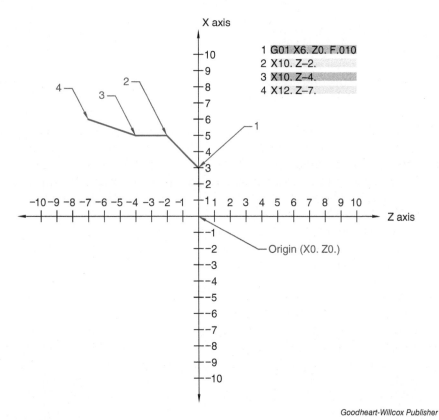

Figure 16-2. Using the G01 command to carry out a series of straight-line movements. The G01 command is a modal command and remains in effect until canceled.

16.3 G-Code Commands

The table in **Figure 16-3** lists the most commonly used G-code commands in lathe programming. It is not a complete list, and some codes are specific to certain controller types. Consult your machine manufacturer's technical manual or website for exact G-codes that are available. Many of these codes will be discussed in greater detail in subsequent chapters.

In a CNC program, multiple G-code commands can be used in the same block of code, as long as they do not have conflicting functions. For example, a block written as **G80 G54 G00 X2. Z–6.** is a valid line of code. Referring to the G-code commands in **Figure 16-3**, this line is easily deciphered:

- **G80.** Cancels any previously active canned cycles.
- **G54.** Designates work coordinate offset 1 as the work coordinate system.
- **G00.** Places the machine in rapid positioning mode.

Commonly Used G-Codes

G-Code	Function	Mode
G00	Rapid positioning	Modal
G01	Linear interpolation in feed mode	Modal
G02	Circular interpolation clockwise (CW)	Modal
G03	Circular interpolation counterclockwise (CCW)	Modal
G04	Dwell (in milliseconds)	Nonmodal
G09	Exact stop check	Nonmodal
G10	Programmable offset setting	Nonmodal
G17	XY plane designation	Modal
G18	XZ plane designation	Modal
G19	YZ plane designation	Modal
G20	US Customary units of input	Modal
G21	Metric units of input	Modal
G28	Machine zero return	Nonmodal
G29	Return from machine zero	Nonmodal
G32	Thread cutting cycle	Modal
G40	Tool nose compensation cancel	Modal
G41	Tool nose compensation left	Modal
G42	Tool nose compensation right	Modal
G50	Spindle speed RPM limit	Modal
G52	Work coordinate system shift	Nonmodal
G53	Machine coordinate system	Nonmodal
G54	Work coordinate offset 1	Modal
G55	Work coordinate offset 2	Modal
G56	Work coordinate offset 3	Modal
G57	Work coordinate offset 4	Modal
G58	Work coordinate offset 5	Modal

Continued

Figure 16-3. G-code commands used in lathe programming. Some machine-specific codes are not listed.

G-Code	Function	Mode
G59	Work coordinate offset 6	Modal
G70	Finishing cycle	Nonmodal
G71	OD/ID stock removal cycle	Nonmodal
G72	End face stock removal cycle	Nonmodal
G73	Irregular path stock removal cycle	Nonmodal
G74	Face grooving cycle; high speed peck drilling cycle	Nonmodal
G75	OD/ID peck grooving cycle	Nonmodal
G76	Multiple pass threading cycle	Nonmodal
G80	Canned cycle cancel	Modal
G81	Drilling cycle	Modal
G82	Spot drilling cycle	Modal
G83	Full retract peck drilling cycle	Modal
G84	Tapping cycle	Modal
G85	Boring cycle—bore in, bore out	Modal
G86	Boring cycle—bore in, rapid out	Modal
G88	Boring cycle—bore in, dwell, rapid out	Modal
G89	Boring cycle—bore in, dwell, bore out	Modal
G92	Threading cycle	Modal
G94	End facing cycle	Modal
G95	Live tooling tapping cycle	Modal
G96	Constant surface speed (CSS on)	Modal
G97	Constant spindle speed (CSS cancel)	Modal
G98	Feed per minute	Modal
G99	Feed per revolution	Modal

Goodheart-Willcox Publisher

Figure 16-3. *(Continued)*

This line will instruct the machine to move to the absolute coordinate position X2. Z–6. in the G54 work coordinate offset in full rapid mode. The G-code commands in this block cancel any canned cycles that may still be active from previous programs, establish the work offset, and define the mode of movement for the cutting tool. The functions of these commands do not conflict with each other.

In CNC programming, G-codes are organized into specific groups by function. The group designation identifies modal G-code commands that cannot be in effect at the same time. Multiple G-codes from different groups can be placed on the same block of code, but no two G-codes from the same group can be specified on the same block. G-codes belonging to the same group are associated with the same function. For example, a block that states **G01 G40 G42 X3.** is invalid. The G40 and G42 commands are both related to the same function: tool nose radius compensation. Tool nose radius compensation commands are used to compensate for the tool nose radius of a lathe tool by offsetting the tool to the left or right of the programmed cutting path. The G40 and G42 commands have conflicting functions. The G40 command cancels tool nose radius compensation and

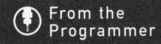

G-Code Entry Format

On many machines, G-code commands beginning with a zero in the digit portion can be shortened to a single digit following the letter G. For example, G00 can be shortened to G0 and G01 can be shortened to G1. This alternate format may not be recognized on older machines.

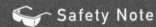

When using the G00 command, be cautious of any possible collisions with this movement. G00 is a modal command, so all movements following it will be in full rapid mode until it is canceled.

G42 initiates tool nose radius compensation right. If conflicting G-codes are used on a single block of code, the machine's alarm will activate.

G-code groups are listed in the table in **Figure 16-4**. Not all of the groups are listed, as some pertain only to milling operations. The number of groups and the group numbers vary by controller type. Generally, G-code groups are numbered from Group 00 to Group 25. Group 00 consists of *nonmodal commands*. A nonmodal command is only active in the block in which it appears and terminates as soon as the function is complete. All other G-code commands are modal and remain in effect until they are canceled, or a new G-code command places the machine in a different mode of operation.

16.3.1 G00: Rapid Positioning

The G00 (or G0) command is only used when the tool is not engaged into the part. This command will move the machine in full rapid movements. This command is commonly used when going to the home position or coming away from the home position. **Figure 16-5** shows a rapid move to the Z–5. position.

Caution should be used when programming movement along multiple axes simultaneously. When a machine is moving in full rapid mode, a block of code with both X and Z coordinates may not necessarily move on a straight line to that position, depending on the machine. Older machines may move at the same rate along each axis to the nearest position first. See **Figure 16-6**.

16.3.2 G01: Linear Interpolation

The G01 (or G1) command is used when entering a piece of material while cutting, and when exiting a cut. This command creates linear movement from the current position to a specified position. *Linear interpolation* refers to determining a straight-line distance by calculating intermediate points between a start point and end point. The G01 command is followed by a feed rate command (F command) to designate the rate at which the cutting tool moves into the material in inches per revolution (ipr). See **Figure 16-7**.

G-Code Groups

Group Number	Description	G-codes
00	Nonmodal G-codes	G04, G09, G10, G28, G29, G52, G53, G70, G71, G72, G73, G74, G75, G76
01	Motion commands	G00, G01, G02, G03, G32, G92, G94
02	Plane selection	G17, G18, G19
06	Unit input	G20, G21
07	Tool nose radius offset	G40, G41, G42
09	Canned cycles	G80, G81, G82, G83, G84, G85, G86, G88, G89
12	Coordinate system	G54, G55, G56, G57, G58, G59
13	Constant surface speed	G96, G97

Goodheart-Willcox Publisher

Figure 16-4. G-code groups.

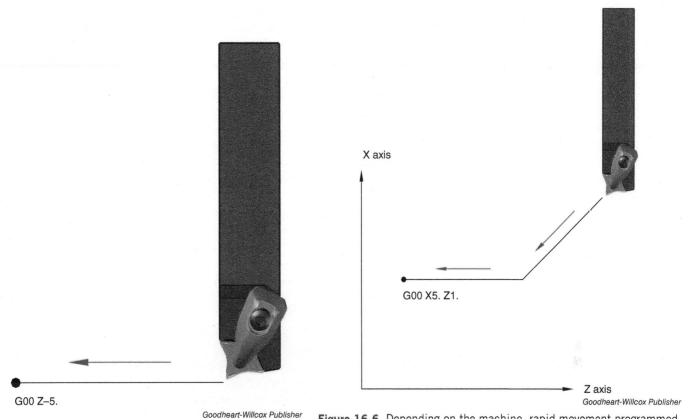

Figure 16-5. The G00 command sets the machine to rapid movement. Shown is a programmed rapid move to the Z–5. position.

Figure 16-6. Depending on the machine, rapid movement programmed along two axes with the G00 command may not occur along a straight line. Shown is a programmed move to X5. Z1. Notice that the movement of the machine was to the closest axis position first.

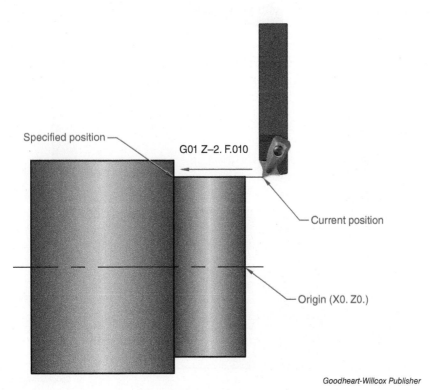

Figure 16-7. The G01 command creates linear movement. Shown is a programmed move to the Z–2. position. A feed rate command (F command) follows the G01 command and specifies the rate at which the cutting tool moves into the material.

When more than one axis of movement is programmed on a single block of code, the G01 command will make a direct linear move to the specified coordinates. The example in **Figure 16-8** shows a move to X7. Z–6.

16.3.3 G02 and G03: Circular Interpolation

The G02 (G2) and G03 (G3) commands produce radial or circular tool movement at a specified feed rate. The G02 command produces a clockwise rotation, and the G03 command produces a counterclockwise rotation. See **Figure 16-9**.

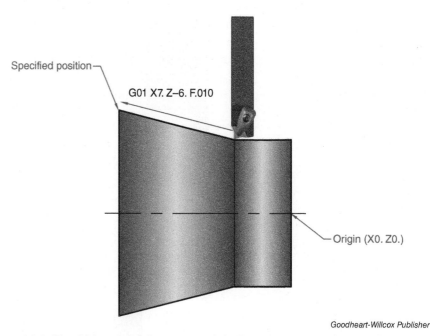

Goodheart-Willcox Publisher

Figure 16-8. The G01 command creates straight-line movement to the specified coordinates.

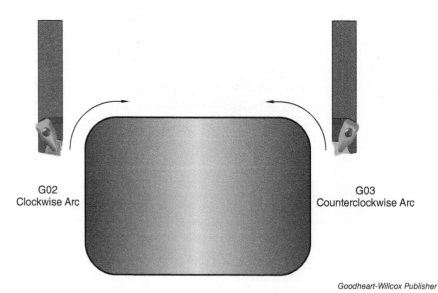

Goodheart-Willcox Publisher

Figure 16-9. The G02 and G03 commands produce motion in a circular path. The G02 command generates a clockwise rotation, and the G03 command generates a counterclockwise rotation.

When programming a G02 or G03 command, it is common to use a G01 command first to position the tool at the start point of the radius or arc, then use the G02 or G03 command to complete the arc. The arcs generated can be partial arcs or full circles, depending on the code. Since these are modal commands, the G01 command will need to follow the G02 or G03 command to return the machine condition to linear interpolation mode. Programming methods using the G02 and G03 commands are explained later in this text.

16.3.4 G20 and G21: Measurement Modes

The G20 and G21 commands set the type of working units used by the machine. CNC machines can read programs in US Customary units or metric units. In the US Customary system, also known as the English (or Imperial) system, the basic unit of linear measurement is the inch. Most prints for manufacturing in the United States are dimensioned in decimal inches. On metric prints, dimensions are commonly specified in millimeters. It can be beneficial to program in one unit format over another. For example, if the print is dimensioned in metric units, it is logical to write the CNC code in the same units.

The G20 command sets the machine to use US Customary units. The G21 command sets the machine to use metric units. Using the G21 command means that the lines of code and specified offsets will be based on metric units. Always make sure to specify the correct code for the unit format. On some machines, the G20 and G21 commands check that the machine control has been set to the correct mode for the program and will produce an alarm if the units are not set correctly.

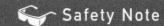

Safety Note

Be aware that if tool offsets are specified while the machine is in inch mode and then the program uses a G21 command, the offsets will not automatically convert. For example, the machine will read a 15″ offset as a 15 mm offset. This is potentially dangerous and can result in a machine crash. Always specify the appropriate unit format for machining and do not use more than one unit format in a program.

16.3.5 G28: Machine Zero Return

The G28 command is used to return the machine to the home position (the machine zero position). Returning the machine to the home position is known as *homing* the machine. The G28 command is commonly used at the end of a program and is normally used prior to a tool change.

The G28 command sends the machine to the home position in rapid mode along the specified axis or axes. For example, a block written as **G28 U0. W0.** sends the machine home along the X and Z axes. In lathe programming, the letters U and W are alternate designations used to specify incremental positioning coordinates along the X and Z axes. Incremental coordinates represent a relative measurement from the current position of the tool, not from the work coordinate system origin. As discussed in Chapter 15, in most cases, X and Z coordinates are specified in lathe programming to command absolute positioning moves. Specifying the U and W codes with the G28 command is a practical example of using incremental positioning in lathe programming. In mill programming, the G91 command is used to activate incremental positioning mode. The U and W codes are used instead of this method in lathe programming.

The G28 command allows the programmer to send the machine home via an intermediate point called the *reference point*. In the **G28 U0. W0.** block, the coordinates U0. W0. designate the amount of incremental movement from the current position to the reference point. Because each coordinate is zero, the reference point is bypassed when the block is read by the machine. In effect, the machine interprets the current position as the reference point and moves directly from the current position to machine home along the specified axes (the X and Z axes). Always exercise caution when using the G28 command and be certain of the current position of the tool. Ensure the tool is in a safe position prior to homing the machine to prevent engagement with the part, workholding, or tailstock.

16.3.6 G40, G41, and G42: Tool Nose Radius Compensation

The G41 and G42 commands are used to offset the tool to the left or right of the programmed cutting path. When programming coordinate positions in a CNC lathe program without tool nose radius compensation activated, the theoretical tip of the tool is being moved to those points. The machine does not account for the radius of the cutting tool being used. To offset the center of the tool and allow the tool edge to follow the correct path, the G41 or G42 command is used. Each command uses the stored radius offset setting for the cutting tool. The G41 command activates tool nose radius compensation left and the G42 command activates tool nose radius compensation right. The terms *left* and *right* refer to the side of the material on which the tool is cutting. If the tool is cutting on the right-hand side of the material, the G42 command is used. The G42 command will offset the tool to the right-hand side of the material. The offset direction of the tool is determined by viewing the tool from behind as it moves in the cutting direction. See **Figure 16-10**.

If the tool is cutting on the left-hand side of the material, the G41 command is used. The G41 command will offset the tool to the left-hand side of the material. The offset direction of the tool is determined by viewing the tool from behind as it moves in the cutting direction. See **Figure 16-11**.

In lathe operations, radii and chamfered features will not be accurate if tool nose radius compensation is not used. Typically, the G42 command is used when machining an outside diameter and the G41 command is used when machining an inside diameter, but this can vary depending on the cutting direction.

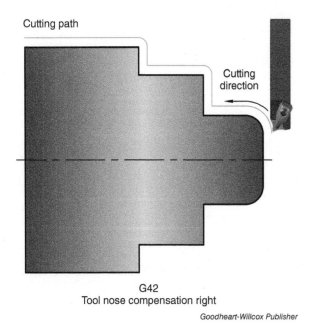

Figure 16-10. The G42 tool nose radius compensation command is used to offset the tool to the right-hand side of the material. This command is used when the tool is cutting an outside diameter and the machining process is on the right side of the material.

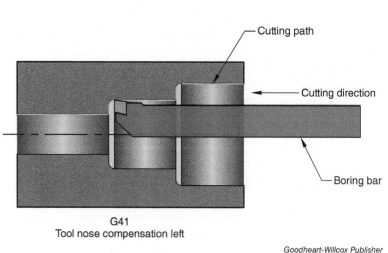

Figure 16-11. The G41 tool nose radius compensation command is used to offset the tool to the left-hand side of the material. This command is used when the tool is cutting an inside diameter and the machining process is on the left side of the material.

The G41 and G42 commands are modal commands and will continue to offset the tool for any additional moves until canceled. The G40 command cancels the G41 or G42 command. Additional explanations and examples of the G41 and G42 commands are provided later in this text.

16.3.7 G70, G71, G72, and G73: Multiple Repetitive Cycles

The G70, G71, G72, and G73 commands are canned cycle commands. A *canned cycle* is an abbreviated machining cycle that automates complex repetitive commands and functions. The G70, G71, G72, and G73 cycles are the most robust and powerful canned cycles used for lathe work. These cycles allow for heavy material removal through a process of machining in repeating patterns, while using only a few lines of programming code. These cycles are explained in more detail in Chapter 19, but an introduction is provided here because of their importance.

The G70 cycle is a profile finishing cycle used in conjunction with the G71, G72, and G73 cycles. The G71, G72, and G73 cycles are commonly used for heavy roughing operations. These cycles greatly simplify programming and increase the efficiency of machining parts with contours made up of multiple diameters.

The G71 cycle is designed to rough cut all the excess material from a piece of stock by producing tool motion in the Z axis direction. It is used in turning and boring operations. A part machined using the G71 cycle is shown in **Figure 16-12**.

Each blue line represents a path traveled by the cutter to machine excess stock. In a point-to-point program written in long form, each pass would require a line of machine code. Using the G71 cycle reduces the program to just a few lines of code and makes editing much simpler. The codes used to define parameters in the following program are discussed in greater detail in the following chapters. The G01 and G03 commands are used to define the contour of the part. The X axis coordinates define the

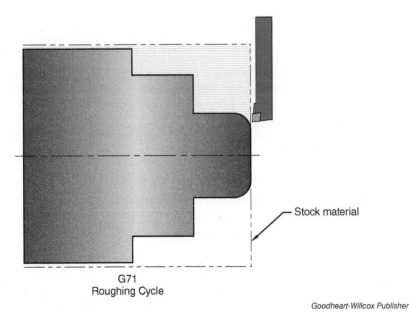

G71 Roughing Cycle

Goodheart-Willcox Publisher

Figure 16-12. The G71 cycle is used to rough cut stock material by cutting in the Z axis direction.

diameters of the contour and the Z axis coordinates define length dimensions. The dimensions and cutting parameters specified in the program define where stock material is to be removed and are used by the machine controller to calculate the roughing passes.

G71 P100 Q110 U.02 W.01 F.01	
N100 G00 X2.9312 S200	Start of part contour
G01 Z-.0005	
G03 X4.9312 Z-1.0313 I-.0313 K-1.0308	
G01 Z-3.4275	
X9.358	
G03 X9.4205 Z-3.4588 K-.0313	
G01 Z-7.1132	
X12.446	
G03 X12.5085 Z-7.1444 K-.0313	
G01 Z-13.6378	
N110 X13.	End of part contour

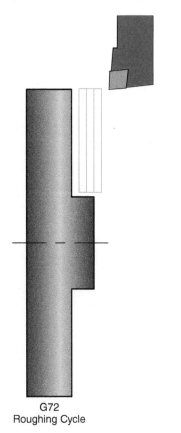

G72
Roughing Cycle

Goodheart-Willcox Publisher

Figure 16-13. The G72 cycle is used to rough cut stock material by cutting in the X axis direction.

The G72 cycle performs a similar operation, but the travel of the cutting tool is in the X axis direction. See **Figure 16-13**. This cycle is used for facing operations and is designed for machining large amounts of material off the face of the part.

The G73 cycle, referred to as the irregular path stock removal cycle, is used for roughing parts with irregular contours. See **Figure 16-14**. This cycle is specifically designed to follow an existing irregular contour, usually on a casting. The tool makes a series of cuts in repeating passes as it reduces the dimensions of the stock.

The G70 cycle is used to create the final finish pass on the part after using the G71, G72, or G73 cycle. Chapter 19 provides additional explanations and examples of these cycles.

16.4 Startup Blocks

Most CNC programmers will begin new programs with a ***startup block***. This block, also referred to as a default block or safety block, cancels any machine conditions left from the previous program or establishes a new starting condition for the current program. Canceling modes that may have been previously active prevents errors and helps ensure the machine will start and operate safely. While this practice can vary from programmer to programmer, a typical startup block could be written as **G80 G18 G40 G20**. The commands are read as follows:

- **G80.** Cancels a previously active canned cycle.
- **G18.** Designates the XZ plane for machining.
- **G40.** Cancels tool nose radius compensation.
- **G20.** Places the machine in inch mode.

G73
Irregular Path Stock Removal Cycle

Goodheart-Willcox Publisher

Figure 16-14. The G73 cycle is used for rough machining operations on parts with irregular contours.

Chapter 16 Review

Summary

- The program address *G* identifies a preparatory command.
- The purpose of a preparatory command is to prepare the machine controller, or preset the machine, into a specific state of operation.
- The G00 command places the machine in rapid positioning mode.
- The G01, G02, and G03 commands are interpolation commands used to move the machine to specific coordinate positions.
- Multiple G-code commands can be used on a single block in a program as long as they do not conflict with each other.
- A modal command stays active until it is canceled or until a subsequent command changes the machine's condition.
- A nonmodal command is only active in the block in which it appears.
- G-code commands are grouped by their use and purpose.
- CNC machines can operate using US Customary or metric units. The G20 or G21 command is used to set the units used by the machine.
- The G41 command offsets the tool to the left-hand side of the part. The G42 command offsets the tool to the right-hand side of the part. The G40 command cancels any tool nose radius compensation in the controller.
- The G70, G71, G72, and G73 cycles are specific CNC lathe canned cycles designed to reduce programming complexity. The G71 cycle is a rough stock removal cycle used in turning and boring operations. The G72 cycle is a rough stock removal cycle used in facing operations. The G73 cycle is used for roughing parts with irregular contours. The G70 cycle is a profile finishing cycle for the G71, G72, and G73 cycles.

Review Questions

Answer the following questions using the information provided in this chapter.

Know and Understand

1. The letter used in a word in CNC programming is referred to as a(n) ____.
   A. symbol
   B. coordinate
   C. address
   D. designator

2. G-code commands are considered ____ commands.
   A. address
   B. compensation
   C. incremental
   D. preparatory

3. The ____ command will initiate full rapid machine travel.
   A. G00
   B. G01
   C. G02
   D. G03

4. *True or False?* To program an angled move while cutting, use the G01 command.

5. To produce radial tool movement in a counterclockwise rotation, the ____ command is used.
   A. G00
   B. G01
   C. G02
   D. G03

6. To move the tool to a position of X5. Z3. in relationship to the origin, use the ____ command.
   A. G00
   B. G18
   C. G20
   D. G40

7. To return the machine to the home position, the _____ command is used.
   A. G18
   B. G28
   C. G41
   D. G42

8. *True or False?* The G80 command is used to cancel tool nose radius compensation.

9. To position a tool cutting on the left-hand side of a piece of material, the correct tool nose radius compensation command is _____.
   A. G40
   B. G41
   C. G42
   D. G43

10. To position a tool cutting on the right-hand side of a piece of material, the correct tool nose radius compensation command is _____.
    A. G40
    B. G41
    C. G42
    D. G43

11. *True or False?* The G43 command cancels the G41 or G42 command.

12. *True or False?* The G20 command sets the machine to use metric units.

Apply and Analyze

Use the illustration below to answer Questions 1 and 2.

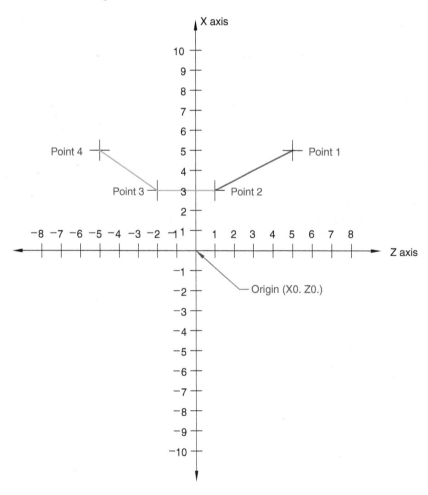

1. What are the correct lines of code to move to Point 1 and then to Point 2 in rapid mode?
2. What are the correct lines of code to move to Point 3 and then to Point 4 in feed mode? Use a feed rate of .010.
3. Identify the correct G-code command to activate tool nose radius compensation based on the cutting direction of the tool.

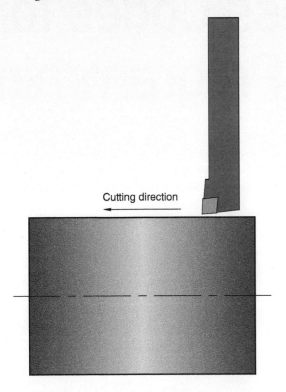

4. Identify the correct G-code command to activate tool nose radius compensation based on the cutting direction of the tool.

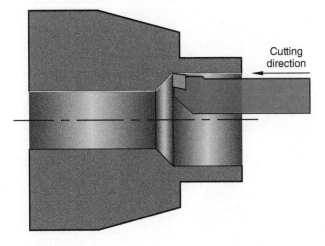

5. Identify and briefly describe the two types of working unit formats and corresponding commands used in CNC programming.
6. Explain the difference between modal and nonmodal commands.
7. Why are G-code commands organized into groups?
8. What is the purpose of the G71 cycle and how does it differ from the G72 cycle?

Critical Thinking

1. Think about the movement that takes place in your daily routine. How many different types of modes do you enter? Consider walking, running, sitting, standing, riding in a vehicle or an elevator, and other types of movement. If you were to assign a G-code to represent each type of movement, how many different codes would it take?
2. Consider the use of G-codes by different machines. Why is it important for G-codes to be standardized for different machine types? How does this simplify machining? Explain the difficulty of using different machine controllers with different G-codes.
3. Depending on the type of machining and programming you may attempt in a machine shop, the most common G-codes used may vary. What are some common G-codes you might use if you were machining? How does the type of machining influence the selection of G-codes?

17 Miscellaneous Functions and Address Codes for Lathe Programming

Chapter Outline

17.1 Introduction
17.2 Use of M-Codes
17.3 M-Codes as Program Functions
  17.3.1 M00: Program Stop
  17.3.2 M01: Optional Stop
  17.3.3 M02, M30, and M99: Program End
17.4 M-Codes as Machine Functions
  17.4.1 Coolant Functions
  17.4.2 Spindle Functions
17.5 Address Codes
  17.5.1 B: Tailstock Movement
  17.5.2 C: Rotary Axis Movement and Automatic Chamfering
  17.5.3 D: Depth of Cut
  17.5.4 E: Precision Feed Rate
  17.5.5 F: Feed Rate
  17.5.6 N: Block Number
  17.5.7 O: Program Number
  17.5.8 P and Q: Cycle Starting Block and Cycle Ending Block References
  17.5.9 S: Spindle Speed
  17.5.10 T: Tool Number
  17.5.11 U and W: Alternate Axis Designations

Learning Objectives

After completing this chapter, you will be able to:

- Explain the purpose of miscellaneous functions.
- Identify machine functions and program functions for turning operations.
- Describe the functions of the stop codes M00 and M01.
- Describe the functions of the end codes M02, M30, and M99.
- Apply techniques for initiating the coolant codes M07, M08, and M09.
- Use the spindle control functions M03, M04, and M05.
- Identify common address codes in CNC turning operations.
- Use address codes in a turning program.

Key Terms

address
address code
flood coolant
M-code
miscellaneous function
mist coolant
program loop
program number
program rewind
turret
word

17.1 Introduction

CNC programs are constructed using words to create blocks of code. As you learned in Chapter 16, a *word* is a letter and number grouping used to execute a command in a program. The letter preceding the number is called an *address*. This chapter looks specifically at the letter addresses that define what the machine controller should do with the numerical data that follows in a block of code.

Chapter 16 covered commonly used G-codes in lathe programming. Programming codes used to control machine functions and program functions are categorized as *M-codes*. The program address M identifies a *miscellaneous function*. Miscellaneous functions are referred to as M-functions or M-codes. Sometimes, M-codes are referred to as machine functions. However, not all M-codes control machine functions, so this is not an accurate definition. Some M-codes relate only to controlling the program process and not a machine function. In this text, M-codes are referred to as miscellaneous functions.

In CNC programming terminology, M-codes and G-codes both fall into a general classification called *address codes*. There are other address codes in addition to M-codes and G-codes in programming. This chapter covers commonly used M-codes and other address codes used in lathe programming.

In a CNC program, the programmer often needs to control machine functions or the program operation outside of machine movements or cutting commands. This is most often done with the functions of M-codes. Examples of machine control functions include turning the spindle on, turning the coolant on, and operating a tailstock movement. Examples of program control functions include program stops, program rewinds, and program optional stops.

Since some M-codes are used to control machine functions, they can vary greatly by manufacturer and control type. Even the same type of lathe can use different M-codes, depending on the functions and features of the machine. This chapter covers the most commonly used M-codes in lathe programming. Not all M-codes are discussed. The table in **Figure 17-1** lists the most commonly used M-codes for lathe operations.

17.2 Use of M-Codes

Unlike G-codes, where multiple codes can be placed on a single block of program code, there can only be one M-code on a single block of code. The M-code can exist on its own line and perform a function. For example:

G00 X2.375 Z.1	
M08	Coolant on
G01 Z–1.625 F.010	

In this example, the machine moves in rapid mode to an absolute X and Z coordinate, and then the coolant is initiated with the M08 code. On the last line, a feed rate move is commanded to the Z–1.625 position at a rate of .010 inches per revolution (ipr). An alternative to these lines of programming could be the following:

G00 X2.375 Z.1 M08
G01 Z–1.625 F.010

Commonly Used M-Codes

M-Code	Description	Controller
M00	Program stop	Fanuc/Haas
M01	Optional program stop	Fanuc/Haas
M02	Program end	Fanuc/Haas
M03	Spindle on clockwise	Fanuc/Haas
M04	Spindle on counterclockwise	Fanuc/Haas
M05	Spindle stop	Fanuc/Haas
M07	Coolant on—mist coolant / Coolant on—shower coolant	Fanuc/Haas
M08	Coolant on	Fanuc/Haas
M09	Coolant off	Fanuc/Haas
M10	Chuck open	Fanuc/Haas
M11	Chuck close	Fanuc/Haas
M17	Rotate turret forward	Fanuc/Haas
M18	Rotate turret reverse	Fanuc/Haas
M19	Orient spindle	Fanuc/Haas
M21	Tailstock forward	Fanuc/Haas
M22	Tailstock retract	Fanuc/Haas
M23	Thread angle pull-out on	Fanuc/Haas
M24	Thread angle pull-out off	Fanuc/Haas
M30	Program end and rewind	Fanuc/Haas
M31	Chip auger forward	Fanuc/Haas
M33	Chip auger stop	Fanuc/Haas
M41	Spindle low gear override	Fanuc/Haas
M42	Spindle high gear override	Fanuc/Haas
M85	Automatic door open	Haas specific
M86	Automatic door close	Haas specific
M88	High pressure coolant on	Haas specific
M89	High pressure coolant off	Haas specific
M98	Subprogram call	Fanuc/Haas
M99	Subprogram return or loop	Fanuc/Haas

Goodheart-Willcox Publisher

Figure 17-1. M-codes commonly used in lathe programming.

Is there a difference in these two programming approaches? Yes—the difference is *when* the coolant is turned on. In the first example, the X and Z positioning is fully completed, and then the coolant is activated. In the second example, the coolant is initiated as the X and Z axis moves are happening.

Not all M-codes can be used in this manner. M-codes that control program processing functions should not be on a line with movement commands. M00, M01, M02, M30, M98, and M99 are examples of functions that should reside either on their own line or on lines with no movement. For example:

G01 X3.25 Z.5 F.015 M00

Safety Note

It is always essential to know how your machine operates and how the controller carries out functions. On many machines, coolant will initiate with axis movement when the M08 function is read by the controller on the same line. However, this is not true of every machine. The Haas controller will complete motion first and then initiate coolant, even if these actions are on the same line of code. Be cautious and verify that the coolant is coming on before the tool starts the cut. If this is not the case, edit the program as needed.

From the Programmer

Sequence of Commands

CNC program lines are not read from left to right, one block at a time, like you would read a sentence. The full line of code is processed at the same time. That is why an X and Z coordinate move on the same line creates a straight-line angle and not two straight lines. The X and Z coordinates are being processed simultaneously. Likewise, depending on the machine controller, coolant is turned on simultaneously when tool motion initiates. This occurs regardless of where the M08 code is placed on the line. The order in which commands are placed within a single line of code does not make a difference. However, most programmers will follow a certain pattern in their methodology and specify commands in a logical order that represents a normal sequence of functions.

This block of code has a feed movement command to an X and Z location, but the M00 code specifies a program stop. Although this block of code will function, it is not recommended or considered the best programming practice. The correct lines of programming should read as follows:

```
G01 X3.25 Z.5 F.015
M00
```

This code allows the machine to complete the X and Z axis movement command, and then perform the program stop function.

M-codes can be modal or nonmodal. Nonmodal M-codes are only active in one block of the program. Modal M-codes stay active until they are canceled or altered. **Figure 17-2** shows partial listings of modal and nonmodal M-codes.

For example, in a line of code that states **G97 M03 S3500**, the spindle is turned on clockwise at 3500 revolutions per minute (rpm). The spindle will remain rotating until the program ends or an M05 (spindle stop) code is initiated in the program body. The M04 function can also change the direction of the spindle to counterclockwise, thus canceling the M03 function. However, it is best practice to use the M05 code to cancel the spindle rotation to prevent possible damage to the spindle.

17.3 M-Codes as Program Functions

As previously discussed, miscellaneous functions can be used to control either program processing functions or machine functions. Program processing functions can be used to stop, pause, or end the program processing sequence. There are multiple applications for these functions, and there are specific places in programs to use them.

Modal M-Codes

M-Code	Description	Codes that Cancel
M03	Spindle on clockwise	M05
M04	Spindle on counterclockwise	M05
M05	Spindle stop	M03, M04
M07	Mist coolant or shower coolant on	M09, M08
M08	Coolant on	M09, M07
M09	Coolant off	M07, M08

Nonmodal M-Codes

M-Code	Description
M00	Program stop
M01	Optional program stop
M02	Program end
M30	Program end and rewind

Goodheart-Willcox Publisher

Figure 17-2. Examples of modal and nonmodal M-codes.

17.3.1 M00: Program Stop

The M00 function is defined as a *mandatory program stop*. Each time the machine controller reads the M00 code, the spindle will stop, any machine motion will stop, and the part program will stop. No control settings will change and the program functions will resume with a cycle start button push. Once the cycle start is engaged, the previous feed rate, coolant condition, and coordinate system will still be active. The spindle will need to be restarted with an M03 or M04 code, but the spindle speed set prior to the M00 code will be restored.

The M00 code is used any time the program needs to completely stop. For example, the program can be stopped to perform an inspection check of a feature that was just machined. See **Figure 17-3**. In this example, the G71 and G70 canned cycles will be used to rough and finish the part. Before proceeding to the G70 finish pass, the program can be stopped with an M00 code to inspect the part dimensions.

O0425	
G20	
(TOOL – 1 OFFSET – 1)	
(OD ROUGH RIGHT – CNMG-432)	
G00 T0101	Tool change
G18	
G97 S180 M03	
G00 G54 X4.25 Z.1 M08	
G50 S3500	
G96 S200	
G71 U.075 R0.	
G71 P100 Q110 U.01 W.005 F.01	
N100 G00 X2. S200	Start of part contour
G01 Z–1.	
X3. Z–3.	
X4.25	
N110 Z–3.8	End of part contour
G00 Z.1	
M00 (CHECK 2.020 DIAMETER BEFORE FINISHING)	
G70 P100 Q110	
G00 Z.1	
M09	
G28 U0. W0. M05	
T0100	
M30	

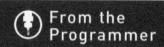

Notes in Programs
Anything written in a program and enclosed by parentheses ( ) is a note and is not read as code. Referring to the previous example, there is a note to inspect the end diameter on the same line as the M00 code. There are also notes used to identify the roughing tool. You can add a note at any place in a program. It is good practice to note part numbers, tool references, and any special machining information for the operator.

Another example of using the M00 code is to flip a part for a secondary operation or to engage the tailstock at a designated point in the program. It is good practice to create a note in the program that explains the purpose of the M00 code when it appears.

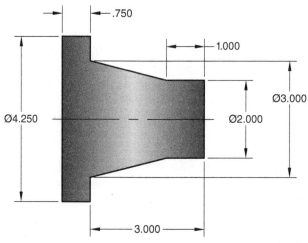

Goodheart-Willcox Publisher

Figure 17-3. A part to be machined with the G71 heavy roughing and G70 finish machining cycles. Before finish machining, the M00 code is used to stop the program to check the part dimensions.

17.3.2 M01: Optional Stop

The M01 function is defined as an *optional program stop*. Similar to the M00 code, the M01 code is added to a program to create a full program stop. The difference is that the function of the M01 code can be turned on or off using the Optional Stop switch on the control panel. When this switch is turned on, the program will stop when an M01 code is read. If the switch is off, the program will proceed without stopping.

The most common use of the M01 code is with tool change commands. Operators and programmers prefer to have the ability to pause at tool changes and ensure the cutting tool is free from chips or debris. An optional program stop also provides the opportunity to make sure the tool is not broken or damaged. The following is the same program from the previous example, but the M01 code is used instead of the M00 code and is specified before the tool change prior to the finish operation.

O0425	
G20	
(TOOL – 1 OFFSET – 1)	
(OD ROUGH RIGHT – CNMG-432)	
G00 T0101	
G18	
G97 S180 M03	
G00 G54 X4.25 Z.1 M08	
G50 S3500	
G96 S200	
G71 U.075 R0.	
G71 P100 Q110 U.01 W.005 F.01	
N100 G00 X2. S200	Start of part contour
G01 Z–1.	
X3. Z–3.	
X4.25	

Continued

N110 Z-3.8	End of part contour
G00 Z.1	
T0100	
G28 U0. W0. M05	Machine home and spindle off
M01 (PREPARE TO TOOL CHANGE FOR FINISH CUT)	
T0202 (OD FINISH RIGHT – VNMG-431)	
G97 S180 M03	
G00 X4.25 Z.1 M08	
G50 S3500	
G96 S200	
G70 P100 Q110	
G00 Z.1	
M09	
G28 U0. W0. M05	
T0200	
M30	

In this case, the operator can stop at all tool changes to inspect the tool if needed. However, if the Optional Stop switch is in the off position, the program will continue without interruption in machining. Although pausing the program is a good practice, the operator should be aware that this adds time to machining and the program should only be stopped if there is a need for such an inspection.

Programmers will also use the M01 code for verifying the program on the machine during setup. The program can be stopped during tool changes or after critical features are machined to make sure the program is operating as expected. Once the program is verified and running as needed, the Optional Stop switch is turned off and production is allowed to take place.

17.3.3 M02, M30, and M99: Program End

At the end of every program, a function is needed to tell the machine that the program is complete. Without this notification, the operator would need to complete at least two resets on the controller to restart the program and continue the machining process.

Older machines often used the M02 code to define the end of a program. This was common when programs were run from 3.5″ floppy drives or external memory sources. At the end of each program, the machine controller had to be reloaded with the file and reactivated. The following example uses the M02 function to terminate the program.

O1020
G20
(TOOL – 6 OFFSET – 6)
(DRILL .5 DIA.)
G00 T0606
G18
G97 S450 M03
G00 X0. Z.25 M08
Z.1

Continued

```
G99 G01 Z-2.3 F.01
G00 Z.25
G28 U0. W0. M05
T0600
M01
(TOOL - 2 OFFSET - 2)
(OD ROUGH RIGHT - 80 DEG. INSERT - CNMG-432)
G00 T0202
G18
G97 S382 M03
G00 X2.5 Z.1 M08
G50 S3000
G96 S250
G71 U.1 R0.
G71 P100 Q110 U.01 W.005 F.005
N100 G00 X1. S200
G01 Z-.5
X1.1875
G03 X1.75 Z-.7813 K-.2813
G01 Z-1.
X1.9375
G03 X2.5 Z-1.2813 K-.2813
N110 G01 Z-2.
G00 Z.1
G18
G70 P100 Q110
G00 Z.1
M09
G28 U0. W0. M05
T0200
M02                           Program stops here
```

The M30 function ends the program and executes a *program rewind*. The term *rewind* is a reference to the past when CNC programs were loaded onto tapes. The M30 code would stop the program and rewind the tape back to the beginning. Using the M30 function is the most common way to end a CNC program. It stops the program at the end and then resets it back to the beginning. By comparison, the M02 function ends the program but does not reset it to the beginning. After the M30 code is executed, the Cycle Start button can be pushed to start the program again. The following is the same program from the previous example, but the M30 function is used instead of the M02 function to return to the beginning.

```
O1020
G20
(TOOL - 6 OFFSET - 6)
(DRILL .5 DIA.)
G00 T0606
G18
G97 S450 M03
G00 X0. Z.25 M08
Z.1
G99 G01 Z-2.3 F.01
G00 Z.25 M09
G28 U0. W0. M05
T0600
M01
(TOOL- 2 OFFSET - 2)
(OD ROUGH RIGHT - 80 DEG. INSERT - CNMG-432)
G00 T0202
G18
G97 S382 M03
G00 X2.5 Z.1 M08
G50 S3000
G96 S250
G71 U.1 R0.
G71 P100 Q110 U.01 W.005 F.005
N100 G00 X1. S200
G01 Z-.5
X1.1875
G03 X1.75 Z -.7813 K-.2813
G01 Z-1.
X1.9375
G03 X2.5 Z-1.2813 K-.2813
N110 G01 Z-2.
G00 Z.1
G18
G70 P100 Q110
G00 Z.1
M09
G28 U0. W0. M05
T0200
M30
```

Program stops here and rewinds to beginning

The M99 function executes a *program loop*. This code is used at the end of a program, but it does not actually stop a program, it loops the program. As the CNC program reaches the end and the M99 code is processed, the program loops back to the start and continues to run. The M99 code is most often used in subprogramming. It is commonly used at the end of a subprogram to loop back into the main program, or at the end of a program with an M01 or M00 code.

17.4 M-Codes as Machine Functions

Miscellaneous functions that control machine operations are much like electrical switches. One M-code turns on a specific machine function and another M-code is used to turn off that function. Most machines come equipped with spare M-codes for special functions. The spare M-codes can be used to operate optional equipment or electrical sources for add-on features.

17.4.1 Coolant Functions

Many machining functions require the use of coolant to reduce friction and heat during the machining process. Machine coolants have become more advanced over time, and delivery methods from the machine have also evolved. Two common delivery methods for coolant are flood delivery and mist delivery. **Flood coolant** is liquid coolant that floods the cutting area. **Mist coolant** is a mix of fluid and air sprayed at the cutting area. Machines can be equipped with a variety of coolant delivery systems. Some machines have the ability to deliver coolant through the spindle. This functionality enables coolant to directly reach the cutting area on deep cuts and remove chips through pressure. Some machines are equipped with a programmable coolant nozzle system. The nozzle can be programmed to automatically move to different positions during machining so that coolant targets the cutting area and flushes away chips. In addition to using liquid coolant to clear chips and debris, chip removal can be accomplished with compressed air. Controllers can use different M-codes to control these delivery methods, but the codes most often used are M07 (mist coolant) and M08 (flood coolant).

M07: Mist Coolant

The M07 code activates the mist coolant function. Mist cooling combines a light air blast with microparticles of coolant and provides a direct spray at the cutting edge. This method minimizes consumption of coolant and reduces condensation loss, making it cost effective.

The M07 function can be activated at any point in the program. Where it is programmed determines the timing of the mist arriving at the cutter or material, as shown in the following example programs.

Example 1:

O1020	
G20	
(TOOL – 6 OFFSET – 6)	
(DRILL .5 DIA.)	
G00 T0606	
G18	
G97 S450 M03	
G00 X0. Z.25 M07	Mist coolant turns on during initial position move
Z.1	
G99 G01 Z–2.3 F.01	
G00 Z.25 M09	
G28 U0. W0. M05	
T0600	
M01	

Example 2:

```
O1020
G20
(TOOL – 6 OFFSET – 6)
(DRILL .5 DIA.)
G00 T0606
G18
G97 S450 M03
G00 X0. Z.25
Z.1 M07                         Mist coolant turns on before drilling
                                cycle begins
G99 G01 Z–2.3 F.01
G00 Z.25 M09
G28 U0. W0. M05
T0600
M01
```

Example 3:

```
O1020
G20
(TOOL – 6 OFFSET – 6)
(DRILL .5 DIA.)
G00 T0606
G18
G97 S450 M03
G00 X0. Z.25
Z.1
G99 G01 Z–2.3 F.01 M07          Mist coolant turns on as drilling cycle
                                begins
G00 Z.25 M09
G28 U0. W0. M05
T0600
M01
```

These examples show three different approaches to initiating coolant for a drilling operation. The differences in these approaches are minor, but they have an influence on the desired delivery of coolant to the cutting edge. Initiating coolant too early can waste coolant and obstruct the operator's view of the machining process, while initiating coolant too late can damage the cutting tool or material. As previously discussed, be aware of how the machine operates and verify that coolant is arriving before cutting starts.

M08: Flood Coolant

The M08 code is the most widely used coolant function. It floods coolant at full pressure. See **Figure 17-4**. The benefit of flooding the coolant is it provides massive coolant supply to the cutter and material, while flushing any chips or debris away from the part being machined.

Dmitrii Pridannikov/Shutterstock.com

Figure 17-4. Flood cooling floods the cutting area with coolant at full pressure.

Flood delivery of coolant can be a useful tool, especially when substantial metal removal is required or when machining an inside diameter where chip evacuation is limited. Flood cooling supplies a high volume of coolant, provides lubrication, removes heat, and clears chips from surfaces. Chips and debris can cause tool breakage or inferior surface finishes.

Flood cooling requires machines with larger coolant tanks and has additional maintenance costs compared to other coolant delivery methods. Coolant loss occurs in flood cooling due to carryout on parts, accumulation on chips, and evaporation. A flood coolant system must be cleaned regularly and machine coolant must be monitored for contaminants such as bacteria and tramp oil. Tramp oil is unwanted oil that migrates into coolant from an outside source, such as machine lubricants and oil-based protective coatings on materials.

The M08 code is used in the same manner as the M07 code in a CNC program. If the M07 code is active and the M08 code is initiated, the mist cooling function will cease and the flood cooling function will activate.

Delivering coolant through the spindle is an option on some machines. Fanuc controls often use the M08 code for this option, while Haas machines use the M88 code. Other controllers also have functions such as M13, which turns on the spindle and coolant simultaneously. Always check the machine's technical manual to verify specific uses of codes in a machine.

M09: Coolant Off

The M09 code is used to turn coolant off. Any code that was used to initiate coolant will be canceled with the M09 code. This code is often found at the end of programs or at tool change locations, although it is not completely necessary. The coolant will "pause" at tool changes and will be shut off at the end of the program with the M30 or M02 code. The following example uses the M09 function before a tool change.

G99 G01 Z-2.3 F.01	
G00 Z.25 M09	Coolant is turned off
G28 U0. W0. M05	
T0600	
M01	
(TOOL – 2 OFFSET – 2)	
(OD ROUGH RIGHT – 80 DEG. INSERT – CNMG-432)	
G00 T0202	
G18	
G97 S382 M03	
G00 X2.5 Z.1 M08	Coolant is restarted
G50 S3000	

The following example shows what happens at the tool change without the M09 code.

G99 G01 Z-2.3 F.01	
G00 Z.25	
G28 U0. W0. M05	
T0600	
M01	
(TOOL – 2 OFFSET – 2)	
(OD ROUGH RIGHT – 80 DEG. INSERT – CNMG-432)	
G00 T0202	Coolant pauses during tool change
G18	
G97 S382 M03	
G00 X2.5 Z.1 M08	Coolant is restarted after tool change
G50 S3000	

The following example shows what happens at the end of a program without the M09 code.

G00 T0202	
G18	
G97 S382 M03	
G00 X2.5 Z.1 M08	Coolant is turned on after tool change
G50 S3000	
G96 S250	
G71 U.1 R0.	
G71 P100 Q110 U.01 W.005 F.005	
N100 G00 X1. S200	
G01 Z-.5	
X1.1875	
G03 X1.75 Z-.7813 K-.2813	
G01 Z-1.	
X1.9375	
G03 X2.5 Z-1.2813 K-.2813	
N110 G01 Z-2.	

Continued

G00 Z.1	
G18	
G70 P100 Q110	
G00 Z.1	
G28 U0. W0. M05	
T0200	
M30	Coolant is turned off at M30

17.4.2 Spindle Functions

Spindle on forward, spindle on reverse, and spindle stop are all M-code machine functions. For virtually every CNC lathe operation, the spindle requires activation to begin rotation of the material. All spindles can rotate in either a clockwise or counterclockwise direction. On a CNC lathe, the direction of clockwise rotation is established by a view looking from the headstock toward the tailstock along the spindle centerline.

M03: Spindle Forward or Spindle Clockwise

In most cases, the M03 code is used to start the spindle. The M03 code initiates forward (clockwise) rotation of the spindle. The M03 code is accompanied by the spindle speed address code (S) and a numerical value to designate the spindle speed in revolutions per minute (rpm). The following block of code will initiate the spindle to rotate clockwise at 3500 rpm.

G97 M03 S3500

Figure 17-5 shows forward (clockwise) rotation of the spindle and the orientation of the lathe tool. In this configuration, the *turret* holding the lathe tool is located on the side of the spindle farthest from the operator. Notice the yellow insert at the cutting edge. The material is being rotated into the cutting tool to create the cut. Standard lathe tools are designed for cutting in this direction. Depending on the orientation of the tool turret, the tooling may face the opposite direction, but the material is still rotating into the cutting edge. See **Figure 17-6**. In this configuration, the turret holding the lathe tool is located on the side of the spindle closest to the operator. Rotation of the spindle is still in a clockwise direction.

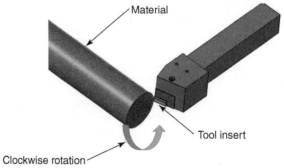

Goodheart-Willcox Publisher

Figure 17-5. The M03 code initiates forward (clockwise) rotation of the machine spindle. Clockwise spindle rotation on a lathe corresponds to a view that looks from the headstock toward the tailstock along the spindle centerline. In this machine configuration, the tool is held in place by a rear turret.

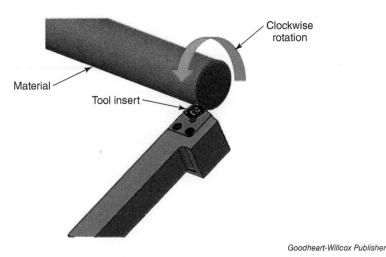

Goodheart-Willcox Publisher

Figure 17-6. The lathe tool in this machine configuration is held in place by a front turret. The tool is oriented so that material rotates into the cutting edge as the spindle rotates clockwise.

The following is an example of using the M03 function to initiate rotation of the spindle in a program. If the M03 function is given on its own block, the machine reverts to the previous spindle speed.

O1020	
G20	
(TOOL – 6 OFFSET – 6)	
(DRILL .5 DIA.)	
G00 T0606	
G18	
G97 S450 M03	Spindle on clockwise to 450 rpm
G00 X0. Z.25 M08	
Z.1	
G99 G01 Z–2.3 F.01	
G00 Z.25 M09	
G28 U0. W0. M05	Spindle off
T0600	
M01	
(TOOL – 2 OFFSET – 2)	
(OD ROUGH RIGHT – 80 DEG. INSERT – CNMG-432)	
G00 T0202	
G18	
M03	M03 function initiated with no speed; will revert to previous speed (450 rpm)
G00 X2.5 Z.1 M08	

M04: Spindle Reverse or Spindle Counterclockwise

The M04 code turns the spindle in reverse, or in a counterclockwise rotation. The M04 code is not a common function, but there are left-hand drills, taps, and cutting tools that require the use of the M04 code. The M04 code is utilized in the same manner as the M03 code, and it must also be accompanied by a spindle speed specification.

```
O1020
G20
(TOOL – 6 OFFSET – 6)
(LEFT HAND DRILL .5 DIA.)
G00 T0606
G18
G97 S450 M04                    Spindle on counterclockwise to 450 rpm
G00 X0. Z.25 M08
Z.1
G99 G01 Z–2.3 F.01
G00 Z.25 M09
```

M05: Spindle Stop

The M05 spindle stop function is used to stop all spindle motion. It is most commonly used before a tool change or at the end of the program. It is not mandatory to stop at these locations, because a tool change and an M30 or M02 program stop will stop the spindle from turning. Most programmers add this code to their normal programming code to shut the spindle off early and safely, before any doors on the machine would open and cause a possible safety concern.

```
O0425
G20
(TOOL – 1 OFFSET – 1)
(OD ROUGH RIGHT – CNMG-432)
G00 T0101
G18
G97 S180 M03
G00 G54 X4.25 Z.1 M08
G50 S3500
G96 S200
G71 U.075 R0.
G71 P100 Q110 U.01 W.005 F.01
N100 G00 X2. S200
G01 Z–1.
X3. Z–3.
X4.25
N110 Z–3.8
G00 Z.1
G70 P100 Q110
G00 Z.1
M09
G28 U0. W0. M05              Stops spindle as machine returns to
                              home position
T0100
M30
```

Remember that different machines and controllers can have a multitude of M-codes. Many machines come with optional M-code functions specifically designed to work as switches to turn external devices on or off. As machines become increasingly sophisticated, the need for additional M-code functions will also increase.

17.5 Address Codes

As previously discussed, an address code is a letter code that is coupled with a number to create a word. The address code defines what the machine controller should do with the numerical data that follows. Address codes are fairly standard between machine controllers and even between lathe and mill applications. However, some address codes are specific to certain machine types and there are additional address codes used for programming multiaxis machines. **Figure 17-7** lists commonly used address codes for lathe operations. Some address codes are used to control more than one parameter in programming. Specific uses of address codes in programs are discussed later in this text.

Commonly Used Address Codes for Lathe Programming

Code	Description	Specific to Lathe
B	Tailstock advance and retract	Yes
C	Rotary axis about Z axis and chamfer distance in automatic chamfering	Yes
D	Depth of cut for G71, G72, and G76 canned cycles	Yes
E	Precision feed rate for threading cycles	Yes
F	Feed rate function	
G	Preparatory command	
I	X axis distance in circular interpolation	
K	Z axis distance in circular interpolation	
L	Loop count for canned cycles	
M	Miscellaneous function	
N	Block number	
O	Program number	
P	Program designator in subprogramming and starting block number in G70, G71, G72, and G73 cycles	Yes
Q	Canned cycle peck amount and ending block number in G70, G71, G72, and G73 cycles	Yes
R	Arc radius in circular interpolation and rapid plane reference in canned cycles	
S	Spindle speed function	
T	Tool number designator	
U	X axis incremental movement	
W	Z axis incremental movement	
X	X axis movement	
Z	Z axis movement	

Goodheart-Willcox Publisher

Figure 17-7. Address codes commonly used in lathe programming.

17.5.1 B: Tailstock Movement

The B address is used to advance and retract a programmable tailstock along the B axis of a CNC turning center. The B axis serves as an axis for forward or reverse tailstock motion. A move in the negative direction (– value) is toward the spindle and a move in the positive direction (+ value) is away from the spindle. See **Figure 17-8**. The B axis is used for optional equipment on most machines.

17.5.2 C: Rotary Axis Movement and Automatic Chamfering

The C address is used to specify rotary axis movement where rotation is around the machine's Z axis (the spindle centerline). As discussed in Chapter 15, the C address references the C axis. The C axis is used with live tooling in lathe operations to machine features on the periphery of the part.

The C address is also used to program distances for 45° chamfered corners in automatic chamfering. This greatly simplifies the process of programming chamfers by eliminating the need to calculate the X and Z positions of the chamfer. For example, adding a C address value to a block with an X or Z linear move creates a 45° chamfer. The following blocks of code are used to create the .125 × 45° chamfer for the part shown in **Figure 17-9**.

```
G00 X0. Z0.
G01 X1.75 C.125 F.010
Z–1.50
```

The initial position of the tool is X0. Z0. In the second block, specifying a C address value of .125 in conjunction with a linear move to X1.75 creates the 45° chamfered corner automatically. The last block makes the transition down the part in the negative Z axis direction.

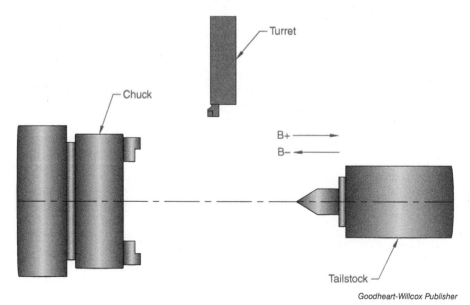

Goodheart-Willcox Publisher

Figure 17-8. A programmable tailstock allows for forward and reverse movement of the tailstock along the B axis in lathe operations.

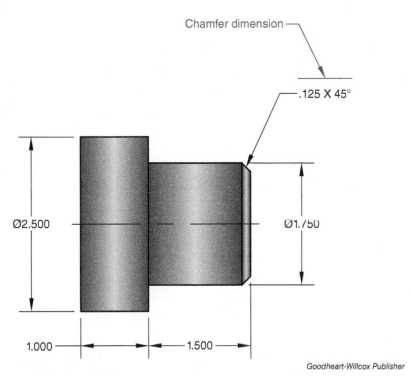

Figure 17-9. Automatic chamfering is used to generate 45° chamfered corners on a part automatically. The C address is used to designate the chamfer distance.

From the Programmer

Automatic Chamfering Addresses

Not all machines use the C address in automatic chamfering. On some machines, the I and K addresses are used to designate automatic chamfer distances. The I address designates a chamfer distance when moving from the Z axis to the X axis. The K address designates a chamfer distance when moving from the X axis to the Z axis.

17.5.3 D: Depth of Cut

The D address is used to designate the depth of cut (DOC) in roughing and threading cycles. It is used in the G71, G72, G73, and G76 canned cycles. The following example shows how the depth of cut is specified in the G71 roughing cycle.

G50 S3600	
G96 S200	
G71 P10 Q20 U.02 W.01 D.05 F.01	The D value represents a .050 DOC radially for each roughing pass
N10 G00 X.2 S200	
G01 Z0.	

In the G76 threading cycle, the D address value designates the depth of the first cut only.

17.5.4 E: Precision Feed Rate

The E address can be used to specify the feed rate in threading cycles. It is programmed in the same manner as the F address, but with exacting precision to allow for threads requiring tight tolerances. The E address is not often used on newer machines because of improvements in tolerance control and repeatability of parts.

17.5.5 F: Feed Rate

The F address is used to command the feed rate that the cutting tool will travel. In lathe programming, the feed rate is usually expressed in inches per revolution (ipr). The feed rate can also be expressed in inches per minute (ipm). The G99 command is used to program the feed rate in inches per revolution. The G98 command is used to program the feed rate in inches per minute. The feed rate specified in the program is modal. The last feed rate commanded will stay active until a new feed rate is designated.

```
G99 G01 Z–2.3 F.01
```

In this block, the tool will travel to Z–2.3 at a feed rate of .01 ipr.

17.5.6 N: Block Number

The N address is used for block numbering in a program. A block number, or sequence number, identifies a block of code with a sequential number. An N address number can be placed in front of any block for reference purposes and searching capabilities.

Block numbers are primarily used for organization and may be omitted to reduce program length. In most cases, block numbers have no purpose other than numbering the lines. However, some canned cycles in lathe programming require certain blocks to be numbered in order to be referenced. In the G70, G71, G72, and G73 roughing and finishing cycles, the N address must be used to specify the block where the roughing or finishing contour operation starts.

17.5.7 O: Program Number

The O address is used to designate the *program number*. It is only used in the opening block of a program. Each program begins with the letter O followed by a four- or five-digit number. It is good practice to use a number that represents a part number or engineering number so that the part created by the program can be easily identified.

It is common to include a program name or description on the same line as the program number. The name or description is enclosed in parentheses. The numbering format used by the programmer should follow the naming standards of the company or organization. Sometimes, a revision number is included as part of the program name, as shown in the following example.

```
O42378 (Part #13-42378 Rev 1)
```

17.5.8 P and Q: Cycle Starting Block and Cycle Ending Block References

The P address is used in the G70, G71, G72, and G73 roughing and finishing cycles to reference the number of the starting block where the roughing or finishing contour operation begins. The P address is followed by a Q address, which references the number of the ending block where the operation is completed. The G70 cycle is a finishing cycle used to create the final finish pass on the part after using the G71, G72, or G73 cycle. In the G70 cycle, the P and Q addresses are restated to tell the machine controller where the finish operation starts and ends.

G71 P100 Q110 D.05 U.02 W.01 F.01	The P100 word indicates the roughing cycle starts at block N100; the Q110 word indicates the cycle ends at block N110
N100 G00 X.2 S200	Start of cycle
G01 Z0.	
X–1.75 C.125	
Z–1.5	
X–3.	
Z–2.75	
N110 X–3.1	End of cycle
G00 Z.05	
G70 P100 Q110	The P100 and Q110 words are restated in the G70 cycle to indicate the finishing cycle starts at block N100 and ends at block N110

17.5.9 S: Spindle Speed

The S address is used to set the spindle speed. In lathe programming, the spindle speed can be programmed in two different ways. As in mill programming, the spindle speed can be programmed in revolutions per minute (rpm). The G97 command is used to program the spindle speed in rpm. As the tool is cutting material, the spindle speed remains fixed for the duration of the machining operation.

In outside diameter and inside diameter machining operations, it is common to program the spindle speed relative to the desired surface speed. This instructs the machine controller to adjust the spindle speed during the cut to allow for changes in the material diameter and maintain the same surface speed. Surface speed is measured in surface feet per minute (sfm) and is commonly referred to as surface footage. Programming the spindle speed in sfm places the machine in constant surface speed mode. The G96 command is used to activate constant surface speed mode. Using constant surface speed in turning operations reduces tool wear and produces better surface finishes.

As material is cut in constant surface speed mode, the diameter of the stock is reduced and the machine increases the spindle speed to maintain the programmed sfm. It is common to program a maximum setting for the spindle speed to prevent an unsafe condition in machining. The G50 command is used for this purpose. In the following example, the spindle speed is set to a limit of 3500 rpm in the first block and constant surface speed is activated in the second block with a value of 200 sfm.

G50 S3500
G96 S200

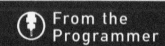

From the Programmer

Spindle Speed Commands
The G96 command is meant for use in turning operations on a lathe. In hole machining operations on a lathe, such as drilling, the G97 command is used to set the spindle to a fixed rpm speed.

17.5.10 T: Tool Number

The T address is used to designate a tool for a tool change. The numerical values following the T address specify the position on the turret where the tool is located and the number of the tool offset stored in the controller. For example, the block **T0101** tells the machine to change to tool #1 and turn on tool offset #1. This programming format is different from mill programming. In lathe programming, there is no M-code used for a tool change. Simply calling the tool will initiate the change.

Before a tool change is initiated, the offset for the previous tool that was in use is canceled, as shown in the following example. The machine must be in a safe position prior to performing the tool change.

T0101	Calls tool #1 and turns on offset #1
...	
...	
...	
T0100	Cancels offset for tool #1
...	
...	
...	
T0202	Calls tool #2 and turns on offset #2

17.5.11 U and W: Alternate Axis Designations

The U and W addresses are used to execute incremental positioning moves in lathe programming. In incremental positioning, coordinate values are measured from the current point to the next point. Incremental coordinates represent a relative measurement from the current position of the tool, not from the work origin. Normally, X and Z axis coordinates are used in lathe programming to specify absolute positioning moves measured from the work origin. The U address is used to specify incremental movement along the X axis and the W address is used to specify incremental movement along the Z axis. For example, the position entry **W–.005** will move the tool .005″ incrementally in the negative Z axis direction. The position entry **U.005** will move the tool incrementally .005″ in the positive X axis direction. In diameter programming, this represents an incremental measurement in diameter. The amount of radial movement is half the change in diameter, or .0025″.

In mill programming, absolute and incremental positioning moves are programmed differently. The G90 and G91 commands set the positioning mode used to locate coordinates. The G90 command sets the positioning mode to absolute positioning and the G91 command sets the positioning mode to incremental positioning. In lathe programming, U and W coordinates are used for incremental movement instead of these commands.

Chapter 17 Review

Summary

- The program address M identifies a miscellaneous function.
- Miscellaneous functions can be used to control machine functions or program processing functions.
- Only one M-code can be placed on a single line of code in a CNC program.
- Miscellaneous codes that control program processing functions can be used to stop, pause, or end the program processing sequence.
- Miscellaneous codes can be used for a variety of machine functions. They act much like electrical switches to turn functions on and off.
- The M00 and M01 codes can be used to create program stops. The M00 code forces the program to stop each time it is encountered. The M01 program stop can be bypassed using a selector switch on the machine control.
- The M02 and M30 codes are used to end a program. The M02 code ends the program while the M30 code stops the program and rewinds to the program beginning.
- The M99 code is used primarily for subprograms to loop back into the main program after the subprogram is complete.
- Coolant functions are controlled with the M07, M08, and M09 codes. The M07 code turns on mist coolant, while the M08 code turns on flood coolant. The M09 code is used to turn off coolant.
- Spindle control functions are commanded with the M03, M04, and M05 codes. The M03 code initiates forward (clockwise) spindle motion. The M04 code initiates reverse (counterclockwise) spindle motion. The M05 code is used to cease all spindle motion.
- An address code is a letter code that is coupled with a number to create a word. The address code defines what the machine controller should do with the numerical data that follows.
- The B address is used to advance and retract a programmable tailstock along the B axis of a CNC turning center.
- The D address is used to designate the depth of cut (DOC) in roughing and threading cycles. It is used in the G71, G72, G73, and G76 canned cycles.
- The F address is used to command the feed rate that the cutter will travel.
- The N address is used for block numbering in a program. The O address is used to designate the program number.
- The S address is used to set the spindle speed. The G97 command is used to program the spindle speed in revolutions per minute (rpm). The G96 command is used to activate constant surface speed mode and program the spindle speed in surface feet per minute (sfm).
- The T address is used to designate a tool for a tool change.
- The U and W addresses are used to execute incremental positioning moves in lathe programming.

Review Questions

Answer the following questions using the information provided in this chapter.

Know and Understand

1. M-codes are used to control ____.
   A. machine movements
   B. cutting commands
   C. machine functions and program functions
   D. work offsets

2. M-codes are considered ____ functions.
   A. preparatory
   B. compensation
   C. interpolation
   D. miscellaneous

3. The ____ command will initiate a full program stop.
   A. M03
   B. M00
   C. M04
   D. M05

4. *True or False?* Multiple M-codes can be used on a single line of a program.

5. The ____ code must be programmed in conjunction with a selector switch on the controller.
   A. M03
   B. M00
   C. M01
   D. M99

6. Which of the following codes is used to loop a program back to the main program or to the program start?
   A. M02
   B. M30
   C. M09
   D. M99

7. The M07 code will activate which of the following functions?
   A. spindle forward
   B. mist coolant on
   C. flood coolant on
   D. coolant off

8. *True or False?* The M03 function initiates spindle forward motion.

9. For the spindle rotation to be stopped, the _____ code is used.
   A. M08
   B. M11
   C. M05
   D. M85

10. *True or False?* The M03 code is accompanied by a spindle speed specification.

11. An address code is a letter code that is coupled with a number to create a _____.
    A. sentence
    B. word
    C. program
    D. block

12. *True or False?* An address code defines what the machine controller should do with the numerical data that follows.

13. On a CNC turning center, which address is used to advance and retract a programmable tailstock?
    A. A
    B. B
    C. C
    D. D

14. *True or False?* The C address is used to designate the depth of cut (DOC) in roughing and threading cycles.

15. *True or False?* The F address is used to command the feed rate that the cutting tool will travel.

16. The _____ address is used in roughing and finishing cycles to reference the number of the starting block where the roughing or finishing contour operation begins.
    A. N
    B. O
    C. P
    D. Q

17. The _____ address is used in roughing and finishing cycles to reference the number of the starting block where the roughing or finishing contour operation is completed.
    A. N
    B. O
    C. P
    D. Q

18. The S address is used to set the _____.
    A. axis of rotation
    B. feed rate
    C. spindle speed
    D. tool number

19. *True or False?* The G96 command is used to program the spindle speed in revolutions per minute (rpm).

20. The _____ addresses are used to execute incremental positioning moves in lathe programming.
    A. W and X
    B. X and Z
    C. U and W
    D. W and Y

Apply and Analyze

1. Referring to the blanks provided in the following program, give the correct M-codes to turn the spindle on clockwise, turn on flood coolant, and turn coolant off before the machine returns to the home position.

G97 S180 _____
G00 X4.25 Z.1 _____
G50 S3500
G96 S200
G70 P100 Q110

Continued

G00 Z.1
_____
G28 U0. W0. M05
T0200
M30

2. Referring to the blank provided in the following program, give the correct M-code to initiate mist coolant after the spindle is turned on and before the machine travels along the Z axis.

G00 T0606
G18
G97 S450 M03
G00 X0. Z.25
Z.1 _____
G99 G01 Z–2.3 F.01
G00 Z.25 M09

3. Referring to the blank provided in the following program, give the correct M-code to initiate an optional stop before the tool change.

G00 Z.1
T0100
G28 U0. W0. M05
_____
T0202

4. Referring to the blanks provided in the following roughing and finishing program, give the correct address codes to indicate the starting and ending blocks for the roughing cycle and activate the finish cycle.

G71 _____ _____ D.05 U.02 W.01 F.01
_____ G00 X.2 S200
G01 Z0.
X1.75 C.125
Z–1.5
X3.
Z–2.75
_____ X3.1
G00 Z.05
_____ P100 Q110

5. Referring to the blanks provided in the following program, give the correct codes to machine the outside diameter. Automatic chamfering is used to create the chamfered corners.

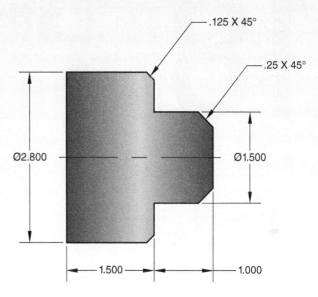

_____ X0. Z0. F.01
X1.5 C.25
Z–1.0
_____ _____
Z–2.5

Critical Thinking

1. Consider the placement of M-codes in a CNC program. What are some possible benefits to starting these functions at different locations inside the program? What are some possible machining failures that could result from specifying an M-code in the wrong location in a program?

2. Why is it important to have spare M-codes available in a CNC machine? What are some add-on equipment features that could be used in a CNC lathe operation?

3. Research CNC multiaxis machines and describe some of the alternative equipment that they use. Is there a need for specific address codes for other types of machines, and what purpose do they serve?

18 Lathe Program Planning

Chapter Outline

18.1 Introduction
18.2 Print Review
18.3 Part Workholding
    18.3.1 Chucks
    18.3.2 Lathe Collet Chuck
    18.3.3 Tailstock
18.4 Tool Selection
18.5 Order of Operations

Learning Objectives

After completing this chapter, you will be able to:

- Explain the importance of preplanning a CNC lathe program.
- Review a print to determine critical features and machining strategy.
- Explain how to establish an appropriate workholding setup based on the final part and program.
- Identify tools needed to create a CNC lathe program.
- Identify features that may reduce the total number of tools required to produce a part.
- Explain how to define the order of operations in a program.

Key Terms

chuck	insert	tailstock
design intent	order of operations	workholding

18.1 Introduction

In this chapter, you will learn how a print can tell you exactly how to machine a part. You will learn how complex operations can be managed by planning the production sequence step-by-step. You will also learn how workholding impacts part production.

Think of the many aspects of machining a part that must be planned and communicated to the machine. What tools are available? How do you control where each tool goes and how fast it gets there, and how do you determine how fast you can machine a piece of material? These are things that are all controlled by the CNC program—and, ultimately, by the programmer.

The most critical, and often overlooked, steps in creating a successful CNC part program occur during the planning stage. Beginning programmers may grab a print and start producing code, adding tools that make sense to them. This approach overlooks the complex issues of part workholding, machine availability, and the part geometry to be machined. An experienced programmer will spend time considering the machines and tooling available, the most efficient workholding method, and the customer's requirements.

It is always a good idea to assemble a team when preparing to program a part and move it to production. A customer representative (perhaps a salesperson or manager), someone from the engineering staff, a setup specialist, and an operator could all provide helpful feedback in the planning process to make a quality finished part. Every program produced and every part manufactured should give end users a product that fits their needs in a cost-effective manner, while giving your company a profitable solution. This type of customer service and smart planning takes a team of highly skilled and organized professionals.

In a machine shop environment, there can be a perception or culture that leaves all of the part production decisions to the programmer. Maybe someone even says, "There just isn't enough time to meet with all these people." Maybe the programmer believes it is easier to work out all of these decisions alone and thinks programs can be created without other input to save time. Consider this in a different way. There is not enough time or money to make scrap parts or pay for the cost of scrap parts. Parts made incorrectly represent wasted time and can lead to missed deadlines or unhappy customers. Mistakes in part production are extremely expensive and can lead to the demise of a machine shop.

Developing a CNC program requires effective planning and decision making. The following general steps are used in the program planning stage to improve efficiency and eliminate unnecessary loss of time. The order of these steps may vary depending on company practice, but this is a typical sequence.

1. Review the print.
2. Establish the part workholding.
3. Identify the required tools.
4. Determine the order of operations.

18.2 Print Review

Interpreting design intent is an important first step in reviewing a print. The term *design intent* refers to the way in which the part is dimensioned and noted to define the features of the part and the part function. For example, if a print is dimensioned exclusively from one point, that location must be critical to the function and orientation of the part. If a tolerance on a feature is ±.001″, that feature is critical. Conversely, if a feature's tolerance is ±.020″, that feature can be considered noncritical. By examining the tolerances of features and the originating points from which tolerances are established, a print reader can gain an understanding of how a part functions, what features are critical in manufacturing, and how the part can be machined.

The part shown in **Figure 18-1** has three features with ±.001″ tolerances. The ±.001″ tolerances are applied to two diameter dimensions and one linear dimension. Of the three diameters, two of them are critical and one is not as critical. See **Figure 18-2**. The critical features should be machined with the same tool and inspected frequently, depending on the number of parts to be produced. The diameter with the ±.020″ tolerance will probably be machined with the same tool as the other two diameters, but it is safe to assume that if the critical diameters meet the tolerance requirements, the noncritical diameter also meets requirements.

> **From the Programmer**
>
> **Print Specifications**
>
> The information on a print provides the specifications needed to manufacture a part. The machinist should make the part to meet the specifications and dimensions provided. Never attempt to measure or scale a drawing to determine a missing dimension or make an assumption about the designer's intent. If necessary, consult with the designer for clarification.

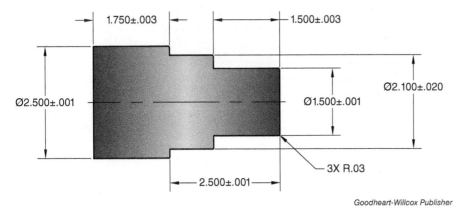

Goodheart-Willcox Publisher

Figure 18-1. A cylindrical part with 1.500″, 2.100″, and 2.500″ outside diameters.

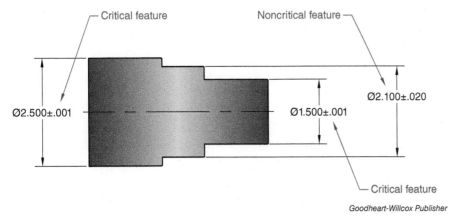

Goodheart-Willcox Publisher

Figure 18-2. Based on tolerance requirements, two of the outside diameter features are critical features and one is not as critical.

> **From the Programmer**
>
> **Roughing and Finishing Passes**
>
> Many programmers assume that the final pass on a critical diameter is the most important cut. Actually, it is the second to last pass that is most important. When trying to hold a close tolerance such as ±.001″, it is most important to determine the appropriate amount of roughing stock to leave. Leaving too little material makes the finish tool rub against the surface and results in the tool not cutting correctly. Leaving too much material causes tool pressure and pushes the tool away, making the ±.001″ tolerance difficult to hold. For the finish pass, try to leave .015″–.020″ material and adjust as needed.

The other critical feature on this print is a length feature with a 2.500″ linear dimension and a ±.001″ tolerance, **Figure 18-3**. This length feature might require a finishing pass that leaves a small amount of material on both faces and then a separate finishing tool to skim cut the faces to hold the tolerance.

The strategies previously discussed are not meant to represent the only way, or the best way, to machine the part. This example serves as an exercise in evaluating a print and developing a machining strategy based on the information specified. Depending on the machine type, material type, and tools available, a different strategy might be used. The most important consideration is to spend the time needed to analyze the part requirements before a program is produced.

Figure 18-4 shows a part similar to the previous example. However, this part has both outside diameter (OD) and inside diameter (ID) features. This part requires additional considerations in program planning. Because the largest inside diameter feature is on the opposite side from the smallest

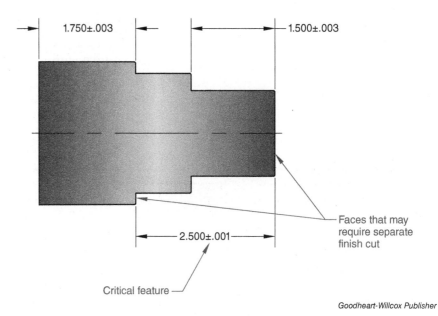

Goodheart-Willcox Publisher

Figure 18-3. Based on tolerance requirements, one of the length features is a critical feature. If necessary, a separate finishing tool can be used to remove a slight amount of material to meet requirements.

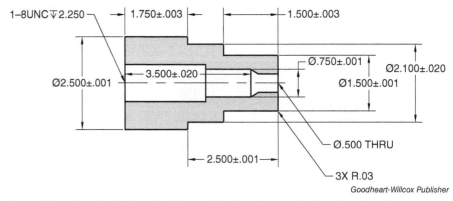

Goodheart-Willcox Publisher

Figure 18-4. A cylindrical part with outside diameter and inside diameter features, including a threaded section and two smaller diameter holes.

outside diameter feature, it makes sense to complete this part in two operations. The ID features can be completed in one operation and then the part will be flipped, or turned 180°, to complete the OD cuts. If needed, these operations can be reversed. If the programmer prefers to complete the OD features first and then flip the part to complete the ID features, it will not affect cycle time or machining efficiency.

A closer examination of the inside features shows that there is a .500″ diameter hole through the part, a .750″ diameter hole with a close tolerance in the middle, and a threaded hole on the end. The same cutting strategy can be used to machine the three outside diameters of the part, but the inside diameter features are more complex and might require a secondary cutting operation or a CNC lathe with dual spindles. These are the types of decisions that can be made in the program planning phase. By carefully examining the print, the print reader gains a much better understanding of the part, which can then inform the best machining process.

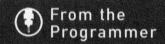

From the Programmer

Interpreting Dimensions

The part shown in **Figure 18-4** has an angled feature at the end of the .750″ diameter hole. This represents the angled shape of the drill tip past the full-diameter portion of the hole. Using an undersized drill (.735″) allows quick metal removal while still leaving some material for the .750″ finish size. But the .750″ diameter must be machined to the full 3.500″ depth, per the dimension on the drawing.

18.3 Part Workholding

Once the requirements of the print are fully understood, the programmer must decide how the part will be held on the machine. **Workholding** refers to any device that is used to secure a workpiece against the forces of machining. Holding the part securely during the machining process is critical for part accuracy and safety, but machinists also need to have access to machine as many features as possible in a single workholding configuration. This reduces setup and overall cycle time.

CNC lathe operations are generally simpler than milling operations because the material is being rotated around the spindle centerline. However, it is critical to understand the complexities and possible machining issues with lathe workholding in the preprogramming phase.

18.3.1 Chucks

A *chuck* is a workholding device that secures a workpiece for machining. On a lathe, a chuck is mounted to the headstock and rotates in the spindle assembly. The most common type of workholding device in a lathe is a three-jaw chuck, **Figure 18-5**. In the CNC environment, three-jaw chucks are almost always hydraulically operated, but some older or less expensive machines may be equipped with chucks that close manually. The three-jaw chuck is a simple and effective workholding device.

Another option for the workholding setup is a four-jaw chuck, **Figure 18-6**. A four-jaw independent chuck is manually operated. This type of chuck allows the jaws to move independently, thus permitting the part workpiece to move off center. A four-jaw chuck provides the ability to machine rectangular parts or create two diameters with different center axes.

The four-jaw chuck is slow to set up and can require the operator to set each workpiece individually in the jaws to produce the required feature. **Figure 18-7** shows an example of machining a rectangular workpiece in a four-jaw chuck. In this setup, the workpiece is centered in the jaws and a hole is machined in the center of the part. This configuration can be faster and more accurate than completing the operation in a CNC milling machine. The issue to consider with this type of workholding configuration is achieving a repeatable part setup in alignment with the spindle centerline.

Copyright Goodheart-Willcox Co., Inc.

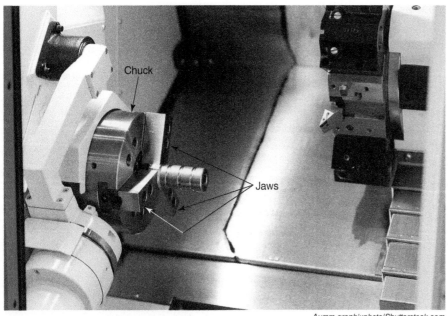

Figure 18-5. A three-jaw chuck is used to secure workpieces for machining. Three-jaw chucks are widely used in CNC lathe operations.

Figure 18-6. A four-jaw independent chuck has four jaws that move independently and are adjusted manually. This provides flexibility in machining and allows operations to be performed on noncylindrical parts.

In **Figure 18-8,** the same rectangular piece of stock is mounted on a four-jaw chuck, but the hole is machined 1/2″ off the center of the part material. This work setup is accomplished by simply moving two jaws to the desired offset distance. This allows the same feature to be drilled and bored off the center of the part.

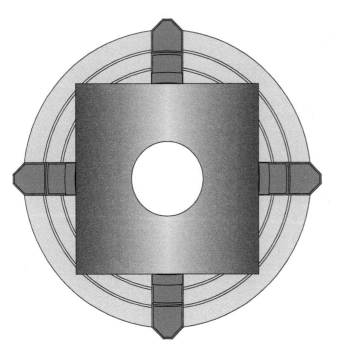

Figure 18-7. A four-jaw chuck allows machining on noncylindrical parts, such as this rectangular workpiece.

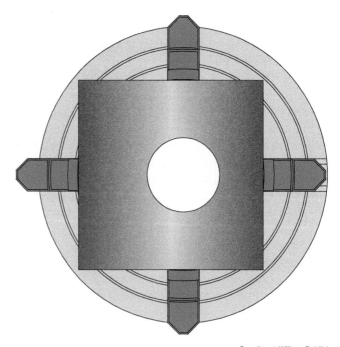

Figure 18-8. Offsetting two of the chuck jaws in a four-jaw chuck setup allows features to be machined off the center of the part.

Another common machining operation on a four-jaw chuck is turning two diameters that are offset, or off center, from each other. See **Figure 18-9**. By moving two of the chuck jaws, the stock material will be offset from the spindle centerline. During machining, the material will rotate off the centerline of the spindle, which allows for turning one diameter feature that is offset from the other diameter feature.

18.3.2 Lathe Collet Chuck

Collets are used in a variety of workholding and toolholding applications in CNC machining. Collet chucks are used to secure workpieces in lathe setups and are similar to other types of workholding chucks, **Figure 18-10**. Collets are also used to hold live tools in lathe operations. On CNC milling machines, collets can be used to hold milling tools or secure round stock in a milling operation.

Collet chucks are mounted to the headstock of a lathe and are typically used on machines equipped with a bar feeder. Collet chucks offer a number of advantages in machining. When closed, a collet makes almost complete contact around the material, supplying tremendous gripping strength and allowing highly concentric rotation of the work. Collets come in many shapes and sizes.

Collet chucks permit quick part changes and can be run at higher speeds than three-jaw or four-jaw chucks. The only disadvantage of collet chucks is that they are limited to smaller stock material sizes, usually 2″ and under.

18.3.3 Tailstock

A *tailstock* is a manually operated or programmable device that provides longitudinal support for a workpiece, **Figure 18-11**. A tailstock is not considered a main workholding device, but a secondary workholding

Goodheart-Willcox Publisher

Figure 18-9. A part with two diameter features on different center axes. This part can be machined by securing the workpiece off center with a four-jaw chuck.

Aumm graphixphoto/Shutterstock.com

Figure 18-10. Collet chucks are used to secure workpieces in lathe setups. Collets provide strong gripping force around the work and allow for shorter setup times.

> **From the Programmer**
>
> **Identifying Workholding Requirements**
>
> The type of part workholding required in machining varies with each job. Knowing how to identify the appropriate workholding method for a specific job is gained through experience. The workholding method used for the part must allow access to areas that require machining and must be repeatable in production. A secure work setup will allow the programmer to maximize cutting speeds and reduce cycle times. Material that is not secured sufficiently poses a serious safety risk and impacts the ability to meet the tolerances specified.

oYOo/Shutterstock.com

Figure 18-11. A tailstock is a secondary device on a lathe used to provide support for long workpieces.

device. The term *secondary* means that it is not the primary workholding of the material, but it supplements, or assists, with the workholding. When machining longer material, a tailstock will provide stability, or support, through the center axis of the workpiece and prevent the work from being forced away by the cutting tool.

The following example explains what happens during the lathe cutting cycle and shows how a tailstock can provide a vital advantage. **Figure 18-12** shows a 2″ diameter bar of stock material that extends out 9″ from the chuck jaws in a turning operation. As the cutter enters the material, at the point farthest away from the chuck, the material is pushed away by cutter pressure. This will create a tapered diameter cut or a poor surface finish.

Using a tailstock will reduce or eliminate deflection of the part that occurs when cutting pressure forces the stock away from the spindle centerline. When machining longer parts, the length to diameter ratio of the work is used to determine when to use a tailstock. The general rule of

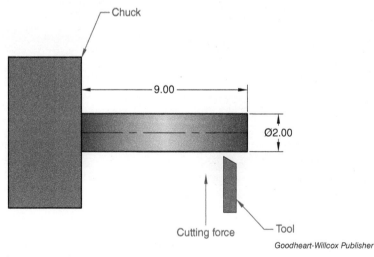

Goodheart-Willcox Publisher

Figure 18-12. A long workpiece without support at the end can be forced away from the spindle centerline by the cutting tool during machining.

thumb is a ratio of 4:1. If the length of stock is four times or greater than the diameter, a tailstock should be used. The length measurement to consider is the unsupported portion of the part that extends away from the chuck jaws. For example, a 2″ diameter bar that protrudes more than 8″ from the chuck jaws exceeds the 4:1 length to diameter ratio and will require a tailstock. See **Figure 18-13**. The diameter measurement to consider is the finish diameter size, not the starting size. Machining a bar down significantly to a smaller size will impact the length to diameter ratio and may require a tailstock.

18.4 Tool Selection

The next important decision in program planning is determining what tools to use. Start by making a list of tools and identify any part surfaces and features that may require a special or uncommon tool. **Figure 18-14** shows a part with a groove located between two outside diameters. The groove is .135″ wide. This feature requires a grooving tool.

A groove is normally machined after the diameter features are completed. A grooving operation requires a tool width smaller than the groove width. See **Figure 18-15**.

To reduce tool changes and cycle time, try to minimize the number of tools used. Always evaluate the possibility of using the same tools for multiple features. For example, the two different outside diameters for the part shown in **Figure 18-15** can be machined in a turning operation using a single roughing tool. Depending on the tolerance requirements for the diameters, they can be finished with the same tool, or a separate finish tool can be used after roughing. If any specialized tooling is required for additional operations, such as the grooving tool, it is best to identify this in the planning phase so that it can be accounted for while programming.

Most tools used in turning operations are carbide inserts. An *insert* is a replaceable cutting tool with a polygonal or round shape that is fastened to a matching toolholder. See **Figure 18-16**. This section is a general introduction to inserts and is not intended to serve as a guide for determining the exact lathe tool for a specific operation. However, there are some general rules that apply in lathe tooling selection. When performing roughing

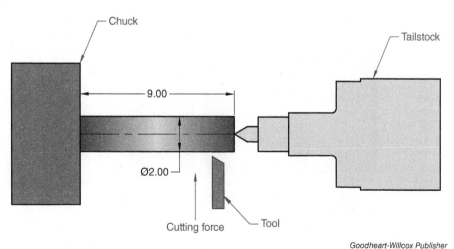

Goodheart-Willcox Publisher

Figure 18-13. When the unsupported portion of a long workpiece exceeds a length to diameter ratio of 4:1, a tailstock is used for end support.

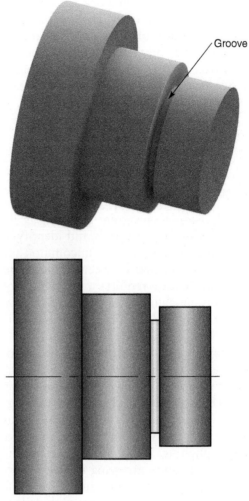

Goodheart-Willcox Publisher

Figure 18-14. A cylindrical part with three outside diameters and a groove. A grooving tool is required for the grooving operation.

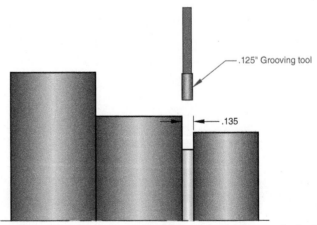

Goodheart-Willcox Publisher

Figure 18-15. The cutting tool used in a grooving operation must have a width smaller than the groove width.

operations, for either outside or inside diameters, an 80° or 85° insert design will allow for deeper cuts and reduced cycle times. The CNMG and WNMG inserts shown in **Figure 18-17** provide substantial support behind the cutting edge and are capable of absorbing a tremendous amount of impact. The names of these tools are based on a standard identification system developed by the International Organization for Standardization (ISO). Insert names are made up of a series of letters and numbers, such as CNMG-432. The four letters in the first part of the name are designations used to identify parameters of the tool. The first letter designation identifies the outside shape. The second letter designation identifies the relief angle between the top and front faces. The third letter designation identifies tolerances for the size dimensions of the insert. The fourth letter designation identifies the insert hole configuration and chipbreaker type. The numbers following the letters are designations used to identify the insert dimensions. In simple terms, the CNMG insert shown in **Figure 18-17** has an 80° diamond shape. The WNMG insert has an 80° trigon shape. The angle refers to the included angle measured at the nose of the tool (the cutting edge).

Inserts are selected for specific machining conditions and requirements. For example, a larger nose radius at the cutting edge will reduce the chance of fracturing, but it will also require more machine horsepower and create more spindle load. The type of material being cut and the capabilities of the machine will dictate the specific type of insert used.

For finishing cuts on the outside of a part, a 35° or 55° insert is commonly selected. The VNMG (35° diamond) and DNMG (55° diamond) inserts shown in **Figure 18-18** are commonly used for finishing cuts. The added clearance behind the cutting edge and the additional support along the length of the insert generally create better surface finishes.

With the limited support directly behind the cutting edge, these types of inserts cannot withstand the larger depths of cut commonly made with CNMG or WNMG inserts. Inserts designed for finishing operations can be used with high spindle speeds and lower feed rates to achieve efficient cycle times and superior surface finishes while maintaining close tolerances.

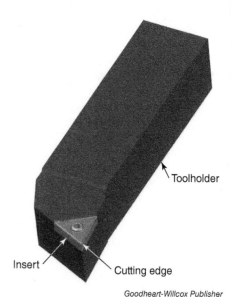

Goodheart-Willcox Publisher

Figure 18-16. A lathe tool with an insert. The insert has a triangular shape with a 60° included angle.

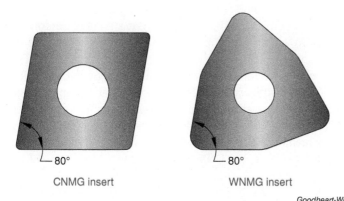

Goodheart-Willcox Publisher

Figure 18-17. Cutting inserts commonly used for roughing operations. The CNMG insert has an 80° diamond shape. The WNMG insert has an 80° trigon shape.

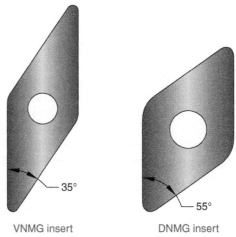

Figure 18-18. Cutting inserts commonly used for finishing operations. The VNMG insert has a 35° diamond shape. The DNMG insert has a 55° diamond shape.

Selection of tooling for machining inside diameter features can be tricky. First, a hole will need to be machined, usually with a drill. Then, the appropriate tools must be selected to fit inside the hole and machine the features to the required dimensions. Inserts similar to those previously discussed are used in boring bar setups, but they are smaller in size. A CCMT (80° diamond) insert is common for boring operations.

The most important considerations in tool selection are minimizing tool changes and the number of tools used when possible. Be deliberate in selecting tools and look for every opportunity to maximize tool removal rates.

From the Programmer

Cutting Insert Data

There are many tooling manufacturers and hundreds of different types of cutting inserts available. Inserts are designed for use with specific materials and machining operations. Deciding which tool style to use for a certain operation is the first step. The next step is to determine which tool coating to use for the type of material you are cutting. The final step is determining which type of chipbreaker is most appropriate. This will help clear chips and remove heat from the cutting edge. However, the chipbreaker only works correctly if programmed at the appropriate speed and feed rate. Your best resources for tooling selection are the technical data and recommendations provided by the tooling manufacturer. Most inserts are manufactured with several cutting edges, making them cost efficient. For example, the CNMG insert shown in **Figure 18-17** has four cutting edges, two on the top and two on the bottom. When one of the cutting edges wears from use, the tool is indexed (moved) to the next edge. This process is repeated until all edges are worn. Keep track of how many parts are produced per cutting edge and note the type of wear that occurs over the life of the tool. This process takes time, but it will help you identify the right insert for the job.

18.5 Order of Operations

To this point in the planning, the print has been reviewed, workholding has been established, and tooling has been selected. These procedures are normally followed specifically in that order, although there might be cases when a slightly different order is taken. The last decision to be made is the sequence in which machining operations will occur, or the *order of operations*. Determining the order of operations is normally more straightforward in lathe programming than in mill programming, but it is often possible to make smart decisions with the machining sequence to translate into reduced cycle times and higher-quality parts. The following example shows how the order of operations can make a significant difference.

Figure 18-19 shows a part requiring outside diameter and inside diameter machining operations. On one end of the part is a threaded shaft, and on the opposite end are two inside diameter features. Assuming the machine is a two-axis machine with a single spindle, this configuration will require the part to be flipped over for a second set of operations.

The most efficient machining strategy may be to secure the workpiece material in a chuck with jaws, or a collet chuck, and machine the large end and inside diameters first. This allows the 2″ outside diameter and the inside diameters to be roughed and finished in one setup. See **Figure 18-20**.

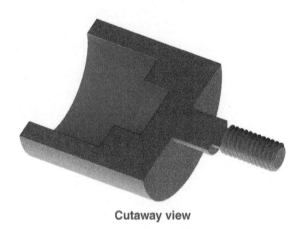

Cutaway view

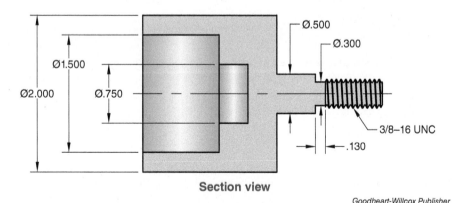

Section view

Goodheart-Willcox Publisher

Figure 18-19. A part with outside diameter features, a threaded shaft on one end, and inside diameter features on the other end. The part requires several operations to rough and finish the outside and inside features. Note: Some dimensions are missing from the drawing, but they are not relevant to the example and are omitted for clarity.

The next operation will require turning the part over and machining the threaded shaft. This allows the part to be held on the 2″ diameter completed in the previous operation, improving concentricity. By clamping on as much of the outside diameter as possible, material deflection will be reduced and surface finish will be improved.

Figure 18-21 shows the part flipped and secured for machining the end with the threaded shaft. One possible strategy for machining this end of the part is to rough the .500″ diameter and the 3/8 (.375″) major thread diameter, leaving some material for the finish pass. Then, the small .300″ diameter groove behind the threads will be machined. Next, the .500″ diameter and .375″ major thread diameter will be finished with the appropriate finishing tool. The last operation will be to machine the threads.

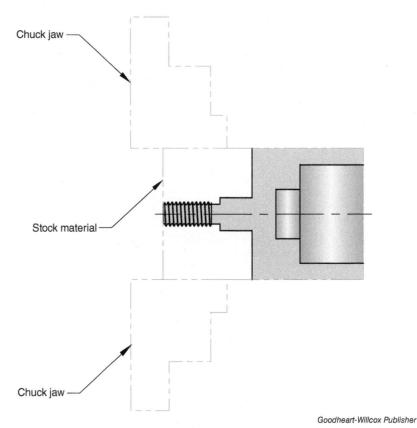

Goodheart-Willcox Publisher

Figure 18-20. The part is secured to the chuck on one end to machine the large end diameter and the inside diameter features. These are the first operations for the part.

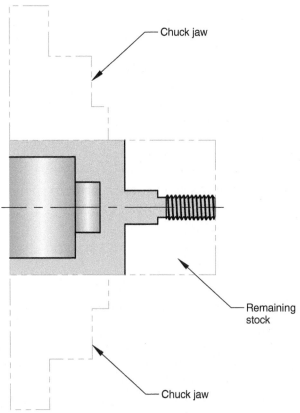

Figure 18-21. The part is flipped and secured to the chuck on the other end to machine the .500″ diameter, the .300″ diameter groove, and the threads. The threading operation is the final operation.

This is not the only possible machining strategy for this part, but it is an efficient approach that minimizes tool changes and cycle time. Efficiency gains in machining should be a primary consideration in program planning. The time saved from planning will save in machine setup, operation, and cycle times. Although it can be an often overlooked step in the programming process, take the extra time and consider all of these factors before beginning a program.

Chapter 18 Review

Summary

- The planning that happens before a CNC lathe program is created is vital to its success. It is necessary to understand all of the customer's requirements and the capabilities of the machinery and available tools.
- Design intent refers to the way in which the part is dimensioned and noted to define critical features of the part and the part function.
- By examining a print, important information on machining procedures can be derived. The dimensioning and tolerancing on the print identify critical features and can help in determining factors such as the work coordinate system origin and workholding.
- Workholding is a critical factor in machining and programming. The material must be securely held while allowing full access to all features that need machining. It is critical to understand the complexities and possible machining issues with lathe workholding in the preprogramming phase.
- Tooling is an important factor to consider before starting the CNC program. Utilizing the same tool for multiple features can reduce cycle times. Specialty tooling needs to be identified before the program goes to the machine for production. When possible, minimize the amount of tools used and the tool changes needed.
- Identify features and operations that can be consolidated to reduce cycle times. A clear understanding of the required machining operations will improve efficiency.
- Before programming begins, consider the order of operations. Having a clear, definitive list of operations in a logical order can greatly reduce cycle times and eliminate mistakes in program operation.
- A clear idea of customer requirements, tooling needed, workholding, and order of operations is likely to reduce the total machine time for any product.

Review Questions

Answer the following questions using the information provided in this chapter.

Know and Understand

1. What is an important factor to consider before writing a CNC mill program?
   A. Customer requirements
   B. Workholding
   C. Tooling
   D. All of the above.

2. *True or False?* A typical sequence in program planning begins with reviewing the print.

3. *True or False?* Examining tolerances on a print can help determine what features are critical in manufacturing.

4. *True or False?* A tailstock is used to secure a length of stock when it extends more than four times the diameter from the chuck jaws in a turning operation.

5. The most common workholding device in a lathe is a ____.
   A. vise
   B. three-jaw chuck
   C. four-jaw chuck
   D. magnetic chuck

6. Which type of workholding device allows for machining off the spindle centerline?
   A. vise
   B. collet chuck
   C. three-jaw chuck
   D. four-jaw chuck

7. *True or False?* Tool selection does not impact machine cycle time.

8. Which insert type has an 80° diamond shape?
   A. DNMG
   B. CCMT
   C. VNMG
   D. CNMG

9. Which insert type has a 55° diamond shape?
   A. DNMG
   B. CCMT
   C. WNMG
   D. VNMG

10. *True or False?* The same tools selected for cutting the outside diameter of a part can always be used for the inside diameter machining.

11. Which of the following is impacted by the order of operations in a program?
    A. Cycle time
    B. Tool usage
    C. Program length
    D. All of the above.

12. Usually, the last step in program planning is to determine _____.
    A. workholding
    B. tolerances
    C. the order of operations
    D. tooling

Apply and Analyze

1. Study the drawing below. What are two possible workholding devices that may be required to complete the machining for this part? Explain your answer.

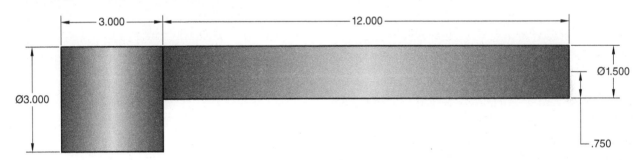

2. Study the drawing below. Other than the outside diameters that will require roughing and finishing tooling, which features will require special tooling?

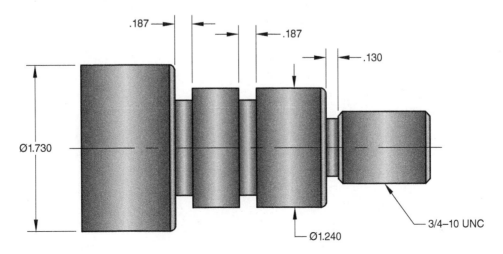

For questions 3–6, match each term with the correct type of workholding.

    A. Three-jaw chuck
    B. Four-jaw chuck
    C. Collet chuck
    D. Tailstock

3.

Aumm graphixphoto/Shutterstock.com

4.

ID1974/Shutterstock.com

5.

oYOo/Shutterstock.com

6.

Herman Hernandez/Shutterstock.com

7. Identify each type of insert as a *roughing* or *finishing* tool.

   A.

   B.

   C.

   D.

Critical Thinking

1. Explain what each team member in production can contribute to program planning. Other than the personnel discussed in this chapter, who else might need to be included? Why is the customer a valuable contributor? What can an operator or setup person contribute? What about sales personnel?

2. Preplanning is an important consideration in most projects. Consider a project you may have in your personal life—perhaps painting a room, repairing a vehicle, or even taking a trip. Write out an explanation of how you might plan for the project. Who else might need to be involved in the planning? What supplies or tools are needed? What are the steps required and in what order should they be completed to make the project successful?

3. This chapter discussed the importance of efficiency and cycle times. What is the impact of not having machines running production to the shop personnel? How does this affect profitability? How can planning prevent inefficiencies? Conduct research and find out what the average hourly shop rates are for machine shops in your area.

19 Lathe Contour Programming

Chapter Outline

19.1 Introduction
19.2 Point-to-Point Programming
   19.2.1 Inside Diameter Contour Programming
19.3 Tool Nose Radius Compensation
19.4 Programming Radial Moves
   19.4.1 Radius Method
   19.4.2 I and K Method
19.5 Programming Angular Moves
19.6 Turning Canned Cycles
   19.6.1 G71 Roughing Cycle
   19.6.2 G70 Finishing Cycle
   19.6.3 G71 and G70 Cycles for Inside Diameter Machining
   19.6.4 G72 Roughing Cycle
   19.6.5 G73 Pattern Repeating Cycle
   19.6.6 G94 Facing Cycle

Learning Objectives

After completing this chapter, you will be able to:

- Create point-to-point contouring programs.
- Explain the use of tool nose radius compensation.
- Differentiate between tool nose radius compensation left and right.
- Create angular moves in two axes.
- Use the I and K method to create arc moves.
- Use the R method to create arc moves.
- Use the G71, G72, and G73 canned cycles to perform roughing operations.
- Use the G70 canned cycle for finishing operations.
- Program a facing operation using the G94 canned cycle.

Key Terms

canned cycle
contouring
facing
I and K method
point-to-point programming
radius method
tool nose radius compensation

19.1 Introduction

Contouring is a machining operation that follows a joined path through a series of points to cut material. A contour is perhaps the most common type of toolpath. Simply think of contouring as cutting straight along a line, around a circle, or around a more complex shape. Contours can be cut around the outside of a shape or on an internal feature, such as an internal bore. The shape can be simple or complex, but the steps to creating the program will be the same.

Creating a contouring program is much like giving directions. First, the tool is positioned in a safe starting location. Then, step-by-step instructions are given to move the tool to create the desired shape. Contouring is often used to create finishing passes. However, with creative use of canned cycles or multiple tools, roughing paths can also be created.

19.2 Point-to-Point Programming

Look at the sample part in **Figure 19-1**. This part is used in creating a simple contouring program. The X0. Z0. part origin is located at the spindle centerline and the front face. This is the most common position for the origin for ease in programming, but it is possible to establish a different location as the origin. Remember that all corresponding points are relative to the origin, so it is critical to identify the origin position before starting a program.

Based on the origin position, the part is oriented in the second quadrant, or the X+, Z– quadrant. In this orientation, the X axis coordinates are always positive (+) and the Z axis coordinates are always negative (–). The coordinate table in **Figure 19-1** shows the X and Z coordinates for each point on the part contour. The coordinates are specified using diameter programming. These coordinates can be used to create a contouring program for a finishing operation. The following program is a simple form of *point-to-point programming* for the finishing operation.

O0526	
G20	
(80 DEG. INSERT – CNMG-432)	
G0 T0101	
G97 S500 M3	
G0 X0. Z.1 M8	Safe move in front of part
G1 Z0. F.01	Move to X0. Z0. origin
X1.5	Point 1
Z–1.5	Point 2
X2.1	Point 3
Z–2.5	Point 4
X2.5	Point 5
Z–4.25	Point 6
X2.65	Safe move away from part
M9	
G28 U0. W0. M5	Return to machine home
T0100	
M30	

From the Programmer

Programming Efficiency

Notice that when the tool moves from Point 1 (X1.5 Z0.) to Point 2 (X1.5 Z–1.5) in the program, X1.5 does not appear on the second line. If the tool is commanded to move along only one axis (the Z axis in this case), the other axis position does not need to be repeated in the program. A move along a single axis can be specified without repeating the axis position that does not change. If both coordinate axis positions are given on the same line for a single axis move, the program will be interpreted correctly by the machine, but this is not necessary. For example, the block of code for the move to Point 2 could also be written as X1.5 Z–1.5. Leaving out code that is not necessary improves efficiency and uses less machine memory.

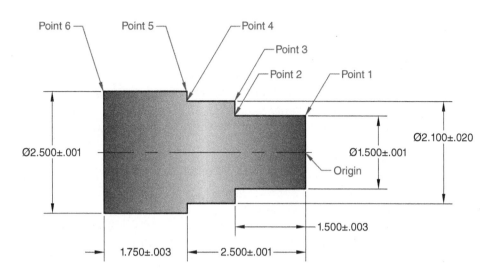

Coordinates	
Origin	X0. Z0.
Point 1	X1.5 Z0.
Point 2	X1.5 Z–1.5
Point 3	X2.1 Z–1.5
Point 4	X2.1 Z–2.5
Point 5	X2.5 Z–2.5
Point 6	X2.5 Z–1.25

Goodheart-Willcox Publisher

Figure 19-1. A part with coordinate positions defined for point-to-point programming.

19.2.1 Inside Diameter Contour Programming

Creating a contouring program for an inside diameter feature is similar to creating a contouring program for an outside diameter feature. Plotting the individual points and programming through the points will create the required toolpath for an inside diameter finish machining operation.

Figure 19-2 shows a part with inside diameter features and the X and Z coordinates for each point on the contour. Be aware of the direction of travel and the entry and exit positioning, as the coordinates will be specified in the opposite X axis direction in comparison to outside diameter contour programming. On this part, the X0. Z0. part origin is located at the spindle centerline and the front face. All of the X axis coordinates are positive (+) and all of the Z axis coordinates are negative (–). However, the X coordinates start with the largest diameter size and become smaller with each positioning move. The contouring program is as follows.

O1902	
G20	
(TOOL – 1 OFFSET – 1)	
(ID BORE – MIN. .75 DIA. BAR)	
G0 T0101	
G97 S500 M3	
G0 X3. Z.1 M8	Safe move in front of part
G1 Z0. F.01	Move to Point 1
Z–.75	Point 2
X2.25	Point 3
Z–1.5	Point 4
X1.	Point 5
Z–4.	Point 6
X.8	Safe move off part
M9	
G0 Z.1	Safe move out of bore
G28 U0. W0. M5	Return to machine home
T0100	
M30	

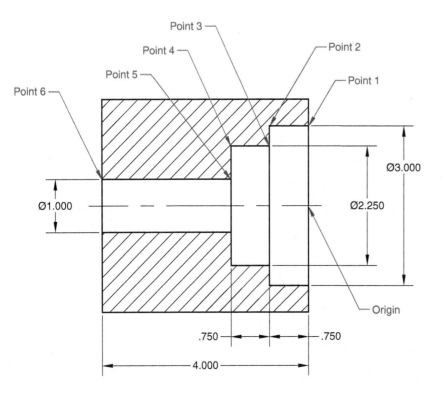

Figure 19-2. A part with coordinate positions defined for point-to-point programming for inside diameter machining.

19.3 Tool Nose Radius Compensation

Tool nose radius compensation is a programmed offset from the center of the cutting tool to the tool's edge along a cutting path. In lathe programming, the center of the cutting tool refers to the center of the tool nose at the tool tip. Tool nose radius compensation enables different offsets to be specified for tools and cutting operations. The G41 and G42 commands are used to activate tool nose radius compensation. These commands compensate for the tool nose radius by offsetting the tool to the left or right of the programmed cutting path. Since these are modal commands, they will stay active until they are turned off in the program. The G40 command is used to cancel the tool nose radius compensation mode when the operation is complete.

In mill programming, the G41 and G42 commands are used to activate cutter compensation mode. These commands offset the cutting tool to the left or right of the cutting path based on the cutting direction of a cylindrical tool. Compensating for the tool radius is similar in lathe programming, but the shape of the cutting tool is different. In lathe operations, the machine control needs to know the orientation of the tool tip in addition to the size of the tool nose radius. The tool tip orientation defines the angular configuration of the cutting tool relative to the workpiece. The tool tip orientation and tool radius are recorded at the machine during setup. These parameters are used by the machine to position the tool correctly and offset it to the left or right of the programmed cutting path when tool nose radius compensation is active. If tool nose radius compensation is not used in a lathe program, angled features and radii will not be finished as programmed. The theoretical tip of a lathe tool is defined where the edges come to a sharp point. In a noncompensated program, the tool tip will follow the programmed positions, but the machine does not account for the radius or orientation of the tool being used. When cutting angles and radius features, the machine must know how to orient the tool correctly and offset the tool radius to create the proper features.

The G41 command activates tool nose radius compensation left. *Left* refers to the side of the material on which the tool is cutting, **Figure 19-3**. The G41 command is often used for internal diameter machining operations, but this can vary depending on the cutting direction. Because of the multiple configurations of lathes, it is necessary to identify the cutting direction relative to the work in order to determine the appropriate command to use for tool nose radius compensation.

The G42 command activates tool nose radius compensation right. In this case, the tool is cutting on the right-hand side of the material, **Figure 19-4**. The G42 command is often used for outside diameter machining operations, but different machine configurations and cutting directions can allow for tool nose radius compensation in either direction. Visualize the cutting direction and determine whether the tool is on the right- or left-hand side of the cut.

When setting tool offsets during machine setup on a lathe, the tool offset page in the controller contains settings for the tool nose radius and tool tip orientation. These settings are entered by the setup person. The settings are used by the controller to calculate the position of the tool when tool nose radius compensation is active. In general, it is best practice to use tool nose radius compensation in most lathe programming applications. For basic parts that only require straight cuts, it is sometimes possible to program tool positions without offsetting the tool. However, many parts have angled or radial features, making it necessary to use tool nose radius compensation.

There are some rules and methodology to follow when activating the tool nose radius compensation feature in a program. Tool nose radius compensation requires an initial activation move to position the tool correctly and establish the offset position relative to the part material. The activation move must be in a clear area away from the part. The distance of this move must be greater than the radius of the tool. Depending on the controller, a move equal to twice the radius of the tool may be required. The activation

> **From the Programmer**
>
> **Walking the Path**
>
> To visualize the cutting direction and determine whether to use the G41 or G42 command, a simple trick is to "walk the path." Imagine walking down the path the tool is following as it is cutting material. You should be facing the same direction as the tool's path of travel. Are you on the right side of the material, or the left? This will determine which command to use for tool nose radius compensation.

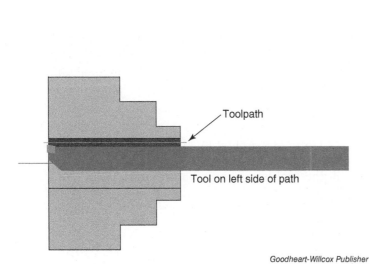

Figure 19-3. The G41 command is used when the tool is cutting on the left side of the material when viewing the tool from behind.

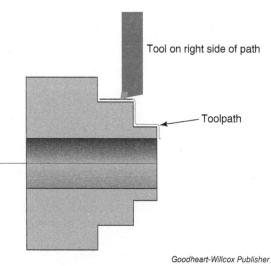

Figure 19-4. The G42 command is used when the tool is cutting on the right side of the material when viewing the tool from behind.

Debugging the Code

Machine Alarms

Machine controls will create alarms for incorrect code. A common alarm that occurs when tool nose radius compensation is not activated correctly is an interference alarm. The interference alarm means that the activation move is smaller than the radius input on the tool offset page. Make sure the travel distance along the X or Z axis is greater than the stored tool nose radius when the G41 or G42 command is initiated.

move must be a straight-line move. This allows the control to *compensate* for the cutter as it is being positioned. Tool nose radius compensation cannot be activated while making an arc move with a circular interpolation command. After completing the operation, tool nose radius compensation must be canceled with a move away from the part. The distance of the exit move must be greater than the radius of the tool.

19.4 Programming Radial Moves

Programming radial movement in a contouring operation is a common requirement in lathe programming. On most parts produced in CNC machining, sharp corners are not acceptable. To address this, many prints include a general note such as *Break Sharp Edges*. This note is a specification to remove all sharp corners and edges where surfaces meet. Creating a radial feature, whether it is a large radius or a simple edge break, is fairly straightforward. The position of the tool is programmed to the starting point of the radius, and then the radius size and direction are specified.

Radial moves are referred to as circular interpolation or arc moves. The G02 command is used to create an arc in the clockwise direction. The G03 command is used to create an arc in the counterclockwise direction. There are two different methods used to program radial moves. These are referred to as the *radius method* and the *I and K method*. Older machine controllers required the I and K method for arc programming, but all models since the early 2000s have incorporated the radius method. Both of these methods are described in the following sections.

19.4.1 Radius Method

The radius method, also called the R method, is the most commonly used arc programming method on newer machines. This method is considered simpler than the I and K method because it requires fewer calculations and less programming input. The information required is the arc start position, the arc direction (G02 or G03 command), the arc end position, and the R value, or radius required.

Figure 19-5 shows an example of a typical CNC lathe part with two different radii. The .100″ radius will be programmed with the G03 command and the .250″ radius will be programmed with the G02 command.

The .100″ radius is an outside corner radius. **Figure 19-6** shows the coordinates required in programming to complete the arc move. The X0. Z0. part origin is located at the front face of the part. Point 1 is located at X1.8 Z0., Point 2 is located at X2. Z–.1, and the arc radius is .100″. When calculating the X axis position for Point 1, remember that the .100″ radius is on both sides of the part. The X axis position is calculated as X2.0 – 2 (.100), or X1.8. The cutter is traveling in a counterclockwise direction, so the G03 command is used.

In the program, the tool is positioned at Point 1. Then, the arc command specifying the direction is given and the arc end position and radius size are given to move the tool to Point 2. The program is as follows.

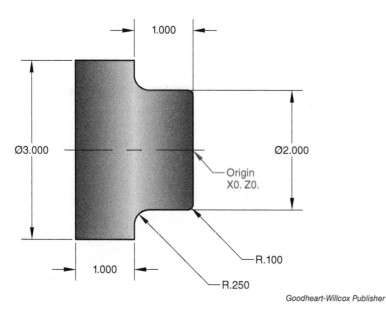

Figure 19-5. A part with two radius features. Arc moves are programmed with the G02 and G03 circular interpolation commands.

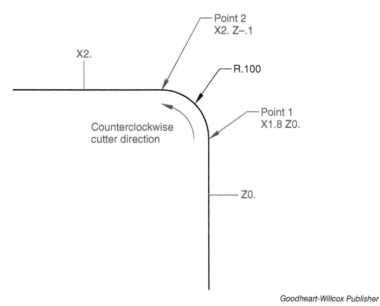

Figure 19-6. The .100" outside corner radius is created with an arc move in the counterclockwise direction using the G03 command.

G0 T0101	
G97 S3600 M3	
G42 G0 X0. Z.1 M8	Activate tool nose radius compensation
G50 S3600	
G96 S200	
G1 Z0. F.01	
X1.8	Feed move to Point 1
G3 X2. Z-.1 R.1	CCW arc to Point 2; .100 radius value
G1 Z-.75	Linear interpolation mode turned on; move to next point

This completes the first radius path. Now study the second radius path and determine the start point, cutter direction, end point, and radius size. **Figure 19-7** shows the coordinates required to complete the arc move for the .250″ inside corner radius.

Point 3 is located at X2. Z−.75. Point 4 is located at X2.5 Z−1. The cutter is traveling in the clockwise direction, so the G02 command is used. From Point 2 in the first radius, a linear move (G01 command) is used to move to Point 3. Then, the second radius is generated to move to Point 4. The entire program is as follows.

G0 T0101	
G97 S3600 M3	
G42 G0 X0. Z.1 M8	Activate tool nose radius compensation
G50 S3600	
G96 S200	
G1 Z0. F.01	
X1.8	Feed move to Point 1
G3 X2. Z−.1 R.1	CCW arc to Point 2; .1 radius value
G1 Z−.75	Linear move to Point 3
G2 X2.5 Z−1. R.25	CW arc to Point 4; .25 radius value
G1 X3.	Linear move to 3.000 diameter
G40 X3.25 Z−.8	Cancel tool nose radius compensation with exit move
M9	
G28 U0. W0. M5	

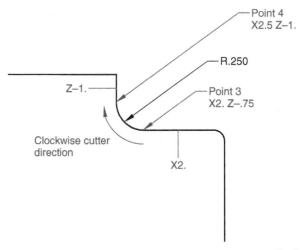

Goodheart-Willcox Publisher

Figure 19-7. The .250″ inside corner radius is created with an arc move in the clockwise direction using the G02 command.

19.4.2 I and K Method

The I and K method is another method for creating arc moves. The I and K addresses are alternate axis designators. The I address is an alternate axis designator for the X axis. The K address is an alternate axis designator for the Z axis. The I and K method was the original means to designate an arc move in lathe programming. On some older machine controllers, this is the only method that works.

When using the I and K method, the X axis and Z axis distances from the arc start point to the arc center point are entered as positive or negative incremental values in the line of code. The R value (radius value) is not used. As with the radius method, the arc start point, arc direction, and arc end point are specified. **Figure 19-8** shows the same part used in the previous example, with the distance from the arc start point to the arc center point added. This example shows the information required to complete the arc move for the .100" outside corner radius.

If the incremental distance to the arc center point along any axis is zero, it does not need to be entered. The following program uses the entry K–.1 to specify the distance along the Z axis to the arc center point. The distance along the X axis to the center point is zero, so an I address is not included. Adding I0. for the X axis distance has no negative impact, but it is not required.

G0 T0101	
G97 S3600 M3	
G42 G0 X0. Z.1 M8	Activate tool nose radius compensation
G50 S3600	
G96 S200	
G1 Z0. F.01	
X1.8	Feed move to Point 1
G3 X2. Z–.1 K–.1	CCW arc to Point 2; .100 distance to center
G1 Z–.75	Linear interpolation mode turned on; move to next point

Figure 19-9 shows the information required to complete the arc move for the second radius. This is the .250" inside corner radius. This arc will require an I address with positive value input, because the center of the

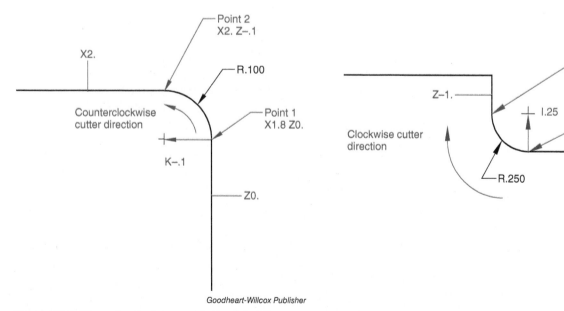

Goodheart-Willcox Publisher

Figure 19-8. The entry K–.1 is used in the I and K method to program the arc move for the outside corner radius.

Figure 19-9. The entry I.25 is used in the I and K method to program the arc move for the inside corner radius.

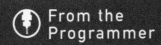

From the Programmer

Arc Programming Methods

When programming arc moves, be aware of the capabilities of the machine. The I and K method will work on virtually every machine, and the R method will work on most newer machines. The R method is simpler by comparison because it does not require calculating the center point distance and direction, so it is the preferred method.

radius is .250" in the positive X axis direction from the arc start point. Movement is in a clockwise direction, so the G02 command is used. The entire program is as follows.

G0 T0101	
G97 S3600 M3	
G42 G0 X0. Z.1 M8	Activate tool nose radius compensation
G50 S3600	
G96 S200	
G1 Z0. F.01	
X1.8	Feed move to Point 1
G3 X2. Z–.1 K–.1	CCW arc to Point 2; .100 distance to center
G1 Z–.75	Linear move to Point 3
G2 X2.5 Z–1. I.25	CW arc to Point 4; .250 distance to center
G1 X3.	Linear move to 3.000 diameter
G40 X3.25 Z–.8	Cancel tool nose radius compensation with exit move
M9	
G28 U0. W0. M5	

19.5 Programming Angular Moves

Angled features such as chamfers and tapers are common in CNC lathe programming. From a simple 45° chamfer to a precision taper, the need to create angled features will occur in almost every program. As previously discussed in this chapter, when a single axis move is programmed, it is not necessary to repeat the X or Z coordinate on a line of code if the axis position does not change. For example, if the machine needs to move from X3. Z0. to X3. Z–1., there is no need to place the X3. coordinate in the subsequent line of code. By simply specifying the Z–1. position in the second line of code, the machine will interpret this input as only moving in the Z axis and not moving in the X axis. However, what will happen if the program commands the machine to move from X3. Z0. to X5. Z–1.? By positioning two axes simultaneously, angular movement is created.

Figure 19-10 shows a 2" diameter part with a 45° chamfer. The X0. Z0. part origin is located on the front face. To program the chamfer, position the tool at the origin, move the tool to the positive X1.5 position, and then move to X2. Z–.25 on the same block of code. The X and Z moves, contained in the same block, will create the angled chamfer.

G0 T0101	
G97 S3600 M3	
G42 G0 X0. Z.1 M8	Activate tool nose radius compensation
G50 S3600	
G96 S200	
G1 Z0. F.01	
X1.5	

Continued

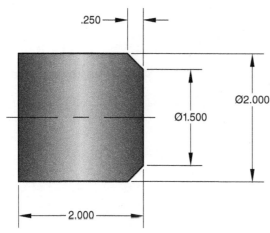

Figure 19-10. A part with a chamfered edge. An angular move is used to program a chamfer.

X2. Z−.25	Angular move
Z−2.	
G40 X2.25 Z−1.8	Cancel tool nose radius compensation with exit move
M9	
G28 U0. W0. M5	

The same process is used to create an extended taper cut. On prints, it is common for tapers to be dimensioned in different ways. Often, one of the taper dimensions must be calculated from the given information for programming purposes. The key point to remember is to calculate the X axis and Z axis starting position and ending position.

Figure 19-11 shows two different dimensioning methods for the same angled cut. The example shown on the left is easily programmed, because the X and Z positions are clearly identified by the dimensions. The example on the right shows the taper dimensioned with an angle instead of a linear distance. However, the X and Z positions can be easily calculated by using a right triangle and a trigonometric function. Refer to the right triangle shown below the drawing. The 60° angle in the drawing is an included angle representing the entire angle measured between the sides of the taper. Thus, a 30° angle is used to solve for the unknown side of the right triangle. This angle represents the angle between one side of the taper and the part centerline. The large end diameter is used to calculate one side of the triangle by subtracting the small end diameter and dividing the result in half. This segment represents the opposite side of the triangle and is equal to 2.309″. The tangent function can then be used to calculate the adjacent side of the triangle. Using a calculator, the formula is as follows.

Calculator formula:

$$\text{Opp} \div \text{Angl TAN} = (\text{Adj})$$

Solution:

$$2.309 \div .5774 = 4″$$

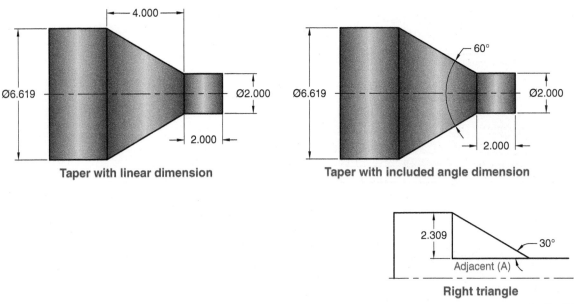

Figure 19-11. Different dimensioning methods are used to define tapered features on prints. The part on the left has a taper 4.000″ in length. The part on the right has the same taper, but it is dimensioned with an angle. To calculate the length for programming, a right triangle is used with the tangent function. The opposite side of the triangle is calculated from the end diameters of the taper. The adjacent side represents the unknown length dimension. The calculation of this segment establishes the Z axis position for the taper in the program.

The length of the taper is 4″. The Z axis position for the angular move is Z–6. The program for the part is as follows.

G0 T0101	
G97 S3600 M3	
G42 G0 X0. Z.1 M8	Activate tool nose radius compensation
G50 S3600	
G96 S200	
G1 Z0. F.01	
X2.	
Z–2.	
X6.619 Z–6.	Angular move
G40 X7. Z–5.8	Cancel tool nose radius compensation with exit move
M9	
G28 U0. W0. M5	

19.6 Turning Canned Cycles

To this point in the chapter, there has only been discussion of creating contouring programs for finishing passes. Finishing operations are important, but most machined parts are made from solid bar stock. Typically, this requires removal of a significant amount of material, not just a finish pass. There is a group of very powerful canned cycles for heavy roughing operations. A *canned cycle* is an abbreviated machining cycle that automates complex repetitive commands and functions. Canned cycles in lathe work are designed to make all of the calculations needed to rough the extra stock off a piece of material and leave just enough material for a finish pass.

The following example shows one way a program can be created for a heavy roughing operation. **Figure 19-12** shows the angled part that was used in a previous example for programming a finish contour. However, when the part is machined, it will most likely be made from 7″ diameter solid bar stock. All of the extra material will have to be machined away before the finish contour machining can be accomplished.

To remove all of the extra material takes a lot of machining and a lot of programming. Even with a very aggressive programmed path that completes depths of cut of .200″ per pass, the resulting toolpath would look like **Figure 19-13**. The lines in gold represent feed rate, or cutting, moves, while the lines in blue represent rapid moves returning back to the front face of the material.

The roughing program to create each cutting pass would be at least 200 lines long. Imagine if there was an error in the program, or if the .200″

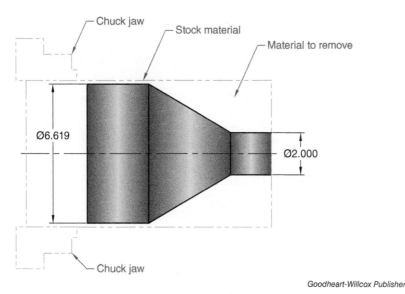

Goodheart-Willcox Publisher

Figure 19-12. The angled part is machined from 7″ diameter solid bar stock. This requires a heavy roughing operation to remove extra material.

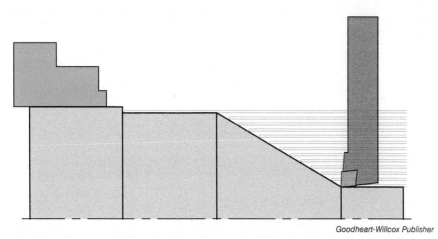

Goodheart-Willcox Publisher

Figure 19-13. A heavy roughing operation requires a complex program to define each pass of the tool when using point-to-point programming.

depth of cut was too aggressive. The program would have to be entirely rewritten, making it longer and increasing the chance of an error. The following is a representation of what the first 100 lines of the program might look like and how complicated programming becomes.

```
O1913
G20
(OD ROUGH RIGHT CNMG-432)
G0 T0101
G18
G97 S117 M3
G42 G0 X6.5388 Z.2 M8
G50 S3500
G96 S200
G1 Z.1 F.01
Z-5.9336
X6.6388 Z-6.0202
Z-9.0313
X6.7802 Z-8.9605
G0 Z.2
X6.4373
G1 Z.1
Z-5.8457
X6.5588 Z-5.9509
X6.7002 Z-5.8802
G0 Z.2
X6.3343
G1 Z.1
Z-5.7565
X6.4573 Z-5.863
X6.5987 Z-5.7923
G0 Z.2
X6.2298
G1 Z.1
Z-5.666
X6.3543 Z-5.7738
X6.4957 Z-5.7031
G0 Z.2
X6.1237
G1 Z.1
Z-5.5741
X6.2498 Z-5.6833
X6.3912 Z-5.6126
G0 Z.2
X6.016
G1 Z.1
Z-5.4808
X6.1437 Z-5.5914
```

Continued

```
X6.2851 Z-5.5207
G0 Z.2
X5.9067
G1 Z.1
Z-5.3862
X6.036 Z-5.4981
X6.1774 Z-5.4274
G0 Z.2
X5.7957
G1 Z.1
Z-5.2901
X5.9267 Z-5.4035
X6.0681 Z-5.3328
G0 Z.2
X5.6832
G1 Z.1
Z-5.1926
X5.8157 Z-5.3074
X5.9572 Z-5.2367
G0 Z.2
X5.5689
G1 Z.1
Z-5.0936
X5.7032 Z-5.2099
X5.8446 Z-5.1392
G0 Z.2
X5.4529
G1 Z.1
Z-4.9932
X5.5889 Z-5.111
X5.7303 Z-5.0403
G0 Z.2
X5.3352
G1 Z.1
Z-4.8913
X5.4729 Z-5.0105
X5.6144 Z-4.9398
G0 Z.2
X5.2158
G1 Z.1
Z-4.7878
X5.3552 Z-4.9086
X5.4967 Z-4.8379
G0 Z.2
X5.0946
G1 Z.1
Z-4.6828
```

Continued

```
X5.2358 Z–4.8052
X5.3772 Z–4.7344
G0 Z.2
X4.9715
G1 Z.1
Z–4.5763
X5.1146 Z–4.7002
X5.256 Z–4.6295
G0 Z.2
```

There is nothing inherently wrong with this program or the technique of programming all of the roughing passes one at a time, but it takes a lot of programming code and is prone to errors. The point of this example is to show that there is a much easier, more efficient method. The G70, G71, G72, G73, and G94 canned cycles are designed to handle repetitive roughing operations. These cycles make lathe programming a much simpler task.

19.6.1 G71 Roughing Cycle

The G71 roughing cycle is used to automate the process of defining where stock is to be removed in a heavy roughing operation. This cycle allows the programmer to define the finish profile positions of a part and specify cutting parameters to remove most of the excess material from the stock workpiece. The finish profile is defined in the program by specifying the X and Z axis positions corresponding to the part contour. The cutting parameters establish the depth of cut, cutting feed rate, and material to leave for a finish pass. A starting position for the tool is defined outside the stock material and the machine makes a series of cutting passes to rough out the material. Cutting occurs in repeating patterns in the Z axis direction. The finish profile dimensions and cutting parameters are used by the machine controller to calculate the passes required to remove material. A part machined using the G71 cycle is shown in **Figure 19-14**. This is the most powerful and most often used canned cycle in CNC lathe programming.

The G71 cycle can be used for outside diameter machining and inside diameter machining. The G70 cycle is used to create the finish cut and is discussed in the next section. These cycles allow for heavy material removal while using only a few lines of programming code.

The following shows the programming format used with the G71 canned cycle. On some machine controls, two blocks are used to define the parameters for the cycle. Each of these blocks begins with the G71 command. This format is shown and explained below.

```
G71 U__ R__
G71 P__ Q__ U__ W__ F__
```

The entries are read as follows:

- **G71.** Roughing canned cycle.
- **U.** On some machines, primarily machines with older Fanuc controllers, the depth of cut is specified on the first block with the U address. This designates the depth of cut, incrementally, for each pass.

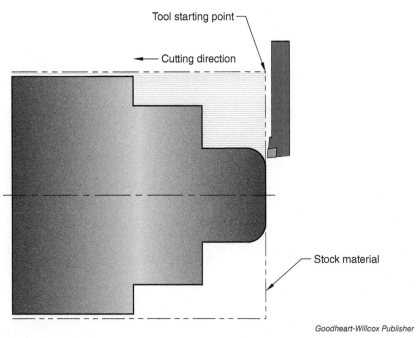

Figure 19-14. The G71 cycle is used to rough cut stock material by cutting in the Z axis direction.

- **R.** The R address designates the retract amount of the tool after each pass. The value specified is the amount the tool will pull away in the X axis after it makes a machining pass. The tool should be lifted off of the material before it moves back in the Z axis for the next cut, preventing it from dragging along the material. A typical designation is R.100. The .100 value is a safe distance, but it can be reduced to .050 or even .020, as long as the tool has enough clearance to retract safely. The R address specification sets a modal parameter in the controller, so the retract amount will stay set at that value until changed.
- **P.** The P address references the number of the block designating the start of the part contour.
- **Q.** The Q address references the number of the block designating the end of the part contour.
- **U.** The U address designates the amount of stock to leave, in diameter, in the X axis for the finish operation.
- **W.** The W address designates the amount of stock to leave in the Z axis for the finish operation.
- **F.** The F address designates the feed rate for the roughing operation.

On some machines, the parameters for the G71 roughing cycle are defined on a single block. The following format is used.

```
G71 P__ Q__ D__ U__ W__ F__
```

The parameters are the same as those defined on the second block in the two-block format, except the D address is used to specify the depth of cut. When using the single-block format, the retract amount is a parameter stored in the machine controller. The single-block format is common on Haas controllers.

Figure 19-15 shows a part with the programmed coordinate positions defining the final profile. The following program will be used to explain the parameters of the G71 cycle in more detail. For clarity, only the X and Z positions that require programming are shown on the drawing. The stock size is 7″ diameter.

```
T0101
G97 S500 M3
G0 X7. Z.1 M8
G50 S3600
G96 S200
G71 P100 Q110 D.100 U.02 W.01 F.015
N100 G0 X0.
G1 G42 Z0. F.01
X2.
Z-2.
X6.619 Z-6.
Z-9.
N110 X7.
G0 Z.1
```

- **G0 X7. Z.1 M8.** The X7. Z.1 position is the starting position of the tool. This is the point from which the roughing passes are calculated by the controller. For this part, the starting position is defined at the stock diameter size of 7″ and .100″ away from the Z0. position. It is important to start at an appropriate position that is away from the stock face and at the initial stock diameter. This is the first position the tool will travel to, and the Z axis position will be the return plane for all of the roughing passes.
- **G71 P100 Q110 D.100 U.02 W.01 F.015.** This block activates the G71 cycle. The parameters are explained as follows.

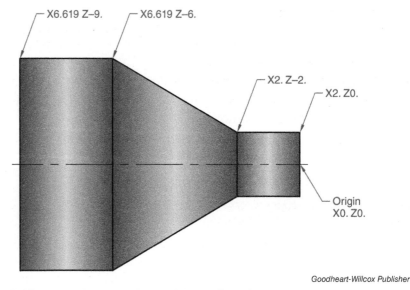

Goodheart-Willcox Publisher

Figure 19-15. A part with coordinate positions defined for an outside diameter roughing program. The positions define the final part profile.

- **P100 Q110.** The P and Q addresses tell the controller the block numbers where the part contour begins and ends. The referenced block numbers begin with the N address. In this example, P100 Q110 tells the controller that the profile of the part is defined between blocks N100 and N110. The P and Q values can be any numbers, but they must correspond to the beginning and ending block numbers in the program that define the part profile.
- **D.100.** This designates a depth of cut of .100″, incrementally, for each pass. Recalling the previous roughing program that was written in long form, it took more than 200 lines of code to complete the program. If the depth of cut was too aggressive for the tool, all of the lines of code would need to be altered, and more lines added to the program. By using the G71 cycle, the D value can be changed to .075 or .050, and the controller will calculate the passes automatically.
- **U.02 W.01.** The U and W addresses are alternate axis designators for the X axis (U) and Z axis (W). These addresses specify the stock allowance to leave for the finish pass. The U address specifies the amount of stock to leave in the X axis direction. The U.02 designation will leave the rough part .020″, in diameter, larger than the programmed part diameter dimensions. The W address specifies the amount of stock to leave in the Z axis direction. The W.01 designation will leave .010″ stock on the stock faces for finishing. As with the depth of cut, this makes for an easy edit if the finishing passes need a little more, or a little less, stock for desired surface finish or tolerance requirements.
- **F.015.** The F address value designates the feed rate used in the roughing cycle. Notice that there is another feed rate specification two lines down (F.01). The second feed rate is not necessary, but it allows the programmer to use a different feed rate in the finishing passes (in the G70 cycle). Some programmers only use one feed rate, but if a separate tool is used to finish the contour, a slower feed rate is often used to produce a better finish.
- **N100 to N110.** The blocks starting at N100 and ending at N110 define the final part profile positions. Notice that there is a rapid move to X0., then a feed move to Z0. where tool nose radius compensation is initiated. Assuming that there is some raw material on the stock face, this will run the roughing program down the face and machine to the Z0. plane. If there is no extra material, the X2. position at the first turned diameter could be specified instead of the X0. position.

By using the G71 cycle, the roughing operation is reduced from more than 200 lines to just nine lines. This greatly simplifies programming and increases efficiency in machining. If changes are required, it is much easier to make adjustments to the parameters defined in the cycle, as opposed to editing a massive program. If needed, the programmer can edit the depth of cut, feed rates, or stock allowances by changing a single parameter value.

19.6.2 G70 Finishing Cycle

Now that all of the roughing is complete, the finishing passes are programmed. The G70 cycle is a profile finishing cycle used in conjunction

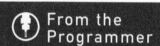

From the Programmer

Type 1 and Type 2 Cycles
The roughing canned cycle used in the previous example is referred to as a Type 1 roughing cycle. This means the programmed path can only go in one axis direction and cannot change direction. The roughing path cannot create undercuts or pockets, and cannot change direction in the course of completing the path. A Type 2 roughing cycle allows the path to change direction. A Type 1 cycle is programmed by defining a single axis move in the starting block referenced with the P address. If your CNC controller supports a Type 2 roughing cycle, program an X and Z move in the starting block in order to perform a Type 2 cycle.

with the G71, G72, and G73 cycles. The G70 cycle will command the machine to go back to the appropriate N numbered blocks and complete the contour to the final dimensions. For the previous roughing example, the block of code is as follows.

```
G70 P100 Q110
```

If the same tool that was used to rough the part is used to finish the part, the program is as follows.

```
T0101
G97 S500 M3
G0 X7. Z.1 M8
G50 S3600
G96 S200
G71 P100 Q110 D.100 U.02 W.01 F.015
N100 G0 X0.
G1 G42 Z0. F.01
X2.
Z-2.
X6.619 Z-6.
Z-9.
N110 X7.
G0 Z.1
G70 P100 Q110
```

The **G70 P100 Q110** block simply tells the controller to machine the final contour and leave no excess material.

It is usually best practice to use a separate finishing tool for the finishing operation. While a roughing insert with a large radius might be used to rough the part, a sharp finishing tool should be used to finish the part. In this case, the program will need to call out a new tool, turn on the spindle, and recall the finish contour lines. The program for the roughing and finishing operations is as follows.

```
T0101 (Roughing Tool)
G97 S500 M3
G0 X7. Z.1 M8
G50 S3600
G96 S200
G71 P100 Q110 D.100 U.02 W.01 F.015
N100 G0 X0.
G1 G42 Z0. F.01
X2.
Z-2.
X6.619 Z-6.
Z-9.
N110 X7.
G0 Z.1
T0202 (Finishing Tool)
```

Continued

```
G97 S500 M3
G0 X7. Z.1 M8
G50 S3600
G96 S200
G70 P100 Q110
G28 U0. W0.
```

19.6.3 G71 and G70 Cycles for Inside Diameter Machining

As easy as it is to use the G71 cycle for machining outside diameter contours, it is just as easy to use it for inside diameter machining, with some minor adjustments. Because a boring bar is not designed to cut into solid material, a hole needs to be drilled first. The starting position of the boring bar will be at the X axis position that coincides with the drill diameter. The other adjustment needed is leaving stock on the inside of the hole for the finishing pass. For inside diameter machining, the stock allowance to leave for the finishing pass in the X axis is programmed with a negative value for the U address (U– value).

Figure 19-16 shows an example of a part with inside diameter features. This part will be rough machined with a boring operation using the G71 cycle. The G70 cycle will be used for finish machining.

Looking at this part more closely, there is a 1.5″ diameter hole going completely through the part. The first operation for this part will be a drilling operation. A 1″ diameter drill will be used. It is important to identify the drill size, as it will be the X axis starting position for the boring bar in the roughing operation.

After the drilling operation is completed, a tool change is programmed and the G71 and G70 cycles are used for the roughing and finish operations. The entire program is as follows.

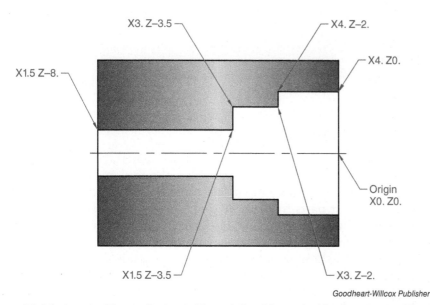

Goodheart-Willcox Publisher

Figure 19-16. A part with coordinate positions defined for an inside diameter roughing and finishing program. The positions define the final part profile.

From the Programmer

Boring Bar Size

Make sure the selected boring bar has a minimum bore diameter smaller than the diameter of the drill used in the drilling operation. Otherwise, the boring bar may collide with the opposite wall when the bar retracts from the hole. If a 1″ diameter drill is used, the minimum bore diameter of the boring bar must be smaller than 1″.

```
O7071
G20
(TOOL – 1 OFFSET – 1)
(DRILL 1.0 DIA.)
G0 T0101
G97 S200 M3
G0 X0. Z.25 M8
Z.1
G1 Z–8.5 F.01                    Drill to Z–8.5 position
G0 Z.25 M9
G28 U0. W0. M5
T0100
M01
(TOOL – 2 OFFSET – 2)
(ID ROUGH MIN. .875 DIA.)
G0 T0202
G97 S509 M3
G0 X1. Z.1 M8                    Starting position is X1. Z.1
G50 S3600
G96 S200
G71 P100 Q110 D.05 U–.02 W.01 F.01    Start of G71 cycle; U– value used for
                                       stock allowance
N100 G0 G41 X4. S200
G1 Z–2.
X3.
Z–3.5
X1.5
Z–8.
N110 G40 Z–8.05                  End of G71 cycle
G0 Z.1
G70 P100 Q110                    Finish part (G70 cycle)
M9
G28 U0. W0. M5
T0200
M30
```

19.6.4 G72 Roughing Cycle

The G71 canned cycle is designed for roughing material in the Z axis direction. As shown in the previous examples, the tool is positioned by the machine to make a number of cutting passes in the Z axis direction. This is appropriate for parts that can be rough machined most efficiently with movement in the Z axis direction. However, many parts lend themselves to movement in the X axis direction.

Figure 19-17 shows a part that has a 12″ outside diameter and a 1.5″ diameter boss that is 1.5″ long. There is 10.5″ of material removal from the large diameter to the boss diameter. A quick calculation shows that if the tool were to cut in the Z axis direction at a depth of .100″, in diameter, it would require 105 roughing passes. Machining in the X axis direction at

.100″ per pass would only require 15 passes. The same amount of material is being removed, but it can be done in a more efficient manner. The cycle time would be shorter, thus saving time and money.

The G72 cycle is similar to the G71 cycle, but the travel of the cutting tool is in the X axis direction. This cycle will be used to machine the part shown in **Figure 19-17**.

The following shows the programming format used with the G72 canned cycle. On some machine controls, two blocks are used to define the parameters for the cycle. Each of these blocks begins with the G72 command. This format is shown and explained below.

```
G72 W__ R__
G72 P__ Q__ U__ W__ F__
```

The entries are read as follows:

- **G72.** Roughing canned cycle.
- **W.** On some machines, primarily machines with older Fanuc controllers, the depth of cut is specified on the first block with the W address. This designates the depth of cut, incrementally, for each pass.
- **R.** The R address designates the retract amount of the tool after each pass. The value specified is the amount the tool will pull away in the Z axis after it makes a machining pass. The tool should be lifted off of the material before it moves back in the X axis for the next cut, preventing it from dragging along the material. A typical designation is R.100. The R address specification sets a modal parameter in the controller, so the retract amount will stay set at that value until changed.
- **P.** The P address references the number of the block designating the start of the part contour.
- **Q.** The Q address references the number of the block designating the end of the part contour.
- **U.** The U address designates the amount of stock to leave, in diameter, in the X axis for the finish operation.
- **W.** The W address designates the amount of stock to leave in the Z axis for the finish operation.
- **F.** The F address designates the feed rate for the roughing operation.

On some machines, the parameters for the G72 roughing cycle are defined on a single block. The following format is used.

```
G72 P__ Q__ D__ U__ W__ F__
```

The parameters are the same as those defined on the second block in the two-block format, except the D address is used to specify the depth of cut. When using the single-block format, the retract amount is a parameter stored in the machine controller. The single-block format is common on Haas controllers.

The following program will be used to explain the parameters of the G72 cycle in more detail. The stock size is 12″ diameter.

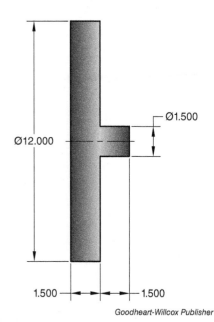

Goodheart-Willcox Publisher

Figure 19-17. A part with a large outside diameter and a boss on the front end. Using the G72 cycle to remove material in the X axis direction is an efficient way to machine the part.

```
O1916
(TOOL – 1 OFFSET – 1)
(80 DEG. INSERT – CNMG-432)
G0 T0101
G97 S64 M03
G0 X12.1 Z.1 M8
G50 S3600
G96 S200
G72 P10 Q20 D.1 U.02 W.01 F.01
N10 G0 G42 X1.5
G1 Z–1.5
X12.
N20 G40 X12.1 Z–1.46
G0 Z–.06
M9
G28 U0. W0. M05
T0100
M30
```

- **G0 X12.1 Z.1 M8.** The X12.1 Z.1 position is the starting position of the tool. This is the point from which the roughing passes are calculated by the controller. For this part, the starting position is defined near the stock diameter size of 12″ and .100″ away from the Z0. position. It is important to start at an appropriate position that is away from the stock face and at the initial stock diameter. This is the first position the tool will travel to, and the X axis position will be the return plane for all of the roughing passes.
- **G72 P10 Q20 D.1 U.02 W.01 F.01.** This block activates the G72 cycle. The parameters are explained as follows.
  - **P10 Q20.** The P and Q addresses tell the controller the block numbers where the part contour begins and ends. The referenced block numbers begin with the N address. In this example, P10 Q20 tells the controller that the profile of the part is defined between blocks N10 and N20. The P and Q values can be any numbers, but they must correspond to the beginning and ending block numbers in the program that define the part profile.
  - **D.1.** This designates a depth of cut of .100″, incrementally, for each pass. If needed, the D value can be changed to a smaller value, such as .075 or .050. The controller will calculate the passes automatically.
  - **U.02 W.01.** The U and W addresses are alternate axis designators for the X axis (U) and Z axis (W). These addresses specify the stock allowance to leave for the finish pass. The U address specifies the amount of stock to leave in the X axis direction. The U.02 designation will leave the rough part .020″, in diameter, larger than the programmed part diameter dimensions. The W address specifies the amount of stock to leave in the Z axis direction. The W.01 designation will leave .010″ stock on the stock faces for finishing. As with the depth of cut, this makes for an easy edit if the finishing passes need a little more, or a little less, stock for desired surface finish or tolerance requirements.

- **F.01.** The F address value designates the feed rate used in the roughing cycle. In this program, it is the only feed rate specified, so the same feed rate will be used in the finish pass.
- **N10 to N20.** The blocks starting at N10 and ending at N20 define the final part profile positions. Notice at the start of the profile that there is a rapid move to X1.5 where tool nose radius compensation is initiated. This is followed by a feed move to Z–1.5 in the next block.

A representation of the roughing toolpath for the G72 cycle is shown in **Figure 19-18**.

19.6.5 G73 Pattern Repeating Cycle

The G73 cycle is designed to remove material by repeatedly following a pattern on an irregular contour. In CNC lathe operations, there are often cases where the part being manufactured is not being made from solid bar stock, but from a casting or a 3D printed blank. In other words, the final shape of the part is similar to the rough stock casting profile or blank, but the casting or blank still needs to be machined to the finish dimensions.

Figure 19-19 shows a drawing for a finished part that needs machining. This part will not be machined from solid bar stock, but rather a casting supplied by a customer. The casting process is not accurate enough for this part, so the design for the casting blank is to leave 1/4″ of material on all surfaces. The rough stock will be similar in appearance to **Figure 19-20**.

When deciding which canned cycle is most efficient to complete this part, it is possible to use the G71 or G72 cycle, but the machine will spend a lot of time during the operation with the tool not engaged in the material. In this situation, the G73 cycle can be used. The G73 cycle is designed for roughing parts that are close in shape to the final part contour. The G73 cycle will step down in incremental amounts and follow the finished part path as programmed. The roughing passes are automatically calculated by the machine based on the programmed path and specified cutting

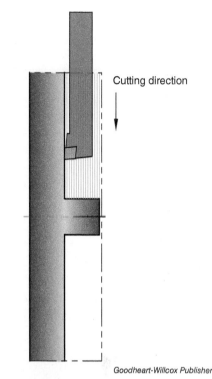

Goodheart-Willcox Publisher

Figure 19-18. In the G72 cycle, the tool removes material by making cuts in repeating passes in the X axis direction.

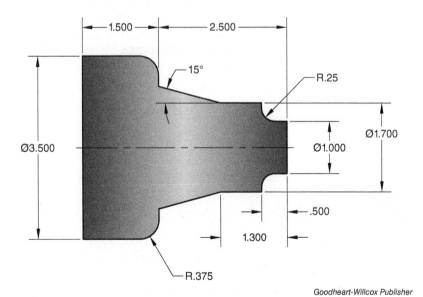

Goodheart-Willcox Publisher

Figure 19-19. A cast part with dimensions defining the final part size.

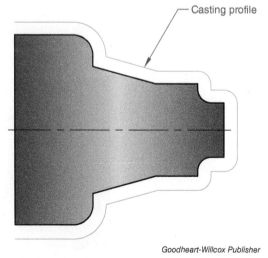

Goodheart-Willcox Publisher

Figure 19-20. The casting has 1/4″ extra material and must be machined to the final part dimensions.

parameters. After roughing is completed, the G70 cycle is used to machine the final contour.

The following shows the programming format used with the G73 canned cycle. In this example, a single block is used to define the cutting parameters. On some machine controls, two blocks are used to define the parameters for the cycle.

G73 P__ Q__ U__ W__ I__ K__ D__ F__

The entries are read as follows:

- G73. Pattern repeating cycle.
- P. The P address references the number of the block designating the start of the part contour.
- Q. The Q address references the number of the block designating the end of the part contour.
- U. The U address designates the amount of stock to leave, in diameter, in the X axis for the finish operation.
- W. The W address designates the amount of stock to leave in the Z axis for the finish operation.
- I. The I address designates the X axis distance and direction from the first cut to the last cut. The distance specified is a radial value.
- K. The K address designates the Z axis distance and direction from the first cut to the last cut.
- D. The D address designates the number of roughing passes.
- F. The F address designates the feed rate for the roughing operation.

Figure 19-21 shows the X and Z coordinates for each point on the contour. The following program will be used to explain the parameters of the G73 cycle in more detail. The program uses the G73 cycle for the roughing operation and the G70 cycle for the finish operation.

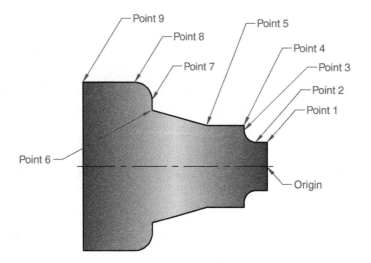

Coordinates	
Origin	X0. Z0.
Point 1	X1. Z0.
Point 2	X1. Z–.25
Point 3	X1.5 Z–.5
Point 4	X1.7 Z–.5
Point 5	X1.7 Z–1.3
Point 6	X2.343 Z–2.5
Point 7	X2.75 Z–2.5
Point 8	X3.5 Z–2.875
Point 9	X3.5 Z–4.

Goodheart-Willcox Publisher

Figure 19-21. A cast part with coordinate positions defined for a roughing and finishing program. The positions define the final part profile.

```
O7373
(INSERT – CNMG-432)
G0 T0101
G18
G97 S218 M03
G0 X4. Z.2 M8
G50 S3600
G96 S200
G73 P100 Q110 U.02 W.01 I.25 K.125 D5 F.01
N100 G0 G42 X1. S200
G1 Z-.25
G2 X1.5 Z-.5 R.25
G1 X1.7
Z-1.3
X2.343 Z-2.5
X2.75
G3 X3.5 Z-2.875 R.375
G1 Z-4.
N110 G40 Z-4.1
G0 Z.1
G70 P100 Q110
M9
G28 U0. W0. M5
T0100
M30
```

- **G0 X4. Z.2 M8.** The X4. Z.2 position is the starting position of the tool. For this part, the starting position is defined near the casting diameter size and .200″ away from the Z0. position. It is important to start at an appropriate position that is away from the stock face and at the initial stock diameter. This is the first position the tool will travel to.

- **G73 P100 Q110 U.02 W.01 I.25 K.125 D5 F.01.** This block activates the G73 cycle. The parameters are explained as follows.
    - **P100 Q110.** The P and Q addresses tell the controller the block numbers where the part contour begins and ends. In this example, P100 Q110 tells the controller that the profile of the part is defined between blocks N100 and N110.
    - **U.02 W.01.** The U and W addresses specify the stock allowance to leave for the finish pass. The U address specifies the amount of stock to leave in the X axis direction. The U.02 designation will leave the rough part .020″, in diameter, larger than the programmed part diameter dimensions. The W address specifies the amount of stock to leave in the Z axis direction. The W.01 designation will leave .010″ stock on the stock faces for finishing.
    - **I.25 K.125.** The I and K addresses are alternate axis designators for the X axis (I) and Z axis (K). These addresses designate the distance and direction from the first cut to the last cut along the corresponding axis. The I address specifies the amount of material to be removed, radially, in the X axis direction. The K address specifies the amount of material to be removed in the Z axis direction.

- **D5.** The D address value specifies the number of roughing passes. In this case, there will be five passes. The value specified must be a positive integer.
- **F.01.** The F address value designates the feed rate used in the roughing cycle. In this program, it is the only feed rate specified, so the same feed rate will be used in the finish pass.
- **N100 to N110.** The blocks starting at N100 and ending at N110 define the final part profile positions. Notice at the start of the profile that there is a rapid move to X1. where tool nose radius compensation is initiated. This is followed by a feed move to Z–.25 in the next block.

A representation of the roughing toolpath for the G73 cycle is shown in **Figure 19-22**.

19.6.6 G94 Facing Cycle

There is one additional canned cycle that is useful for reducing program length in lathe work. Sometimes, stock material is too long in the Z axis direction and cuts are made in the X axis direction to reduce the stock length. To remove material in this manner, a facing operation is used. *Facing* refers to machining the front face of a part to remove any excess material. This can be done by writing a program in long form, but the G94 facing cycle simplifies the process, especially if multiple passes are required.

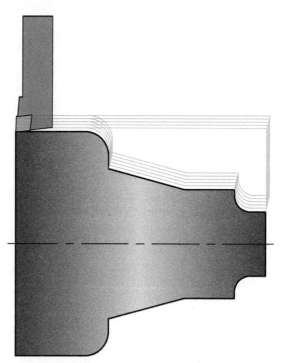

Goodheart-Willcox Publisher

Figure 19-22. In the G73 cycle, the tool removes material by making a series of cuts in repeating passes that follow the part contour.

For example, if the stock material is 1/2″ too long, a simple program can be written to remove the excess stock. The following is an example of using the G94 cycle in a facing operation.

O1234	
T0101	
G97 S1000 M3	
G0 X6. Z.6	Starting position at stock diameter
G94 X0. Z.5 F.02	Initiate facing cycle and face down to X0. Z.5
Z.4	Face at Z.4
Z.3	Face at Z.3
Z.2	Face at Z.2
Z.1	Face at Z.1
Z0.	Face at Z0.
G0 X7. Z2.	Clearance move
G28 U0. W0.	
M30	

To use this cycle, the tool is positioned at a starting point that coincides with the stock diameter, or a larger diameter, and is located away from the face as required. The G94 cycle is then activated and the tool is programmed to an XZ position to remove material. Then, by simply adding the next Z axis coordinate position, the controller automatically pulls off the face, repositions the tool at the starting position, and faces at the new Z axis position. There is no finishing cycle required. This is an ideal cycle for removing material quickly in facing operations.

Chapter 19 Review

Summary

- Creating a lathe contouring program requires programming the tool to each X and Z coordinate location in sequential order. Laying out the coordinates before programming will simplify the process.
- Tool nose radius compensation is used to offset the tool to the correct positions by compensating for the radius on the end of the cutting tool. Without tool nose radius compensation active, angled features and radii will not be machined as programmed.
- Tool nose radius compensation commands are used to offset the cutting tool to the left or right of the cutting path. The G41 command activates tool nose radius compensation left and the G42 command activates tool nose radius compensation right.
- Angular moves are created by programming X and Z axis moves on the same line of the program.
- The radius method (R method) is a simplified method to create arc movement on a contour toolpath. By programming the start point, arc direction, end point, and radius size, the controller will calculate the path.
- To use the I and K method, the arc start position, arc direction, arc end position, and incremental distances to the arc center point are specified.
- The G71 canned cycle is used for removing heavy material while cutting in the Z axis direction.
- The G72 canned cycle is used for removing heavy material while cutting in the X axis direction.
- The G73 canned cycle is best utilized when machining a casting or stock that is preformed to the final shape of the part.
- The G70 canned cycle captures the final toolpath and creates the finish operation as defined by the G71, G72, or G73 cycle.
- The G94 canned cycle is used to remove excess material from the part face. By programming a series of Z axis movements, the program repeats the X axis distance and direction as programmed.

Review Questions

Answer the following questions using the information provided in this chapter.

Know and Understand

1. Contouring refers to an operation that machines which two-dimensional shapes?
   A. Straight lines
   B. Radii
   C. Angles
   D. All of the above.

2. *True or False?* Contouring is an operation that follows a joined path of points to cut material.

3. Which command is used to activate tool nose radius compensation left?
   A. G40
   B. G41
   C. G42
   D. G43

4. When programming a counterclockwise arc, which command is utilized?
   A. G01
   B. G02
   C. G03
   D. G43

5. When using the I and K method for arc programming, what does the K address designate?
   A. The start point of the arc in the X axis.
   B. The end point of the arc in the Z axis.
   C. The X axis distance and direction from the arc start point to the arc center point.
   D. The Z axis distance and direction from the arc start point to the arc center point.

6. When using the R method for arc programming, what does the R address designate?
   A. The start point of the arc in the X axis.
   B. The end point of the arc in the Z axis.
   C. The size of the radius.
   D. The tool nose radius compensation amount.

7. *True or False?* The G71 canned cycle is used for cutting material in the X axis direction.

8. In a two-line G71 program, what does the R address designate in the first line?

    A. Depth of cut.
    B. Cutting feed rate.
    C. Radius size.
    D. Tool retract between cuts.

9. Which address is not used to specify cutting parameters in the G71 canned cycle?

    A. X
    B. F
    C. U
    D. P

10. *True or False?* The G72 canned cycle is used for cutting material in the X axis direction.

11. In the G71, G72, and G73 cycles, the P and Q addresses reference which other address in the program?

    A. F
    B. N
    C. U
    D. W

12. Which canned cycle is best suited for a cast part?

    A. G71
    B. G72
    C. G73
    D. G94

13. In the G73 canned cycle, what does the D address designate?

    A. Depth of cut.
    B. Direction of cut.
    C. Retract amount.
    D. Number of passes.

14. The G70 canned cycle is used as a finishing cycle for which roughing cycle?

    A. G71
    B. G72
    C. G73
    D. All of the above.

15. In the G71 cycle, the F address specified on the same line as the G71 command controls the _____.

    A. Roughing feed rate.
    B. Stock allowance in the X axis.
    C. Finish feed rate.
    D. Depth of cut.

Apply and Analyze

1. Referring to the drawing and coordinate table shown, plot the X and Z positions for a finish contour machining program.

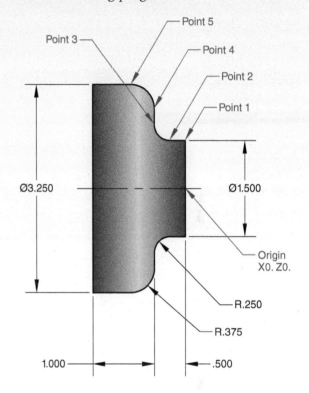

Coordinates	
Origin	X0. Z0.
Point 1	
Point 2	
Point 3	
Point 4	
Point 5	

2. Referring to the drawing given for Question 1, and the blanks provided below, give the correct codes that are missing for the G71 roughing cycle.

O0804
(INSERT – CNMG-432)
G0 T0101
G97 S235 M3
G0 X3.3 Z.1 M8
G50 S3600
G96 S200
G71 _____ _____ D.05 U.02 W.01 F.01
N100 G0 _____ X1.5 S200
G1 _____
_____ X2. Z–.5 R.25
G1 X2.5
G3 _____ _____ R.375
G1 Z–1.5
N110 _____ Z–1.6
G0 Z.1
M9
G28 U0. W0. M5
T0100
M30

3. Referring to the given drawing and the blanks provided below, give the correct codes that are missing for the G72 roughing cycle and the finishing canned cycle.

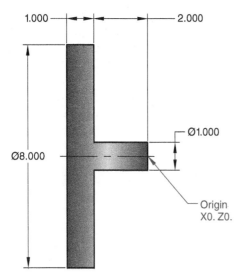

O0930
G0 T0101 (OD ROUGH RIGHT – CNMG-432)
G97 S95 M3
G0 X8.1 Z.1 M8
G50 S3500
G96 S200
G72 _____ _____ D.05 U.02 W.01 F.01
N100 G0 G42 X1. S200
_____ Z–2.
_____
N110 _____ X8.1
G0 Z.1
G70 _____ _____
M9
G28 U0. W0. M5
T0100
M30

4. Identify the canned cycle used to generate each toolpath as the G71, G72, G73, or G94 cycle.

A.

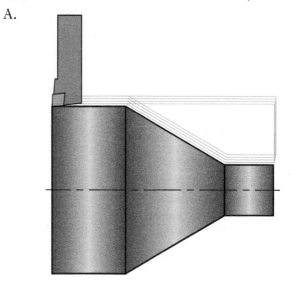

B.

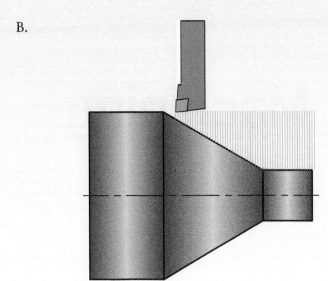

5. Referring to the blanks provided below, give the correct codes that are missing for the facing canned cycle. Assume that each successive pass of the tool removes .05" of material from the stock face.

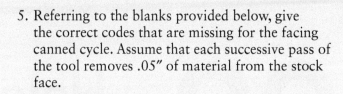

```
O1234
G0 T0101
G97 S1000 M3
G0 X5. Z.5
____ X0. Z.25 F.02
____
____
____
Z0.
G0 X5. Z.6
G28 U0. W0.
M30
```

C.

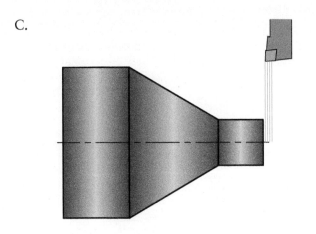

Critical Thinking

1. If an inside diameter roughing operation requires tool nose radius compensation left to be active and the G42 command is used in the program, how would this affect the final part?

2. Why is it important to evaluate the shape and orientation of a part when determining the best way to remove material in rough machining? How does this improve efficiency and reduce program length?

3. Identify a cylindrical part that contains angled or curved features and requires a Type 2 roughing cycle in programming. Explain why it would be necessary to change the direction of the cutting path in machining.

D.

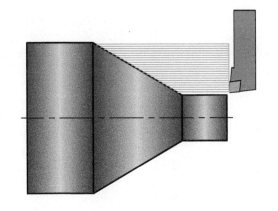

Copyright Goodheart-Willcox Co., Inc.

429

20 Hole Machining on a Lathe

Chapter Outline

20.1 Introduction

20.2 Hole Drilling Cycles

   20.2.1 G80: Cancel Canned Cycles

   20.2.2 G81: Standard Drilling and Spot Drilling Cycle

   20.2.3 G82: Spot Drilling and Counterboring Cycle

   20.2.4 G83: Peck Drilling Cycle

   20.2.5 G84: Right-Hand Tapping Cycle

   20.2.6 G85: Boring and Reaming Cycle

20.3 Machining Holes with Live Tooling

Learning Objectives

After completing this chapter, you will be able to:

- Describe different types of hole drilling operations.
- Identify canned cycles used for machining holes on a CNC lathe.
- Explain how to program drilling operations using canned cycles.
- Identify the proper canned cycle for the machining required.
- Identify applications for multiaxis drilling operations.
- Program three-axis drilling operations.

Key Terms

axial drilling
boring
canned cycle
counterbore
countersink

live tool
multiaxis machine
peck drilling
radial drilling
reamer

reaming
spot drill
spotface
spring pass

20.1 Introduction

Hole machining is a common process in lathe work, and is almost always required on parts with inside diameter features. There are specific canned cycles available to streamline the programming process for hole making operations on a lathe. Knowing when to use these cycles and how to optimize them will increase machining efficiency and reduce cycle times.

Tooling plays a significant role in deciding how to machine a part. Although drill bits made from high-speed steel can be utilized on a CNC lathe, most machine shops using CNC machines have converted to solid carbide drills and carbide inserted tooling. The capabilities of carbide tools have significantly reduced costs and cycle times and paved the way for precision drilling and deep hole drilling cycles. In the preprogramming phase, consider the cost of tooling and evaluate the total cycle time when planning out the machining process.

20.2 Hole Drilling Cycles

A standard two-axis lathe is a machine that holds and rotates the workpiece while the cutting tool is fixed in place. This requires drilling operations to be programmed with the tool positioned at X0. so that it aligns with the Z axis (the spindle centerline). This requirement is for two-axis machine configurations. On multiaxis machines, operations can be performed off the center of the part with live tooling, as discussed later in this chapter.

Figure 20-1 shows a drilling operation on a two-axis lathe. The way a drill *works*, or drills holes, is by cutting on center along the Z axis. The tool must be positioned at X0. to align with the part center. If a drill were to be programmed to drill off center at a position of X.500, for example, the drill would simply not cut. In fact, it would result in a catastrophic tool failure and cause severe damage to the work material. On a CNC mill, the drill works at any location because the tool is rotating. On a CNC lathe, the material is rotating and the drill point must be programmed at the center of rotation, or the X0. position. This requires all drilling cycles to be initiated from the X0. position and at a safe location along the Z axis, such as Z.1.

A drilling operation can be programmed in longhand form by commanding a series of rapid moves and feed moves with the G00 and G01 commands. For example, the part shown in **Figure 20-1** has a 1″ diameter

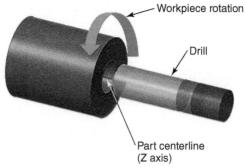

Goodheart-Willcox Publisher

Figure 20-1. On a standard two-axis lathe, a drilling operation must be programmed with the tool at the X0. position. This aligns the tool with the part center along the Z axis.

through hole. The part is 3″ long. To drill the hole correctly, the drill must travel an extra distance past the part so that the entire drill diameter clears the end of the workpiece. The extra distance to travel is the length of the drill tip extending past the full diameter of the tool. The drill tip length is calculated by multiplying the drill diameter by the constant value .3. This value is used for drills with tips that form an angle of 118°. The following formula is used to calculate the drilling depth for programming.

$$\text{Programmed depth} = \text{hole depth} + (.3 \times \text{drill diameter})$$
$$= 3 + (.3 \times 1)$$
$$= 3.3$$

The drilling depth is 3.3″. The program to drill the hole is as follows.

O0824	
G20	
(1″ Drill)	
G0 T0101	
G97 S300 M3	
G0 X0. Z.1 M8	Safe move in front of part
G1 Z−3.3 F.01	Feed move to final drill depth
Z.1	Safe move away from part
M9	
G28 U0. W0. M5	Return to machine home
T0100	
M30	

Often, drilling operations require the tool to drill only to partial depths to remove chips and allow coolant into the hole. When drilling deep holes, it is common to use a *peck drilling* operation. In peck drilling, the drill feeds down to a partial depth and then retracts to remove chips, then re-enters the material to continue cutting. The process of programming moves for a peck drilling operation becomes more complicated, as shown in the following program.

O0824	
G20	
(1″ Drill)	
G0 T0101	
G97 S300 M3	
G0 X0. Z.1 M8	Safe move in front of part
G1 Z−.5 F.01	Feed move to first drill depth
Z.1	Safe move away from part
G1 Z−1.	Feed move to second drill depth
Z.1	
G1 Z−1.5	Feed move to third drill depth
Z.1	
G1 Z−2.	Feed move to fourth drill depth
Z.1	
G1 Z−2.5	Feed move to fifth drill depth

Continued

Debugging the Code

G96 and G97 Commands

The G96 spindle speed command is meant for use in turning operations on a lathe. The G96 command is used to activate constant surface speed mode, as discussed in Chapter 17. This allows the machine controller to adjust the spindle speed to allow for changes in the material diameter and maintain the same surface speed for material removal. In drilling operations, the X axis position of the tool is programmed at X0. and the G97 command is used to program the spindle speed to an exact speed in revolutions per minute (rpm).

Z.1	
G1 Z–3.	Feed move to sixth drill depth
Z.1	
G1 Z–3.3	Feed move to final drill depth
Z.1	
M9	
G28 U0. W0. M5	Return to machine home
T0100	
M30	

Canned cycle commands simplify this process and reduce the opportunity for programming errors. A *canned cycle* is an abbreviated machining cycle that automates complex repetitive commands and functions. **Figure 20-2** shows some of the canned cycles available for hole making processes. These cycles are discussed in the following sections. Canned cycle commands are modal and must be canceled when the operation is completed. The G80 command is used to cancel canned cycles.

20.2.1 G80: Cancel Canned Cycles

The G80 command is often used in the startup block at the beginning of a program to ensure that no canned cycle is still active from the previous program. Canned cycle commands are modal and will remain active until they are shut off. The G80 command should always be used directly after the last hole is machined in a canned cycle.

20.2.2 G81: Standard Drilling and Spot Drilling Cycle

Before using a drilling canned cycle, the tool must be positioned at a safe starting location. On a two-axis machine, the starting location is at X0. and a Z axis position that is safely in front of the work material. The activation block for the canned cycle will specify the final depth along the Z axis. The cycle will feed to the final depth at the programmed feed rate. The R address is used in the activation block to set the retract plane, also called the rapid plane or R plane. This is the position the tool rapids to before the drilling cycle begins and after the drilling cycle is complete.

Hole Machining Canned Cycles

Command	Description	Application
G80	Canned cycle cancel	Cancels any active cycle; used at the end of all canned cycles
G81	Standard drilling cycle	Shallow drilling and spot drilling
G82	Spot drilling and counterboring cycle	Spotface drilling, counterboring, countersinking, and spot drilling
G83	Deep hole drilling cycle	Full retract peck drilling
G84	Tapping cycle	Standard tapping cycle for right-hand taps
G85	Boring and reaming cycle	Bore cycle—feed in and feed out

Goodheart-Willcox Publisher

Figure 20-2. Canned cycle commands for hole machining.

The G81 canned cycle is designed to simplify the task of drilling shallow holes or spot drilling prior to drilling. A *spot drill* is a tool that creates a starting position for a drilling operation. The G81 cycle will feed down to the programmed Z depth and then rapid retract before starting the next drilling cycle. This cycle is used when the depth of the hole is less than four times the diameter of the drill. This cycle should also be used for solid carbide drills, as carbide will chip or shatter if peck drilling cycles are used. The following shows the programming format used with the G81 canned cycle.

```
G81 Z__ R__ F__
```

The entries are read as follows:

- **G81.** Drilling cycle.
- **Z.** Final drilling depth.
- **R.** Z position to rapid to before the drilling cycle begins and after the drilling cycle is complete.
- **F.** Feed rate in inches per revolution.

The following is an example of using the G81 canned cycle for a drilling operation.

G0 X0. Z1.	Position at hole location
G81 Z–1.25 R.1 F.01	Complete canned cycle, drilling 1.25″ deep
G80	Cancel canned cycle

20.2.3 G82: Spot Drilling and Counterboring Cycle

The G82 canned cycle is specifically designed for programming a short dwell at the hole bottom. The dwell time allows the tool to make at least one additional revolution and leave no unwanted tool marks on the surface finish. This cycle is used for spot drilling operations that require a short dwell at the hole bottom and is particularly useful when performing a counterboring, spotfacing, or countersinking routine. A *counterbore* is an enlarged cylindrical recess at the top of a smaller hole, used to provide a flat surface for a fastener. A *spotface* is a bearing surface similar to a counterbore, but not as deep. A *countersink* is an enlarged conical recess at the top of a smaller hole, used to allow the top of a fastener to sit flush with the part face. Special tools are used to machine these features. Programming a short dwell time in drilling operations for these features allows for accurate machining at the hole bottom with a clean finish.

Usually, the G82 cycle is programmed when a slow spindle speed is utilized. The following shows the programming format used with the G82 canned cycle.

```
G82 Z__ R__ P__ F__
```

The entries are read as follows:

- **G82.** Spot drilling cycle.
- **Z.** Final drilling depth.

- **R.** Z position to rapid to before the drilling cycle begins and after the drilling cycle is complete.
- **P.** Dwell time at the bottom of the hole in milliseconds. A number entered without a decimal point represents milliseconds. A number entered with a decimal point represents seconds.
- **F.** Feed rate in inches per revolution.

The following is an example of using the G82 canned cycle for a drilling operation.

G0 X0. Z1.	Position at hole location
G82 Z–1.25 R.1 P20 F.010	Complete canned cycle, drilling 1.25″ deep
G80	Cancel canned cycle

20.2.4 G83: Peck Drilling Cycle

The G83 canned cycle is a peck drilling cycle used for deep hole drilling. Peck drilling is used when the depth of the hole is four times greater or more than the diameter of the drilling tool. See **Figure 20-3**. For example, a .500″ diameter drill that has to drill a hole more than 2″ deep needs a pecking cycle to remove chips and allow coolant down to the cutting surface. This cycle can be used at shallower depths, but it is not intended for that purpose. The cycle will feed down in incremental steps and fully retract between steps out of the hole. Avoid using any peck drilling cycle with solid carbide drills. Carbide is brittle by nature and the re-entry into the cut after a pecking cycle can cause tool failure. The following shows the programming format used with the G83 canned cycle.

G83 Z__ R__ Q__ F__

KPixMining/Shutterstock.com

Figure 20-3. Peck drilling is used for drilling deep holes to allow for chip removal and coolant delivery to the cutting edge.

The entries are read as follows:

- **G83.** Full retract peck drilling cycle.
- **Z.** Final drilling depth.
- **R.** Z position to rapid to before the drilling cycle begins and after the drilling cycle is complete.
- **Q.** Incremental amount of motion along the Z axis for each peck.
- **F.** Feed rate in inches per revolution.

The following is an example of using the G83 canned cycle for drilling.

G0 X0. Z1.	Position at hole location
G83 Z–2. R.1 Q.3 F.01	Complete canned cycle, drilling 2″ deep
G80	Cancel canned cycle

20.2.5 G84: Right-Hand Tapping Cycle

Tapping operations are used to produce threaded holes. Tapping is one way to create internal threads on a lathe. A tapping operation requires a spot drill, followed by a drill, and lastly the tap. The G84 canned cycle is designed to perform tapping cycles with right-hand taps only. This cycle synchronizes the spindle speed and feed rate and automatically reverses the spindle at the final Z depth to allow the tap to exit properly. See **Figure 20-4.** The feed rate calculation is based on the threads per inch (TPI) of the tap. The following shows the programming format used with the G84 canned cycle.

G84 Z___ R___ F___

> **From the Programmer**
>
> **Tapping on a Lathe**
>
> It is possible to tap holes on a lathe using the G84 cycle, but consider alternatives to this method and pick your spots. Often, a tapping cycle is an easier method to thread smaller holes (holes smaller than 1/2″ in diameter). However, when possible, a better option for a threading operation is to use a threading canned cycle. Internal threads will be more accurate and controllable with an inside diameter threading cycle. More importantly, it is fairly easy to break a tap in a hole on a lathe. Threading canned cycles are covered later in this text. When machining internal threads, consider all of the options available.

Dovzhykov Andriy/Shutterstock.com

Figure 20-4. A tapping canned cycle automates the tapping operation when creating a tapped hole on a lathe.

The entries are read as follows:

- **G84.** Right-hand tapping cycle.
- **Z.** Final tapping depth.
- **R.** Z position to rapid to before the tapping cycle begins and retract after the tapping cycle is complete.
- **F.** Feed rate in inches per revolution. Calculated as 1 ÷ threads per inch (TPI).

The following is an example of using the G84 canned cycle for tapping holes.

G97 M3 S500 (3/8–16 Tap)	Spindle on to 500 rpm
G0 X0. Z1.	Position at hole location
G84 Z–.75 R.1 F.0625	Complete canned cycle, tapping .75″ deep
G80	Cancel canned cycle

The feed rate defines the amount of movement in the tool in one revolution of the spindle. Feed rate on a lathe is expressed in inches per revolution (ipr) when using the G99 command, and the tap needs to advance one full thread with each revolution of the spindle. For example, a tap designation of 3/8–16 indicates a 3/8″ diameter tap with 16 threads per inch. The feed rate is calculated by dividing 1″ by the number of threads per inch.

$$\text{Feed rate} = 1 \div \text{threads per inch}$$
$$= 1 \div 16 = .0625$$

20.2.6 G85: Boring and Reaming Cycle

Boring and reaming operations are used to machine previously drilled holes to a finish size. *Boring* is an operation used to enlarge an existing hole to a final size. *Reaming* is an operation used to enlarge an existing hole to a final size and produce a smooth finish. Reaming uses a special tool called a *reamer*. A reamer is considered a *hard tool*, meaning it is purchased in an exact size and that is the only size that it cuts. For example, if the hole to be machined is dimensioned as .3748″–.3752″, a .3750″ reamer could be a suitable tool to produce the hole at a functional size and hold the tolerances required. It is important to note that a reamer cannot machine a hole like a drill does. A reamer only machines a predrilled hole. The hole must be drilled before the reamer enters the material. A normal practice is to drill the hole .010″–.015″ smaller than the reamed hole. In the .3750″ example, a drilled hole approximately .360″ to .365″ would be recommended. A reamer, by design, only follows a drilled hole. This means that a reaming operation will not improve geometric tolerance requirements for characteristics such as concentricity or perpendicularity.

The G85 canned cycle is the preferred cycle for boring and reaming operations. This cycle will feed down to the final Z depth and then feed back to the retract plane using the programmed feed rate. This feeding in both directions creates a fine surface finish and allows for a spring pass while the tool travels back to the retract plane. A *spring pass* is a final pass that removes any material that remains as a result of tool deflection on the initial cutting pass. The final cutting pass will not leave any scratches or tool marks on the surface finish. The following shows the programming format used with the G85 canned cycle.

G85 Z__ R__ F__

The entries are read as follows:

- **G85.** Feed in and feed out cycle.
- **Z.** Final hole depth.
- **R.** Z position to rapid to before the cycle begins and retract after the cycle is complete.
- **F.** Feed rate in inches per revolution.

The following program is an example of using the G85 canned cycle.

G0 X0. Z1.	Position at hole location
G85 Z-.75 R.1 F.005	Complete canned cycle
G80	Cancel canned cycle

 From the Programmer

Hole Machining
Creating a hole is a fairly simple operation in lathe work. Choosing the right tool and selecting the right operation are the most important decisions. Most parts machined on a lathe do not simply require making a hole with a drill, but will require finishing operations such as threading or inside diameter machining with a boring bar. Consider the drilling operation as a rough machining operation that will probably require finish machining.

The application of the correct drilling canned cycle is an important decision in lathe programming. When drilling a hole to a depth more than four times the diameter, a full retract G83 peck drilling cycle should be used. Solid carbide drills should not be used in peck drilling cycles. For reamers or similar tools that require the tool to feed out of the hole, the G85 canned cycle should be considered.

20.3 Machining Holes with Live Tooling

To this point in the chapter, hole machining operations with the tool positioned at X0. and traveling through the part center have been discussed. This configuration must be used for machining holes on a standard two-axis lathe. But it is also possible to perform hole machining operations *off center* in conjunction with one or more additional machine axes. In reference to lathe work, a ***multiaxis machine*** is a machine with more axes of travel than a standard two-axis machine. In the modern machining environment, many machine shops have turned to three-axis and four-axis lathes.

Figure 20-5 shows an example of a typical part that is well-suited for machining on a three-axis lathe. During the preprogramming phase, there might be discussions about what machine, or machines, will be most efficient in manufacturing this specific part. An argument could be made that

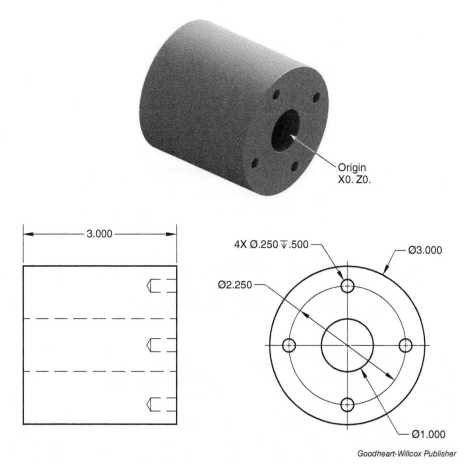

Figure 20-5. A part with four .250" diameter holes equally spaced on a 2.250" diameter bolt circle. The holes can be machined on a lathe equipped with a third axis.

this is a mill part where soft jaws could be machined to hold the outside diameter and all of the holes are drilled in a single operation. If a three-axis lathe is not available, this part could be faced, drilled, and bored in a lathe. A second operation would then be required in a CNC mill to create the four .250" holes. Multiple setups and excessive part handling probably make this approach the least efficient way. However, if a multiaxis lathe is available, this part is a perfect application.

A three-axis lathe has the standard X and Z machine axes, exactly like the two-axis lathe, but it has an additional axis referenced as the C axis. The C axis is a rotational axis used to rotate the spindle to a specific angle. This allows the programmer to lock the spindle in a position designated by the angle. Because the third axis is a rotational axis, the programmed coordinates are expressed in degrees. Angles are measured from 0°. When looking directly at the face of the chuck or collet nose on the machine, the 3 o'clock position is designated as 0°. See **Figure 20-6.** When looking in this viewing direction, positive rotation is counterclockwise.

The third axis on a three-axis lathe is used in conjunction with live tooling. A *live tool* is a driven tool used to perform drilling or milling operations while the workpiece remains in orientation with the main spindle of the lathe. Live tooling can be used for hole machining and milling operations. This expanded capability is used to reduce secondary operations and improve efficiency. A typical application for a lathe live tool is to machine features on the part's face or outside diameter. An angular

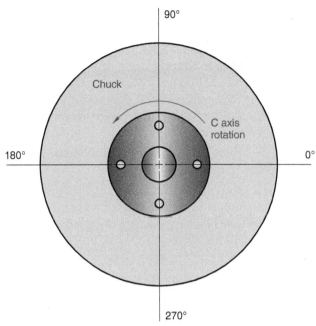

Figure 20-6. Angle positions for the C axis on a lathe with 0° oriented at the 3 o'clock position.

direction is specified to rotate the spindle about the C axis to the specific angle, and the live tool is positioned with X and Z coordinates to perform the operation.

The following program is used for a C axis drilling operation. This program will be used to explain the format used for C axis programming. The G81 canned cycle is used for drilling. Notice the similarities between standard two-axis programming and C axis programming and the M-codes that are used specifically for C axis programming.

| O2005 |
| (TOOL – 2 OFFSET – 2) |
| (.250 DRILL) |
| (C AXIS FACE DRILL) |
| T0202 |
| M154 |
| M8 |
| G97 P2500 M133 |
| G0 Z.25 C0. |
| X2.25 |
| G98 G81 Z–.650 R.1 F4. |
| C90. |
| C180. |
| C270. |
| G0 G80 Z1. X5. |
| M9 |
| M155 |
| M135 |
| G28 U0. W0. |

- **M154.** The M154 function engages the C axis motor.
- **G97 P2500 M133.** The G97 command places the machine in constant spindle speed mode. This programs the speed in revolutions per minute (rpm). The P address value sets the speed of the drill. In this case, the speed is 2500 rpm. Normally, the S address is used to set the speed, but the S address will command the spindle and that is not what is desired. The M133 function turns on forward motion of the live tooling motor.
- **G0 Z.25 C0.** This block specifies a rapid move to Z.25 and orients the spindle to the 0° position.
- **X2.25.** This is a rapid move that positions the tool at X2.25. This position is the diameter of the bolt circle where the first hole in the pattern is drilled. This represents a shift from the X0. position programmed to drill on center.
- **G98 G81 Z−.650 R.1 F4.** The G98 command is used to program the feed rate in inches per minute. In this case, the material is not rotating and the feed rate specifies the distance the cutting tool will travel per minute. The G81 command activates the same drilling canned cycle used for drilling on center. The final hole depth is Z−.650.
- **C90.** After the first hole is drilled, the C axis will rotate 90° and the second hole is drilled.
- **C180.** After the second hole is drilled, the C axis will rotate 90° and the third hole is drilled.
- **C270.** After the third hole is drilled, the C axis will rotate 90° and the fourth hole is drilled.
- **M155.** The M155 function disengages the C axis motor.
- **M135.** The M135 function stops the live tooling motor.

Figure 20-7 shows the toolpath for the drilling operation. This type of operation is referred to as *axial drilling* because the travel of the tool is perpendicular to the part face and parallel to the Z axis direction. Live tooling operations can also be completed on the outside diameter of the

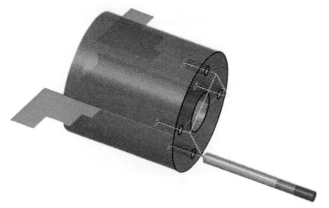

Goodheart-Willcox Publisher

Figure 20-7. A representation of the toolpath for an axial drilling operation. The cutting direction is perpendicular to the face of the part and parallel to the Z axis.

part. A drilling operation is referred to as *radial drilling* when the travel of the tool is perpendicular to the part diameter and parallel to the X axis direction. There are similar drilling canned cycles used for each type of operation, but different programming codes are used and programming formats can vary significantly among different types of controllers.

Figure 20-8 shows a part with four .250″ diameter holes machined on the outside diameter of the part. The holes extend from the outside diameter to the inside diameter. A radial drilling canned cycle will be used to machine the holes. In radial drilling, the travel of the tool is in the X axis direction, not the Z axis direction.

The following is the program used for the radial drilling operation and an explanation of the programming format. The G243 peck drilling canned cycle is used. This cycle is common on Haas controllers. In this program, the G18 command is used to designate the XZ work plane. On some controllers, it may be necessary to designate the YZ work plane with the G19 command in a radial drilling cycle. The Y axis is a third axis used in multiaxis lathe work and is perpendicular to the X and Z axes.

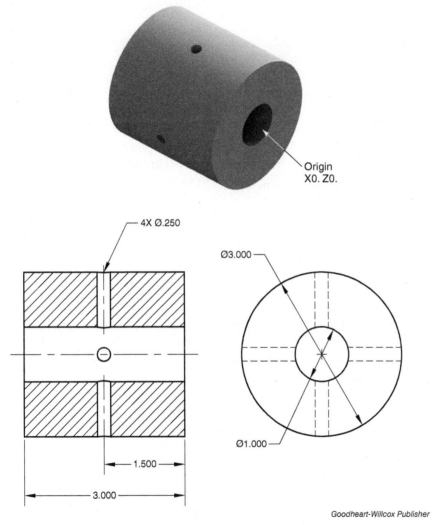

Goodheart-Willcox Publisher

Figure 20-8. A part with four .250″ diameter holes equally spaced on the circumference of the part. A radial drilling operation is used to machine the holes.

```
O2007
G20
(TOOL - 2 OFFSET - 2)
(.250 DRILL)
(C AXIS CROSS DRILL)
T0202
M154
M8
G97 P3600 M133
G98
G18
G0 X3.5 Z-1.5 C0.
G243 X.6 R3.2 Q0.2 F3.42
C90.
C180.
C270.
G00 G80 X5. Z3.
M9
M155
M135
G28 U0. W0.
M30
```

- **M154.** The M154 function engages the C axis motor.
- **G97 P3600 M133.** The G97 command places the machine in constant spindle speed mode. This programs the speed in revolutions per minute (rpm). The P address value sets the speed of the drill. In this case, the speed is 3600 rpm. Normally, the S address is used to set the speed, but the S address will command the spindle and that is not what is desired. The M133 function turns on forward motion of the live tooling motor.
- **G98.** The G98 command is used to program the feed rate in inches per minute.
- **G18.** The G18 command activates the XZ work plane.
- **G0 X3.5 Z-1.5 C0.** This block specifies a rapid move to X3.5 Z-1.5 and orients the spindle to the 0° position.
- **G243 X.6 R3.2 Q0.2 F3.42.** The G243 command activates the peck drilling cycle for radial drilling on the C axis lathe. The X.6 designation is the final X axis hole depth. The R3.2 position is the X axis position to rapid to before the cycle begins and after the cycle is complete. The Q address specifies the incremental peck amount (.200″). The F address specifies the feed rate in inches per minute.
- **C90.** The cycle is repeated at 90°.
- **C180.** The cycle is repeated at 180°.
- **C270.** The cycle is repeated at 270°.
- **M155.** The M155 function disengages the C axis motor.
- **M135.** The M135 function stops the live tooling motor.

The following program is used for the same operation, but the programming format is common on Fanuc controllers. Depending on the controller type, the G87 cycle is used for peck drilling in radial drilling operations.

G0 T0202
M23
M8
G0 X3.5 Z−1.5
C0.
G97 S2500 M51
G87 X.6 R3.2 Q0.2 F3.42
C90.
C180.
C270.
G80
M9
G28 U0. W0. M5
T0200

- **M23.** The M23 function engages the C axis motor.
- **G97 S2500 M51.** This block places the machine in constant spindle speed mode, sets the speed of the drill to 2500 rpm, and turns on forward motion of the live tooling motor.
- **G87 X.6 R3.2 Q0.2 F3.42.** This block activates the peck drilling cycle and sets the cycle parameters.

Figure 20-9 shows the toolpath for the drilling operation.

The ability to learn and use multiaxis drilling techniques on a CNC lathe is an invaluable skill. Multiaxis machining reduces setups and machine cycle times, improving the overall performance of a machine shop. The ability to program multiaxis drilling operations is a required skill for the CNC lathe programmer.

From the Programmer

Multiaxis Programming Formats

Check your machine's programming manual for the correct G- and M-codes for C axis programming. In multiaxis programming, there can be significant differences in programming formats and controllers. Haas and Fanuc machines have variations in code from controller to controller. As machines have evolved, multiaxis operations have become highly specialized and the specific G- and M-codes used in programming have changed.

Goodheart-Willcox Publisher

Figure 20-9. A representation of the toolpath for a radial drilling operation. The cutting direction is perpendicular to the outside diameter of the part and parallel to the X axis.

Chapter 20 Review

Summary

- Many lathe parts with inside diameter features require drilling operations prior to finish machining. The ability to program hole drilling operations in lathe work is a required skill.
- On a standard two-axis lathe, a drill must be programmed at the X0. position to complete a drilling operation properly. This aligns the drill with the Z axis (the spindle centerline).
- Drilling operations on a CNC lathe are simplified through the use of canned cycles. Canned cycles allow for minimal parameter input and complete control of the toolpath.
- Drilling canned cycles are programmed with a Z axis final hole depth, an R address specifying the R plane the tool rapids to before and after the cycle is complete, and a feed rate. Peck drilling cycles are programmed with a Q address that defines the peck amount.
- Holes drilled off the center axis of a part, or from the outside diameter of the part toward the inside diameter, must be completed on a multiaxis lathe. Multiaxis machines use live tooling to reduce secondary operations and improve efficiency.
- Axial drilling refers to lathe live tooling operations that machine in the Z axis direction. Radial drilling refers to lathe live tooling operations that machine in the X axis direction.
- Axial drilling and radial drilling operations can be accomplished on a multiaxis lathe with some special programming considerations. It is best to consult the programming manual for your machine to identify specific codes and operations that apply to the machine.

Review Questions

Answer the following questions using the information provided in this chapter.

Know and Understand

1. *True or False?* To complete a drilling operation on a two-axis lathe, the tool must be programmed at the X0. position.

2. *True or False?* A drilling operation can be programmed in long form by commanding a series of moves with the G00 and G01 commands.

3. In a _____ operation, the drill feeds down to a partial depth and then retracts to remove chips, then re-enters the material to continue cutting.
   A. reaming
   B. peck drilling
   C. threading
   D. turning

4. Which of the following G-codes is programmed to cancel a drilling canned cycle?
   A. G85
   B. G83
   C. G81
   D. G80

5. Which canned cycle would be used to program a 3/4–10 hole tapping cycle?
   A. G85
   B. G84
   C. G83
   D. G81

6. In the G83 canned cycle, what does the letter Q represent?
   A. The depth of each peck.
   B. The clearance plane before the cycle starts.
   C. The tool retract plane.
   D. The programmed dwell between drilling paths.

7. In the G85 canned cycle, what will the tool do after moving to the final Z depth?
   A. Return to machine home.
   B. Rapid back out.
   C. Feed back out.
   D. None of the above.

8. When using a 1/2–13 tap in the G84 tapping cycle, what is the programmed feed rate?

   A. F13.
   B. F.077
   C. F.013
   D. F.010

9. *True or False?* When possible, it is best to machine internal threads with a tap and tapping cycle to produce the most accurate threads.

10. What are the benefits of using a multiaxis lathe to complete drilling operations?

    A. Fewer machine setups.
    B. Less handling of parts.
    C. Reduced cycle times.
    D. All of the above.

11. *True or False?* When completing axial drilling or radial drilling operations, a lathe with a minimum of three axes is required.

12. On a multiaxis CNC lathe, the _____ axis is used to control the spindle orientation.

    A. C
    B. X
    C. Y
    D. Z

13. In an axial drilling operation, the codes P2000 M133 command what action?

    A. Spindle orientation at 133°.
    B. Main spindle forward at 2000 rpm.
    C. Live tooling motor forward at 2000 rpm.
    D. Tool dwell for 2000 milliseconds.

14. On a Haas machine controller, the M154 and M155 functions control which of the following?

    A. Live tooling forward and reverse.
    B. C axis motor engage and disengage.
    C. Main spindle forward and reverse.
    D. None of the above.

Apply and Analyze

1. Referring to the given drawing and the blanks provided below, give the correct codes that are missing for the standard drilling cycle on a two-axis lathe. This part requires drilling a .750″ diameter through hole. The part is 1.500″ in length.

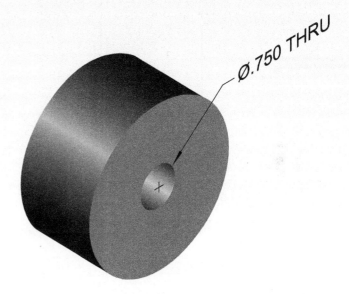

O1111
(TOOL – 2 OFFSET – 2)
(.750 DRILL)
T202
M8
_____ S200 M3
G0 _____ Z.25
_____ Z-1.875 R.1 F.01
G28 _____ _____ M9
M5
T200
M30

2. Referring to the given drawing and the blanks provided below, give the correct codes that are missing for the C axis peck drilling cycle. This part requires an axial drilling operation to drill three .375" diameter through holes equally spaced on a 1.400" diameter bolt circle.

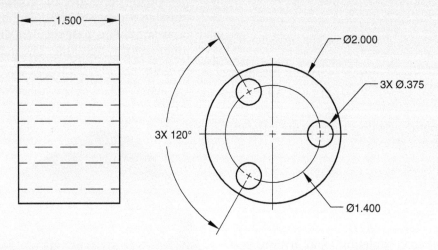

O2222	X1.4
(TOOL – 1 OFFSET – 1)	____ Z–1.613 R.1 Q0.1 F4.22
(3/8 DRILL)	____
(C AXIS FACE DRILL)	____
T0101	G0 G80 X5.
M154	M9
M8	____
G97 P713 ____	M135
G98	G28 U0. W0.
G0 Z.25 C0.	M30

3. Referring to the given drawing and the blanks provided below, give the correct codes that are missing for the standard drilling cycle and tapping cycle on a two-axis lathe. This part requires drilling a 21/32″ through hole and then tapping a 3/4–10 threaded hole. The part is 1.500″ in length.

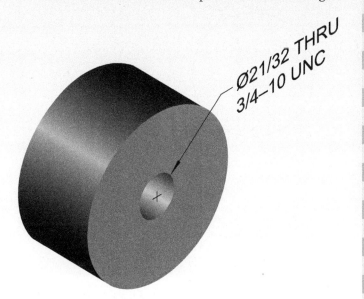

O3333
(TOOL – 1 OFFSET – 1)
(.65625 DRILL)
T101
____ S200 M3
G0 ____ Z.25 M8
____ Z–1.8281 R.1 F.01
M9
G28 U0. W0. M5
T100
M01
(TOOL – 2 OFFSET – 2)
(3/4–10 TAP)
T202
____ S200 M3
G0 ____ Z.25 M8
Z.1
____ R.1 Z–1.8281 ____
G80
G0 Z.25 M3
M9
G28 U0. W0. M5
T200
M30

4. Referring to the given drawing and the blanks provided below, give the correct codes that are missing for the radial drilling cycle. A Haas controller is used.

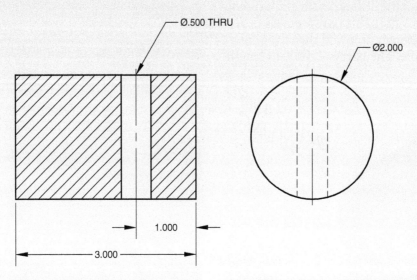

O4444
(1/2 DRILL)
(C AXIS CROSS DRILL)
T0101
____
M8
____ P534 M133
G98
G__
G0 Z–1. C__

X2.5
____ X–2.3 ____ Q0.2 F4.28
G80
G0 X3.
M9
____
____
G28 U0. W0.
M30

Critical Thinking

1. Identify some common parts with drilled holes that were machined on a lathe. Consider different types of valves, automotive parts, and any hydraulic or pneumatic driven components. Make a list and describe the possible machining processes used to produce the parts.

2. The most common machines purchased today for use in CNC shops are multiaxis machines. What are some of the major advantages to multiaxis lathe work? Explain why multiaxis machines would add value to any shop and how they increase efficiency.

3. Over the last few years, in addition to improvements in CNC machine equipment, CNC tooling has become significantly more advanced. Search online for information about carbide drills and insertable carbide drills. How has precision tooling affected cycle times and efficiency?

21 Programming Grooves and Parting Off Operations

Chapter Outline

21.1 Introduction

21.2 Planning Grooving Programs
   21.2.1 Tool Offsets
   21.2.2 Groove Width and Shape
   21.2.3 Groove Depth

21.3 Programming Straight Wall Grooves

21.4 Programming Chamfers

21.5 G75 Grooving Cycle

21.6 Tapered Wall Grooves

21.7 Full Radius Grooves

21.8 Parting Off

Learning Objectives

After completing this chapter, you will be able to:

- Describe different applications for grooving.
- Identify straight wall, tapered wall, and full radius grooves.
- Explain the relationship of the tool offset to the programmed positions in a grooving operation.
- Identify the proper tool width and shape prior to machining a groove.
- Prepare programs for machining straight wall grooves and tapered wall grooves.
- Explain programming strategies used for parting off operations.

Key Terms

bar feeder	chamfer	parting off
bar puller	grooving	tool offset

Chapter opening photo credit: Dmitry Kalinovsky/Shutterstock.com

21.1 Introduction

The previous chapters on CNC lathe programming have covered contouring, boring, and hole machining. These are all very common operations used in producing the vast majority of lathe parts. There are many other types of CNC lathe operations that might not be used on all parts, but they are common in lathe work. One of these operations is groove machining. The term *grooving* usually refers to a process of machining a narrow three-sided cavity.

There are three main types of grooves in CNC lathe work: outside diameter (OD) grooves, inside diameter (ID) grooves, and face grooves. See **Figure 21-1**. Grooves can also be described as straight wall, tapered wall, or full radius grooves. See **Figure 21-2**.

There are many applications for grooves in machined parts. Often, the required application defines tolerance requirements and the groove shape. Grooves can serve as recesses for O-ring seals, **Figure 21-3**, and are commonly used to install fasteners such as snap rings and retaining rings. Grooves can also be used as thread reliefs or simply to remove unnecessary material for weight reduction. On a print, grooves can be defined by a military standard, an American National Standards Institute (ANSI) standard, or an ASTM International specification. Grooves can also be defined by fully dimensioning each feature. It is important to have a complete understanding of the specified tolerances and customer requirements before machining takes place. It is always important to understand the print requirements for a part, but grooves can be especially tricky and critical to the finished part functionality. For example, the failure of an O-ring seal caused the explosion of a solid rocket booster in the Space Shuttle *Challenger*, resulting in the loss of that shuttle and seven crew members.

21.2 Planning Grooving Programs

Before programming can begin for a grooving operation, there are several aspects of the machining process to consider. Factors such as the tool offset, groove width, groove depth, and insert shape all affect the program and the final part dimensions. It is important to understand these factors before the program is written.

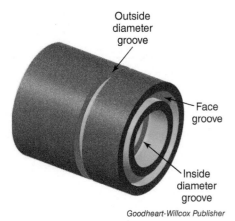

Goodheart-Willcox Publisher

Figure 21-1. Grooves can be machined on the outside diameter, inside diameter, or end face of a part.

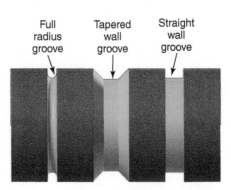

Goodheart-Willcox Publisher

Figure 21-2. Grooves can be defined by straight walls, tapered walls, or a full radius.

Urri/Shutterstock.com

Figure 21-3. O-ring seals are used in mechanical assemblies and are held in place by grooves when assembled.

21.2.1 Tool Offsets

Later in this text, setting tool offsets will be covered in more detail, but it is important to understand how the tool offsets factor into the grooving program. *Tool offsets* represent the distance the tool travels from machine home to the X0. Z0. part origin. The most common procedure for setting tools on a lathe is to manually bring the tool insert over and touch the face of the part and then designate that coordinate position as the Z axis tool offset. For the X axis tool offset, the tool is moved to a known diameter and then the size of that diameter is input to the controller to record the offset. This process tells the machine the distance from the machine home position to the programmed part zero location. The machine controller uses the tool offset settings to calculate the position of the tool during machining. This setup process is used for each tool, including grooving inserts. See **Figure 21-4**.

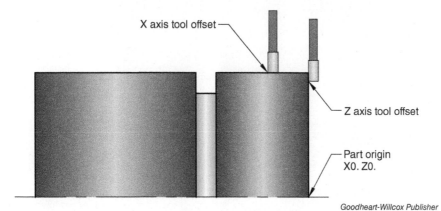

Goodheart-Willcox Publisher

Figure 21-4. During machine setup, the X axis and Z axis tool offsets for each tool are established. This is achieved by touching off the tool on the part face to set the Z axis position, then touching off the tool on the outside diameter of the part to set the X axis position.

Copyright Goodheart-Willcox Co., Inc.

However, unlike a contouring operation, where the tool is machining only at the tool tip, the grooving insert is removing material with the entire width of the tool. This must be accounted for in programming. It is also necessary to determine how the groove is dimensioned and account for the side of the tool used to establish the tool offset. For example, if the starting location for the right side of a groove is Z–1.138 and the groove is .250″ wide, the tool cannot be programmed to the Z–1.138 location. This is because the left side of the tool is normally used to set the tool offset and that side is positioned by the machine at the cutting location. In addition, the entire width of the tool, not just one end, will be removing material. See **Figure 21-5**.

To locate the tool properly for cutting the right-hand side of the groove, you will need to add the starting Z axis position and the tool width together. In this example, the width of the tool insert is .125″. The calculation is as follows.

1.138 + .125 = 1.263 (negative Z axis position)

In **Figure 21-6**, the position of the tool is programmed at Z–1.263 for cutting the right end of the groove. Before programming, consider the required Z axis positions for cutting and where the tool offsets are established during setup.

21.2.2 Groove Width and Shape

Grooving inserts come in a wide variety of shapes and sizes. Identifying the finish size requirements of the groove is the first step in selecting the correct insert for the job. Many grooves will be standard sizes, such as .125″ (1/8″), .187″ (3/16″), or .250″ (1/4″), and inserts will also come in those

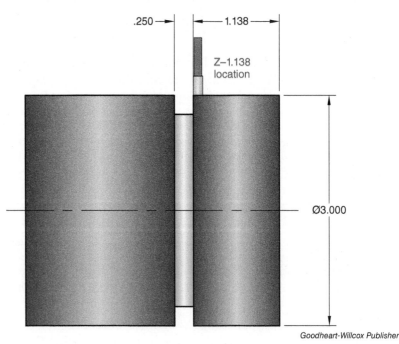

Goodheart-Willcox Publisher

Figure 21-5. If the groove is dimensioned from the right side of the groove, the dimension cannot be used directly in programming. Cutting will occur along the entire width of the tool. If the tool offset is established on the left side of the tool, the machine positions that side of the tool at the cutting location.

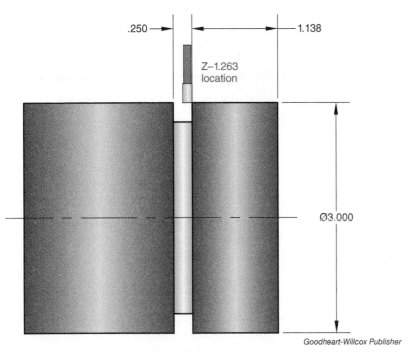

Figure 21-6. The dimension on the print is added to the width of the grooving insert to establish the programmed Z axis position for cutting.

sizes. But the best practice is to use a smaller groove tool than the final width of the groove. One trick is to consider using a metric width for the insert. For example, a .125″ wide groove can be machined with a 3 mm (.118″) wide insert. This allows for a roughing plunge in the groove and then a finish pass, where light cuts will result in better surface finish and more precise sizes.

Another factor can be a small radius dimension on the groove floor. Even straight wall grooves often require a small radius where the vertical wall meets the horizontal floor. In this case, check the insert corners for the radius size, or select the matching radius insert.

21.2.3 Groove Depth

Grooving inserts and toolholders are nonconventional tools that have a limit to the amount of depth travel in the X axis direction. Grooving inserts, just like many other types of inserts, come in a variety of styles and shapes. It is critical to identify the tool used for grooving and ensure it is capable of completing the operation safely.

Figure 21-7 shows a top-clamped insert that can only machine grooves to a radial depth of .219″. Depending on the desired groove depth, this tool may be adequate, but there are cases when deeper grooves are required.

For deeper grooves, a grooving tool such as the one shown in **Figure 21-8** can be used. This tool can machine grooves more than 1″ in radial depth. This toolholder style is generally made for wider insert sizes, approximately .125″ wide or larger. This type of tooling can also create chatter, or tool marks, depending on the application of the tool. The key point to remember is to make sure the tool can penetrate deep enough to reach the desired depth, without causing a crash.

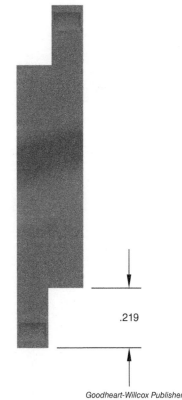

Figure 21-7. A grooving tool with an insert used for machining small grooves.

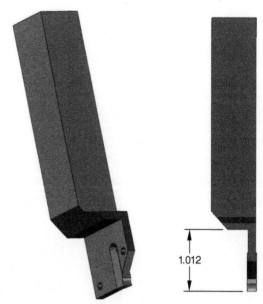

Figure 21-8. A grooving tool designed for machining deep grooves .125" wide or larger.

21.3 Programming Straight Wall Grooves

Once the tooling requirements are identified and the required dimensions are determined from the print, the program can be written. **Figure 21-9** shows a fairly simple part with a straight wall groove on the outside diameter. The outside diameter is 3" and the diameter at the bottom of the groove is 2.7". The groove is .135" wide and the right-hand edge is located 1" from the face. Whenever possible, do not use the same size insert as the groove width. For this example, a .125" wide groove tool will be used.

The four points to be programmed for the finish pass are shown in **Figure 21-10**. Before the tool goes to those locations, a roughing pass can be programmed at Z–1.130, the middle of the groove, and down to X2.710, removing most of the stock in the groove. The programmed Z

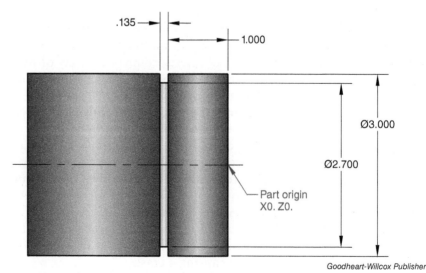

Figure 21-9. A part with a .135" wide straight wall groove on the outside diameter of the part.

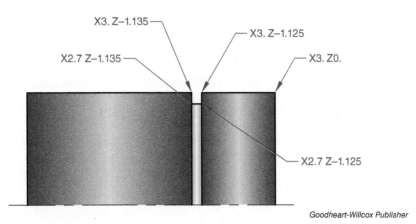

Figure 21-10. Points used in programming for machining the straight wall groove.

axis tool position at the middle of the groove is determined by subtracting the insert width from the groove width, dividing the difference in half, and offsetting by that amount from the left end of the groove. Once the roughing is completed, a finish pass can be used to finish each wall and the groove floor. The program is as follows.

O0921	
G20	
G0 T0101	
(.125 Groove Tool)	
G97 S500 M3	
G0 X3.1 Z.1 M8	Safe move in front of part
Z-1.130	Move to roughing location
G1 X2.710 F.005	Roughing plunge
X3.1	
Z-1.125	Move to finish pass location
X2.700 F.01	
Z-1.135	
X3.05	
G0 X3.5 Z.1	
M9	
G28 U0. W0. M5	Return to machine home
T0100	
M30	

Take note of the differences in programming in this grooving operation. The first is programming the spindle in constant spindle speed mode with the G97 command. This programs the spindle to an exact speed in revolutions per minute (rpm). A grooving cycle can be programmed in constant surface speed mode using the G96 command, but the tool is very narrow and can be fragile. It is beneficial to calculate the rpm speed and have it stay constant. The second difference in programming is an added feed rate after the roughing pass. The first groove is machined at .005" inches per revolution (ipr), and then the finish passes move to .010" ipr. The first pass is engaging the entire tool width, while the second pass is only removing about .005" of material. The finish pass can be accelerated without damaging the tool.

Occasionally, more than one roughing pass is required. If using the same .125″ wide tool to machine a groove that is .185″ wide, a second roughing pass would be created, again leaving between .005–.010″ of material for the finish pass. See **Figure 21-11**. The program is as follows.

O0526	
G20	
G0 T0101	
(.125 Groove Tool)	
G97 S500 M3	
G0 X3.1 Z.1 M8	Safe move in front of part
Z–1.130	Move to roughing location
G1 X2.710 F.005	First roughing plunge
X3.1	
Z–1.180	
X2.710	Second roughing plunge
X3.05 F.01	
Z–1.125	Move to finish pass location
X2.700	
Z–1.185	
X3.05	
G0 X3.5 Z.1	
M9	
G28 U0. W0. M5	Return to machine home
T0100	
M30	

Machining multiple grooves on a single part is also a common operation. See **Figure 21-12**. Multiple grooves can be programmed one at a time, by roughing and then finishing each groove, or the programmer may decide to complete the roughing plunge in each groove and then go back

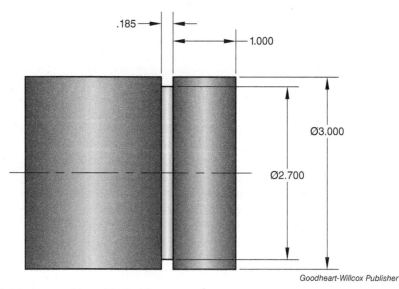

Goodheart-Willcox Publisher

Figure 21-11. A part with a .185″ wide straight wall groove. The groove is machined by programming two roughing passes and a finish pass.

Pixel B/Shutterstock.com

Figure 21-12. A grooving operation used to machine multiple grooves on a single part.

to each location and finish. If rapid moves are used between grooves, the cycle time will be very similar for both types of toolpaths.

Internal grooves are completed using the same process. The insert fits into a special boring bar holder, **Figure 21-13**. The same concerns exist in programming internal grooves relative to the groove depth, groove width, and the roughing and finish operations. The added concern is the boring bar holder. Make sure that the bar can safely enter and exit the workpiece bore diameter, and the depth of groove does not exceed the clearance of the insert or bar.

21.4 Programming Chamfers

Creating small edge breaks on sharp corners can be critical to the functionality of a part. A *chamfer* is a beveled edge machined to relieve a sharp

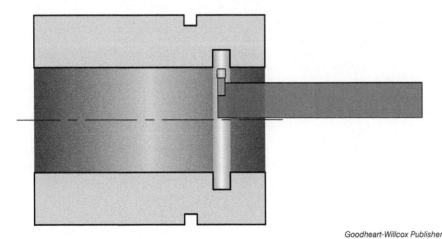

Goodheart-Willcox Publisher

Figure 21-13. An internal grooving operation must allow for clearance between the tool and the inside diameter features.

corner. It is often necessary to machine chamfers on grooves to smooth edges. For example, leaving a small sharp edge on a groove wall can cause an O-ring seal to be cut or damaged in assembly. A sharp corner on a part can also cause assembly personnel to cut their hands when handling parts. Although the part is *technically* made to the print specifications, that simple oversight can cause a machine failure or a safety issue at assembly.

Programming a small chamfer on a straight wall groove is similar to any other chamfering operation. Just be aware of where the tool is in relation to the part and that cutting occurs on both sides of the tool. On a print, chamfers are typically noted as edge breaks. In **Figure 21-14**, the part in the previous example has .015″ edge breaks specified on the groove edges.

Figure 21-15 shows an enlarged view of the chamfers and a table with X and Z coordinate positions for the final pass. After the roughing pass is made, the tool can be positioned at Point 1 and programmed through the remaining points. Remember that the Z axis positions are accounting for the .125″ wide groove tool.

The following program is similar to the previous grooving program with a roughing plunge in the center of the groove, followed by a safe move off the part, and then a move to Point 1. The tool then moves to the remaining points to complete the final pass.

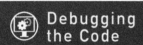

Debugging the Code

Safe Movement

Notice in this program that feed rate moves (G01 command moves) were used even when exiting the cut. However, these are very small moves and the cycle time saved by rapid moves (G00 command moves) would be miniscule. Safety is more important than speed. Never sacrifice a safe move for a fast move, especially when there is little benefit.

O1020	
G20	
G0 T0101	
(.125 Groove Tool)	
G97 S500 M3	
G0 X3.1 Z.1 M8	Safe move in front of part
Z–1.130	Move to roughing location
G1 X2.710 F.005	Roughing plunge
X3.1	
Z–1.110	Point 1 Z axis position
X3.0	Point 1 X axis position
X2.970 Z–1.125	Angle move to Point 2
X2.700	Point 3
Z–1.135	Point 4
X2.970	Point 5
X3.00 Z–1.150	Angle move to Point 6
X3.05	Safe move off part
G0 X3.5 Z.1	
M9	
G28 U0. W0. M5	Return to machine home
T0100	
M30	

The same technique for programming edge breaks, or chamfers, can also be used for programming a radius on a groove wall. Follow through with the same steps, but use the G02 and G03 circular interpolation commands for creating the radii.

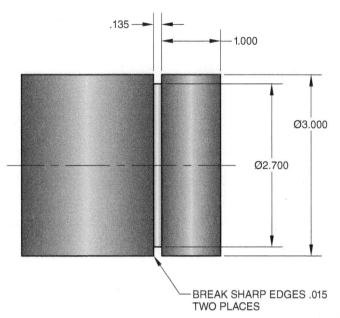

Figure 21-14. A part with a straight wall groove and edge breaks on the sides.

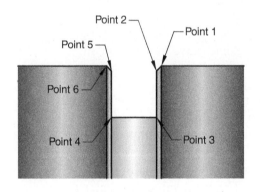

Point	X Position	Z Position
Point 1	X3.000	Z–1.110
Point 2	X2.970	Z–1.125
Point 3	X2.700	Z–1.125
Point 4	X2.700	Z–1.135
Point 5	X2.970	Z–1.135
Point 6	X3.000	Z–1.150

Goodheart-Willcox Publisher

Figure 21-15. An enlarged view of the part with plotted points and the programmed coordinate positions used in finish machining.

21.5 G75 Grooving Cycle

The G75 canned cycle is specifically for rough grooving on a CNC lathe. This can be a great help, especially with wide grooves. The cycle is simple to use and it is easy to make quick edits for cut widths or depths. The G75 canned cycle also has the ability to perform peck cutting. As with a peck drilling cycle, this means the tool will feed down to a partial depth and then make a small retraction to shear off the chips, and then re-engage the cut.

Figure 21-16 shows a part with a single straight wall groove. This groove is much wider than the grooves in the previous examples. A .125″ wide groove tool will be used to make a series of repetitive cuts to rough out material. The position at the right side of the groove for programming is Z–.625 (.500 + .125 tool width), the position at the left side of the groove is Z–1.250, and the finish groove diameter is X2.000. The following program is used to rough the groove in .050″ steps along the Z axis, leaving .005″ on the walls and floor of the groove for the finish pass. This program is used to explain the parameters of the G75 cycle in more detail. The two-block format used in this program is common on Fanuc controllers.

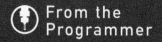

Becoming a Great Programmer

There are three types of programmers: bad programmers, good programmers, and great programmers. Bad programmers have to have their programs edited at the machine and every operator hates that. Good programmers develop good programs, but they always require secondary operations or they are slow and not thought out well. Great programmers produce efficient programs that make good parts right off the machine. They pay attention to all of the little details, like deburring parts on the machine. Their parts come out looking great, feeling great, and machined to the print specifications. There are many good programmers, but you should strive to be a *great* programmer.

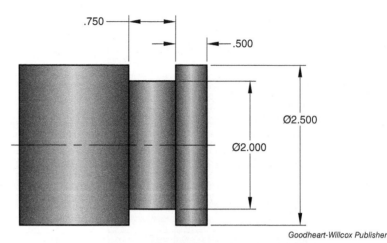

Figure 21-16. A part with a .750″ wide straight wall groove on the outside diameter of the part.

G0 T0404	Tool 4
G97 S500 M3	Spindle on to 500 rpm (constant spindle speed)
G0 X2.7 Z-.63 M8	Rapid move to starting position
G75 R.02	First G75 block
G75 X2.01 Z-1.245 P.07 Q.05 F.003	Second G75 block
G0 X2.7	Rapid move to safe position

- **G0 X2.7 Z-.63 M8.** This block is critical because it defines the first plunge location of the tool in the X and Z axes. The part is 2.5″ in diameter, so the X2.7 position is a safe distance off the part. The Z-.630 position is the Z-.625 finish position plus .005″ to leave stock for the finish pass.
- **G75 R.02.** The first G75 block includes the R address to establish a retract amount between pecks. The R.02 value designates a retract distance of .020″.
- **G75 X2.01 Z-1.245 P.07 Q.05 F.003.** The second G75 block defines the completion of the cycle. The entries are explained as follows.
    - **X2.01.** This is the final X axis diameter position for the groove. Remember to leave material for a finish pass. In this case, .010″ of material (.005″ per side) is left for the finish pass.
    - **Z-1.245.** This is the final Z axis position for the left side of the groove, leaving .005″ for the finish pass.
    - **P.07.** The P address value is the peck amount in the X axis direction. The tool will travel down .070″ and then retract .020″, and then feed down another .070″ until the final X axis depth is reached.
    - **Q.05.** The Q address value represents the Z axis stepover amount between plunging cuts. The first plunge is at Z-.630, the second plunge is at Z-.680, and so on, until the final Z axis position is reached.
    - **F.003.** The F address value designates the feed rate in inches per revolution.

Note: On older Fanuc controllers, the P and Q address values cannot use a decimal point. In such a case, the decimal is moved four places to the right. For example, a peck value of .070″ is expressed as P700 and a stepover amount of .050″ is expressed as Q500.

To this point in the program, the G75 cycle has only roughed the groove. To finish the groove, manually program the finish points. Just remember to be aware of the tool positions and create safe positioning moves. The final program with the finish pass is as follows.

G0 T0404	
G97 S500 M3	
G0 X2.7 Z-.63 M8	
G75 R.02	
G75 X2.01 Z-1.245 P.07 Q.05 F.003	
G0 X2.7	Move to safe position after roughing
Z-.625	Tool position at right side of groove
G1 X2. F.005	Feed move to X2.
Z-.750	Finish to final Z axis position
X2.7	Feed out of part
G0 X3. Z1.	Safe move away from part

The G75 cycle used on Haas controllers is very similar. It uses a single-block format to define the cycle. The P and Q addresses are replaced with I and K, as shown in the following example.

G0 T0404
G97 S200 M3
G0 X2.7 Z-.63 M8
G75 X2.01 Z-1.245 I.07 K.05 R.02 F.003
G0 X2.7

The following is an explanation of the cycle and the parameters used in programming.

- **G0 X2.7 Z-.63 M8.** Before the G75 cycle is activated, program the tool to the X and Z axis position for the first plunge cut.
- **G75 X2.01 Z-1.245 I.07 K.05 R.02 F.003.** A single G75 block defines the completion of the cycle. The entries are read as follows.
    - **X2.01.** This is the final X axis diameter position for the groove. Remember to leave material for a finish pass. In this case, .010″ of material (.005″ per side) is left for the finish pass.
    - **Z-1.245.** This is the final Z axis position for the left side of the groove, leaving .005″ for the finish pass.
    - **I.07.** The I address value is the peck amount in the X axis direction. The tool will travel down .070″, retract, and then feed down another .070″ until the final X axis depth is reached.
    - **K.05.** The K address value represents the Z axis stepover amount between plunging cuts. The first plunge is at Z-.630, the second plunge is at Z-.680, and so on, until the final Z axis position is reached.

- **R.02.** The R address sets the retract amount, or clearance amount, between pecking moves. A Haas controller can store this value as a default setting in Setting 22.
- **F.003.** The F address value designates the feed rate in inches per revolution.

The G75 cycle can also be used for creating multiple grooves on a single diameter, as long as the grooves all have the same depth along the X axis, the grooves are the same distance apart, and the tool is the same size as the groove. Since it is not best practice to plunge cut a groove with a tool of the same size, this method will not be covered. However, check your machine's programming manual for additional options for the G75 grooving cycle.

21.6 Tapered Wall Grooves

A tapered wall groove refers to any type of groove that has angled walls. Grooves with tapered walls can have different shapes. For example, some grooves have one wall that is vertical and an opposite wall that is tapered. The angle can be as small as 2° or it can be larger, such as 60°. The approach to programming is similar to programming straight wall grooves, with the addition of the angular cut for the side walls. CNC lathe programming is meant to be creative and there are many variables to consider, such as machine type, material type, tooling, machine horsepower, and controller type. This section will cover some of the possible ways to program tapered wall grooves, but there are many other ways, so be creative and determine which method works for a particular part.

Figure 21-17 shows a part with a tapered wall groove that has a 45° angle on each end. There are enough dimensions provided to complete the tapered groove. To complete the part profile, the G70 and G71 cycles would be used in programming. The main part profile would be completed before machining the groove. However, before the groove can be programmed, some calculations are required to establish all of the X and Z axis coordinates.

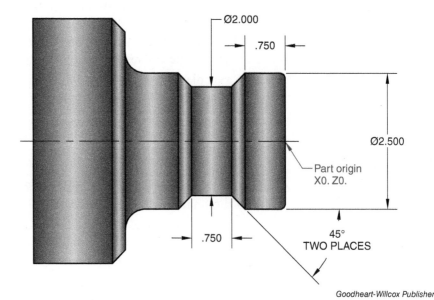

Goodheart-Willcox Publisher

Figure 21-17. A part with a tapered wall groove on the outside diameter of the part.

An enlarged view of the part is shown in **Figure 21-18**. The Z axis coordinates are not specifically defined by the dimensions provided. However, to plot the points, a right triangle can be created to determine the distance between Points 1 and 2. Because the groove angle is 45°, the sides of the triangle are equal in length. As shown in **Figure 21-18**, the sides are .250″. Remember that when defining the X axis coordinates, the 2.500 and 2.000 dimensions represent distances in diameter. Thus, the radial distance from Point 1 to Point 2 represents half the difference between the 2.500 and 2.000 diameters (2.500 – 2.000 = .500 and .500 ÷ 2 = .250). The same distance exists between Points 3 and 4 on the left side of the groove. The table in **Figure 21-18** shows the X and Z coordinate positions for each of the points.

Now that the four programming positions are calculated, there are two more positions that are helpful to determine. These are the middle positions on each angle. In **Figure 21-19**, these points are identified as Points 2 and 5. If the center of the groove is roughed out, and the machine tries to cut all of the material along the angle in one finish pass, it is highly likely that the tool will break. To avoid this, the tool can be programmed to plunge cut to the middle position on each angle. Point 2 represents half the distance between Points 1 and 3 in each axis direction. This distance is .125″. Point 5 represents the same distance between Points 4 and 6. The table in **Figure 21-19** shows the coordinate positions for Points 2 and 5. Remember that the X axis positions represent diameter measurements.

Now the program can be written. The cutting tool will be a .125″ wide grooving insert, so Points 1, 2, and 3 will be shifted by Z–.125 to allow for cutting on the right-hand side of the insert. The program could be written in longhand form, but it is simpler to use the G75 cycle. The G75 cycle will be used to rough between Points 3 and 4, and then additional plunge moves will be added at Points 2 and 5. This will be fully explained after the G75 cycle is completed. The following program is used for initial roughing, leaving .005″ of material for the finish pass. The single-block format common on a Haas controller is used.

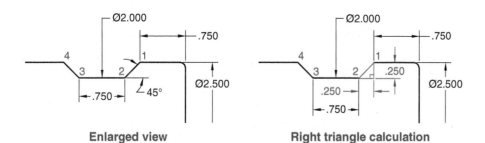

Point	X Position	Z Position
Point 1	X2.500	Z–.750
Point 2	X2.000	Z–1.000
Point 3	X2.000	Z–1.750
Point 4	X2.500	Z–2.000

Coordinates

Goodheart-Willcox Publisher

Figure 21-18. To define the Z axis positions for the groove, a right triangle can be used to establish the side lengths corresponding to the 45° angle. The resulting .250 distance is used to establish the coordinates for Points 2, 3, and 4.

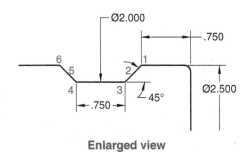

Figure 21-19. Two additional points (Points 2 and 5) are plotted at the middle position on each angle of the groove to allow for roughing at those locations.

```
G0 T0404
G97 S200 M3
G0 X2.7 Z-1.13 M8
G75 X2.01 Z-1.745 I.07 K.05 R.02 F.003
G0 X2.7
```

The G75 cycle will remove the material in the center of the groove. The material adjacent to the angles remains in place. As shown in **Figure 21-20**, it is apparent that if the .125″ grooving tool attempts to machine the angles, the tool will most likely fail. At this point, the tool should be moved back to Points 2 and 5 to remove the excess material. In the following program, the roughing passes are added and a finish pass is completed.

G0 T0404	
G97 S200 M3	
G0 X2.7 Z-1.13 M8	
G75 X2.01 Z-1.745 I.07 K.05 R.02 F.003	
G0 X2.7	
Z-1.000	
G1 X2.255 F.003	Point 2
G0 X2.7	
Z-1.875	
G1 X2.255 F.003	Point 5
G0 X2.7	
Z-.875	Move to Point 1 for finish pass
G1 X2.500 F.005	
X2.0 Z-1.125	
Z-1.750	
X2.5 Z-2.000	
G0 X2.7	
G28 U0. W0.	
M30	

Another possible strategy, especially for deep grooves with extended angles, is to use two separate G75 canned cycles. See **Figure 21-21**. In the first G75 cycle, roughing is completed using Points 2 and 5 as boundaries. The blue lines in **Figure 21-21** represent the path for the first cycle. Then, a second G75 cycle is used to complete roughing between Points 3 and 4.

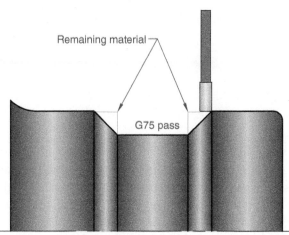

Figure 21-20. After the initial roughing, the material in the center of the groove has been removed and excess material remains along the angles. Machining the angles in one pass will most likely cause the tool to break.

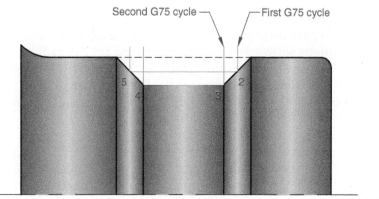

Figure 21-21. Two separate G75 cycles can be used in the program for roughing the groove. The path programmed in the first G75 cycle is represented by the blue lines. The path programmed in the second G75 cycle is represented by the green lines.

The green lines in **Figure 21-21** represent the path for the second cycle. The groove is completed with the same finishing pass used previously.

G0 T0404	
G97 S200 M3	
G0 Z–1.005 X2.7 M8	
G75 X2.255 Z–1.87 I.07 K.05 R.03 F.003	First G75 roughing cycle
G0 X2.7	
G0 Z–1.13	
G75 X2.01 Z–1.745 I.07 K.05 R.02 F.003	Second G75 roughing cycle
G0 X2.7	
Z–.875	Move to Point 1 for finish pass
G1 X2.500 F.005	Start finish pass
X2.0 Z–1.125	
Z–1.750	

Continued

```
X2.5 Z–2.000
G0 X2.7
G28 U0. W0.
M30
```

21.7 Full Radius Grooves

A full radius groove is a groove that has no flat on the groove bottom, but a 180° radius. The width of the groove is the same size as the diameter of the grooving tool. Full radius grooving operations are straight plunging, but some may require a roughing operation as needed. This type of operation is often used for parts that require O-ring seals.

A full radius groove is a perfect application for the G75 pecking canned cycle. Because this groove is often just directly plunged, the chip breaking feature will help prevent a tool failure. The G75 cycle can be used for machining a single full radius groove or multiple grooves spaced at equal distances.

Figure 21-22 shows a part that has a threaded end with a straight wall groove. This part also has a full radius groove, likely an O-ring seal groove. The dimensions shown are specific to creating the full radius groove. The groove radius is .0625″, so the groove is .125″ wide (.0625 × 2 = .125). The groove is dimensioned from the face of the part to the groove centerline. Because the center of the cutting tool is aligned with the center of the feature, and the tool offset measurement is to the center of the tool, it is not necessary to program from the tool edge. During machine setup, the setup person will touch off one side of the tool and then offset in the negative Z axis direction the distance to the tool centerline (.0625″ in this case). The order of operations for this part will be to rough and finish the outside diameter (using the G70 and G71 cycles), machine the straight wall groove, thread the end, and then machine the full radius groove. The following program creates a single full radius groove and is used to explain the G75 cycle in more detail. The two-block format used in this program is common on Fanuc controllers.

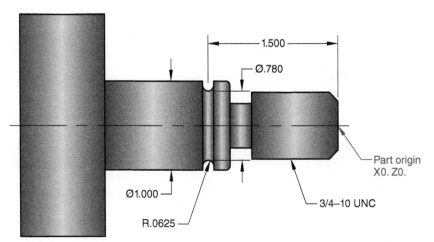

Goodheart-Willcox Publisher

Figure 21-22. A part with a single straight wall groove and a single full radius groove on the outside diameter of the part.

```
T0404
G97 S800 M3
G0 X1.1 Z-1.5 M8
G75 R.015
G75 X.78 Z-1.5 P.04 Q0. F.003
G0 X1.3
G28 U0. W0.
```

- **G0 X1.1 Z-1.5 M8.** The starting position of the tool is at the groove center.
- **G75 R.015.** The first G75 block sets the retract amount between pecks.
- **G75 X.78 Z-1.5 P.04 Q0. F.003.** The second G75 block designates the X axis finish diameter and the Z axis end position. In this case, the Z axis end position is the same as the initial position. This is because there is only one groove. The P.04 value is the peck amount. The Q0. value is used to specify that there is no stepover amount and a single plunge cut is made.

It is common to have multiple full radius grooves on a single part. As long as the final X axis position is the same for all grooves and the grooves are equally spaced, the G75 cycle can be used for a multiple grooving operation. **Figure 21-23** shows a part with two full radius grooves. The program is as follows.

```
T0404
G97 S800 M3
G0 X1.1 Z-1.5 M8
G75 R.015
G75 X.78 Z-1.75 P.04 Q0.25 F.003
G0 X1.3
G28 U0. W0.
```

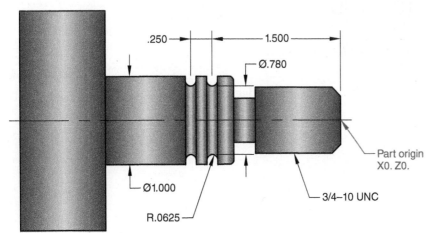

Figure 21-23. A part with two full radius grooves on the outside diameter of the part.

> **From the Programmer**
>
> **Grooving Operations**
>
> In lathe programming, you will encounter many opportunities to create grooves on parts. The part size, groove size, part material, and groove shape can influence the best approach for creating the groove. Consider all of the available options in machining and write the program for repeatability and efficiency.

- **G0 X1.1 Z–1.5 M8.** The starting position of the tool is at the center of the first groove.
- **G75 R.015.** The first G75 block sets the retract amount between pecks.
- **G75 X.78 Z–1.75 P.04 Q0.25 F.003.** The second G75 block designates the X axis finish diameter and the Z axis position of the second groove. The P.04 value is the peck amount. The Q0.25 value specifies that there is .250″ stepover between grooves.

If the part has more grooves, the program can be altered as needed. **Figure 21-24** shows a part with four grooves. The grooves are equally spaced and have the same final X axis location. The program is as follows.

| T0404 |
| G97 S800 M3 |
| G0 X1.1 Z–1.5 M8 |
| G75 R.015 |
| G75 X.78 Z–2.25 P.04 Q0.25 F.003 |
| G0 X1.3 |
| G28 U0. W0. |

The only change to this program from the previous program for two grooves is the Z axis position specified in the second G75 block. The Z–2.25 position is the location of the last groove. This method only works if the grooves are equidistant and have the same final X axis depth.

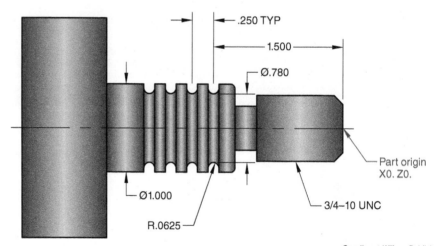

Goodheart-Willcox Publisher

Figure 21-24. A part with four full radius grooves on the outside diameter of the part.

21.8 Parting Off

Parting off refers to a process in which a machined part is separated from the rest of the bar stock being held in the lathe chuck or collet. Parting off is a final operation that occurs after the part is completed. This is a common operation with high production parts less than 2″ in diameter. A special tool holder and insert are used to perform a parting operation. CNC lathes equipped with bar feeders or bar pullers are best suited for parting off. A *bar puller* is a programmable accessory tool mounted to a turret that grips bar stock and pulls it to a specific length through the machine spindle. A *bar feeder* is a separate unit that holds bar stock and automatically loads it into the lathe for machining. Bar feeders allow for longer production runs. As with other types of machining operations, parting off requires the appropriate tooling and a suitable programming strategy. The programming process is similar to that used for grooving operations.

Figure 21-25 shows a small part with a 1″ outside diameter and a 1″ length. The part also has a 1/2″ diameter bore and chamfers on all edges. Although this is a part with close tolerances, with a ±.001″ tolerance on each dimension, it is a relatively simple part to program and produce. However, the customer needs 10,000 pieces and the parts need to be produced in a cost-efficient manner. The programmer decides to produce these parts from 1 1/8″ bar stock on a CNC lathe with a bar feeder.

Figure 21-26 shows a setup for the part. The bar is extending out far enough so the part can be completely machined and there is still enough clearance to use a parting tool on the left-hand side of the part to separate the finished part from the bar stock.

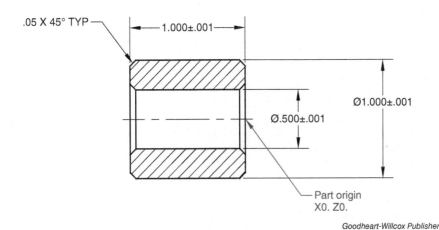

Goodheart-Willcox Publisher

Figure 21-25. A small 1″ diameter part with 45° chamfers on all edges.

From the Programmer

Efficiency in Machining

The *hard* part of machining for the part shown in **Figure 21-26** is completing the internal diameter chamfer on the back side, or left side, of the part. Many programmers would just skip programming that chamfer and leave it up to a secondary operation on another machine, but that adds time, money, and the opportunity to scrap the part. This is a good opportunity to use a grooving operation. One way to complete the internal machining would be to use an internal diameter grooving tool after the bore is completed and machine the chamfer on the back end. When the part is parted off, it is ready for packaging and shipping to the customer.

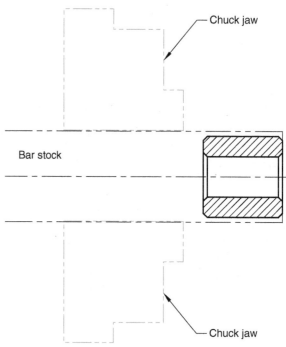

Figure 21-26. A work setup for a part machined from round bar stock on a lathe equipped with a bar feeder.

Although the G75 cycle can be used for parting off this part, it makes sense to create a program in longhand form. Also, a clearance cut should be used to aid the cutter in making a smooth, load-free cut. This clearance pass is made .005″ behind the part in the Z axis direction and goes down to a diameter position between the outside diameter and inside diameter, or approximately X.750. To program the Z axis position, the tool offset must be considered. The tool offset could be set from either side of the tool, but for consistency, the tool will be offset from the left side, just as with a grooving tool. For a .125″ wide parting tool, the programmed Z position for the final pass will be Z–1.125. The program is as follows.

T0505	
G97 S500 M3	
G0 X1.2 Z–1.13 M8	Z–1.13 is the final Z axis location plus an allowance of .005″
G1 X.75 F.005	Clearance cut
X1.1	Feed back to clearance position above part
Z–1.075	Start of outside diameter chamfer
X1.0	
X.90 Z–1.125	Complete chamfer
X.45	Feed to inside diameter bore
G0 X1.2 Z0.	
G28 U0. W0.	

Depending on the application, multiple parts can be made from a single bar pull. There are multiple programming strategies to consider in this scenario. For example, multiple parts can be programmed through the use of a work shift or work offsets. A work shift is a programmed shift of the active work coordinate system to a different location for machining. A work shift can be programmed with the G10 command. On some controllers, multiple work offsets can be used for machining multiple parts. The G54–G59 work offset designations can be used for this purpose. Multiple parts can also be machined through the use of subprograms or by simply making the required math calculations and programming to the desired coordinates. Use creativity and develop a system that will work with the available machine. When planning a work setup for multiple parts, make sure there is sufficient room between parts to part off the first piece and machine the second piece. **Figure 21-27** shows a part setup that provides enough clearance for a .125″ wide parting tool to remove the first part and enough material for a facing operation for the second part.

When parting off parts without an inside diameter bore, it is best practice to use a parting tool with an angled face. See **Figure 21-28**. The angled face on the insert allows the finished part to detach first and not leave any extra material on the back face. A parting tool with a straight insert will leave a small piece of material on the back face of the part, which will require removal in a secondary operation or by hand sanding.

One additional consideration in parting off is the method used for catching the part. CNC lathes can be purchased with an option for a programmable parts catcher. There are specific machine functions used to activate the operation of a parts catcher on a lathe. On a Haas controller, the M36 and M37 functions are commonly used to activate and deactivate a parts catcher. On a Fanuc controller, the M73 and M74 functions are commonly used. Without a programmable parts catcher, there will need to be some consideration for preventing the part from falling to the bottom of the machine and damaging the exterior.

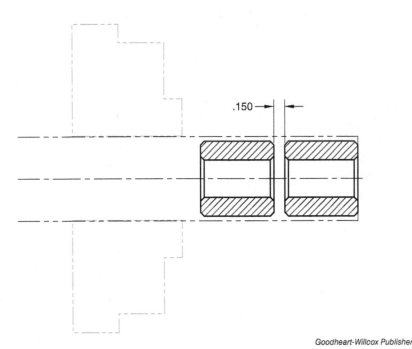

Goodheart-Willcox Publisher

Figure 21-27. A work setup for machining multiple parts on a lathe equipped with a bar feeder. Sufficient clearance between parts is necessary for parting off the first part and machining the second part.

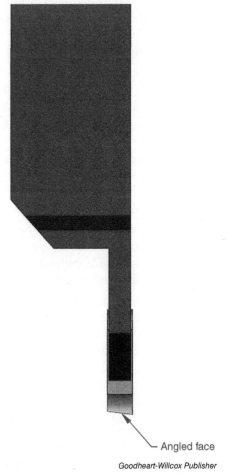

Goodheart-Willcox Publisher

Figure 21-28. A parting tool insert with an angled face cuts without leaving extra material when parting off parts that do not have a machined bore.

Chapter 21 Review

Summary

- Grooves have many applications on machined parts. Often, the required application defines tolerance requirements and the groove shape.
- Grooves can have straight or tapered walls. They can have horizontal floors or full radius floors. On some parts, there can be a combination of these configurations, depending on the application of the groove.
- Tool offsets on a lathe represent the distance the tool travels from machine home to the X0. Z0. part origin. The Z axis offset is usually established by touching the left edge of the tool to the face of the part and then recording that coordinate position on the controller. To set the X axis offset, the tool is moved to a known diameter and then the size of that diameter is input to the controller. Be aware that the grooving tool will remove material with the entire width of the tool, so program accordingly.
- Whenever possible, select a tool width smaller than the finish groove size. This allows for a roughing plunge followed by a finishing pass to better ensure a superior surface finish and dimensional accuracy.
- Grooving inserts come in a variety of styles. Identify the correct shape for the intended groove and verify there is adequate tool clearance to achieve the desired groove depth.
- Straight wall grooves can be programmed in longhand form. For deeper grooves, or wider grooves, the G75 canned cycle can be used to reduce program length and programming errors.
- When creating tapered wall grooves, the G75 canned cycle can be used for roughing. This is followed by a finish pass programmed in longhand form.
- Parting off operations usually involve a clearance cut before a final pass. This allows the parting tool to produce a more accurate cut, with an improved surface finish.
- Parting off operations can be used with a bar feeding system, improving production times and efficiencies. Depending on the application, multiple parts can be machined by completing parts, parting off, and then machining second or third pieces one at a time.

Review Questions

Answer the following questions using the information provided in this chapter.

Know and Understand

1. Grooving is usually considered to be a process of machining a narrow cavity with _____ sides.
   A. 2
   B. 3
   C. 4
   D. There is no set number of sides.

2. *True or False?* Grooves can serve as recesses for O-ring seals and are commonly used to install fasteners such as snap rings and retaining rings.

3. A groove can be described as a straight wall, tapered wall, or _____ groove.
   A. curved
   B. vertical
   C. horizontal
   D. full radius

4. *True or False?* In machining a groove, the best practice is to use a larger groove tool than the final width of the groove.

5. The Z axis tool offset for a groove tool is usually established using the _____ of the tool.
   A. left side
   B. center
   C. right side
   D. top side

6. If machining a groove .265″ wide, which insert width should be selected?
   A. .125″
   B. .187″
   C. .250″
   D. .312″

7. In most grooving operations, the spindle is programmed in constant spindle speed mode with the _____ command.
   A. G96
   B. G97
   C. G98
   D. G99

8. *True or False?* Any groove shape can be machined on an inside diameter or outside diameter, as long as there is sufficient tool clearance.

9. The Z axis starting position specified before the G75 block tells the controller where the _____ is located.
   A. initial groove plunge
   B. second groove plunge
   C. finish diameter
   D. clearance plane

10. *True or False?* The parameters defined in the G75 canned cycle are identical in programs created for Haas and Fanuc controllers.

11. In the G75 canned cycle, what does the Q or K address represent?
    A. Retract amount
    B. Starting position, incrementally
    C. Peck amount in X axis
    D. Z axis stepover amount

12. In the G75 canned cycle, what does the I or P address represent?
    A. Retract amount
    B. Starting position, incrementally
    C. Peck amount in X axis
    D. Z axis stepover amount

13. When programming a tapered wall groove, the G75 cycle is used for _____.
    A. face cutting
    B. finish machining
    C. rough machining
    D. thread cutting

14. If a full radius groove is .250" wide, what size radius is on the groove floor?
    A. .125"
    B. .250"
    C. .500"
    D. .750"

15. In a parting off operation, a clearance cut should be made before the final cut at approximately _____ from the finish surface.
    A. .005"
    B. .050"
    C. .125"
    D. .250"

Apply and Analyze

1. Study the following drawing. Identify the missing X and Z coordinates used in the finish pass for the groove. The groove tool is .125" wide.

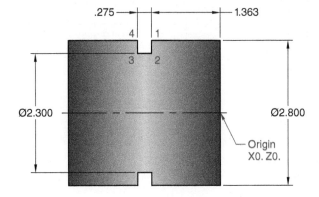

Coordinates	
Origin	X0. Z0.
Point 1	
Point 2	
Point 3	
Point 4	

2. Referring to the drawing given for Question 1, and the blanks provided below, give the correct codes that are missing. The G75 roughing cycle is for a Fanuc controller. The groove tool is .125" wide. The roughing pass should leave .005" of material for the finish pass.

```
G0 T0404
G97 S500 M3
G0 X2.9 _____ M8
_____ R.02
G75 X2.310 _____ .05 _ .05 F.003
G0 X2.9
_____
G1 _____ F.005
_____
X2.9
G0 X3. Z1.
```

3. Study the following drawing. Identify the missing X and Z coordinates used in the finish pass for the groove. The groove tool is .250" wide. Points 2 and 5 are at the midpoints along the angles.

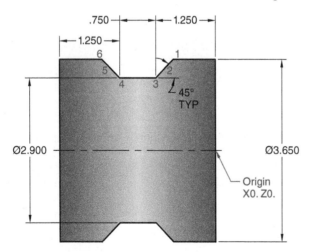

Coordinates	
Origin	X0. Z0.
Point 1	
Point 2	
Point 3	
Point 4	
Point 5	
Point 6	

4. Referring to the drawing given for Question 3, and the blanks provided below, give the correct codes that are missing for the G75 roughing cycles for a Haas controller. Two cycles are used for roughing and the deepest part of the groove is roughed first. The groove tool is .250″ wide. The roughing passes should leave .005″ of material for the finish pass.

| G0 T0404 |
| G97 S500 M3 |
| G0 X3.85 _____ M8 |
| G75 X2.910 _____ .05 _.10 F.003 |
| G0 X3.85 |
| _____ |
| G75 X3.285 _____ .05 _.100 F.003 |
| G0 X3.85 |
| Z–1.125 |
| X3.65 |
| _____ Z–1.500 |
| _____ |
| _____ |
| G0 X4. Z1. |

5. Referring to the given drawing and the blanks provided below, give the correct codes that are missing for the parting off operation. A clearance cut is made first and should leave .005″ of material for the final pass. The parting tool is .125″ wide.

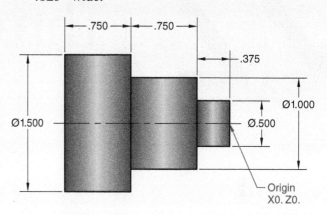

| T0505 |
| G97 S500 M3 |
| G0 X1.7 _____ M8 |
| _____ X.75 F.005 |
| X1.7 |
| _____ |
| X0. |
| G0 X1.7 Z0. |
| G28 U0. W0. |

Critical Thinking

1. Make a list of parts that require a grooving operation in machining. What industries commonly use parts with grooves to create recesses for seals or thread reliefs?

2. The programming strategy for grooving often involves a canned cycle followed by writing out code in longhand form. Explain the importance of plotting out X and Z coordinates when preparing a program for a grooving operation.

3. Parting off operations are often used in manufacturing smaller parts. Estimate the cycle times of machining parts one at a time and analyze the potential cost savings in machining parts from bar stock. Explain why producing parts from bar stock increases efficiency and adds value to your company.

22 Threading

Chapter Outline

22.1 Introduction
22.2 Thread Terminology
22.3 Programming Threads
    22.3.1 G32: Single Pass Threading Cycle
    22.3.2 G76: Multiple Pass Threading Cycle
    22.3.3 G78: Multiple Pass Threading Cycle
    22.3.4 G92: Thread Cutting Cycle
22.4 Tapered Threads
22.5 Multi-Start Threads

Learning Objectives

After completing this chapter, you will be able to:

- Identify different applications for threading operations.
- Describe the major elements of threads.
- Explain the programming methods used in the G32, G76, G78, and G92 threading canned cycles.
- Program a single pass thread cutting cycle using the G32 canned cycle.
- Program a multiple pass threading cycle using the G76 canned cycle.
- Program tapered threads using threading canned cycles.
- Explain techniques for programming multi-start threads.

Key Terms

ball screw	pitch diameter	thread
lead	multi-start threads	thread angle
major diameter	nominal size	thread depth
minor diameter	single-point threading	thread form
pitch	spindle synchronization	threads per inch (TPI)

22.1 Introduction

Thread cutting is one of the most common machining operations on a lathe. Threads have a variety of applications in industry. Threads are often used to fasten parts together and can be used in mechanical assemblies to transmit motion. Programming threading operations requires precise calculations and a detailed understanding of the elements that make up threads. Before programming methods and specific threading operations are discussed, it is important to have an understanding of common terminology associated with threads. A *thread* is a helical groove machined along a cylindrical or conical surface. A *thread form* is the basic profile shape of the thread. Examples of thread forms are 60° threads, Acme threads, buttress threads, and pipe threads. Threads can be machined as internal or external threads and can be straight or tapered. The programming techniques used for thread machining operations are similar for different types of threads.

22.2 Thread Terminology

There are a number of elements that define the form and shape of a thread. Identifying the target dimensions of these elements is necessary in selecting the correct tooling and programming a threading operation. The following terms apply to screw thread. See **Figure 22-2**. Screw thread is the most common type of thread, but the nomenclature is the same for other thread types.

- *Pitch*. The distance from a point on one thread to another point at the same location on the next thread, measured parallel to the axis of the cylinder.
- *Major diameter*. The largest diameter of the thread. On external threads, the major diameter is measured from one crest to the other crest in the opposite direction.
- *Minor diameter*. The smallest diameter of the thread. On external threads, the minor diameter is measured from one root to another root in the opposite direction.
- *Thread depth*. The distance between the major diameter and minor diameter of the thread, measured perpendicular to the axis of the cylinder. Also called the *thread height*. This is a radial distance. To calculate the thread depth, subtract the minor diameter from the major diameter, then divide the difference by two.
- *Pitch diameter*. The theoretical diameter measured at the point where the width of the ridge is equal to the width of the groove. This measurement is also known as the *effective diameter* and determines whether two threaded parts can be successfully mated together. Properly mated parts will demonstrate equal distance between the thread walls when in contact. Although the pitch diameter is not needed for programming, it is often defined in the dimensioning to ensure mating parts will properly fit as designed.
- *Thread angle*. The included angle between the sides of the thread. In V-shaped screw threads, this angle is 60°. Other thread forms will have different profiles or angles. This angle is critical to tool selection.
- *Lead*. The distance the thread advances along the axis of the cylinder in one revolution of the part.

From the Programmer

Ball Screw Assemblies

Although threads are widely used in fastening applications, threaded parts can be used for many other purposes. For example, threads are commonly used in mechanical assemblies to produce linear motion through rotational motion. Threads perform this type of action and play an integral role in the operation of CNC machines. On CNC lathes and mills, threaded ball screw assemblies allow for precise positioning of the machine axes. A *ball screw* is a long threaded shaft mated with a ball nut in a ball screw assembly. See **Figure 22-1**. As the screw is rotated, ball bearings roll continuously in the thread grooves between the screw and nut and the nut moves in a linear direction along the screw. This configuration greatly reduces friction and provides a highly efficient way to convert rotational motion to linear motion. Ball screw assemblies are driven by servo motors, which monitor the amount of shaft rotation and allow the controller to calculate precise movement of the machine axes.

Figure 22-1. A ball screw assembly consists of a ball screw and a movable ball nut.

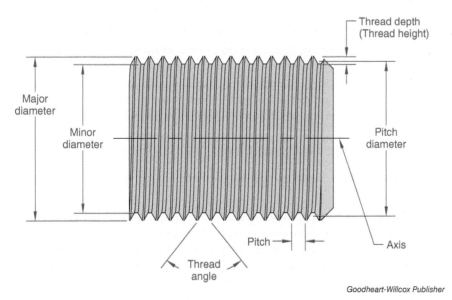

Figure 22-2. Elements of threads. The major diameter is the largest diameter of the thread. The minor diameter is the smallest diameter of the thread.

- *Threads per inch (TPI).* The total number of complete revolutions of the thread in exactly 1″. The threads per inch of the thread is used to calculate the pitch for inch-based threads.

On a print, screw threads are defined by the major diameter and number of threads per inch. For example, a print may define a thread as 3/8–16, 1/2–13, or 5/8–11. The first number represents the major diameter and the second number represents the number of threads per inch. The major diameter in the designation is a *nominal size*. A *nominal size* is a designated value used for general identification purposes. Usually, for an external thread measurement, the measured size is smaller than the nominal size. For example, the thread designation 1/2–13 specifies a major diameter of 1/2″. The allowable limits of size for the major diameter of this thread are .4876–.4985.

A thread note on a print also typically includes the thread series and thread class. See **Figure 22-3**. Inch-based threads are classified in the Unified system. Threads in the Unified coarse thread series are designated as UNC. Threads in the Unified fine series are designated as UNF. Threads in the Unified extra fine series are designated as UNEF. Fine threads have more threads per inch than coarse threads of the same diameter. For example, a UNC 3/8 thread has 16 threads per inch. A UNF 3/8 thread has 24 threads per inch. A UNEF 3/8 thread has 32 threads per inch. The thread class identifies the class of fit of the thread. The class of fit designates the amount of tolerance for the thread. The higher the number of the class, the tighter the tolerance of the thread. The thread class includes a letter designation to identify external or internal threads. For external threads, the letter A is designated. For internal threads, the letter B is designated.

Metric screw threads are specified slightly differently. In a thread note, the letter *M* designates metric thread. The thread note specifies the major diameter and pitch. The pitch is the distance between threads. For example, the thread note M8 × 1.25 specifies that the major diameter is 8 mm and the pitch is 1.25 mm.

22.3 Programming Threads

CNC canned cycles are designed to make the programmer and operator's job easier by automating repetitive functions. In most cases, a canned cycle reduces the number of lines of code and simplifies editing. The threading canned cycles for the CNC lathe are also designed to automate repetitive functions, but they are different from other cycles in that more calculations and programming input are required. Threading canned cycles are required to machine single-point threads. *Single-point threading* refers to a process in which a formed insert makes a series of passes to create the desired thread. See **Figure 22-4**. It would be very rare, and not recommended, to try to use a threading die or a tap to cut threads on a CNC lathe. The threading canned cycles are the best option in lathe threading operations.

The reason that the canned cycles are so important in threading operations is the tool must enter the material at the exact same point on each cutting pass. To visualize the cutting entry, think of the end view of the part as a full circle and divide it into degrees. The tool must enter the material at the 90° position each time. If the tool were to enter at 90° on the first pass and 180° on the next pass, the threads would be *cross-threaded* and unusable. Each cut must enter at the same rotational position. The CNC lathe does this by orienting the spindle rotation to the tool feed rate. This process is referred

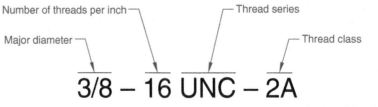

Goodheart-Willcox Publisher

Figure 22-3. A thread note for inch-based threads designates the major diameter, number of threads per inch, thread series, and thread class.

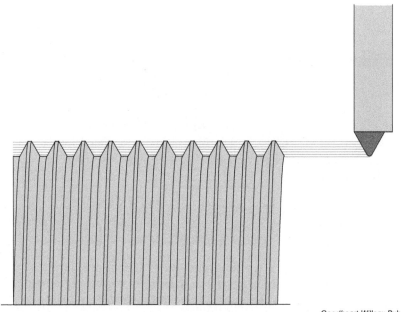

Goodheart-Willcox Publisher

Figure 22-4. In single-point threading, an insert tool makes a series of cutting passes to produce the required dimensions of the thread.

to as *spindle synchronization*. It is an amazing process, but the servo motors controlling the spindle and axial movements make it possible. The threading canned cycles automatically activate spindle synchronization in the controller.

22.3.1 G32: Single Pass Threading Cycle

The G32 cycle is the simplest and least versatile of the threading cycles. Although it allows the programmer to machine tapered threads and change angles during a single operation, it only allows for programming one machining pass at a time. It is a cycle that the programmer might use for a very odd or specialized part. The following example uses a simple part with straight threads to show how the G32 command can be used to program a thread cutting operation.

Figure 22-5 shows a part with a threaded section 1″ in length. The X0. Z0. part origin is located at the front face and part centerline. The minor diameter of this thread is .300″. Before the threads can be machined, the roughing and finishing passes on the 3/8″ (.375″) outside diameter must be completed. A 60° threading tool will be positioned at a safe position in front of the part and at the diameter of the threading pass. Then the G32 cycle is used to synchronize the spindle to the tool feed rate and complete a threading pass. The program is as follows.

T0202	
G97 S500 M3	Spindle on to 500 rpm (constant spindle speed)
G0 X.4 Z.1	Safe starting position
X.300	Diameter of first pass
G32 Z-1. F.0625	Threading pass completed
G1 X.4 Z-1.01	Tool retract
Z.1	

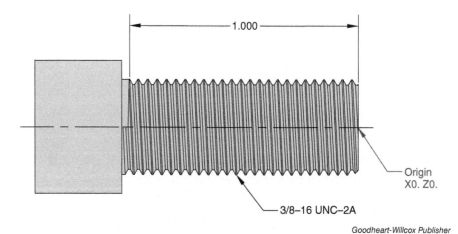

Figure 22-5. A threaded fastener with straight 60° coarse threads.

This is a very simple program in which the tool is activated (T0202), the spindle is turned on clockwise, and then the tool is positioned at a safe starting position. From there, the tool is positioned at the cutting diameter and the G32 cycle is activated. Note that after the G32 block, the tool must be retracted with a G01 or G00 command move. Unlike other threading cycles, the G32 cycle will not automatically retract the tool. The following explains two of the blocks in more detail.

- **X.300.** This is the diameter position of the first cut (in this case, it is the minor diameter of the thread).
- **G32 Z–1. F.0625.** The G32 command initiates single-pass thread cutting and synchronizes the spindle to the feed rate. The .0625 feed rate is calculated from the number of threads per inch (1 ÷ 16 = .0625). The Z–1. position is the final Z axis position of the thread.

The reason this cycle is not often used is that it only makes a single pass. In this example, the most likely result would be a broken threading tool. Attempting to thread from .375″ diameter to .300″ diameter in a single pass will not be successful. To create this thread properly with the G32 command, it is necessary to repeat the command in the program to make an appropriate number of cuts at smaller thread depths. In this case, the program is prone to errors because multiple lines are required and there are better threading canned cycles available.

However, the G32 cycle does have some unique characteristics that other cycles do not have. This cycle differs from other thread cutting cycles in that the thread taper angle and/or lead can vary continuously throughout the entire thread.

A threaded part with a uniform taper is shown in **Figure 22-6.** A taper angle is programmed by adding the final X diameter position to the final Z axis position on the G32 block. This moves the thread cutting tool in an angular direction. Note that only one pass is programmed in this example. Normally, multiple passes are required to cut the thread.

T0202	
G97 S500 M3	
G0 X.4 Z.1	
X.298	Diameter of first pass

Continued

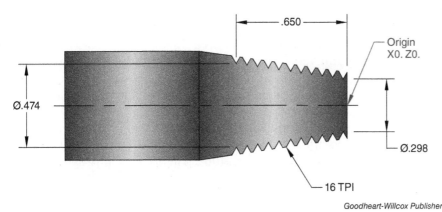

Goodheart-Willcox Publisher

Figure 22-6. A part with threads machined at an angle. Note: The only dimensions shown are those relevant to creating the program.

G32 X.474 Z–.65 F.0625	Angular move
G1 X.5 Z–.7	Tool retract
Z.1	

Tapered threads can be programmed with other canned cycles, but with the G32 command, it is possible to change direction multiple times. This capability is unique to the G32 command. **Figure 22-7** shows a part with a threaded profile that alternates between straight and tapered thread. The thread starts with a taper, then follows a straight path, and then tapers again. A single pass for this type of thread can be programmed using three lines, as shown in the following program. Note that only one pass is programmed and multiple passes would normally be required to cut the thread.

T0202	
G97 S500 M3	
G0 X.4 Z.1	
X.298	Diameter of first pass
G32 X.474 Z–.65 F.0625	Angular move
Z–1.325	
X.617 Z–1.7	Angular move
G1 X.75 Z–1.8	Tool retract
Z.1	

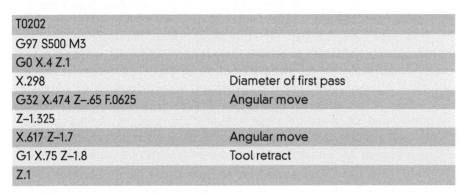

Goodheart-Willcox Publisher

Figure 22-7. A threaded part with an irregular shape made up of tapered and straight threads.

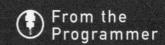

Tapered Thread Cutting

Can you think of an application where a thread changes angle along a part? This is not a common application for threads. However, a part may have helical grooves or a similar shape that must follow this type of pattern. Just remember that this type of thread cutting path is available when applying the G32 command and use creativity as needed.

22.3.2 G76: Multiple Pass Threading Cycle

The G76 cycle is the most commonly used thread cycle, because of its ease of programming and versatility. Both straight and tapered threads can be programmed with the G76 cycle, either on the outside or inside diameter of a part. There are some differences between the Fanuc and Haas cycles, so both will be discussed.

The part shown in **Figure 22-8** has a thread relief groove at the end of a threaded section with 1/2–13 threads. The major diameter of the thread is .500″ and the minor diameter is .406″. The following program is used to describe the components of the G76 cycle in more detail. This program uses a two-block format and is common on Fanuc controllers.

T202	
G97 S500 M3	Spindle on to 500 rpm (constant spindle speed)
M8	
G0 X.7 Z.2	Safe starting position
G76 P010030 Q10 R20	First G76 block
G76 X.406 Z–.9875 P400 Q50 F.077	Second G76 block
G0 X1. Z1.	Safe move away from part

The first G76 block establishes the following cutting parameters.

- **G76 P010030.** The P address value consists of six digits. The P address code will not accept decimals. The first two digits (01) represent the number of spring passes made by the tool. Spring passes are additional cuts made at the final diameter to remove any material that remains as a result of tool deflection. The more precise the thread, the more spring passes to consider in machining. In this case, there is only one spring pass. The second two digits (00) represent the length of the pullout angle at the end of the thread (Z–.9875). Because there is a thread relief, this will be zero. If there is no thread relief, an angle may be added to ensure the tool pulls out on an angle and does not damage the last thread turn. The third two digits (30) represent the tool infeed angle. The standard thread angle is 30° per side. Other options for this are 00 and 29. When the angle is set at 0°, the tool will plunge straight in on the center of the tool. When the angle is set at 29°, the tool will plunge in at 29° and only cut with the front edge of the insert.
- **Q10 R20.** The Q and R address codes will not accept decimals. The Q address defines the minimum depth of cut in .0001″ increments. Thus, Q10 designates a minimum cutting depth of .001″. The controller will divide the passes based on decreasing removal rates, so not setting a minimum cutting depth can cause the controller to take excessive amounts of passes and add substantial cycle time. The R address defines the depth of the final cut in .0001″ increments. The R20 value specifies the final cutting pass will be .002″ deep.

The second G76 block establishes the following parameters. The parameters in this block define the size and shape of the thread.

- **G76 X.406 Z–.9875.** The X.406 position is the final X axis diameter of the thread. This is the minor diameter of the thread for external threads. The Z–.9875 position is the final Z axis position for threading. The part in **Figure 22-8** has a thread length of .925″,

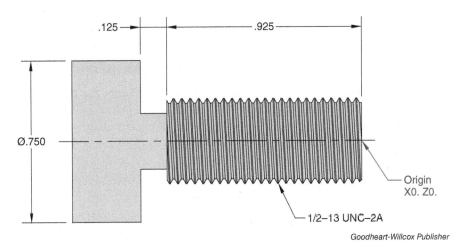

Figure 22-8. A threaded fastener with straight 60° coarse threads and a thread relief.

but also a thread relief groove length of .125″. The thread is programmed to go to the end of the .925 dimension, plus half of the thread relief groove. This will ensure there is a complete thread at the Z−.925 position.

- **P400 Q50 F.077.** The P and Q address codes will not accept decimals. The P address value represents the thread depth, or thread height, radially. The value is defined in .0001″ increments. The thread depth is calculated as the major diameter minus the minor diameter, divided by two. In this case, the thread depth is .04″: (.500 − .406) ÷ 2 = .04. Notice that the major diameter was never stated in the program. The controller calculates cutting passes and depths from the minor diameter, the thread depth, and the depth of the first cut. Each successive cut decreases in depth after the first cut. This is to account for the amount of surface area cut by the tool. The Q address defines the depth of the first cut in .0001″ increments. Thus, the Q50 value designates a .005″ depth for the first cutting pass. The F address designates the feed rate. This is critical for producing the correct thread pitch. The feed rate is programmed in inches per revolution and is calculated as 1 ÷ threads per inch (TPI).

The G76 cycle used with a Haas controller has a similar programming format. The Haas format uses less coding, but the same amount of control exists. After the programming format is explained, there will be more discussion on how this is accomplished.

G97 S500 M3	Spindle on to 500 rpm (constant spindle speed)
G0 X.7 Z.2	Rapid to safe starting position
M8	
G76 X.406 Z−.9875 K.040 D.005 F.077	Begin G76 cycle
G28 U0. W0. M9	Return to machine home, coolant off

The G76 cycle used with the Haas controller has a single-block format. Just as in the previous example, the tool is positioned at a safe starting location, the spindle is turned on, and then the G76 cycle is initiated. The parameters are explained as follows.

- **G76 X.406 Z-.9875.** The X.406 position is the final X axis diameter of the thread. This is the minor diameter of the thread for external threads. The Z-.9875 position is the final Z axis position for threading.
- **K.040 D.005 F.077.** The K address value represents the thread depth, or thread height, radially. The thread depth is calculated as the major diameter minus the minor diameter, divided by two. In this case, the thread depth is .04″: (.500 − .406) ÷ 2 = .04. The D address defines the depth of the first cut. The F address designates the feed rate. The feed rate is programmed in inches per revolution and is calculated as 1 ÷ threads per inch (TPI).

It is easy to notice the similarities between the two different formats. However, there are several parameters in the Fanuc cycle that do not appear to be in the Haas cycle. These parameters are placed in the machine control settings by Haas because it is believed that they will not routinely change from thread to thread. The following settings are for use with the G76 cycle.

- Setting 86 controls the depth of the final cut.
- Settings 95 and 96 control the chamfer pullout distance and angle.
- Setting 99 controls the minimum depth of cut for the thread.

All of the control of the threading operation is still in place, but the variables are stored by the machine controller under specific settings. This simplifies the programming procedure while still allowing access to the desired thread cutting parameters.

22.3.3 G78: Multiple Pass Threading Cycle

The G78 threading cycle is only applicable to certain Fanuc controls. Generally, the newer machine controls, from the 21T series controller to later, will use the G78 cycle. The G78 cycle looks and acts just like the G76 cycle. Fanuc moved to a system called *G-code System B*, which utilizes the G78 cycle. The G78 cycle is also seen in the Siemens and Hurco controller systems. The G78 cycle is virtually the same as the G76 cycle and uses a two-block format, as shown in the following program. This program is for the part shown in **Figure 22-8**.

T202	
G97 S500 M3	Spindle on to 500 rpm (constant spindle speed)
M8	
G0 X.7 Z.2	Safe starting position
G78 P010030 Q10 R20	First G78 block
G78 X.406 Z-.9875 P400 Q50 F.077	Second G78 block
G0 X1. Z1.	Safe move away from part

The first G78 block establishes the following cutting parameters.

- **G78 P010030.** The P address value consists of six digits. The first two digits (01) represent the number of spring passes made by the tool. In this case, there is only one spring pass. The second two digits (00) represent the length of the pullout angle at the end of the thread (Z–.9875). Because there is a thread relief, this will be zero. The third two digits (30) represent the tool infeed angle. The standard thread angle is 30° per side.

- **Q10 R20.** The Q address defines the minimum depth of cut in .0001" increments. Thus, Q10 designates a minimum cutting depth of .001". The R address defines the depth of the final cut in .0001" increments. The R20 value specifies the final cutting pass will be .002" deep.

The second G78 block establishes the following parameters.

- **G78 X.406 Z–.9875.** The X.406 position is the final X axis diameter of the thread. This is the minor diameter of the thread for external threads. The Z–.9875 position is the final Z axis position for threading.

- **P400 Q50 F.077.** The P address value represents the thread depth, or thread height, radially. The value is defined in .0001" increments. In this case, the thread depth is .04". The Q address defines the depth of the first cut in .0001" increments. The Q50 value designates a .005" depth for the first cutting pass. The F address designates the feed rate. The feed rate is programmed in inches per revolution and is calculated as 1 ÷ threads per inch (TPI).

> **Debugging the Code**
>
> **Thread Chamfer Pullout**
> In the given G76 and G78 programs, there was no programmed move for the tool pullout at the end of the Z axis positioning move. Unless programmed otherwise, at the end of the thread, an automatic chamfer is cut by the controller before reaching the Z axis target position. The default for this chamfer is one thread lead at 45°. This is controlled by a setting in the machine controller.

22.3.4 G92: Thread Cutting Cycle

The G92 cycle is used for simple thread cutting operations. Straight threads are simply made by specifying the X axis position (minor diameter), the final Z axis position, and the feed rate. It is possible to program multiple threading passes by specifying the X axis positions of additional passes. This can add more programming control over the threading passes, but it requires more calculations and lines of code than a multiple pass threading cycle.

Figure 22-9 shows a part that has a 1" diameter thread with 8 threads per inch. The minor diameter of this thread is .851". The following program shows the entry format for the G92 cycle. The first cutting pass is defined on the same block used to activate the cycle. Then, additional passes are programmed by specifying the X axis diameter positions.

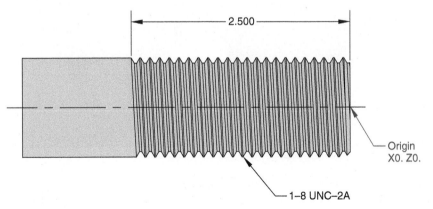

Goodheart-Willcox Publisher

Figure 22-9. A fastener with a threaded section 2.5" in length made up of straight 60° coarse threads.

T202	
G97 S500 M3	Spindle on to 500 rpm (constant spindle speed)
M8	
G0 X1.1 Z.2	Safe move to starting position
G92 X.980 Z–2.500 F.125	First pass (feed rate = 1 ÷ 8 TPI)
X.950	Second pass
X.920	Third pass
X.910	Fourth pass
X.900	Fifth pass
X.890	Sixth pass
X.883	Seventh pass
X.876	Eighth pass
X.869	Ninth pass
X.862	10th pass
X.857	11th pass
X.852	12th pass
X.851	Last pass
G28 U0. W0.	
M30	

As you can see, this takes more lines of programming and some additional thought as to defining the depth of cut, but it is a very straightforward programming method. There are some calculations for programming decreasing depths of cut, but they vary significantly according to material type and tool insert style. The main point to understand is the impact of the surface area of the cut as the tool goes deeper into the thread.

Figure 22-10 shows that the deeper the insert goes into the threading cut, the more surface area that is engaged. What might be surprising in this example is that the first cut is at .010″ deep and the second cut is at .020″ deep. By doubling the depth of cut, the tool contact surface area is increased by three times. In programming multiple passes using the G92 cycle, the depth of cut must be reduced as the X axis position decreases. Failure to adjust the cutting depth can result in a catastrophic tool failure and cause damage to the threads.

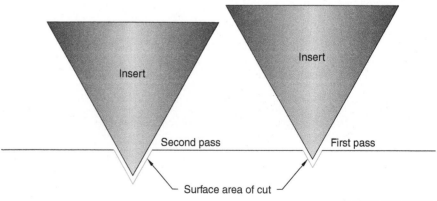

Goodheart-Willcox Publisher

Figure 22-10. Increasing the cutting depth of the tool with each pass in a threading operation results in more engagement of the surface area. After the first cut, the depth of cut must be reduced in subsequent passes to cut threads properly.

22.4 Tapered Threads

Most of the threading examples discussed in this chapter have been for parts with straight threads. However, not all threads are straight. For example, American Standard taper pipe threads (designated NPT) have a 1.79° taper. The specific reason for the taper in a pipe thread is to create a sealed fit between pipes or pipe fittings. There are other types of threads that have standard tapers and some thread designs that will call for a specific thread angle, but taper pipe thread is the most common. All of the threading cycles are capable of producing tapered threads, but there are some different programming techniques to create them.

Since the G32 cycle is used for single-line programming for thread cutting, each cutting pass would have to be calculated and programmed individually. The G76 and G92 cycles are better options for tapered threads. The G76 cycle is easy to program, but it is necessary to make some calculations to complete the program.

Figure 22-11 shows the thread calculations for a standard 3/8 NPT thread. Notice that a 3/8 NPT thread is not 3/8" diameter. The 3/8 designation is a nominal size. The starting diameter for the thread is the major diameter at the end (.656"), the minor diameter is .568", the Z axis end position is –.408, and the thread depth (thread height) is .044". These are the dimensions needed to write the program. The G76 cycle is used in the following program to create the threads. Remember that the outside diameter would be machined first, before the threading cycle is initiated. This program uses the two-block format common on a Fanuc controller.

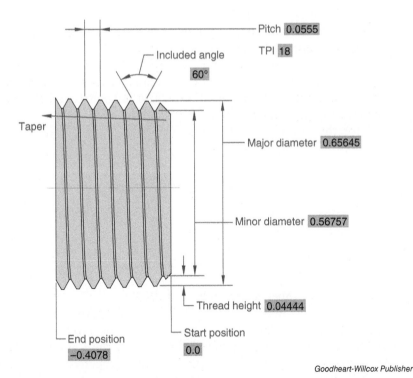

Goodheart-Willcox Publisher

Figure 22-11. Taper pipe threads have a 1.79° taper. Before programming tapered threads, calculations must be made to establish the required parameters.

T202	
G97 S500 M3	Spindle on to 500 rpm (constant spindle speed)
M8	
G0 X.8 Z.2	Safe move to starting position
G76 P010030 Q10 R20	First G76 block
G76 X.568 Z–.408 R–.013 P440 Q50 F.0555	Second G76 block
G0 X1. Z1.	Safe move away from part

The following is an explanation of the cycle and the parameters used in programming.

- **G76 P010030 Q10 R20.** The parameters in the first block are the same as those used in straight threading G76 cycles.
- **G76 X.568 Z–.408 R–.013 P440 Q50 F.0555.** The second block defines how the cycle is completed. The entries are read as follows.
  - **X.568 Z–.408.** The X axis position is the minor diameter at the start of the thread. The Z axis position is the final Z position of the thread.
  - **R–.013.** The R address value represents the taper amount in the X axis. This is a radial value that represents the difference in taper height between the start and end of the thread. The specified value is negative when the taper increases in diameter on external threads, as shown in **Figure 22-11**.
  - **P440.** The P address value represents the thread depth. This is a radial value defined in .0001″ increments.
  - **Q50.** The Q address defines the depth of the first cut in .0001″ increments.
  - **F.0555.** The F address designates the feed rate. The value is calculated as 1 ÷ threads per inch (1 ÷ 18 = .0555).

The taper amount in the X axis must be calculated and can be determined using a trigonometric function. See **Figure 22-12**. As previously discussed, taper pipe threads have a 1.79° taper. The length of the thread in the Z axis is .408″. Using this information, a right triangle can be used to solve for the height of the opposite side. The tangent function is used to calculate the taper distance, which represents the shift in the X axis. Using a calculator, the formula is as follows.

Calculator formula:

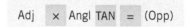

Solution:

$$.408 \times .0313 = .013″$$

The amount of taper for the thread is .013″.

When using the G76 cycle with a Haas controller, the I address is used to specify the amount of taper in the X axis. The taper is specified as a radial value. To create the tapered thread, add the I address as shown in the following program. The specified value is negative when the taper increases in diameter on external threads, as shown in **Figure 22-11**.

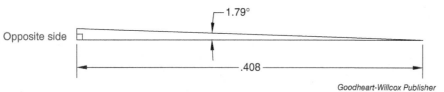

Figure 22-12. To calculate the programmed taper amount for the thread, a right triangle is used with the tangent function. The length and angle of the thread establish two sides of the triangle. The opposite side represents the taper height.

T202	
G97 S500 M3	Spindle on to 500 rpm (constant spindle speed)
G0 X.8 Z.2	Safe move to starting position
M8	
G76 X.568 Z-.408 I-.013 K.044 D.005 F.0555	
G28 U0. W0. M9	Return to machine home, coolant off

The parameters in the G76 block are explained as follows.

- **G76 X.568 Z-.408.** The X axis position is the minor diameter at the start of the thread. The Z axis position is the final Z position of the thread.
- **I-.013 K.044 D.005 F.0555.** The I address value represents the taper amount in the X axis. The K address value represents the thread depth. The D address value defines the depth of the first cut. The F address designates the feed rate. The value is calculated as 1 ÷ threads per inch (1 ÷ 18 = .0555).

It is possible to create the same 3/8 NPT thread using the G92 canned cycle. The R address is used to specify the taper amount. This will shift the cutting tool in the X axis on every pass.

T202	
G97 S500 M3	Spindle on to 500 rpm (constant spindle speed)
M8	
G0 X.67 Z.2	Safe move to starting position
G92 X.650 Z-.408 R-.013 F.0555	First pass
X.640	Second pass
X.630	
X.620	
X.610	
X.600	
X.590	
X.580	
X.570	
X.568	Final pass
G0 X1. Z1.	
G28 U0. W0.	Return to machine home

From the Programmer

Thread Forms

There are many thread forms and many different applications for threads. The same principles used for programming straight and tapered threads can be applied to other types of threads. Some of the many thread forms used in industry include the Acme, British Whitworth, buttress, square, and knuckle forms. Different thread forms can be programmed with threading canned cycles, but they will require specific cutting tools. When creating threads, always take the time needed to examine the print and understand the customer requirements. Identify the tooling that will be required and determine the thread parameters that will be needed in programming.

22.5 Multi-Start Threads

The threading examples discussed in this chapter have been for single-start threads. *Multi-start threads* are made up of two or more intertwined threads running parallel to one another. Intertwining threads allow the lead of a thread to be increased without changing the pitch. The *lead* is the distance the thread advances in one revolution. Sometimes, multi-start threads are called *multi-lead threads*. Single-start threads have a lead distance that is equal to the pitch. A double-start thread will have a lead that is double that of a single-start thread. A triple-start thread will have a lead that is three times longer than that of a single-start thread. These threads are common in applications where less stress is introduced in the thread, but faster rotation is desired. The ball screws on CNC machines are multi-start threads because the threads provide faster rotation and engagement of the mating nut.

As previously discussed, lathe threading cycles are designed to automatically activate spindle synchronization. This allows the tool to enter the material at the same rotational position each time to begin cutting. The starting angular position is 0° (or 360°). The orientation of the spindle by the machine controller can be altered to the programmer's advantage when cutting multi-start threads.

Figure 22-13 shows a threaded part with a note specifying multi-start threads. The note specifies four-start thread, or four starts, and 16 threads per inch. The simplest method to create the program is to use the G76 canned cycle. The address code Q will be added to alter the starting angular position of the cycle. The minor diameter of 1–16 thread is .9234", which will be used in programming. The thread pitch is calculated as 1 ÷ 16, or .0625. This would normally be used for the feed rate, but because this is multi-start thread, the pitch needs to be multiplied by the number of starts (4) to calculate the lead. The lead is used instead of the pitch when calculating the feed rate for multi-start threads. In this case, the lead is equal to 4 × .0625, or .25.

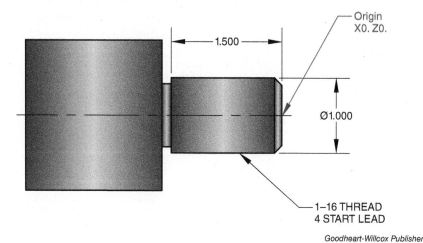

Goodheart-Willcox Publisher

Figure 22-13. A part with multi-start threads specified in a thread note. The threads are four-start threads.

The following is the G76 line used to cut the first start of the thread, with an explanation of the cutting parameters.

```
G76 X.9234 Z-1.5 Q360000 K.0383 D.005 F.25
```

- **G76 X.9234 Z-1.5.** The X axis position represents the minor diameter of the thread. The Z axis position is the final Z position of the thread.
- **Q360000.** The Q address sets the spindle orientation for this cycle. The value must be expressed without a decimal point. The angular position is designated as 360° to prevent confusion in expressing the value 0°.
- **K.0383.** The K address value represents the thread depth. This is calculated as the major diameter minus the minor diameter, divided by two: (1 − .9234) ÷ 2 = .0383.
- **D.005.** The D address value defines the depth of the first cut.
- **F.25.** The F address designates the feed rate. This is the lead of the thread, which is equal to four times the pitch for four-start threads: 4 × .0625 = .25.

To program the other three starts, repeat the G76 line and alter the Q address for each angular position: 270°, 180°, and 90°. Running the G76 canned cycle at four equal angle synchronizations will create the four-start thread.

```
G76 X.9234 Z-1.5 Q360000 K.0383 D.005 F.25
G76 X.9234 Z-1.5 Q270000 K.0383 D.005 F.25
G76 X.9234 Z-1.5 Q180000 K.0383 D.005 F.25
G76 X.9234 Z-1.5 Q90000 K.0383 D.005 F.25
```

This is a very complicated operation for a threaded part, made simple by using a threading canned cycle. This program can be used for many different multi-start thread configurations. Just remember to multiply the pitch by the number of starts to calculate the feed rate and divide the angular positions equally for cutting. This is achieved by dividing 360° by the number of thread starts.

Chapter 22 Review

Summary

- Threading is a common machining operation on a CNC lathe. Single-point threading is a process in which a formed insert makes a series of passes to create the desired thread.
- Spindle synchronization is a process used in lathe threading cycles to orient the spindle rotation to the tool feed rate.
- The largest diameter of the thread is known as the major diameter. The smallest diameter of the thread is the minor diameter. The pitch diameter is the theoretical diameter measured at the point where the width of the ridge is equal to the width of the groove.
- The thread pitch is the distance from a point on one thread to another point at the same location on the next thread. The thread depth is the distance between the major diameter and minor diameter of the thread.
- Threading canned cycles are designed to automate repetitive functions, but they require more calculations and programming input than other canned cycles.
- The G32 canned cycle allows for programming one machining pass at a time. It can be used in special operations where spindle synchronization is required.
- The G76 cycle for a Fanuc controller uses a two-block format to define parameters for completion of the canned cycle. The G78 cycle is similar to the G76 cycle and is used in newer machine controls.
- The G76 cycle for a Haas controller uses a single-block format. Some of the additional programming parameters are controlled by machine settings.
- The G92 cycle is used to program multiple threading passes by specifying the X axis positions of additional passes. It requires more calculations and lines of code than a multiple pass threading cycle.
- Any of the threading canned cycles can be used to create tapered threads. The amount of taper must be calculated for programming.
- Multi-start threads are used in applications where faster rotation is desired. Lathe threading cycles simplify the process of creating multi-start threads.

Review Questions

Answer the following questions using the information provided in this chapter.

Know and Understand

1. Which of the following is a type of thread form?
   A. Acme
   B. buttress
   C. pipe thread
   D. All of the above.

2. *True or False?* To create threads on a lathe, the spindle must be synchronized with the feed rate of the tool.

3. The distance between a point on one thread and the same point on the next thread is known as the _____.
   A. major diameter
   B. thread angle
   C. minor diameter
   D. pitch

4. To calculate the thread depth, subtract the _____ diameter from the _____ diameter and divide by two.
   A. major, minor
   B. major, pitch
   C. minor, major
   D. pitch, major

5. To calculate the feed rate for a threading cycle, divide _____ by _____.
   A. 1, the threads per inch
   B. the threads per inch, 1
   C. 1, the thread depth
   D. the pitch, the thread depth

6. Which threading canned cycle would be used if the threads transitioned from straight to tapered?
   A. G92
   B. G78
   C. G32
   D. G76

7. When using the G76 cycle on a Fanuc controller, the first two digits in the code P020030 represent what function?
   A. A .002″ depth for the first cut.
   B. Two spring passes.
   C. The amount of material for the final pass.
   D. The length of the pullout angle at the end of the thread.

8. *True or False?* The Haas G76 cycle requires two lines of programming code to control the thread form and function.

9. In the Haas G76 cycle, the address code K represents what parameter?
   A. Thread depth.
   B. Feed rate.
   C. Amount of material for the final pass.
   D. Depth of the first cut.

10. *True or False?* The G76 and G78 cycles are identical, but the G78 cycle is used in the Haas controller.

11. In the G92 canned cycle, each _____ axis move must be programmed individually.
    A. K
    B. I
    C. Z
    D. X

12. To create tapered threads on a machine with a Fanuc controller, the letter address _____ is used to specify the taper amount.
    A. R
    B. I
    C. K
    D. L

13. To create tapered threads on a machine with a Haas controller, the letter address _____ is used to specify the taper amount.
    A. R
    B. I
    C. K
    D. W

14. To create multi-start threads, the pitch is multiplied by _____ to calculate the feed rate.
    A. 2
    B. 4
    C. the number of starts
    D. the thread depth

15. To create multi-start threads, what letter address is used in the G76 cycle to orient the position of the spindle?
    A. Q
    B. K
    C. D
    D. S

Apply and Analyze

1. Identify the names of the designations labeled A, B, C, and D in the following thread note.

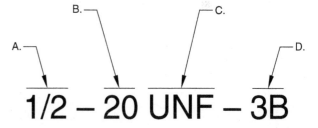

2. Referring to the drawing and the blanks provided below, give the correct codes that are missing for the final diameter pass in the G32 threading cycle.

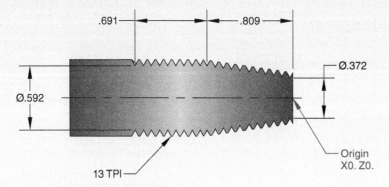

T0202
G97 S500 M3
G0 X.6 Z.1
X____
G32 X____ Z____ F____
Z____
G1 X.65 Z−1.8

3. Referring to the blanks provided below, give the correct codes that are missing in the G76 threading cycle for cutting 5/8–18 threads 2″ long using a Haas controller. The minor diameter is .555″.

G97 S300 M3
G0 X.8 Z.2
M8
G76 X____ Z____ ____ D.005 ____
G28 U0. W0. M9

4. Referring to the blanks provided below, give the correct codes that are missing in the G76 threading cycle for cutting 3/4–16 threads 1.5″ long using a Fanuc controller. The minor diameter is .673″.

T202
G97 S350 M3
M8
G0 X.950 Z.2
____ P010030 Q10 R20
____ ____ ____ P____ Q50 ____
G0 X1. Z1.

5. Referring to the blanks provided below, give the correct codes that are missing in the G76 threading cycle for cutting 5/8–11 multi-start threads 4″ long with three starts. The minor diameter is .513″.

| G76 X____ Z____ Q____ K____ D.005 F____ |
| G76 X____ Z____ Q____ K____ D.005 F____ |
| G76 X____ Z____ Q____ K____ D.005 F____ |

Critical Thinking

1. Make a list of items in your home that are fastened with threads. Look for items and assemblies with screws, bolts, and nuts. For parts with external threads, identify whether the thread has a relief groove or ends at a shoulder. Explain the parameters that would need to be defined to machine the thread.

2. Consider the threading canned cycles covered in this chapter and how they can be used for different applications. Which canned cycle do you think is the most efficient? Explain your answer and identify a threading application that would be suitable for the cycle.

3. Research tapered threads and multi-start threads online and identify how they are used in industry. What industries rely on these types of threads and their engineering applications to complete work?

23 | Live Tooling

Chapter Outline

23.1 Introduction

23.2 Special Considerations with Live Tooling
- 23.2.1 Live Tooling M-Codes
- 23.2.2 Live Tooling G-Codes

23.3 Radial Hole Machining Canned Cycles
- 23.3.1 G241: Radial Drilling Cycle
- 23.3.2 G242: Radial Spot Drilling Cycle
- 23.3.3 G243: Radial Peck Drilling Cycle
- 23.3.4 G245: Radial Boring Cycle—Feed In and Feed Out
- 23.3.5 G246: Radial Boring Cycle—Feed In and Spindle Stop

23.4 Slot Milling

23.5 CAM Software Programs

Learning Objectives

After completing this chapter, you will be able to:

- Explain the different machining applications for a three-axis CNC lathe.
- Describe the functions of M-codes for live tooling operations.
- Explain how C axis motion is programmed.
- Identify work planes used on a CNC lathe.
- Differentiate between axial and radial machining operations.
- Program drilling canned cycles for live tooling operations.
- Explain how slots are created by using C axis engagement with live tooling.
- Explain how complex part profiles can be machined on a three-axis lathe by using a computer-aided manufacturing (CAM) system.

Key Terms

axial drilling
live tool
live tooling lathe
mill-turn machine
radial drilling

23.1 Introduction

The primary focus in this text has been two-axis lathes with operations programmed in the X and Z axes. However, there are many more options for CNC lathe machines, including three-axis lathes with live tooling (using the X, Z, and C axes) and four-axis lathes (using the X, Y, Z, and C axes). There are also lathes with dual spindles, Swiss machines that can have upwards of nine axes of programmable orientations, and full mill-turn machines. For most of these machine types, programming is almost always done using a computer-aided manufacturing (CAM) system. These machines are capable of complex operations that go far beyond any hand programming operation. This chapter will cover the basics of live tooling CNC lathes, with the focus on using the X, Z, and C axes.

Basic machine functions evolved from the same starting point and are fairly similar on different machine types. It is easy to recognize the similarities in programming a three-axis mill or a two-axis lathe in either the Haas or Fanuc programming language. In fact, in most cases, a program from a Haas controller can run directly on a Fanuc controller. But once the machine type starts to get more complex or newer in design and functionality, the differences become much more apparent. With that in mind, this chapter will specifically address the Haas programming language. The processes may be identical, or very similar to, processes used on Fanuc controller machines, but many of the G-codes or M-codes will vary.

A *live tooling lathe* is a machine equipped with powered tools that can perform various operations while the workpiece remains in orientation with the main spindle. The live tooling is an attachment that mounts to the tool turret and provides a secondary spindle for drills or end mills. Do not confuse a *mill-turn machine* with a live tooling lathe. A mill-turn machine looks like a lathe with milling attachments, but it has full lathe capabilities with full milling capabilities. It has a separate milling spindle with the full horsepower of a milling machine. See **Figure 23-1**.

Dmitry Kalinovsky/Shutterstock.com

Figure 23-1. A mill-turn machine with full turning and full milling capability.

On a live tooling lathe, a special tool holder contains the *live tool* and mounts directly to the turret. See **Figure 23-2**. Live tooling can be either belt or gear driven and can have speeds as high as 6000 revolutions per minute (rpm), but a live tool spindle has much lower torque or horsepower than the main spindle. This limits the ability of live tools to complete deep cuts or heavy roughing operations. However, this does not mean they do not serve a vital role in machining. Secondary operations such as drilling and slotting can be easily accomplished in the same setup used for primary operations, saving substantial manufacturing time.

One might ask, what is a good candidate for a live tooling lathe operation? The answer is, any part that can be completed in less operations, both quickly and efficiently. Consider a simple lathe part that has a single cross hole drilled through the part. See **Figure 23-3**. Prior to the availability of live tooling lathes, this part would be turned on a lathe, then taken to a milling machine, or placed in a drill fixture, and completed in a second operation. Not only is this time-consuming and inefficient, it increases the opportunity to produce a part out of tolerance, resulting in a scrap part.

23.2 Special Considerations with Live Tooling

As discussed in Chapter 20, a three-axis lathe has the standard X and Z machine axes and a third axis referenced as the C axis. The C axis is a rotational axis used to rotate the spindle to a specific angle. This is one of the functions needed to program live tooling. Programming the C axis with the C address will orient the spindle to the desired angular position, but the live tooling motor also needs to be engaged and commanded. There are also specific codes and work plane designations used to control the live tooling. The following sections cover M-codes and G-codes required to successfully program live tooling.

weerasak saeku/Shutterstock.com

Figure 23-2. A lathe equipped with a live tooling attachment for a milling operation.

Goodheart-Willcox Publisher

Figure 23-3. A turned part with a hole drilled on the outer diameter can be produced in a single setup on a live tooling lathe.

23.2.1 Live Tooling M-Codes

M-codes can be used to control machine functions or program functions. M-codes used for machine functions in live tooling operations are used to control operation of the live tooling motor, the main spindle brake and spindle orientation, and the C axis motor.

M133, M134, and M135: Live Tooling Motor Functions

The M133 function turns on forward motion of the live tooling motor to a specific speed in rpm. For example:

M133 P3000	Live tooling spindle forward motion at 3000 rpm

Notice that the P address is used instead of the S address, which is used to set the main spindle speed. The controller must be able to distinguish between drives.

The M134 function turns on reverse motion of the live tooling motor to a specific speed in rpm. The P address is used to set the speed. For example:

M134 P2000	Live tooling spindle reverse motion at 2000 rpm

The M135 function stops the live tooling motor. This function is commonly used when the operation is completed and will automatically occur at a tool change.

M14 and M15: Clamp and Unclamp Main Spindle

The M14 function activates the spindle brake. This will lock the spindle in place. This is needed for heavier machining cuts and will not allow simultaneous spindle rotation with live tooling. The M15 function unclamps the spindle brake. This will automatically occur at a tool change or when the spindle is positioned to a new angle.

M19: Spindle Orientation

The M19 function is used to position the spindle to a specified angle. The M19 function specified without an angle will orient the spindle at 0°. The P or R address code can be used to orient the spindle to the desired angular position in degrees. The C address code works in the same way.

M154 and M155: C Axis Motor Functions

The M154 function engages and turns on the C axis motor. The C axis motor produces precise rotation that is synchronized with movement of the X and/or Z axis. Spindle speeds up to 60 rpm can be programmed. This is used when it is desired to rotate the spindle simultaneously while the cutter is in motion. For example, simultaneous motion of the C axis with X or Z axis movement makes it possible to cut a rotating slot. The M155 function disengages the C axis motor.

23.2.2 Live Tooling G-Codes

The G17, G18, and G19 commands are used to designate the work plane in CNC lathe work. Normally, in two-axis lathe programming, these commands are not used, because all machining is done in the XZ plane. However, the work plane commands will be necessary in certain live tooling operations and in multiaxis operations. When programming machining operations on a four-axis CNC lathe (using the X, Y, Z, and C axes), designating the correct work plane will be critical.

Figure 23-4 shows the CNC lathe work planes. The G17 command is used to designate the XY plane for machining. The G18 command is used to designate the XZ plane for machining. The G19 command is used to designate the YZ plane for machining. This chapter focuses on live tooling operations programmed using the XZ work plane. Depending on the controller and machining operation, it may be necessary to specify a different work plane when using certain live tooling canned cycles.

23.3 Radial Hole Machining Canned Cycles

Canned cycles used in live tooling operations are similar to other lathe canned cycles. There are several canned cycles specifically for hole making live tooling operations. They are designed to simplify programming and automate repetitive functions.

There is a distinction between live tooling operations occurring in the X axis direction and those occurring in the Z axis direction. Drilling operations occurring in the X axis direction are called *radial drilling* operations and drilling operations occurring in the Z axis direction are called *axial drilling* operations. See **Figure 23-5**. When these operations are occurring in a three-axis lathe with C axis motion in conjunction with X and Z axis motion, each operation can only be accomplished on a plane that is coplanar with the machine centerline. The canned cycles discussed in the following sections are used for radial drilling cycles. Axial drilling can be accomplished with the same lathe canned cycles used for drilling on the Z axis centerline.

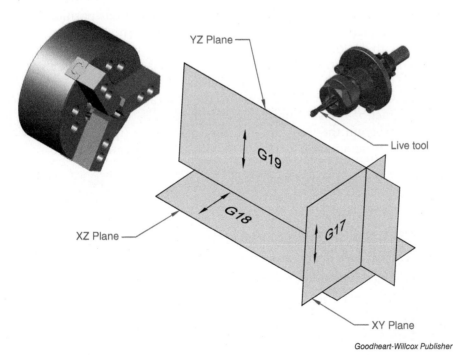

Goodheart-Willcox Publisher

Figure 23-4. The work planes used in CNC lathe work are designated with the G17, G18, and G19 commands.

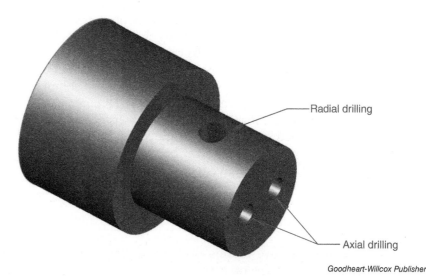

Figure 23-5. In a radial drilling operation, the travel of the tool is parallel to the X axis direction. In an axial drilling operation, the travel of the tool is parallel to the Z axis direction.

The part shown in **Figure 23-6** will be used to demonstrate a drilling canned cycle created for a Haas controller. The X0. Z0. part origin is located at the front face and part centerline. The part has two holes located on the 1″ outside diameter. The holes are .250″ diameter, .500″ deep, and 90° apart. The holes are located .500″ from the part origin in the negative Z axis direction.

23.3.1 G241: Radial Drilling Cycle

The G241 drilling cycle is similar to the G81 standard drilling cycle. In a radial drilling operation, the tool must be positioned at a safe starting location away from the part in the X and Z axes. The X axis position is near the outside diameter of the part at a safe distance. The Z axis position is aligned with the location of the hole along the Z axis. An angular direction is specified to rotate the spindle about the C axis to the desired angle for the hole location. Then, the drilling cycle is performed.

The G241 cycle will feed down to the programmed X depth and then rapid retract before starting the next drilling cycle. The G98 command is used to program the feed rate in inches per minute (ipm). The following shows the programming format used with the G241 canned cycle.

G241 X__ C__ R__ F__

The entries are read as follows:

- **G241.** Radial drilling canned cycle.
- **X.** Final hole depth along the X axis.
- **C.** Angular position for the C axis.
- **R.** X diameter position to rapid to before the cycle begins and after the cycle is complete.
- **F.** Feed rate in inches per minute.

> **From the Programmer**
>
> **Hole Orientation**
>
> The part shown in **Figure 23-6** has two holes located 90° apart. In this case, it does not matter if they are programmed using the angular coordinates C0. and C90., or C90. and C180., or any other angular coordinates, as long as they are 90° apart. However, other parts may have additional radial or axial features that are oriented to the first set of holes. Be careful in programming when parts contain multiple features. Make sure to consider all of the features so they remain oriented as per the print specifications.

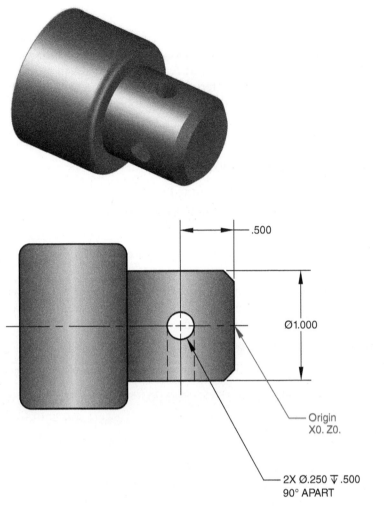

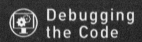

Figure 23-6. A part with two cross holes drilled on the outside diameter. The holes can be machined in a live tooling operation after the turning operation is completed.

The following program is an example of using the G241 canned cycle. This program is used for the part shown in **Figure 23-6**.

T1212	Tool #12
M154	Engage the C axis motor
M133 P2500	Live tool forward at 2500 rpm
G98	Drilling feed rate programmed in ipm
G0 X1.2 Z–.5	Rapid to safe starting position
G241 X0. C0. R1.1 F20.	Initiate cycle, drill to X0. at 0°, rapid plane return to X1.1, feed rate at 20 ipm
X0. Z–.5 C90.	Drill second hole to X0. Z–.5 location, at 90°
G0 G80 Z1.	Cancel canned cycle
M135	Stop live tooling motor
G28 U0. W0.	Return to machine home

Debugging the Code

Interpreting Dimensions

In **Figure 23-6**, notice that the outside diameter where the hole is drilled is 1.000". The print specifies a hole depth of .500", but the program goes to the X0. location. This is the centerline of the part. Use the information specified on the print to determine how the depth is dimensioned and program accordingly. The .500" depth is measured down from the 1" outside diameter and represents a radial dimension. This means that the dimension is measured from one side of the part, or radially, and is not a direct diameter measurement.

23.3.2 G242: Radial Spot Drilling Cycle

The G242 drilling cycle is similar to the G82 spot drilling cycle and is designed for programming a short dwell at the hole bottom. This cycle is used for spot drilling operations that require a short dwell at the hole bottom and is useful for hole chamfering operations. The following shows the programming format used with the G242 canned cycle.

```
G242 X___ C___ R___ P___ F___
```

The entries are read as follows:

- **G242.** Radial spot drilling canned cycle.
- **X.** Final hole depth along the X axis.
- **C.** Angular position for the C axis.
- **R.** X diameter position to rapid to before the cycle begins and after the cycle is complete.
- **P.** Dwell time at the bottom of the hole. A number entered without a decimal point represents milliseconds. A number entered with a decimal point represents seconds.
- **F.** Feed rate in inches per minute.

The following program is an example of using the G242 canned cycle.

T1212	Tool #12
M154	Engage the C axis motor
M133 P2500	Live tool forward at 2500 rpm
G98	Drilling feed rate programmed in ipm
G0 X1.2 Z-.5	Rapid to safe starting position
G242 X0. C0. R1.1 P.5 F20.	Initiate cycle, drill to X0. at 0°, X1.1 rapid plane position, dwell for 1/2 second, feed rate at 20 ipm
X0. Z-.5 C90.	Drill second hole to X0. Z-.5 location, at 90°
G0 G80 Z1.	Cancel canned cycle
M135	Stop live tooling motor
G28 U0. W0.	Return to machine home

23.3.3 G243: Radial Peck Drilling Cycle

The G243 drilling cycle is similar to the G83 peck drilling cycle. The cycle will feed down in incremental steps and fully retract between steps out of the hole. The following shows the programming format used with the G243 canned cycle.

```
G243 X___ C___ R___ Q___ F___
```

The entries are read as follows:

- **G243.** Radial peck drilling canned cycle.
- **X.** Final hole depth along the X axis.
- **C.** Angular position for the C axis.

- **R.** X diameter position to rapid to before the cycle begins and after the cycle is complete.
- **Q.** Incremental amount of motion along the Z axis for each peck.
- **F.** Feed rate in inches per minute.

The following program is an example of using the G243 canned cycle.

T1212	Tool #12
M154	Engage the C axis motor
M133 P2500	Live tool forward at 2500 rpm
G98	Drilling feed rate programmed in ipm
G0 X1.2 Z-.5	Rapid to safe starting position
G243 X0. C0. R1.1 Q.1 F20.	Initiate cycle, drill to X0. at 0°, X1.1 rapid plane position, peck .100 deep, feed rate at 20 ipm
X0. Z-.5 C90.	Drill second hole to X0. Z-.5 location, at 90°
G0 G80 Z1.	Cancel canned cycle
M135	Stop live tooling motor
G28 U0. W0.	Return to machine home

23.3.4 G245: Radial Boring Cycle—Feed In and Feed Out

The G245 cycle is similar to the G85 boring canned cycle used in milling. This cycle will feed down to the final X axis depth and then feed back to the retract plane using the programmed feed rate. The following shows the programming format used with the G245 canned cycle.

G245 X___ C___ R___ F___

- **X.** Final hole depth along the X axis.
- **C.** Angular position for the C axis.
- **R.** X diameter position to rapid to before the cycle begins and retract after the cycle is complete.
- **F.** Feed rate in inches per minute.

The following program is an example of using the G245 canned cycle.

T1212	Tool #12
M154	Engage the C axis motor
M133 P2500	Live tool forward at 2500 rpm
G98	Boring feed rate programmed in ipm
G0 X1.2 Z-.5	Rapid to safe starting position
G245 X0. C0. R1.1 F20.	Initiate cycle, bore to X0. at 0°, X1.1 R plane position, feed in and feed out
X0. Z-.5 C90.	Bore second hole to X0. Z-.5 location, at 90°
G0 G80 Z1.	Cancel canned cycle
M135	Stop live tooling motor
G28 U0. W0.	Return to machine home

23.3.5 G246: Radial Boring Cycle—Feed In and Spindle Stop

The G246 cycle is similar to the G86 boring canned cycle used in milling. This cycle will feed down to the final X axis depth. Then, the spindle stops and the tool rapids back to the retract plane. The following shows the programming format used with the G246 canned cycle.

G246 X___ C___ R___ F___

- X. Final hole depth along the X axis.
- C. Angular position for the C axis.
- R. X diameter position to rapid to before the cycle begins and retract after the cycle is complete.
- F. Feed rate in inches per minute.

T1212	Tool #12
M154	Engage the C axis motor
M133 P2500	Live tool forward at 2500 rpm
G98	Boring feed rate programmed in ipm
G0 X1.2 Z–.5	Rapid to safe starting position
G246 X0. C0. R1.1 F20.	Initiate cycle, bore to X0. at 0°, X1.1 R plane position, feed in and rapid out
X0. Z–.5 C90.	Bore second hole to X0. Z–.5 location, at 90°
G0 G80 Z1.	Cancel canned cycle
M135	Stop live tooling motor
G28 U0. W0.	Return to machine home

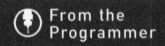

Contour Programming

This text is primarily concerned with manual programming, but it is mathematically possible to create a contour milling toolpath on a three-axis lathe. Normally, this will be accomplished using a CAM system. This text serves as a guide for the codes and format used in programming, but when the part requires multiaxis machining, a CAM system is going to provide the best options for creating an efficient program.

23.4 Slot Milling

Slot milling is a common operation performed with a live tooling lathe. It might seem odd to say *milling* in a lathe operation, but it is a milling operation. An end mill will be used to create slots, but remember that live tooling attachments have limited speed and horsepower, so milling has its limitations on a three-axis lathe. Also consider that the live tool does not have the ability to move from side to side in the slot (in the Y axis), so in this case a 1/4″ slot is most likely machined with a 1/4″ end mill. High precision slots might require roughing with a smaller diameter end mill first, and then finish machining with a separate tool. Use the tolerances specified on the print to help make the machining decisions.

Figure 23-7 shows the part used in earlier examples with a simple slot added. The slot is .250″ wide. A 1/4″ end mill can be used for cutting the slot, but depending on tolerances and cycle time, a 3/16″ end mill could be used to rough the slot before finishing. Be cognizant of the axis positioning and cutting direction when determining the machining strategy. The tool will plunge to a specified depth in the X axis and traverse along the Z axis. The program is as follows.

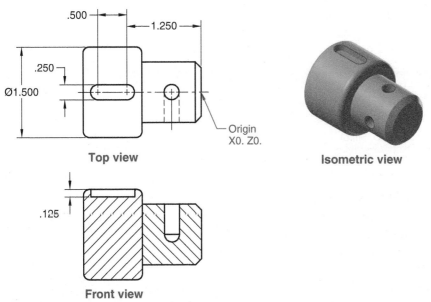

Figure 23-7. A part with a straight slot on the large outside diameter. The slot can be milled in a live tooling operation. Note: Only the dimensions required to define the slot are included.

O2307	
G20	
(TOOL – 11 OFFSET – 11)	
(1/4 FLAT END MILL)	
T1111	Tool #11
M154	Engage the C axis motor
M8	
G97 P3500 M133	Live tool forward at 3500 rpm
G98	Feed rate programmed in ipm
G18	XZ plane selection
G0 Z–1.25 C90.	Rapid to Z–1.25 and rotate C axis to 90°
X1.6	Safe position in X axis
G1 X1.25 F6.42	Plunge to X1.25 position at feed rate
Z–1.75 F28.	Feed to Z–1.75
X2.	Safe pull out
M9	
M155	Disengage C axis motor
M135	Stop live tooling motor
G0 G28 U0. W0.	Return to machine home
M30	

Slots can also be angled, **Figure 23-8**. Completing some math calculations will establish the Z axis positions for programming. The X axis depth positions can be constant or varying, depending on the print dimensions. For this example, the slot is angled at 15° and at 90° from the first programmed slot. The Z axis positions are calculated using a right triangle and the cosine trigonometric function. The calculation is made based on the 15° angle of the slot and the .250″ distance from the slot center to the center of the radius. This calculation is used for the two Z axis positions

Figure 23-8. A part with an angled slot on the large outside diameter. The slot is angled at 15° and at 90° from the straight slot. The slot is machined by moving the tool in the Z axis with simultaneous rotation of the C axis.

at the radius centers. The X axis depth will remain at X1.25, for ease of program interpretation. In the following program, movement of the tool in the Z axis is used in conjunction with the rotation of the C axis.

O2309	
G20	
(TOOL – 11 OFFSET – 11)	
(1/4 FLAT END MILL)	
T1111	Tool #11
M154	Engage the C axis motor
M8	
G97 P3500 M133	Live tool forward at 3500 rpm
G98	Feed rate programmed in ipm
G18	XZ plane selection
G0 Z–1.2585 C172.5	First location in Z axis, –7.5° from 180° center
X1.6	Safe position in X axis
G1 X1.25 F10.	Plunge to X1.25 position at feed rate
X1.25 Z–1.7415 C187.5 F28.	Move to X and Z location while rotating C axis to 187.5°
G0 X2.	
M9	
M155	
M135	
G28 U0. W0	
M30	

By moving the tool in the X and C axis, Z and C axis, or the X, Z, and C axes, simultaneously, a variety of slots can be created. Establish coordinates defining the tool movement as needed. Stay aware of the tool center point location and program operations based on the requirements of the print.

It is also possible to program axial slots. Remember that on a three-axis machine, the tool can only travel along the X and Z axes, so an axial slot can only traverse along the X axis. Straight slots or curved slots can be machined in this manner. Axial arcs can be created using the C axis to rotate the part through the cutting tool. Although these are milling operations, there are limitations to the speeds and feeds used based on the specific machine tooling being utilized.

A straight axial slot can be programmed by positioning the tool at the required X axis position, feeding the tool into the material in the Z axis direction, and then feeding to the final X axis position. See **Figure 23-9**. The part shown has a slot on the front face that is .125" deep. The program for milling the slot is as follows.

T1212	Tool #12
M154	Engage the C axis motor
M8	
G97 P4500 M133	Live tooling forward at 4500 rpm
G98	

Continued

Completed part

Tool movement

Goodheart-Willcox Publisher

Figure 23-9. A part with a straight axial slot on the front face. The tool is programmed to cut to the .125″ depth in the Z axis on one end before cutting to the opposite end.

G18	
G0 Z.25 C0.	Safe position in Z axis, rotate C axis to 0°
X.3125	Position tool in X axis
Z.1	
G1 Z–.125 F6.16	Feed down in face in Z axis
X–.3125 F18.	Feed across face in X axis
G0 Z.25	
M9	
M155	
M135	

An axial slot along an arc can be programmed by positioning the tool at the required X axis position, feeding the tool into the material in the Z axis direction to the cutting depth, and then creating rotation in the C axis. See **Figure 23-10**. The part shown has a slot milled on the front face that is .125″ deep. The cutting motion is programmed by rotating the C axis from the 0° position to –180°. The program is as follows.

(1/8 FLAT END MILL)	
T1313	
M154	
M8	
G97 P4278 M133	Live tooling forward at 4278 rpm
G98	
G18	
G0 Z.25 C0.	Safe position in Z axis, rotate C axis to 0°
X.625	Position tool in X axis
Z.1	
G1 Z–.125 F6.	Feed down in face in Z axis

Continued

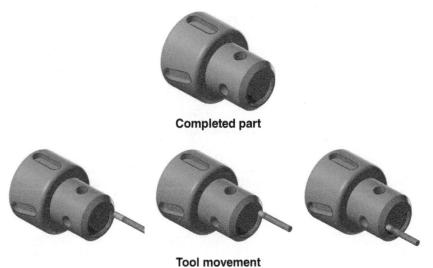

Figure 23-10. A part with an axial slot along an arc on the front face. The tool is programmed to cut to the .125″ depth in the Z axis on one end before the C axis is rotated 180° to complete the cut.

C–180. F12.	Rotate C axis to –180° at programmed feed rate
G0 Z.25	
M155	
M135	

23.5 CAM Software Programs

Manually programming drilling and slot milling operations on a machine with live tooling can add tremendous versatility to a modern machine shop. More complicated part configurations and features can be machined using a three-axis lathe, but CAM software is usually required to complete the more complex mathematical calculations.

Figure 23-11 shows a part with a raised hexagonal (six-sided) feature and a rectangular boss at the front end. The raised hexagonal feature has

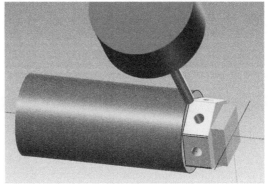

Figure 23-11. CAM software simplifies programming for parts requiring complex calculations in machining. This part has a six-sided raised feature with holes machined on the flat ends.

cross holes machined on the flat ends. It is possible to program this part manually for machining on a three-axis lathe. But, without the additional axis of movement that a four-axis machine provides, the hexagonal profile is machined with small moves in the X or Z axis, while simultaneously rotating the C axis. The math calculations can get overwhelming quickly without the help of a CAM system.

Figure 23-12 shows a part similar to **Figure 23-11**, but with a single raised feature made up of six sides, or flats, that might serve as wrench flats. To machine this part, a 1/2″ diameter end mill is used to cut in the axial direction. The following is a program created in CAM software to machine one of the flat profiles.

Goodheart-Willcox Publisher

Figure 23-12. A part with a raised feature made up of six flat sides. The flats can be machined in a milling operation by cutting in an axial direction.

```
T0606
M154
M8
G97 P2292 M133
G98
G18
G0 G54 Z-.75 C-63.5
X2.998
Z-.9
G1 Z-1. F10.
X2.9077 C-60.709 F27.5
X2.8256 C-57.778
X2.7538 C-54.791
X2.6888 C-51.598
X2.6329 C-48.283
X2.5865 C-44.86
X2.5498 C-41.345
X2.5231 C-37.758
X2.5064 C-34.12
X2.5 C-30.457
X2.5039 C-26.792
X2.5179 C-23.152
X2.542 C-19.56
X2.576 C-16.039
X2.6197 C-12.609
X2.6727 C-9.285
X2.7115 C-7.216
X2.7538 C-5.209
X2.8269 C-2.171
X2.9084 C.73
X2.9523 C2.136
X2.998 C3.5
G0 Z-.75
M9
M155
M135
```

Figure 23-13 shows the tool simulation generated by the CAM software. Although it appears that the tool is moving in a straight line across the flat, the program is made up of a series of X and C axis movements to keep the tool in a flat orientation across the face. Also consider that this program does not include any depth cuts or roughing passes, meaning the programmed toolpath could be considerably longer.

More complicated machining paths are required for parts with greater complexity. Figure 23-14 shows a part with an irregular cam lobe machined on the front end. This part is programmed using a series of X axis movements combined with rotation of the C axis. The following program shows only the final pass to create the cam lobe surface profile.

T0606	X2.0127 C160.662
M154	X2.0764 C165.337
M8	X2.1467 C169.79
G97 P0 M133	X2.2228 C174.026
G98	X2.2749 C176.767
G18	X2.3252 C179.467
G0 G54 Z-.5 C70.314	X2.3738 C182.13
X2.4047	X2.4206 C184.759
Z-.9	X2.4655 C187.358
G1 Z-1. F0.	X2.5083 C189.929
X2.3181 C74.104	X2.5491 C192.474
X2.236 C78.093	X2.5879 C194.996
X2.1589 C82.292	X2.6244 C197.498
X2.0876 C86.709	X2.6588 C199.982
X2.0228 C91.347	X2.6908 C202.448
X1.9652 C96.204	X2.7206 C204.899
X1.9156 C101.269	X2.748 C207.337
X1.8748 C106.524	X2.773 C209.763
X1.8432 C111.94	X2.7956 C212.178
X1.8216 C117.482	X2.8158 C214.584
X1.8102 C123.105	X2.8335 C216.982
X1.8094 C128.758	X2.8486 C219.373
X1.819 C134.389	X2.8613 C221.758
X1.839 C139.948	X2.8714 C224.139
X1.8689 C145.388	X2.879 C226.516
X1.9082 C150.672	X2.884 C228.89
X1.9564 C155.77	*Continued*

Goodheart-Willcox Publisher

Figure 23-13. A milling operation used to machine one of the flats of the raised feature. The toolpath is defined by a series of X and C axis movements.

X2.8864 C231.263	X1.8924 C348.667
X2.8863 C233.636	X1.9213 C353.962
X2.8836 C236.009	X1.9591 C359.115
X2.8783 C238.384	X2.0055 C364.099
X2.8705 C240.761	X2.0596 C368.895
X2.8601 C243.142	X2.121 C373.492
X2.8472 C245.528	X2.1887 C377.884
X2.8318 C247.92	X2.2623 C382.073
X2.8139 C250.318	X2.3409 C386.063
X2.7934 C252.725	X2.424 C389.862
X2.7706 C255.141	X2.4652 C391.745
X2.7453 C257.568	X2.5023 C393.643
X2.7177 C260.007	X2.5354 C395.553
X2.6877 C262.46	X2.5643 C397.473
X2.6554 C264.928	X2.5891 C399.403
X2.6208 C267.413	X2.6096 C401.34
X2.584 C269.917	X2.6259 C403.282
X2.5451 C272.442	X2.6379 C405.229
X2.504 C274.99	X2.6456 C407.179
X2.461 C277.563	X2.649 C409.131
X2.4159 C280.165	X2.648 C411.083
X2.369 C282.797	X2.6428 C413.034
X2.3202 C285.464	X2.6332 C414.983
X2.2698 C288.17	X2.6193 C416.928
X2.1957 C292.339	X2.6011 C418.868
X2.1273 C296.711	X2.5787 C420.802
X2.0654 C301.288	X2.552 C422.728
X2.0105 C306.065	X2.5213 C424.644
X1.9633 C311.031	X2.4864 C426.548
X1.9246 C316.168	X2.4475 C428.44
X1.8948 C321.451	X2.4047 C430.314
X1.8746 C326.844	G0 Z–.5
X1.8641 C332.308	M9
X1.8636 C337.797	M155
X1.8731 C343.265	M135

Goodheart-Willcox Publisher

Figure 23-14. A part with a raised cam lobe. The profile can be programmed in CAM software using a series of X and C axis movements.

This is a basic programming strategy that can be easily implemented with other complex parts that have irregular profiles.

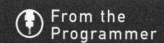

Multiaxis Programming

Using a multiaxis lathe to produce parts can be a major step in the success of a machine shop. Creating complex parts in a single operation increases cost efficiency and part quality by reducing the number of part setups. The ability to program parts in multiaxis operations is a valuable skill for any CNC lathe programmer.

Chapter 23 Review

Summary

- The use of a three-axis lathe, with live tooling, can provide versatility and efficiency to any machine shop. The ability to reduce operations increases productivity and part quality.
- A live tooling lathe is a machine equipped with powered tools that can perform various operations while the workpiece remains in orientation with the main spindle.
- Different machine types and controllers use a variety of specialized M-codes to command live tooling. These M-codes initiate the live tooling, direct its drive in forward or reverse, and engage and disengage the C axis motor. The correct use of these codes in the specific machine being programmed is vital to correct operation of the live tooling.
- The C axis in a CNC lathe is the rotary axis of the machine spindle. The spindle can be locked into place and programmed in degrees for live tooling operations.
- The G17, G18, and G19 commands are used to designate the work plane on a CNC lathe.
- Operations performed in the X axis direction are referred to as radial operations. Operations performed in the Z axis direction are referred to as axial operations.
- Standard drilling canned cycles, such as the G81, G82, and G83 cycles, can be used for axial drilling operations. In radial drilling operations, the G241, G242, and G243 cycles are adaptive cycles specifically used for drilling in the X axis direction.
- Slot milling operations can be programmed on a live tooling lathe using C axis positioning and an end mill.
- As parts increase in complexity, CAM software is usually necessary in programming to complete complex mathematical calculations.

Review Questions

Answer the following questions using the information provided in this chapter.

Know and Understand

1. With the addition of live tooling and a C axis, a CNC lathe is considered a _____ machine.
    A. two-axis
    B. three-axis
    C. four-axis
    D. None of the above.

2. *True or False.* A live tooling lathe can also be called a mill-turn machine.

3. *True or False.* In radial drilling, the travel of the tool is parallel to the Z axis direction.

4. If drilling two holes directly across from each other, what are the possible C axis rotation angles?
    A. 0° and 180°
    B. 90° and 270°
    C. 45° and 225°
    D. All of the above.

5. The M154 command tells the machine to perform which of the following tasks?
    A. C axis motor engage
    B. Live tooling forward
    C. C axis motor disengage
    D. Live tooling reverse

6. The M134 command tells the machine to perform which of the following tasks?
    A. C axis motor engage
    B. Live tooling forward
    C. C axis motor disengage
    D. Live tooling reverse

7. Which radial drilling canned cycle allows a programmable dwell?
    A. G242
    B. G243
    C. G245
    D. G246

8. When drilling a deep hole that requires peck drilling, which radial drilling cycle is most appropriate?
   A. G242
   B. G243
   C. G245
   D. G246

9. *True or False?* The G245 cycle is most similar to the G85 canned cycle used in milling.

10. When machining a slot on the outside diameter of a machined part using the XZ plane, which work plane command should be used?
    A. G17
    B. G18
    C. G19
    D. None of the above.

11. *True or False?* When initiating the live tool motor, the G96 command should be used to activate constant surface speed mode.

12. When creating an angled slot on the outside diameter of a part on a three-axis lathe, which axes are moved simultaneously?
    A. Z and C
    B. X and Z
    C. C and Y
    D. X and Y

13. When creating an axial slot, which axis will be used to make the initial feed move into the part?
    A. X
    B. C
    C. Z
    D. Y

14. When creating a radius arc slot on a part face, which axis will be the drive axis when feeding?
    A. X
    B. C
    C. Z
    D. Y

15. *True or False?* When creating complicated features, such as a cam lobe, CAM software is usually required for calculating the feed moves.

Apply and Analyze

1. Identify the axis designations for the following three-axis CNC lathe.

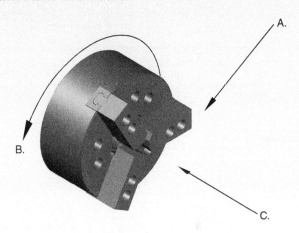

For *questions 2–8, match each code with the corresponding function, cycle, or setting.*

2. C axis motor engage
3. Radial peck drilling cycle
4. Live tooling reverse motion
5. XZ plane selection
6. Radial drilling cycle
7. Live tooling speed
8. Live tooling forward motion

A. M133
B. M134
C. M154
D. G18
E. G241
F. G243
G. P3500

9. Referring to the blanks provided below, give the correct codes that are missing in the radial drilling cycle to drill holes located at 45° and 90°.

| T1212 |
| M____ |
| M____ P2500 |
| G98 |
| G0 X1.2 Z-.5 |
| G____ X0. C45. ___1.1 ___20. |
| ____ |
| G0 G80 Z1. |
| M135 |

10. Referring to the drawing and the blanks provided below, give the correct codes that are missing for the slot milling operation.

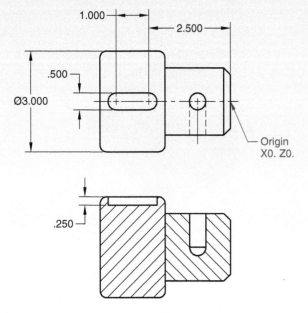

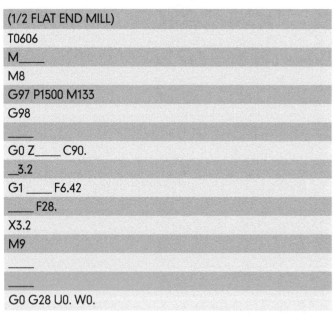

(1/2 FLAT END MILL)
T0606
M____
M8
G97 P1500 M133
G98
____
G0 Z____ C90.
____ 3.2
G1 ____ F6.42
____ F28.
X3.2
M9
____
____
G0 G28 U0. W0.

11. Referring to the drawing and the blanks provided below, give the correct codes that are missing for the face slot milling operation.

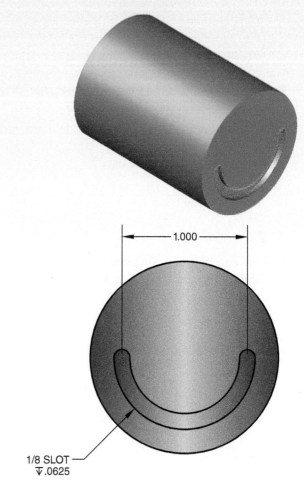

1/8 SLOT
▽ .0625

(1/8 FLAT END MILL)
T1313
M____
M8
G97 P4278 ____
G98
____
G0 Z.25 C0.
____ .5
Z.1
G1 ____ F6.
____ F12.
G0 Z.25
____
____

Critical Thinking

1. The ability to eliminate secondary operations, such as cross hole drilling or slotting operations, has many advantages. Explain why costs are reduced and why quality improves with three-axis machining operations.

2. There are many different types of parts that are well-suited for turning, but would require three-axis machining. What are some parts or tools that would benefit from multiaxis lathe operations? In what industries might you find parts made using three-axis turning operations? Consider industries that use pneumatic or hydraulic components and how they may require air or fluid to be pressurized for specific engineering functions.

3. Cams are often associated with automobile parts, but they have many other applications for any type of machinery requiring a timed action to be repetitively performed. What are some examples of parts with cams and how could they be produced on a three-axis machine?

24 Lathe Setup

Chapter Outline

24.1 Introduction
24.2 Tool Installation
24.3 Tool Offsets
    24.3.1 Tool Length Offset
    24.3.2 Tool Diameter Offset
24.4 Work Offsets
24.5 Setup Sheets
    24.5.1 Header
    24.5.2 Rough Stock
    24.5.3 Tool Data
    24.5.4 Workholding and Fixturing
    24.5.5 Special Instructions
24.6 Setup Sheet Formats
24.7 Efficiency in Production

Learning Objectives

After completing this chapter, you will be able to:

- Explain the importance of a proper CNC lathe setup.
- Describe how lathe tools are correctly installed in the tool turret.
- Describe the process for establishing the X0. and Z0. planes for tool offsets.
- Explain how to input the X and Z axis tool offsets on a Fanuc or Haas machine controller.
- Explain how to use work offsets to create multiple parts.
- Identify the components that are included on a setup sheet.
- Describe different documentation formats used for setup sheets.

Key Terms

setup
setup sheet
tool offsets
work offset
work shift

24.1 Introduction

This section of the text has covered common methods and techniques for programming operations on a CNC lathe. At this point, cutting strategies have been determined, tooling has been selected, and the program has been planned and written. Now, the program moves to the shop floor and the initial setup of the machine begins.

There are many steps involved in the initial work setup. *Setup* is the phase of CNC machining that involves setting up the workholding for the part, setting up the tooling, and making a test run of the program before the first part is machined. The setup process is just as important as any other aspect of CNC lathe work. Although the programmer has determined the types of tooling that will be used, the setup personnel will need to securely mount the tools in the machine, secure the material appropriately, and establish the tool and work offsets. All of these steps have to be accomplished before the program can be tested. In order to produce parts successfully and efficiently, the work setup must be completed correctly.

24.2 Tool Installation

Before tool offsets can be established, the cutting tools and toolholders must be mounted in the machine. Lathe tools should be firmly secured to the toolholder in the turret when installed. Depending on the machine configuration and turret style, tools can be oriented in multiple configurations. Regardless of the turret style, there are some common rules to follow in properly installing any tooling in a CNC lathe.

The first rule is to reduce the distance from the clamping surface to the cutting surface. By reducing the length the tool extends outside of the clamping zone, the rigidity of the toolholder is increased. The longer the tool extends from the clamping zone, the greater the chance for vibration and tool failure. See **Figure 24-1**.

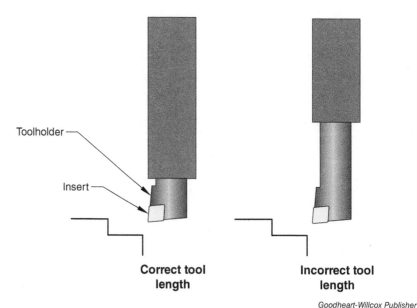

Goodheart-Willcox Publisher

Figure 24-1. The length of the tool extending from where it is mounted must be kept to a minimal distance to prevent vibration or tool failure.

The second rule is to install tools in the correct alignment. See **Figure 24-2**. There may be certain workpieces that appear to require or benefit from a tool placed at an angle. If that is the case, then the wrong tool has been selected to perform the operation. The geometry of the cutting edge of an insert is specifically designed for performance of that insert. If the toolholder is not installed in its correct alignment, the insert geometry will not function as designed. Furthermore, a toolholder that is not in alignment is not correctly secured, so the toolholder could possibly move during a cutting operation. This will cause a catastrophic tool failure and possible damage to the CNC lathe.

24.3 Tool Offsets

Setting up tools in a turning center is much different from the tool setup process in a machining center. First, there are only two axes to work with (the X and Z axes). Second, all lathe parts turn around a common centerline, and this presents an advantage in programming and machining. The most common method for setting up tools, and the one that will be discussed in this chapter, is to use the face of the finished part as the Z axis zero plane (Z0.) and the spindle centerline as the X axis zero plane (X0.). **Figure 24-3** shows a sample part and the tool located at the X0. Z0. position.

Tool offsets represent the distance the tool travels from machine home to the X0. Z0. part origin. During setup, the tool offset settings must be defined for each tool relative to each axis. This process tells the machine the distance from the machine home position to the programmed part zero location. The machine controller uses the tool offset settings to calculate the position of the tool during machining. The Z axis offset represents the tool length offset and the X axis offset represents the tool diameter offset, as discussed in the following sections.

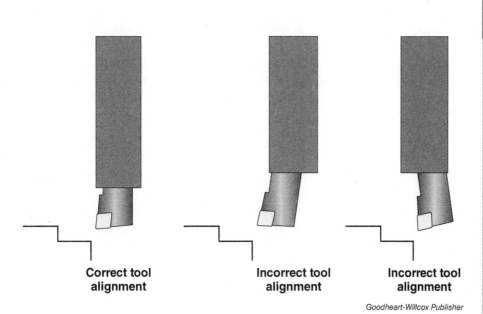

Goodheart-Willcox Publisher

Figure 24-2. The tool must be aligned correctly when mounted to ensure proper performance when cutting.

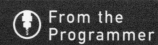

Part Origin and Tool Setup

Although it is common to program from the face of the part and establish that location as the Z0. plane for defining tool offsets, it is possible to program from any location. The key point to remember is that the same Z axis origin point (Z0.) that is utilized in the part program must be established as the Z axis tool plane during setup. For example, the program can be written from the "back" of the part, but the tools must also be located and measured from that plane.

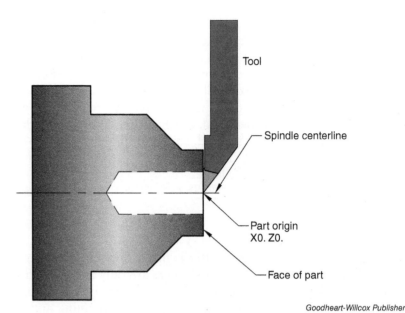

Goodheart-Willcox Publisher

Figure 24-3. The most common position for the X0. Z0. part origin in lathe programming is at the front face of the finished part and the spindle centerline. This position is used to define the X axis and Z axis tool offsets.

24.3.1 Tool Length Offset

The tool length offset establishes a position for each tool from the machine home position to the intersection of the tool face and part face. The tool length offset defines the Z axis offset for the tool. It will tell the machine the distance from the machine home position to the programmed part zero location. The practical application for setting the Z axis offset is to make a light cut on the part material face to establish the part length. The Z axis position where the tool is touching the material face is defined as Z0. See **Figure 24-4**.

Without changing the Z axis position, press the appropriate key on the controller to record the position. First, navigate to the machine's *Tool Offset* page and move the cursor to the corresponding tool number. On a Haas controller, press the *Z Face Measure* key. See **Figure 24-5**. This will record the current Z axis position for the active tool. Each subsequent tool will need to be touched to the same face to set its Z axis offset.

Setting the Z axis offset on a Fanuc controller is similar. Position the tool at the finished face of the part. Navigate to the machine's *Tool Offset* page and move the cursor to the corresponding tool number. Type in Z0. and press the *Measure* key. See **Figure 24-6**. This will record the machine's current Z axis location in reference to the tool being measured.

There is an alternative to using the face of the part to establish the Z0. tool plane. Some machine shops prefer to use the face of the chuck to establish the tool plane. There are some benefits to this method, although it is not as commonly utilized. The main benefit is that it reduces setup time, and the more tools that are used in completing a part, the more time that will be saved. This is assuming that the same tools are used, or that the tools mounted in the tool turret remain in the same tool number location at all times. The setup process is identical to the previously discussed method, but there is one additional step.

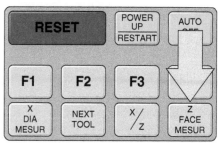

Figure 24-5. The Z axis tool offset for the current tool is recorded by pressing the *Z Face Measure* key on a Haas controller.

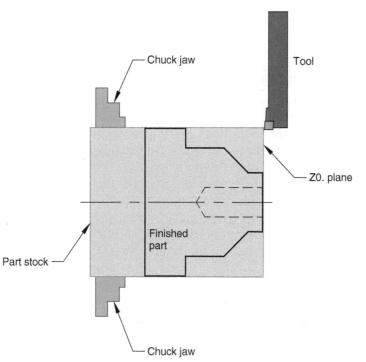

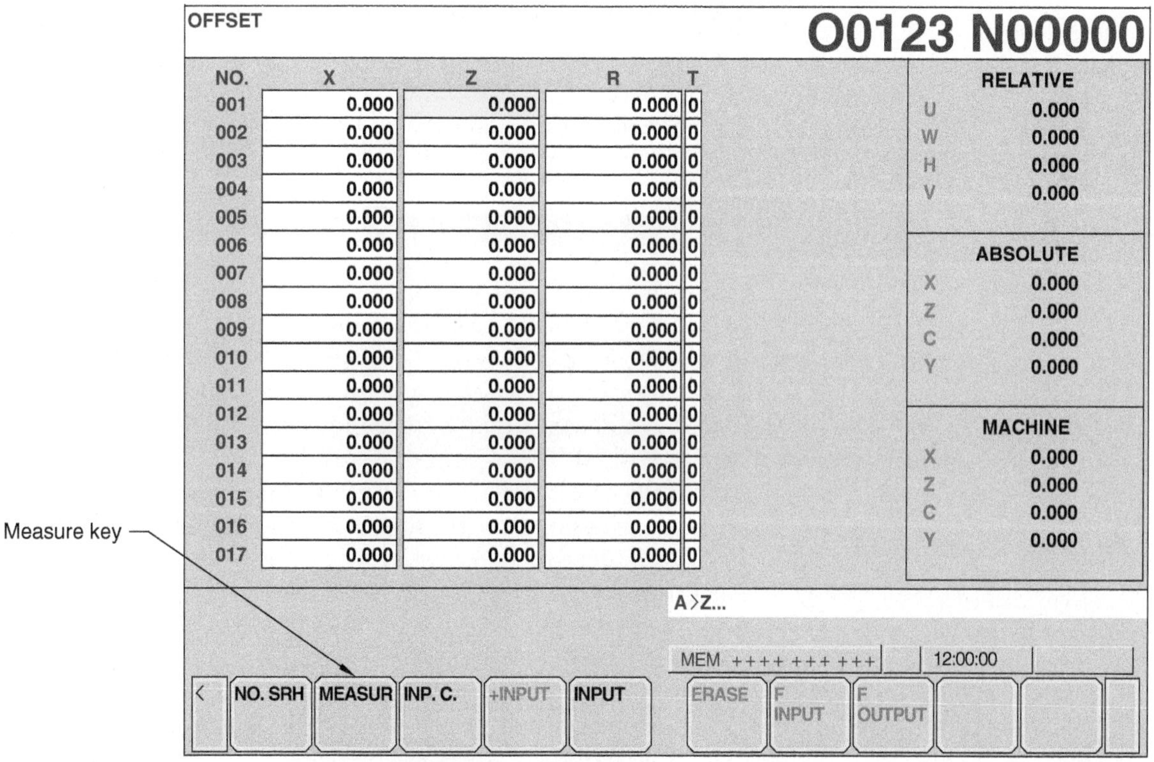

Figure 24-4. The Z axis tool offset location is established by making a small cut on the part face.

Figure 24-6. On a Fanuc controller, the Z axis tool offset for the current tool is recorded by pressing the *Measure* key.

> **Debugging the Code**
>
> **Offset Measurement**
>
> Is there a benefit to setting the tools off the face of the chuck? There can be, if the tools are going to stay in the machine for other jobs. All of the tools can be left in place and the only measurement needed for the next part is the distance from the chuck face to the face of the part. It will make no difference if you are using a G10 work shift or the G54–G59 work offsets. The end result will be the same.

As shown in **Figure 24-7,** the distance from the chuck face to the programmed part zero location must be accounted for in the offset. There are two different ways to shift the tool offset plane to the programmed part zero location. On newer machine controllers, a work offset can be used to shift the Z axis position of the work plane. A *work offset* is a coordinate setting that establishes the location of the work origin relative to the machine home position. During setup, coordinates for the work offset are entered in the corresponding work offset setting in the machine. Although there are more work offsets available, most programs are written with one of the G54–G59 work offset designations. On older controllers, it was possible to add a *work shift* to the program to *shift* the Z axis position of the work plane from the face of the chuck to the front of the part. A work shift is a programmed shift of the active work coordinate system to a different location for machining. A work shift can be programmed with the G10 command. Whether using a work offset or a work shift, simply touch off all of the tools at the chuck face and set the tool offsets. Then, measure from the chuck face to the face of the part. This can be done with one of the tools and the controller positioning screen, or with an indicator. In either case, this measurement should represent a positive Z axis value and is used for the Z axis setting of the work offset or work shift. Using work offsets and work shifts is discussed in further detail later in this chapter.

24.3.2 Tool Diameter Offset

The tool diameter offset defines the X axis offset for the tool. As previously discussed, the spindle centerline on a lathe represents the X axis zero plane

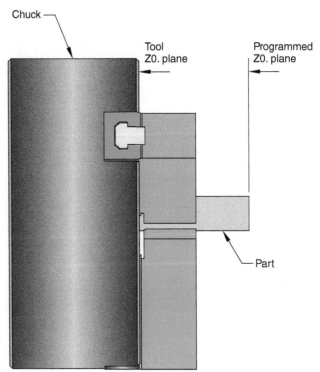

Goodheart-Willcox Publisher

Figure 24-7. When tool offsets are established from the face of the chuck, the Z axis setting of the offset must account for the measured distance between the chuck and the programmed part zero location.

(X0.) for each tool. To establish the X axis offset for a tool during setup, a light cut is made on the outside diameter of the part. Without moving the position of the tool in the X axis, the diameter of the small cut is measured. See **Figure 24-8**.

To record the X axis position, navigate to the machine's *Tool Offset* page, highlight the correct tool number, and press the appropriate key on the controller. On a Haas controller, press the *X Diameter Measure* key. See **Figure 24-9**. When the key is pressed, a prompt will appear on screen asking for the diameter of the part. Enter the measured diameter. The machine controller will calculate the distance from the current position to the spindle centerline and use this distance to calculate the corresponding offset value. The offset will be automatically entered on the *Tool Offset* page for the active tool. The offset represents the distance from the machine home position to the part zero location at the spindle centerline. Each subsequent tool will need to be touched to the same outer diameter to set its X axis offset. This process can be performed for inside diameter (ID) tools by drilling a hole in the part center, making a cut with a boring bar, and measuring and recording that diameter.

On a Fanuc controller, the same procedure is followed. Make a skim cut on the diameter, leave the tool at that X axis location, and measure the machined diameter. Then navigate to the *Tool Offset* page, highlight the correct tool number, and type in "X" followed by the measured diameter. See **Figure 24-10**.

24.4 Work Offsets

As previously discussed, a work offset is used to establish the location of the work origin relative to the machine home position. When CNC turning centers were in their infant stages, they did not have the same type of work coordinate system as CNC mills. The work coordinates were just the tool offsets established from the machine home position. As programmers became more creative, they saw opportunities to make multiple parts

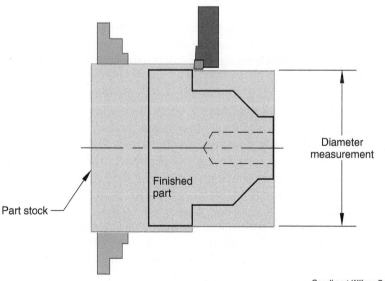

Figure 24-8. The X axis tool offset location is established by making a small cut on the part diameter.

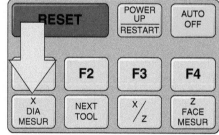

Figure 24-9. The X axis tool offset for the current tool is recorded by pressing the *X Diameter Measure* key on a Haas controller.

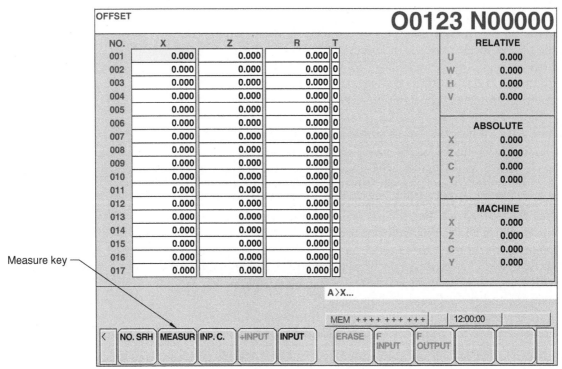

Figure 24-10. On a Fanuc controller, the X axis tool offset for the current tool is recorded by pressing the *Measure* key.

in a single setup. Controller manufacturers implemented some additional features to handle this, such as a work shift. On older controllers, the G10 command allowed the programmer to use a work shift to make a single part and then shift all the tools down the Z axis to make a second part. For example, when the code **G10 W–1.** is initiated, *all* of the tool offsets will shift 1″ in the negative Z axis direction. The W address designates an incremental measurement for the shift. At the end of the program, the G10 command must be used again to reset the tool offsets by specifying the code **G10 W1.** This shifts the tool offsets 1″ in the positive Z axis direction.

Most newer machines have implemented a simpler way to define the work coordinate system by using work offsets, similar to machining centers. In most cases, work offsets will not be used in lathe work unless making multiple parts, but it is possible to use work offsets to solve other machining situations. **Figure 24-11** shows how work can be set up to make multiple parts from round bar stock using the G54, G55, and G56 work offsets.

If the controller is capable of using multiple work offsets, they are a much better alternative to defining a work shift with the G10 command. Work offsets reduce the chance of programming errors and can be easily used in subprogramming routines. As shown in **Figure 24-11**, the G54 work offset is used in programming the first part and there are two additional parts with different starting points. The G55 and G56 work offsets are used for the additional parts. Since the X0. point is established at the spindle centerline, the X axis coordinate for each work offset can be left at X0., while the Z axis coordinate can be shifted for the second and third parts. Because the work coordinates for each work offset are stored by the machine, there is no need to cancel these shifts, just recall the G54 work offset at the beginning of the program in the next operation.

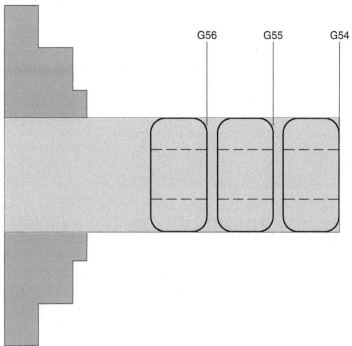

Goodheart-Willcox Publisher

Figure 24-11. Different work offsets can be defined to machine multiple parts in a single work setup.

24.5 Setup Sheets

Setup sheets serve a key purpose in work setup by providing an efficient way to exchange information between the programmer and the setup person. A *setup sheet* is a document that shows detailed information for setting up the part on the machine. A setup sheet lists items such as general information about the part, the type of stock to be machined, the tooling required, and the type of workholding used. Setup sheets vary in format and there can be a variety of opinions about what information needs to be included. The most important consideration is that all of the information required to make the setup easy and seamless is provided.

Setup sheets provide specific information about the work setup and tooling. Different machine shops will have normal operating procedures in place defining *how* parts are machined. The purpose of the setup sheet is to answer all of the unknown questions. The following sections discuss the general components that make up setup sheets and the standard information that most setup sheets should contain.

24.5.1 Header

The area at the top of a setup sheet is called the header. The header contains information such as the part name, part number, part revision, program number, program revision, customer name, and date. The date should indicate the date of completion for the program, and that date should also be included at the beginning of the written program.

24.5.2 Rough Stock

The material to be used for the machining operation should be included on the setup sheet. In some cases, the rough stock will be a previously machined part, a casting, or a saw cut piece of raw material. The rough stock section in the documentation should indicate the size and material type of the stock used for the operation. Usually, a drawing of the rough stock is included in the rough stock section. The drawing should identify the location of the work origin for the part.

24.5.3 Tool Data

The tool data section describes the tooling to be used. Each tool used in machining should be listed and detailed with information required for setup. The information should include the insert type, insert tip radius, toolholder, and tool number location in the tool turret. Any special tool offsets or nonstandard tool configurations should be noted in this section. In a multiaxis operation, define each tool by specifying the axial or radial orientation and the amount the tool must protrude from the live tooling holder.

24.5.4 Workholding and Fixturing

The method used to secure the part material must be clearly defined. If using a specific fixture or soft jaws, define the required device by number or type. If the material is simply held in a three-jaw chuck or a collet, define the amount of material that will protrude past the chuck jaws or collet face for clearance. If a tailstock or follower is required, note that in this section. Every workholding requirement needs to be completely defined. Proper workholding and material stability are the greatest determining factors of successful part completion.

24.5.5 Special Instructions

Any important information that is not covered elsewhere on the setup sheet should also be included. Special instructions might be needed for specific dimensions that need inspection, specific places where the program may pause, and anything else that does not fall under normal operations and requires special attention.

24.6 Setup Sheet Formats

Setup sheets vary in the amount of information provided. The amount of detail included depends on the information needed to completely describe the work setup and tooling requirements. This section presents different examples of setup sheets to illustrate common formats and the options available to the programmer.

Setup sheets can be created using the tools of a computer-aided manufacturing (CAM) program, or they can be created using standard word processing or spreadsheet software. **Figure 24-12** shows an example of a setup sheet created as a spreadsheet document in Microsoft Excel®. This setup sheet has a simple format and includes basic information about the rough stock, the tooling and workholding used, and the part orientation and work offset location.

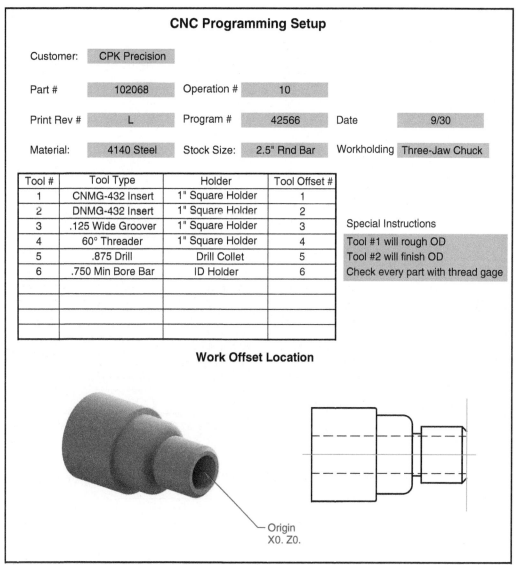

Figure 24-12. A setup sheet generated in spreadsheet software.

As shown in **Figure 24-12**, the part is made from 2.5″ round stock. The part is mounted in a three-jaw chuck and six different tools are used in machining. The first two tools are CNMG and DNMG inserts used to rough and finish the outside diameter. This is detailed in the special instructions next to the tooling data. Then, a grooving tool is used to machine the thread relief groove. A single-point threading tool is used to cut the threads for the threaded section on the small end diameter. The part also has a 1″ bored hole machined with a .875″ drill and a boring bar with a minimum bore diameter of .750″. The tooling data specifies the toolholder type and the tool offset number. At the bottom of the setup sheet, the part orientation and designated work offset are illustrated. The work origin is located at the front part face and the spindle centerline.

Depending on the type of job being processed, a list of tooling may be the only program documentation. **Figure 24-13** shows a tool list with a breakdown of the six tools required. The tool list provides detailed information about each tool, including an illustration of the specified toolholder with dimensions. Although this type of documentation is often used, it is

TOOL LIST

T0101: General Turning Tool—OD ROUGH RIGHT—80 DEG.

TOP TURRET:	Yes
ACTIVE SPINDLE:	Left
SPINDLE SPEED:	200 CSS
FAST FEED:	.01 inch/rev
SLOW SPEED:	.005 inch/rev
OFFSET:	1

HOLDER:	DCGNR-164D
LENGTH:	6.0
WIDTH:	1.25
ORIENTATION:	Vertical ANGLE: 0.0
HAND:	Right

INSERT:	CNMG-432
SHAPE:	C (80° diamond)
RADIUS:	.0313
MATERIAL:	Carbide

T0202: General Turning Tool—OD 55 DEG. RIGHT

TOP TURRET:	Yes
ACTIVE SPINDLE:	Left
SPINDLE SPEED:	200 CSS
FAST FEED:	.01 inch/rev
SLOW SPEED:	.005 inch/rev
OFFSET:	2

HOLDER:	MDJNL-164C
LENGTH:	5.0
WIDTH:	1.25
ORIENTATION:	Vertical ANGLE: 0.0
HAND:	Right

INSERT:	DNMG-432
SHAPE:	D (55° diamond)
RADIUS:	.0313
MATERIAL:	Carbide

T0303: Grooving Tool—OD GROOVE CENTER—NARROW

TOP TURRET:	Yes
ACTIVE SPINDLE:	Left
SPINDLE SPEED:	200 CSS
FAST FEED:	.0025 inch/rev
SLOW SPEED:	.002 inch/rev
OFFSET:	3

HOLDER:	
LENGTH:	6.0
WIDTH:	0.5
ORIENTATION:	Vertical ANGLE: 0.0
HAND:	Right

INSERT:	GC-4125
SHAPE:	Single End (Square)
RADIUS:	.01
MATERIAL:	Carbide

T0404: Threading Tool—OD THREAD LEFT

TOP TURRET:	Yes
ACTIVE SPINDLE:	Left
SPINDLE SPEED:	200 RPM
FAST FEED:	20.0 inch/min
SLOW SPEED:	0.0 inch/min
OFFSET:	4

HOLDER:	
LENGTH:	5.0
WIDTH:	1.0
ORIENTATION:	Vertical ANGLE: 0.0
HAND:	Left

INSERT:	NTC-3L10E
SHAPE:	UN/NPT 60°
RADIUS:	.0124
MATERIAL:	Carbide

T0505: Drilling Tool—DRILL .75 DIA.

TOP TURRET:	Yes
ACTIVE SPINDLE:	Left
SPINDLE SPEED:	200 RPM
FAST FEED:	.01 inch/rev
SLOW SPEED:	0.0 inch/min
OFFSET:	5

HOLDER:	7/8" Drill
LENGTH:	8.75
WIDTH:	2.0
ORIENTATION:	Horizontal ANGLE: 0.0
HAND:	Right

DRILL:	Drill
DIAMETER:	.875
TIP ANGLE:	118.0
MATERIAL:	Carbide

T0606: Boring Bar—ID ROUGH MIN. .75 DIA.—75 DEG.

TOP TURRET:	Yes
ACTIVE SPINDLE:	Left
SPINDLE SPEED:	200 CSS
FAST FEED:	.01 inch/rev
SLOW SPEED:	.005 inch/rev
OFFSET:	6

HOLDER:	A20-MWLNR3
LENGTH:	14.0
WIDTH:	.765
ORIENTATION:	Horizontal ANGLE: 0.0
HAND:	Right

INSERT:	5-16 IC Triangle
SHAPE:	E (75° Diamond)
RADIUS:	.0313
MATERIAL:	Carbide

Goodheart-Willcox Publisher

Figure 24-13. A setup sheet consisting of a tool list showing the details for each tool required.

the least desirable format. Much of the vital information needed for the setup person is not included. The benefit of this format is that it is short and simple, with good detail of the tooling and toolholders being used.

Another common setup sheet format is shown in **Figure 24-14**. This is perhaps the most compact but complete example of a setup sheet. This format provides general information about the program, including an illustration of the part orientation and the programmed part zero location. Also included are the tooling data with toolholder information and a breakdown of cycle times by operation. Although more information and illustrations can be added, this format is generally considered to be fairly comprehensive.

There are cases when it is necessary to provide even more information about the setup requirements. Sometimes, too much information is the right amount of information. For some jobs, the program documentation can be many pages long and can have an extensive amount of data, including explanations of each operation, tool lists, toolpath representations, and cycle times. Most CAM systems are capable of generating multiple views of parts, workholding, and toolpaths. This helps ensure that the setup person and operator have enough information for every step of the process. Do not be hesitant to add photos or in-depth explanations of each operation.

When a CAM system or other software programs are not available to create the setup documentation, a simple layout sheet with photographs and tooling information will be helpful. No matter how setup sheets are created, they should be stored in a secure location, either digitally on a computer network or in a binder.

> **From the Programmer**
>
> **Setup Sheets in Machining**
>
> The more information the setup person and machine operator have, the better the program will be. Often, the programmer will not take the time to build a detailed, proper setup sheet. This can significantly delay the setup and operation of the job. Take the time and complete the process. Consider the setup sheet as another part of the program and provide one with every program.

24.7 Efficiency in Production

Setup sheets serve a key purpose in the production of parts. By providing an important means of communication between the programmer and machinist, setup sheets allow for greater efficiency. Sometimes, programmers assume that the setup person or operator can "figure out" where the program is going or what it is doing. Obviously, this can be a bad assumption. Making sure that the program is executed exactly as it was designed, and programmed, is critical to the success of the operation. The proper transfer of information from the programmer's desk to the machine shop floor can ensure an efficient setup and limited down time for the machine.

A more detailed setup sheet can improve efficiency by allowing the setup person to build the tool and holder assemblies before the machine has completed its current job. In addition, any workholding or fixturing devices can be prepared in advance. Working ahead keeps the machine making product for as much of the available time as possible. The period of transition from finishing one part, or operation, to the next part is a great opportunity to minimize machine down time. Depending on the work being machined, this can save several minutes or hours of total operation time.

There are three vital steps to the completion of a functional part: the program, the machine setup, and the operation. Each of these steps is just as important as the other. A well-written program will not run with an incorrect setup, and a great setup will not result in good parts without successful operation of the machine. All three of these phases work hand in hand. Proper attention to detail during programming, setup, and operation will help maximize the efficiency of a machine shop.

CNC Programming Setup Sheet Report

GENERAL INFORMATION

PROJECT NAME:	CAM Sample Part
CUSTOMER NAME:	CPK Precision Aerospace
PROGRAMMER:	R. Calverley
DRAWING:	102068 REVISION: L
DATE:	Thursday, October 28
TIME:	5:29 PM

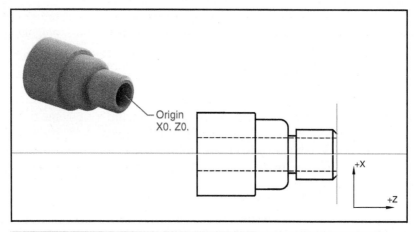

TOTAL CYCLE TIME: 0 HOURS, 17 MINUTES, 7 SECONDS

OPERATION LIST

OP #	OPERATION NAME	TOOL #	MIN-Z	MAX-Z	CYCLE TIME
1	1 - Lathe Canned Rough	1			00:06:27
2	2 - Lathe Canned Finish	2			00:01:17
3	3 - Lathe Groove (Chain)	3			00:04:45
4	4 - Lathe Thread	4			00:00:35
5	5 - Lathe Drill	5			00:02:26
6	6 - Lathe Finish	6			00:01:34

Continued

Figure 24-14. A setup sheet with illustrations of the part to be machined, cycle times for each operation, and the tooling data.

TOOL LIST SORTED: ASCENDING

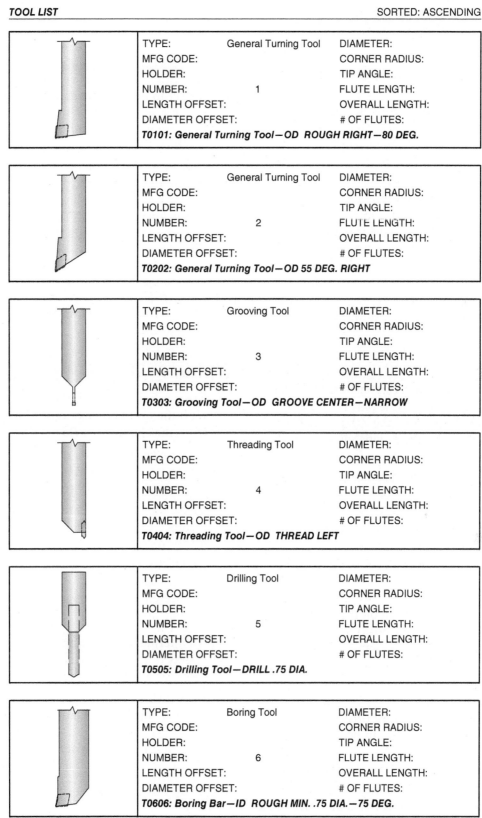

TYPE: General Turning Tool	DIAMETER:
MFG CODE:	CORNER RADIUS:
HOLDER:	TIP ANGLE:
NUMBER: 1	FLUTE LENGTH:
LENGTH OFFSET:	OVERALL LENGTH:
DIAMETER OFFSET:	# OF FLUTES:

T0101: General Turning Tool—OD ROUGH RIGHT—80 DEG.

TYPE: General Turning Tool	DIAMETER:
MFG CODE:	CORNER RADIUS:
HOLDER:	TIP ANGLE:
NUMBER: 2	FLUTE LENGTH:
LENGTH OFFSET:	OVERALL LENGTH:
DIAMETER OFFSET:	# OF FLUTES:

T0202: General Turning Tool—OD 55 DEG. RIGHT

TYPE: Grooving Tool	DIAMETER:
MFG CODE:	CORNER RADIUS:
HOLDER:	TIP ANGLE:
NUMBER: 3	FLUTE LENGTH:
LENGTH OFFSET:	OVERALL LENGTH:
DIAMETER OFFSET:	# OF FLUTES:

T0303: Grooving Tool—OD GROOVE CENTER—NARROW

TYPE: Threading Tool	DIAMETER:
MFG CODE:	CORNER RADIUS:
HOLDER:	TIP ANGLE:
NUMBER: 4	FLUTE LENGTH:
LENGTH OFFSET:	OVERALL LENGTH:
DIAMETER OFFSET:	# OF FLUTES:

T0404: Threading Tool—OD THREAD LEFT

TYPE: Drilling Tool	DIAMETER:
MFG CODE:	CORNER RADIUS:
HOLDER:	TIP ANGLE:
NUMBER: 5	FLUTE LENGTH:
LENGTH OFFSET:	OVERALL LENGTH:
DIAMETER OFFSET:	# OF FLUTES:

T0505: Drilling Tool—DRILL .75 DIA.

TYPE: Boring Tool	DIAMETER:
MFG CODE:	CORNER RADIUS:
HOLDER:	TIP ANGLE:
NUMBER: 6	FLUTE LENGTH:
LENGTH OFFSET:	OVERALL LENGTH:
DIAMETER OFFSET:	# OF FLUTES:

T0606: Boring Bar—ID ROUGH MIN. .75 DIA.—75 DEG.

Goodheart-Willcox Publisher

Figure 24-14. *(Continued)*

Chapter 24 Review

Summary

- Machine setup involves setting up the workholding for the part, setting up the tooling, and making a test run of the program before the first part is machined. This process is critical to producing parts successfully.
- Lathe cutting tools and toolholders must be mounted correctly in the machine. Lathe tools should be firmly secured to the toolholder in the turret when installed. As with all tools, the amount of exposed tooling should be minimized to reduce vibration and prevent tool failure.
- The most common method for setting up tools is to make a skim cut on the face of the material and use that face as the Z axis zero plane (Z0.). The lathe spindle centerline always serves as the X axis zero plane (X0.).
- An alternate technique to establishing the Z0. tool plane on the front face of the part is to use the face of the chuck to define the Z0. plane. There are benefits to this method, depending on the type and quantity of parts being produced.
- Tool offsets represent the distance the tool travels from machine home to the X0. Z0. part origin.
- The Z axis tool offset is set by recording the Z axis position where the tool is touching the material face after making a light cut on the part material face. The X axis tool offset is set by making a skim cut on the material diameter, measuring that diameter, and using the X axis diameter measuring function on the controller to record the X axis diameter position.
- Newer controllers have the ability to use the G54–G59 work offsets. When machining a single piece, this is not required, but for machining multiple parts at a time, it can be a huge benefit. The X axis coordinate for each work offset is left at X0., while the Z axis coordinate can be shifted for additional parts.
- In controllers without the G54–G59 work offsets, a work shift can be programmed with the G10 command to shift the Z axis position of the work plane.
- Communication between the programmer and the setup person is usually achieved through the use of a setup sheet. This communication is vital in relaying the full and correct information to the machine shop floor.
- A setup sheet lists items such as general information about the part, the type of stock to be machined, the tooling required, and the type of workholding used. A complete setup sheet is essential to a successful production of quality parts.

Review Questions

Answer the following questions using the information provided in this chapter.

Know and Understand

1. *True or False?* CNC lathe tools should be installed with the tool body extending as far as possible from the clamping surface.

2. *True or False?* A lathe tool installed incorrectly can cause vibration in machining, or even a catastrophic tool failure.

3. The Z axis tool offset tells the controller the distance from _____ to the programmed part zero location.
    A. the tool face
    B. the chuck face
    C. the spindle centerline
    D. machine home

4. An alternative to setting the Z0. position for tool offsets to the part zero location is to set all the tools off a common plane, such as the _____.
    A. chuck face
    B. machine home position
    C. chuck jaws
    D. centerline of the part

5. On all CNC lathes, the X axis tool offset is in relation to the _____.
   A. finished part face
   B. material diameter
   C. chuck face
   D. spindle centerline

6. To establish the X axis tool offset, make a skim cut and then measure the _____ to record the offset on the controller's tool offset page.
   A. distance from home
   B. part length
   C. part diameter
   D. part radius

7. On a Haas controller, which key is pressed to set the Z axis tool offset?
   A. Z face measure
   B. Tool offset enter
   C. Auto measure
   D. Tool offset set

8. *True or False?* A Fanuc controller has the same key entry as a Haas controller for setting tool offsets.

9. *True or False?* When programming work offsets, the Z axis coordinate is normally left at Z0. for each work offset, while the X axis coordinate can be shifted for additional parts with different starting points.

10. A setup sheet should contain which of the following items?
    A. Tool data
    B. Workholding method
    C. Stock information
    D. All of the above.

11. *True or False?* It is important to list the insert type for each tool on the setup sheet.

12. What is an important component to include in the header section of the setup sheet?
    A. Part number
    B. Insert type
    C. Toolholder size
    D. Tool numbers

13. In which section of the setup sheet would a setup person locate the type of chuck or collet to be used for a particular part program?
    A. Workholding and fixturing
    B. Tool data
    C. Special instructions
    D. Header

14. Which phase of the manufacturing process is the most important?
    A. Programming
    B. Setup
    C. Operation
    D. Each phase is equally important.

Apply and Analyze

1. In certain cases, a machine shop may decide to set tool offsets in relation to a reference plane on the machine rather than the finished part face. Explain how this is usually accomplished and how this approach can allow for more efficiency in production.

2. A setup sheet should be provided with every program to convey details to the setup person. Why is it important to completely describe the work setup requirements?

Critical Thinking

1. Programming, setup, and operation are three different career paths in the manufacturing profession. An operator is on the machine floor making parts on a consistent basis. A setup person is going between machines to install tooling and is probably making the first part run. The programmer may only be creating programs or working on a CAM system. Which one of these roles might be a good fit for you? Explain the skills you will need to be successful on the job.

2. As discussed in this chapter, the format used for setup sheets can vary considerably. Setup sheets can range from a single page with notes to reports made up of multiple pages. What type of setup sheet do you think is most adequate for general use? Explain your answer.

3. The setup sheet is a form of communication. It relays all of the information from the programmer to the shop floor. What are some possible machine failures or part failures that can occur if the programmer and setup person are not communicating with each other?

Section 3

Subprogramming, Probe Programming, and Macros

25	Main Programs and Subprograms	29	On Machine Verification
26	Subprogramming Techniques	30	Macros
27	Probing for Work Offsets	31	Additional Tips and Tricks
28	Probing Inside of the Program		

Aumm graphixphoto/Shutterstock.com

25 Main Programs and Subprograms

Chapter Outline

25.1 Introduction
25.2 Main Programs and Subprograms
25.3 Mill Subprogramming
25.4 Lathe Subprogramming

Learning Objectives

After completing this chapter, you will be able to:

- Describe the difference between main programs and subprograms.
- Explain how a subprogram is called from a main program.
- Explain the difference between the M30 and M99 functions.
- Identify the unique codes required to create subprograms.
- Explain how multiple subprograms can be used in the body of a main program.

Key Terms

bar feeder
bar puller
main program
program
subprogram
work shift

25.1 Introduction

To this point in the text, there has been extensive discussion about the different types of CNC machines and the programming of machining centers and turning centers. The previous two sections provided basic coverage of mill and lathe programming, with specific information on machining operations performed on CNC mills and lathes. This section covers more advanced programming techniques. In this chapter, subprogramming is introduced. The subprogramming techniques that are discussed, in most cases, can be utilized on a CNC mill or a lathe, but they are readily available on almost all machines.

CNC machinists and programmers are creative and want to identify faster and more efficient ways to machine parts. *Efficiency is directly correlated to time.* Everything that is accomplished in manufacturing is based on time. Multiaxis machines are used to reduce setup *time*. Computer-aided manufacturing (CAM) systems are used to reduce programming *time*. Production costs are quoted by adding material cost plus *time*. The amount of time it takes to machine a part determines how efficient and profitable a business is in manufacturing. The customer will demand that every part is 100% in compliance with the print and will award the contracts to the facility that can produce the parts in the least amount of time, at the least cost. The use of subprograms can definitely reduce programming time, and in many cases, can reduce cycle times and eliminate errors.

25.2 Main Programs and Subprograms

The term *program* has been used in this text many times. A **program** is a series of commands and designated functions specified in combination with axis positions to control the machine's movement and operation. CNC machines will not do anything unless instructed by the program. For example, the spindle will not turn on clockwise unless the M03 function is programmed. To this point, the programs in this text have been presented one at a time with a defined beginning and end. However, what if it were possible to "leave" a program to perform a series of commands and then re-enter at a specific place in the program? This is accomplished by using a subprogram. The initial program in which machining starts is called the **main program**, and the secondary program is called the **subprogram**. This process is diagrammed in **Figure 25-1**.

There are several rules that must be followed to use this technique. First, there must be a main program where the program begins and ends. The machining process cannot start in the main program and end in the subprogram. The specific way in which this happens will be discussed later in this chapter. Second, the subprogram must have a program number that is in the machine controller, and can be *called* by the main program. In other words, the machine must have a direction to where the subprogram can be found. The M98 function is used to call a subprogram. The P address is used in conjunction with the M98 function to reference the number of the subprogram. Third, the subprogram cannot end with the M30 function, but must be returned back to the main program with the M99 function. This is because the program cannot end in the subprogram, and must re-enter the main program. The M30 function is used to end the main program. It will rewind the main program back to its beginning.

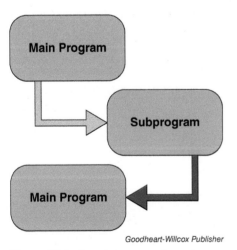

Goodheart-Willcox Publisher

Figure 25-1. A main program branches out to a subprogram. Once completed, the subprogram must return back to the main program.

The purpose of a subprogram is to simplify programming by reducing program length and programming time, while reducing the potential for errors. Subprograms are most often used to complete repetitive types of machining in multiple locations on parts. They are used in cases where it is possible to shorten programs and make the programming process more efficient.

25.3 Mill Subprogramming

Subprograms are commonly used in applications that involve repetitive actions, such as drilling operations for multiple holes. Hole patterns and other types of patterned features are common on milled parts. **Figure 25-2** shows a typical part with a pattern of holes oriented in a straight line. The holes are 1/2″ in diameter. For ease of instruction, this example will just be used to spot drill and drill the holes.

The part is 1″ thick with 1/2″ holes drilled through the part. There are seven holes equally spaced 1″ apart. Based on the print dimensions, it makes sense to use the top-left corner of the part as the work origin. In the following program, the G54 work offset is used to designate the work coordinate system and the G81 canned cycle is used in two operations to spot drill and drill the holes.

```
O1425 (SUBPROGRAM SAMPLE)
(T1|3/4 SPOT DRILL|H1|D1|DIA. – .75)
(T2|1/2 DRILL|H2|D2|TOOL DIA. – .5)
G20
G0 G17 G40 G49 G80 G90
T1 M6
G0 G90 G54 X1. Y–1. S2037 M3
G43 H1 Z1.
M8
G99 G81 Z–.26 R.1 F32.
X2.
```

Continued

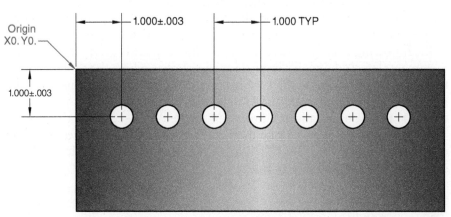

Figure 25-2. A rectangular part with a pattern of holes. A subprogram can be used to define spot drilling and drilling operations for the holes.

```
X3.
X4.
X5.
X6.
X7.
G80
M5
G91 G28 Z0. M9
M01
T2 M6
G0 G90 G54 X1. Y-1. S1910 M3
G43 H2 Z1.
M8
G99 G81 Z-1.15 R.1 F15.3
X2.
X3.
X4.
X5.
X6.
X7.
G80
M5
G91 G28 Z0. M9
G28 X0. Y0.
M30
```

In this program, tool #1 is a 3/4″ spot drill. It travels down to a depth of Z-.26 to create a small edge break and establish a starting location for the drill. Tool #2 is the drill and it travels down to a depth of Z-1.15. This depth allows the drill tip to extend completely through the 1″ part.

This is a straightforward program used for spot drilling and drilling. However, a subprogram can be used to simplify the programming. In this example, the G91 command, which is used to activate incremental positioning mode, will be used with looped cycles (repeated cycles). As discussed in the milling section of this text, the G91 command is normally not used but can be programmed when it is appropriate to specify coordinates relative to the current position. Incremental positioning moves must be programmed carefully because they can cause confusion or operator error. In the following example, incremental positioning is necessary. When using incremental positioning, it's important to visualize where the tool is in relation to the part and how the subprogram can be utilized. In this case, the main program will activate the G54 work offset and position the tool at the Y-1. position, because all holes are located at Y-1. along the Y axis. Then, two subprograms can be used to machine the seven holes by using one subprogram with a spot drill and another subprogram with the 1/2″ drill. The main program is as follows.

```
O1425 (SUBPROGRAM SAMPLE)
(T1|3/4 SPOT DRILL|H1|D1|DIA. - .75)
(T2|1/2 DRILL|H2|D2|TOOL DIA. - .5)
```

Continued

G20	
G0 G17 G40 G49 G80 G90	
T1 M6	
G0 G90 G54 X0. Y–1. S2037 M3	
G43 H1 Z1. M8	
M98 P3535 L7	L address used to loop on a Haas controller; K address used to loop on a Fanuc controller
G80	Cancel drilling canned cycle
T2 M6	
G0 G90 G54 X0. Y–1. S1910 M3	
G43 H2 Z1. M8	
M98 P3636 L7	
G80	Cancel drilling canned cycle
G91 G28 Z0.	
M30	

Before the subprograms are shown and explained, analyze what is accomplished in the main program.

G20	Inch mode
G0 G17 G40 G49 G80 G90	Startup block
T1 M6	Tool 1 (spot drill)
G0 G90 G54 X0. Y–1. S2037 M3	G54 work offset; rapid move to Y–1.; X axis moves will be in incremental mode in the subprogram

These four blocks are programmed in the same manner as any standard milling program. This sequence includes a change to tool #1, and then the tool is positioned at the Y–1. location of the holes. The starting X axis position is X0., and the tool will then move incrementally 1″ to each hole in the subprogram.

G43 H1 Z1. M8	Turn on tool length offset; position safely in Z axis; coolant on
M98 P3535 L7	Go to program #3535 and repeat 7 times
G80	Cancel drilling canned cycle

G43 H1 turns on the tool length offset for tool #1. The move to Z1. safely positions the tool 1″ above the part. This Z axis position can be lower, but the rapid plane will be set in the subprogram, so this is a safe position. Coolant is turned on in this block, but it could be turned on earlier or in the subprogram, depending on programmer preference. The M98 function tells the controller to go to the subprogram. The P address references the program number of the subprogram. The specified program has to be written and stored in the machine controller in order to be initiated. The L7 designation commands the machine to complete the subprogram seven times before moving forward. This is called a *loop*. Haas controllers use the L address to perform loops, while Fanuc controllers use the K address.

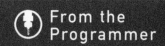

From the Programmer

Programming Modes

It is important to keep track of where the tool is positioned and what modes are active in subprogramming. In the second half of the program, the programmer is *re-calling* absolute positioning mode (with the G90 command) and the required work offset (with the G54 command). This is important because the subprogram is in incremental positioning mode and will only work if the tool starts at the X0. Y–1. location. This can be programmed in different ways, so take special notice of the active modes and programmed locations in the main program and subprogram.

T2 M6	Tool 2 (1/2″ drill)
G0 G90 G54 X0. Y–1. S1910 M3	
G43 H2 Z1. M8	
M98 P3636 L7	
G80	Cancel drilling canned cycle

After the first subprogram is complete, the controller returns to the main program. The second half of the main program is very similar to the first half. It will complete a tool change, set the G54 work offset, position the tool at X0. Y–1., turn on the spindle, and turn on the tool length offset for tool #2.

The two subprograms will be limited to just four lines. This simplifies editing and reduces errors. The spot drilling subprogram (O3535) is as follows.

O3535 (3/4″ Spot Drill Subprogram)
G91 X1.
G99 G81 Z–.26 R.1 F32.
M99

The program number must be stated, as is the case with any program. The first block of the subprogram specifies an incremental move of 1″ in the positive X axis direction. In this example, the tool started in the main program at the X0. position. If the G91 X1. block was programmed after the G81 drilling cycle, the first drilling cycle would happen at the X0. Y–1. location, and this is not what is desired. The next block contains the typical entries for the G81 drilling cycle. The M99 function is programmed in the following block. The M99 function is used at the end of all subprograms, as it returns the subprogram back to the main program. Because the L7 loop count was used in the main program, the controller will complete the subprogram seven times before it moves on in the main program.

After the seven loops, the controller returns to the main program, where the G80 command cancels the drilling cycle. This is followed by a tool change, tool positioning to the starting point, and entry into the second subprogram (P3636). The second subprogram is very similar to the first subprogram, and is written as follows.

O3636 (1/2″ Drill Subprogram)
G91 X1.
G99 G81 Z–1.15 R.1 F15.3
M99

This subprogram is identical to the first subprogram, except for the hole depth and feed rate.

Writing a main program and two subprograms to create a part with seven holes might not seem to be that efficient. However, if this part had 16 holes, only the loop count would need to be altered with the L address and the same program would work. See **Figure 25-3**.

O1425 (SUBPROGRAM SAMPLE)
(T1|3/4 SPOT DRILL|H1|D1|DIA. – .75)

Continued

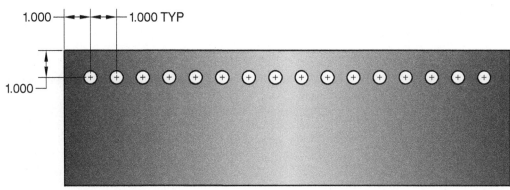

Figure 25-3. A part with a straight-line hole pattern made up of 16 holes spaced equally.

(T2|1/2 DRILL|H2|D2|TOOL DIA. – .5)	
G20	
G0 G17 G40 G49 G80 G90	
T1 M6	
G0 G90 G54 X0. Y–1. S2037 M3	
G43 H1 Z1. M8	
M98 P3535 L16	L address used to loop on a Haas controller; K address used to loop on a Fanuc controller
G80	Cancel drilling canned cycle
T2 M6	
G0 G90 G54 X0. Y–1. S1910 M3	
G43 H2 Z1. M8	
M98 P3636 L16	
G80	Cancel drilling canned cycle
G91 G28 Z0.	
M30	

The more repetitions or sets of patterns in a designed part, the more likely they are to benefit from some type of subprogramming strategy. For example, imagine a similar part with a pattern of 64 holes. See **Figure 25-4**. There are multiple ways to program this part, but placing three additional repositioning blocks between subprograms is an easy solution for this application.

O1425 (SUBPROGRAM SAMPLE)
(T1|3/4 SPOT DRILL|H1|D1|DIA. – .75)
(T2|1/2 DRILL|H2|D2|TOOL DIA. – .5)
G20
G0 G17 G40 G49 G80 G90
T1 M6
G0 G90 G54 X0. Y–1. S2037 M3
G43 H1 Z1. M8
M98 P3535 L16
G80

Continued

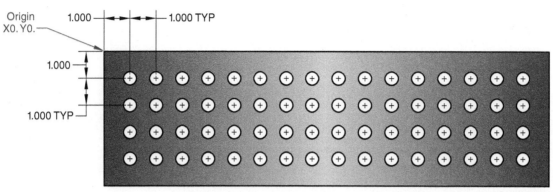

Figure 25-4. A part with a hole pattern made up of 64 holes with equal spacing along the X and Y axes.

G0 G90 G54 X0. Y-2.	Reposition tool for spot drill subprogram
M98 P3535 L16	
G80	
G0 G90 G54 X0. Y-3.	Reposition tool for spot drill subprogram
M98 P3535 L16	
G80	
G0 G90 G54 X0. Y-4.	Reposition tool for spot drill subprogram
M98 P3535 L16	
G80	
T2 M6	
G0 G90 G54 X0. Y-1. S1910 M3	
G43 H2 Z1. M8	
M98 P3535 L16	
G80	
G0 G90 G54 X0. Y-2.	Reposition tool for drill subprogram
M98 P3535 L16	
G80	
G0 G90 G54 X0. Y-3.	Reposition tool for drill subprogram
M98 P3535 L16	
G80	
G0 G90 G54 X0. Y-4.	Reposition tool for drill subprogram
M98 P3535 L16	
G80	
G91 G28 Z0.	
M30	

The program for the 64-hole pattern weaves in and out between the subprogram and the main program. This can be a common approach in subprogramming. Thinking creatively and understanding how to maximize machine efficiency allows the programmer to identify many other subprogramming applications. Subprogramming is not just used with hole patterns, but with all types of repetitive patterns. **Figure 25-5** shows a part with a set of four pockets. The main program will reference the center of each pocket and then a subprogram will be used to machine the pockets.

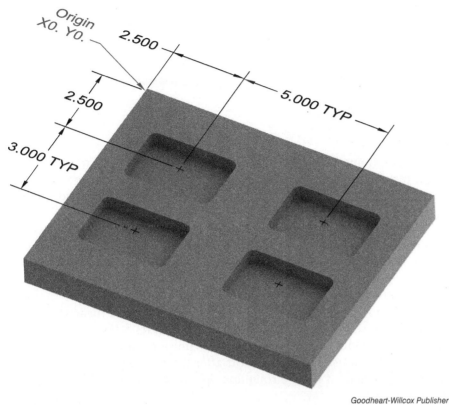

Goodheart-Willcox Publisher

Figure 25-5. A part with a pattern of closed pockets. The pockets can be milled by using the same subprogram for each pocket.

O1020 (POCKET SUBPROGRAM)	
(T1|1/2″ END MILL)	
G20	
G0 G17 G40 G49 G80 G90	
T1 M6	
G0 G90 G54 X2.5 Y–2.5 S3820 M3	
G43 H1 Z1. M8	
M98 P1968	Subprogram for first pocket
G0 G90 G54 X7.5 Y–2.5	
M98 P1968	Subprogram for second pocket
G0 G90 G54 X2.5 Y–5.5	
M98 P1968	Subprogram for third pocket
G0 G90 G54 X7.5 Y–5.5	
M98 P1968	Subprogram for fourth pocket
G91 G28 Z0. Y0.	
M30	

Program O1020 is the main program. Program 1968 is used as the subprogram to machine each pocket. The pockets are programmed with incremental positioning, with the entry and exit moves defined in the subprogram.

Many other types of operations performed on a CNC mill are conducive to subprogramming. Repetitive operations should always be considered for subprograms. Often, subprograms can be saved and used for similar part operations. This is a common strategy that reduces setup time and programming time.

25.4 Lathe Subprogramming

Subprogramming techniques can often be applied to parts machined on a CNC turning center. The techniques used are similar to those used in mill programming. Drilling, outside diameter turning, threading, and grooving operations are often simplified through subprograms.

The most common application of subprogramming in lathe work is making multiple parts from bar stock. **Figure 25-6** shows a part that can be a good candidate for subprogramming. This is a 3″ diameter part that is relatively short, at only 1.75″ long. With the right machine, it is possible to make more than one part at a time.

The following program is used for outside diameter turning and parting off operations. Tool #1 (a CNMG-432 insert) is used for roughing, tool #2 (a CNMG-432 insert) is used for finishing, and tool #3 is a .250″ wide cutoff tool used for parting off after the part is completed. The G71 canned cycle is used for the roughing operation and the G70 cycle is used for finishing.

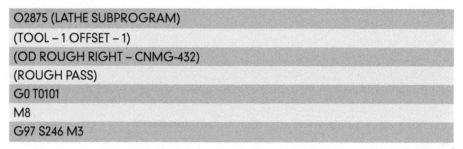

```
O2875 (LATHE SUBPROGRAM)
(TOOL – 1 OFFSET – 1)
(OD ROUGH RIGHT – CNMG-432)
(ROUGH PASS)
G0 T0101
M8
G97 S246 M3
```

Continued

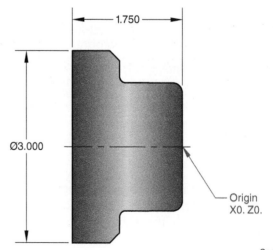

Figure 25-6. A typical part that can be produced from round stock in a turning center. Using subprograms allows for machining multiple parts in one setup. Note: For ease in viewing, the part is not fully dimensioned.

```
G0 X3.1 Z.05
G50 S3000
G96 S200
G71 U.1 R.1
G71 P100 Q102 U.02 W.01 F.01
N100 G0 X-.2 S200
G1 Z0.
X1.6875
G3 X2. Z-.1563 K-.1563
G1 Z-.9063
G2 X2.1875 Z-1. I.0938
G1 X2.6875
G1 X3. Z-1.1563
G1 Z-1.85
N102 X3.1
G0 Z.05
M9
G28 U0. W0. M5
T0100
M01
(OD FINISH RIGHT – CNMG-432)
(FINISH PASS)
G0 T0202
M8
G97 S246 M3
G0 X3.1 Z.05
G50 S3600
G96 S200
G70 P100 Q102
G0 Z.05
M9
G28 U0. W0. M5
T0200
M01
(TOOL – 3 OFFSET – 3)
(OD CUT OFF .250 WIDE – GCP-4250)
G0 T0303
M8
G97 S200 M03
G0 X3.4 Z-2.
G1 X3.2 F.004
X-.02
X.18
G0 X3.2
M9
G28 U0. W0. M05
T0300
M30
```

A creative programmer might look at this program and see that two parts can be made at a time by extending the bar stock out a little farther. This way, every time the operator opens the door, two parts are complete, reducing the amount of time per piece. **Figure 25-7** shows the work setup for producing two parts at a time. It appears that the stock is extended far out from the chuck jaws, but this is 3″ diameter material only sticking out approximately 5″ from the grip of the jaws. This is a safe length to diameter ratio.

The programmer knows that the part finishes at 1.75″ long and it is parted off with a .250″ wide cutoff tool. This means that the second part must be more than 2″ behind the first part. It will also be necessary to leave some stock on the second part face to machine that face clean and flat, so the second part will be started at Z–2.050 from the face of the first part.

As discussed in Chapter 24, before CNC turning centers became more advanced, work offsets were not available to define the work coordinate system. Work offsets were not needed because tool offsets were used to establish the work coordinates and the X0. position was always the spindle centerline. Earlier machines used a different method to create multiple parts. This was called a *work shift*. On older controllers, the G10 command allowed the programmer to use a work shift to make a single part and then shift all the tools down the Z axis to make a second part. It would be highly uncommon to shift the X axis, usually only the Z axis. For the setup shown in **Figure 25-7**, the block **G10 W–2.050** is used to shift the work plane in the negative Z axis direction. The W address designates an incremental measurement for the shift. At the end of the program, the G10 command must be used again to *undo* the shift.

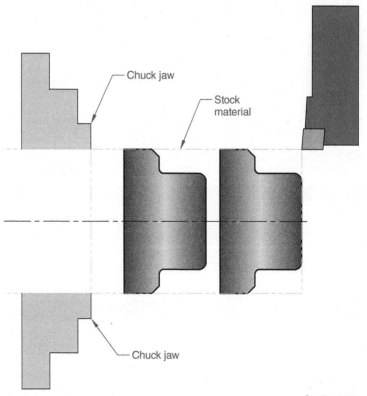

Goodheart-Willcox Publisher

Figure 25-7. Two parts can be produced in the same work setup by extending the bar stock farther from the chuck jaws while maintaining a safe length to diameter ratio.

The following is the main program and subprogram for producing the two parts. After the subprogram is called to machine the first part, the work shift is programmed. The second part is then machined and the work shift is reset.

O1111 (MAIN PROGRAM)
G28 U0. W0.
M98 P2875
G10 W–2.050
M98 P2875
G10 W2.050
M30

The only edit needed in the subprogram is changing the M30 function to the M99 function.

Using the subprogram with the work shift eliminates the need to calculate all of the axis positions needed for the second part. This approach achieves repeatability for the two parts.

It is common to produce multiple parts on a lathe through the use of a bar puller or bar feeder. A *bar puller* is a programmable accessory tool mounted to a turret that grips bar stock and pulls it to a specific length through the machine spindle. A *bar feeder* is a separate unit that holds bar stock and automatically loads it into the lathe for machining. Many lathe parts are produced from round bar stock. Depending on the machine size and configuration, a bar of stock material can be fed through the spindle and into the chuck jaws or collet system. After a part is complete, the bar can be advanced by pulling or pushing the bar to a position where a new part can be machined. In this case, a subprogram can be used to advance the bar and to machine the final part.

There are some different programming options for machining parts repeatedly on a lathe. A main program can be used in conjunction with a subprogram to machine the part with a second subprogram for the bar advance. That program must use the M00 program stop function or the M30 function to prevent the bar from feeding all the way out of the chuck and creating an unsafe workholding hazard. The second option is to create a loop for the machining subprogram and place the bar advance subprogram inside that loop. By calculating the bar length and part length, the number of loops can be determined and programmed. The different bar advancing options can be used to program the part used in the previous example. Refer to **Figure 25-6**.

The first example is a main program that goes to a machining subprogram, followed by a subprogram for the bar advance.

O1111 (MAIN PROGRAM)	
G28 U0. W0.	
M98 P2875	Machining subprogram
M98 P3235	Bar advance subprogram
M00	
M99 (or M30)	

The second option is a main program that loops a machining subprogram and another subprogram to advance the bar stock.

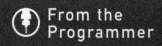

From the Programmer

Using Subprograms

The correct use and application of subprograms can significantly improve efficiency and reduce overall machining time. How subprograms are constructed and implemented will take some creativity and experience, but continue to look for applications that can benefit from this technique. Reducing programming time, setup time, and machining time will provide a decided advantage to any machine shop.

O1111 (MAIN PROGRAM)	
G28 U0. W0.	
M98 P2875 L22	Machining subprogram goes into bar advance subprogram
M30	

This program looks different because there is only one subprogram. Actually, inside subprogram 2875 is a block that reads **M98 P3235**. This block calls another subprogram before the controller re-enters the main program. A diagram of the program path illustrating this process is shown in **Figure 25-8**.

The L22 designation is given to repeat the program 22 times. The loop count is calculated from the length of the bar and the total length of the part. This will allow the machine to run virtually hands free, while 22 parts are machined.

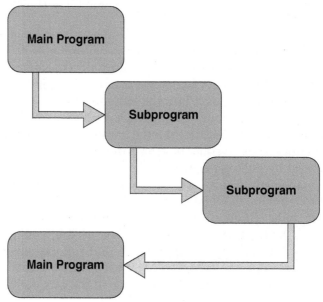

Goodheart-Willcox Publisher

Figure 25-8. A main program that branches out to a subprogram containing an additional subprogram. The controller returns to the main program after each subprogram is completed.

Chapter 25 Review

Summary

- The initial program where the machining cycle is started is called the main program. Any program that branches out from the main program is called a subprogram.
- A subprogram is called, or initiated, from the main program using the M98 function. This is followed by the P address and subprogram number. For example, the block **M98 P1234** will initiate program 1234 as a subprogram.
- The M99 function is used in the final block of a subprogram, commanding the controller to re-enter the main program.
- Subprograms are most often used to complete repetitive types of machining in multiple locations. The use of subprograms reduces program length and programming time, while reducing errors.
- Incremental positioning mode can be used in subprogramming to program repetitive operations by specifying incremental moves. The G91 command is used to activate incremental positioning. It is important to always be aware of the modes that are active in subprogramming.
- Subprograms can be called from inside another subprogram. As a programmer, keep track of where the program is and where it is headed.
- Subprograms can be used in CNC lathe applications for multiple part programming and bar advancing operations. The same rules that apply to mill subprogramming apply to lathe subprogramming.

Review Questions

Answer the following questions using the information provided in this chapter.

Know and Understand

1. The _____ is the initial program where the beginning of the cycle is initiated.
   A. subprogram
   B. main program
   C. macroprogram
   D. microprogram

2. *True or False?* A subprogram is best used when there are a variety of machining operations needed for irregular features.

3. Which code is used to end the main program?
   A. M99
   B. M98
   C. G91
   D. M30

4. Which code is used before the subprogram number, to send the main program into the subprogram?
   A. M99
   B. M98
   C. G91
   D. M30

5. *True or False?* A main program can be ended while in a subprogram using the M30 function.

6. Many subprograms require use of the _____ command to program incremental movements.
   A. G10
   B. G28
   C. G90
   D. G91

7. In CNC mill programming, what type of operation can be used in a subprogram?
   A. Drilling
   B. Tapping
   C. Pocketing
   D. All of the above.

8. On a Haas controller, which letter address is used to create a loop?
   A. K
   B. L
   C. M
   D. R

9. *True or False?* The G10 command is used to create work shifts.

10. *True or False?* If a G10 command is used in a program, the related designation must be undone at the end of the program by repeating the G10 command.

Apply and Analyze

1. Referring to the blanks provided below, give the correct codes that are missing for the milling program. The program uses subprograms to spot drill and drill 12 holes. The main program number is 0718, the spot drilling program number is 0824, and the drilling program number is 1020.

| O____ (SUBPROGRAM SAMPLE) |
| (T1|3/4 SPOT DRILL|H1|D1|DIA. – .75) |
| (T2|1/2 DRILL|H2|D2|TOOL DIA. – .5) |
| G20 |
| G0 G17 G40 G49 G80 G90 |
| T1 M6 |
| G0 G90 G54 X0. Y–1. S2037 M3 |
| G43 H1 Z1. M8 |
| ____ P____ L____ |
| G80 |
| T2 M6 |
| G0 G90 G54 X0. Y–1. S1910 M3 |
| G43 H2 Z1. M8 |
| ____ P____ L____ |
| G80 |
| G91 G28 Z0. |
| M30 |

2. Referring to the drawing and the blanks provided below, give the correct codes that are missing for the main program and the subprogram. There will be just one subprogram and it is used only to drill.

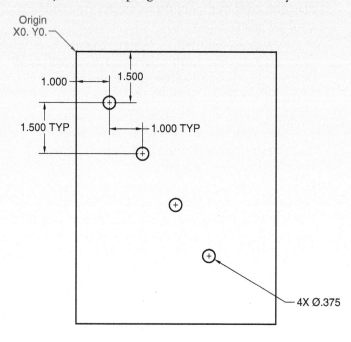

Main Program

| O0526 (MAIN PROGRAM) |
| (T1 3/8 DRILL – DIA. – .375) |
| G20 |
| G0 G17 G40 G49 G80 G90 |
| T1 M6 |
| G0 G90 G54 X__ Y__ S3217 M3 |
| G43 H1 Z1. M8 |
| ____ P____ ____ |
| G80 |
| G91 G28 Z0. |
| ____ |

Subprogram

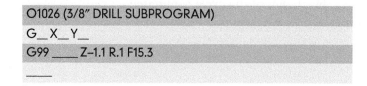

3. Referring to the drawing and the blanks provided below, give the correct codes that are missing for the main program to machine three parts on a lathe. The program uses subprograms to machine each part. The distance between the part faces is 1.140″. The main program number is 0804 and the subprogram number is 0731.

```
O____
G28 U0. W0.
____P____
G__ W____
____P____
G__ W____
____P____
G__ W____
M30
```

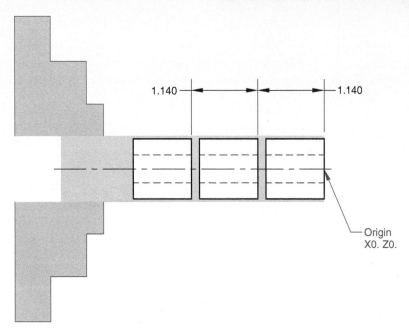

Critical Thinking

1. What are the benefits of repetitive processes in machining? How can repetitive processes be optimized in subprogramming?

2. Other than the operations discussed in this chapter, what are some other operations that can be used in a subprogram? Make a list and explain how the operations can be optimized with subprograms.

3. Subprogramming can be used in manufacturing many types of parts. However, certain industries lend themselves to a more widespread use of subprogramming. What are some industries that make use of the same types of parts, but in various sizes or with small differences between parts? List three industries that make use of parts where subprogramming techniques would be useful.

26 Subprogramming Techniques

Chapter Outline

26.1 Introduction
26.2 The Work Coordinate System
26.3 Using Work Offsets for Subprograms
26.4 Lathe Subprogramming
26.5 Using Subprograms for Multiple Parts and Fixtures
26.6 Layering Subprograms
    26.6.1 Layering Subprograms for Standard Operations
26.7 Contouring Subprograms
26.8 Using Subprograms in Multiaxis Machining

Learning Objectives

After completing this chapter, you will be able to:

- Explain how to use work offsets with subprograms.
- Explain how multiple work offsets can be used to simplify mill programming.
- Describe how work offset locations can be calculated from print dimensions.
- Explain how multiple work offsets can be used to simplify lathe programming.
- Describe techniques for using work offsets in conjunction with subprograms in multiple part setups.
- Explain why subprogramming is particularly useful in multiaxis machining.

Key Terms

impeller
layering
work coordinate system (WCS)
work envelope
work offset

26.1 Introduction

In Chapter 25, basic subprogramming methods were introduced. As you learned in that chapter, there are specific codes and rules that apply when calling a subprogram from a main program. The examples in Chapter 25 were based on the use of incremental positioning to program basic machining operations for patterned features. In many cases, particularly when making shifts in more than one axis, there are other alternatives to building the main program and subprogram. It is important to note that either approach can be used successfully, and deciding which one to use might just be a matter of programmer preference. This chapter discusses the use of work coordinate systems in subprogramming for repetitive machining motions, multiple parts, and multiaxis machining.

26.2 The Work Coordinate System

As discussed in previous chapters, the *work coordinate system* (*WCS*) is the system of point specification that defines where the zero point of the part physically sits within the limits of the machine. Applications in previous chapters were primarily shown using a single work offset (the G54 work offset) to designate the distance from machine home to the part zero location. A *work offset* is a coordinate setting that establishes the location of the work origin relative to the machine home position. Fortunately for the programmer, there are many work offsets, and they can be used in a multitude of ways to make subprogramming more efficient.

Before work offsets are used in a subprogram, it is important to note the different types and number of work offsets that can be utilized. Almost every CNC machine, no matter what manufacturer or age, can use the standard six work offsets. These are the G54, G55, G56, G57, G58, and G59 work offsets. After these, it may depend on the age and type of machine as to what work offsets are available. Haas machine controllers can use additional work offset designators from G110 through G129. Most Haas machines built after 2008 can also use the designators G154 P1 through G154 P99, enabling an additional 99 offsets. Most Fanuc controllers have additional work offset settings using the commands G54.1 P1 through G54.1 P48, with newer controls going all the way to G54.1 P300. As always, consult the programmer's manual for the available work offsets on the machine, but be aware that all modern machines provide many work offsets.

26.3 Using Work Offsets for Subprograms

Now that you have an understanding of the work offsets that are available for use, the following example shows how work offsets can be used to the programmer's advantage. **Figure 26-1** shows one of the example parts from Chapter 25. This part has a pattern of 64 holes arranged in four rows.

In the Chapter 25 example, the top-left corner was established as the G54 work offset origin, and then the program moved to the first row of holes at the X0. Y–1 position. From there, the subprogram was used to shift the movement 1″ incrementally 16 times using the block **G91 X1**. A different strategy will be used for this example. It is important to note that neither of the programs is "better," or faster, they are just different strategies. In this program, the G54 work offset will still be assigned at the

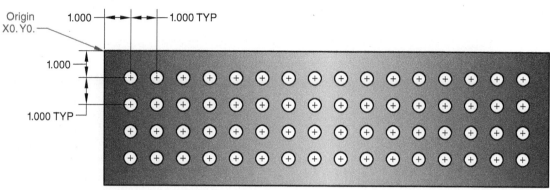

Figure 26-1. A part with a hole pattern made up of 64 holes.

top-left corner, but the first hole of each row will then be used to establish individual work offsets (using the G55–G58 work offsets). See **Figure 26-2**.

First, study the structure of the main program. The main program uses a subprogram for spot drilling and drilling each row of holes. The machine settings for each work offset will be discussed after the part program and subprogram. Notice that the main program is using the G90 command to activate absolute positioning mode, which is generally preferred. This is the mode used in the subprogram.

O1425 (Main Program)	
(T1|3/4 SPOT DRILL|H1|D1|DIA. – .75)	
(T2|1/2 DRILL|H2|D2|TOOL DIA. – .5)	
G20	
G0 G17 G40 G49 G80 G90	
T1 M6	3/4" spot drill
G0 G90 G55 X0. Y0. S2037 M3	Locate at first hole, first row
G43 H1 Z1. M8	
M98 P3535	Spot drill subprogram
G56 X0. Y0.	Locate at first hole, second row
M98 P3535	Spot drill subprogram
G57 X0. Y0.	Locate at first hole, third row
M98 P3535	Spot drill subprogram
G58 X0. Y0.	Locate at first hole, fourth row
M98 P3535	Spot drill subprogram
T2 M6	(1/2" drill)
G0 G90 G55 X0. Y0. S1910 M3	Locate at first hole, first row
G43 H2 Z1. M8	
M98 P3536	Drill subprogram
G56 X0. Y0.	Locate at first hole, second row
M98 P3536	Drill subprogram
G57 X0. Y0.	Locate at first hole, third row
M98 P3536	Drill subprogram
G58 X0. Y0.	Locate at first hole, fourth row
M98 P3536	Drill subprogram
G91 G28 Z0.	
M30	

From the Programmer

Programming Modes

You may notice that there is no G80 command in the main program. That is because it is used in the subprogram. It is programmed that way so the move to the next work offset will not result in running the spot drilling or drilling canned cycle twice. It cannot be stressed enough—keep track of what modes are active and what the program is doing and utilize the main program and subprograms to your full advantage.

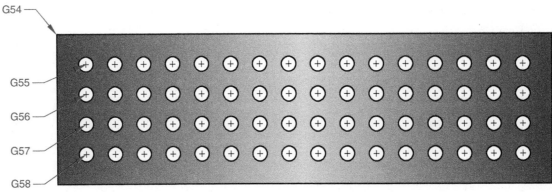

Figure 26-2. Each row of holes can be machined by establishing a work offset at the first hole center in each row.

The following shows the two subprograms. Notice they are fully self-contained and use only absolute positioning moves. This is because the G90 command is used in the main program and remains active in the subprograms.

O3535 (3/4″ Spot Drill Subprogram)
G99 G81 Z−.26 R.1 F32.
X1.
X2.
X3.
X4.
X5.
X6.
X7.
X8.
X9.
X10.
X11.
X12.
X13.
X14.
X15.
G80
M99

O3536 (1/2″ Drill Subprogram)
G99 G83 Z−1.15 R.1 Q.3 F15.3
X1.
X2.
X3.
X4.
X5.

Continued

```
X6.
X7.
X8.
X9.
X10.
X11.
X12.
X13.
X14.
X15.
G80
M99
```

The two subprograms are identical except for the programmed drilling canned cycle. The main program positions the tool at the work offset X0. Y0. position. The first block of the subprogram is the drilling canned cycle (G81 or G83), initiating a drilling operation at the current location. The G80 command is used at the end of the canned cycle to prevent a drilling cycle at the next move to a new work offset. It is also possible to use a looped incremental move command inside the subprogram to complete the same drilling cycle. Once the concept is understood, some experimentation with the main program and subprograms will lead to the best implementation of a specific machining strategy and toolpath.

For clarification, the following discussion is provided to explain the settings on the machine controller's work offset page and how the coordinates for each work offset can be determined. For the part shown in **Figure 26-2**, the top-left corner of the part was set as the origin location for the G54 work offset. The G55 work offset is exactly +1.000″ in the X axis direction and –1.000″ in the Y axis direction. The G56, G57, and G58 work offsets are all located –1.000″ in the Y axis direction, respectively.

To establish the G54 work offset location, the intersection of the X and Y axes at the top-left corner was measured to be at the machine coordinates X–9.250 Y–10.550. When that position is known, the remaining work offset locations can be calculated. The G55 work offset is at X–8.250, or G54 X + 1.000″, and at Y–11.550, or G54 Y–1.000″. Each of the remaining G56, G57, and G58 work offset positions is at the same X axis location, and will move –1.000″ in the Y axis, respectively. See **Figure 26-3**.

The following example shows another part with patterned features that can be machined using repetitive toolpaths in subprograms. **Figure 26-4** shows one of the example parts from Chapter 25. This is a part with multiple pockets. By using subprograms and individual work offsets, the pockets can be programmed with absolute positioning moves, simplifying the programming process.

Because the drawing is dimensioned from the top-left corner to the center of each pocket, the work offsets should be set up with the same strategy. The top-left corner of the part will be used to establish the G54 work offset and the center of each pocket will represent the location of an alternate work offset. See **Figure 26-5**.

Debugging the Code

Work Offset Designation

Is there a reason to keep setting the G54 work offset at the top-left corner of the part and set the other work offsets relative to the G54 offset in numerical order? No, it is just an easier way to keep track. Any work offset number can be used to designate any position. Just make sure the machine controller has a defined location to know where the machining is supposed to occur. Keep your programming code recognizable and easy to understand for the setup person and machine operator.

<< WORK PROBE		WORK ZERO OFFSET		WORK PROBE >>
G CODE	X AXIS	Y AXIS	Z AXIS	
G52	0.	0.	0.	
G54	-9.250	-10.550	0.	
G55	-8.250	-11.550	0.	
G56	-8.250	-12.550	0.	
G57	-8.250	-13.550	0.	
G58	-8.250	-14.550	0.	
G59	0.	0.	0.	
G154 P1	0.	0.	0.	
G154 P2	0.	0.	0.	
G154 P3	0.	0.	0.	

Goodheart-Willcox Publisher

Figure 26-3. The work offset page for the machine controller shows the stored settings for each work offset. The G54 work offset is established by measuring a location on the part. The remaining work offsets used in the program (G55–G58) are calculated from the G54 work offset location.

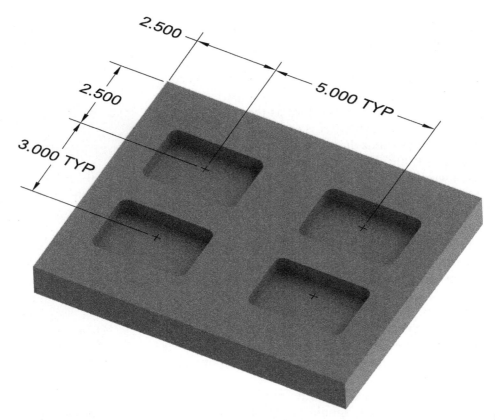

Goodheart-Willcox Publisher

Figure 26-4. A part with a pattern of closed pockets. The pockets can be milled by using the same subprogram for each pocket with a work offset established at the pocket center.

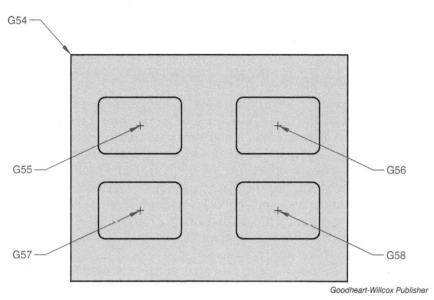

Figure 26-5. The center of each pocket is used to establish a work offset for programming absolute positioning moves.

The following shows the main program with the subprograms used for machining.

O1020 (POCKET SUBPROGRAM)	
(T1|1/2″ END MILL)	
G0 G90 G55	
M98 P1968	Subprogram for first pocket
G0 G90 G56	
M98 P1968	Subprogram for second pocket
G0 G90 G57	
M98 P1968	Subprogram for third pocket
G0 G90 G58	
M98 P1968	Subprogram for fourth pocket
G91 G28 Z0.	
G91 G28 Y0.	
M30	

Notice in the main program there are no startup block, positioning, or tool change commands. In this case, all of that information will be contained in the subprogram.

26.4 Lathe Subprogramming

Subprogramming techniques used in lathe programming are similar to those used in mill programming. As discussed in Chapter 25, on older machine controllers, the G10 command was used to program a work shift in order to shift the tool offsets along the Z axis to create multiple parts. Most newer lathes provide the option to use the G54–G59 work offsets. Not all lathes have work offsets, but almost all machines produced after 2005 will have this capability.

There are two different ways to use work offsets in lathe programming, and both involve manipulating the Z axis position. The first method is to

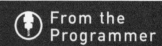

From the Programmer

Development of Work Offsets

The G54–G59 work offsets were first introduced on the Fanuc controller in 1980–81. Previously, the machine had only one offset from machine home and many programs were written in incremental mode only. Work offsets made programming much easier, especially for multiple parts and part fixtures. For a long time, there did not seem to be a need for this type of world coordinate system on a lathe. But around 2005, machine builders started adding the G54 command to lathes to assist in multiaxis applications and multiple part production. Now it's a staple of machine controllers.

set all tool offsets from the face of the finished part and leave the G54–G59 work offset settings at X0. Z0., basically bypassing the work offsets. The second method is to set the tool offsets to a known plane, such as the chuck face or a tool setter, and measure back to the face of the finished part. See **Figure 26-6**. That measurement is then entered as a positive (+) number in the Z axis column of one of the G54–G59 designations on the controller's work offset page. In **Figure 26-6**, the G54 work offset is designated.

Regardless of the method used to set tool offsets, work offsets can be leveraged in a number of ways for use in lathe subprograms. The most common use of work offsets on a lathe is to make multiple parts from a length of bar stock, but there are many other instances where they are used. For example, multiaxis parts, or families of parts where the inside or outside diameter may repeat, can lend themselves to subprogramming. Keep in mind that subprogramming can be used for many applications other than multiple part production.

Figure 26-7 shows three parts machined from an extended piece of bar stock. The first piece is designated as G54. Be aware that G54 is an arbitrary designation, meaning that the first piece could be programmed using any work offset designator, such as G55, G59, or G54.1 P1. It does not matter as long as the programmed location accounts for the tool offsets. This is where a repetitive process and a good setup sheet will help clarify the programming strategy. In this example, the parts finish at 1.000″ long, with a .125″ cutoff operation, and .015″ of remaining material to face on the second part. The work offsets will be measured at Z–1.140″ per position. The program is as follows.

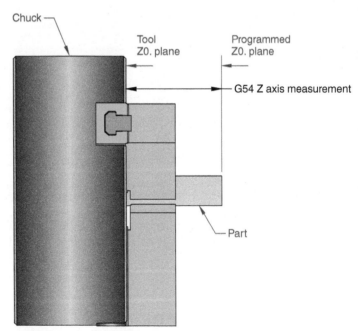

Goodheart-Willcox Publisher

Figure 26-6. When tool offsets are established from the face of the chuck, the Z axis setting of the work offset must account for the measured distance between the chuck and the work offset location.

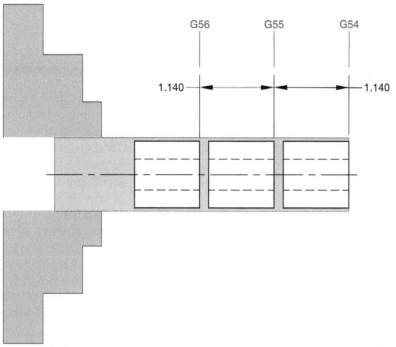

Goodheart-Willcox Publisher

Figure 26-7. A work setup for multiple parts on a lathe with work offsets programmed to set the work plane on the finished face of each part.

O3292 (MAIN PROGRAM)	
G28 U0. W0.	
G54	
M98 P3293	First subprogram
G55	
M98 P3293	Second subprogram
G56	
M98 P3293	Third subprogram
G28 U0. W0.	
M30	

Note the following items from this program:

1. The program starts and ends by sending the machine home with the G28 command. This is a safety move to make sure the machine is in a safe position to start the programmed movements.

2. The work offset commands (G54–G56) are on blocks by themselves. There are no tool length offsets established, so these blocks just activate the designated work offset without making a move.

3. Nothing is commanded in the main program, so all tool changes, spindle activations, and safe entries and exits need to be programmed in the subprogram.

When using multiple work offsets in either lathe or mill programming, be aware that changing to one of the G54–G59 work offset designations moves all the tools to that commanded work offset. Changing the offset for a tool will change the tool position in all subprograms where that tool is utilized.

26.5 Using Subprograms for Multiple Parts and Fixtures

To this point, the text has covered the use of subprograms for different features on the same workpiece. In mill programming, a more common application is using subprograms on multiple parts and parts held by fixtures. The *work envelope*, or operating space, of a CNC milling machine is large enough to accommodate more than one vise or a fixture that can hold more than one part at a time. Using the majority of the work envelope can improve machining efficiency. Work offsets can be used with subprograms to simplify programmed operations in multiple part setups.

Figure 26-8 shows an example of a setup for relatively small parts. The programmer realizes that 10 parts can be easily held by a fixture on the work table, providing a more efficient cycle time. If the programmer attempts to create the program with only one work coordinate system, the setup person and operator will have to figure out how to align the 10 parts perfectly in the X and Y axes each time before running the cycle. This is possible with a complicated fixture or even a series of vises that are aligned perfectly, but the chances of repeating this successfully with each set of parts are very small.

By simply designating a point on the part where part zero will be established, writing the program to that point, and establishing work offsets at those points, the entire process is simplified. Part zero can be established at any location, as long as the programmed part zero and work offset zero are at the same location. For example, the top-left corner, top-right corner, or center of the part can be used for part zero, and that same location needs to be used for the work offset zero point. The following program uses independent work offsets and subprograms to produce the parts.

O6866 (MAIN PROGRAM)	
G91 G28 Z0.	
G90 G54	
M98 P2025	First subprogram
G90 G55	

Continued

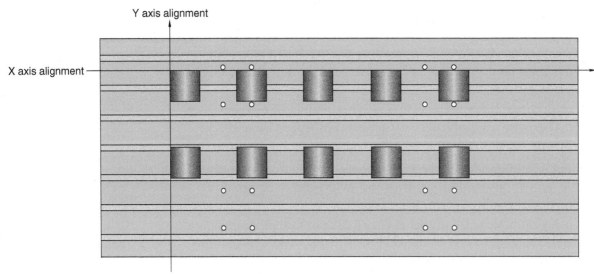

Goodheart-Willcox Publisher

Figure 26-8. A setup of parts in which multiple work offsets can be used to define the work coordinate system for each part.

M98 P2025	Second subprogram
G90 G56	
M98 P2025	Third subprogram
G90 G57	
M98 P2025	Fourth subprogram
G90 G58	
M98 P2025	Fifth subprogram
G90 G59	
M98 P2025	Sixth subprogram
G90 G110	Move to secondary work offset
M98 P2025	Seventh subprogram
G90 G111	
M98 P2025	Eighth subprogram
G90 G112	
M98 P2025	Ninth subprogram
G90 G113	
M98 P2025	Final subprogram
G91 G28 Z0.	
G28 Y0.	
M30	

Note the following items from this program:

1. The main program is setting absolute positioning mode with the G90 command and calling the work offset (G54–G59 and G110–G113).

2. The reason the main program commands the work offset is that in this case, the subprogram cannot command the work coordinate system. If it did, the subprogram would overwrite the WCS and result in machining on the wrong part.

3. In this program, there is no positioning move with the work offset (for example, G54 X0. Y0.), but this could be added if it is appropriate for the subprogram.

4. This type of operation is usually performed with parts mounted in vises or on a fixture. See **Figure 26-9**.

Pixel B/Shutterstock.com

Figure 26-9. A fixture setup for machining multiple parts on a CNC milling machine.

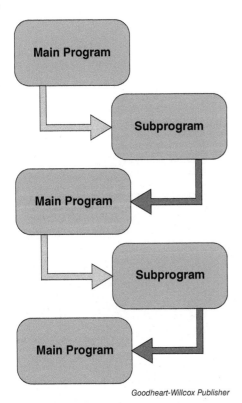

Figure 26-10. A main program calls a subprogram. Once completed, the subprogram must return back to the main program.

26.6 Layering Subprograms

The previous examples in this chapter have covered techniques for transitioning from a main program into a subprogram, and then returning to the main program. This programming path is shown in **Figure 26-10**.

Although this is the most common approach to subprogramming, it is not the only way to use subprograms. It is possible to go from a subprogram into another subprogram. The number of times this can be accomplished will depend on the machine controller. At a minimum, all machine controllers can go four subprograms deep, and the newest controllers and PC-based controllers can go as deep as 12 subprograms. The process of nesting subprograms is called *layering*. See **Figure 26-11**.

This is not a common practice in programming, but there are some powerful applications to make programming more efficient. Machine shops that have a family of similar products can use layered subprograms. For example, a machine shop that manufactures hydraulic valves might have 10 different valve configurations with common forms. Instead of programming all of the forms individually, they can be saved as individual subprograms. Another common application for layered subprograms is probing cycles. These will be covered in later chapters, but using a subprogram to call a probing cycle to establish a WCS inside a subprogram is common. Macro programming is another technique used in layered subprograms. Macros are discussed later in this text.

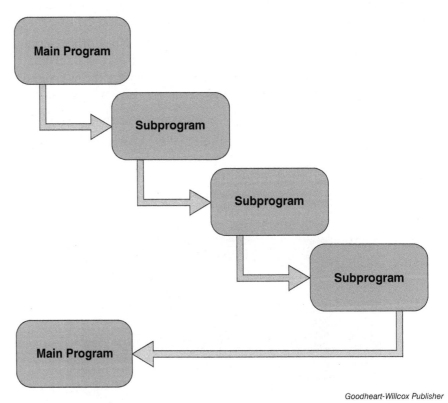

Figure 26-11. Subprograms can be layered to allow for transitioning from a main program into a subprogram, and then transitioning into another subprogram.

26.6.1 Layering Subprograms for Standard Operations

There are many common machining operations that can benefit from layered subprograms. Operations as simple as drilling and tapping a hole can be turned into a set of subprograms for ease in programming. In a standard 1/4–20 tapping operation, there are three things that should happen every time: spot drill the hole to a depth that chamfers the hole, drill the hole with a #7 drill, and tap the hole with a 1/4–20 tap. Because this is the way the programmer machines these tapped holes, every time, this program can be turned into a single operation with layered subprograms.

This programming strategy can be demonstrated by first considering a single hole operation and then expanding it to multiple hole locations for added time savings and efficiency. **Figure 26-12** shows a part with a 1/4–20 tapped hole through the part. The part is 1/2″ thick. For this example, the work origin is established at the top-left corner of the part.

The following is a very short main program. Its only purpose is to get into the first subprogram (the spot drilling program) and then exit the program.

| O0615 (Main Program to Drill and Tap 1/4–20 Holes) |
| (Subprograms are O0616, O0617, and O0618) |
| G20 |
| G0 G17 G40 G49 G80 |
| G91 G28 Z0. |
| M98 P0616 (Spot Drill Subprogram) |
| G91 G28 Z0. |
| M30 |

> **From the Programmer**
>
> **Developing Subprograms**
>
> For this approach to be really beneficial, a few assumptions are being made. First, this operation is being done on a regular basis, with many different parts. Secondly, the machine is going to be set up with three tools in the same location every time—for example, T1, T2, and T3. In later chapters, macro variables are discussed. These can be used in the main program for specifying hole depths or drilling locations to really make this operation easy.

The difference between this main program and previous examples is that only one subprogram is being initiated from the main program. Also notice that there is no specified work offset in the main program. It is possible to specify the work offset in the main program, but for this example it will stay in the subprogram only. The following is the spot drilling subprogram (O0616).

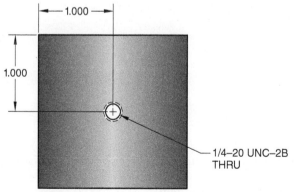

Goodheart-Willcox Publisher

Figure 26-12. A part with a 1/4–20 tapped hole. The hole requires a spot drilling, drilling, and tapping operation.

O0616 (Spot Drill Subprogram)
G0 G90 G54 X1. Y–1.
T1 M6
G43 H1 Z1. M3 S2250
G81 Z–.140 R.1 F5.
G80
M98 P0617 (#7 Drill Subprogram)
M99

In this subprogram, all of the tool data, spindle engagement, and drilling cycle programming is included. Notice that the next subprogram is initiated before the M99 function. Although this subprogram is moving into another subprogram, the M99 function is still required. The main program is "stepping down" through subprograms, but when the last one is complete, the controller will step back up through the subprograms. In each case, the next line after returning to the previous subprogram must have the M99 function to send it "up" to the previous program until it reaches the main program again.

The drilling subprogram is very similar.

O0617 (#7 Drill Subprogram)	
G0 G90 G54 X1. Y–1.	
T2 M6	
G43 H2 Z1. M3 S3350	
G83 Z–.65 R.1 Q.1 F9.	Drill travels .65 deep to drill through the material
G80	
M98 P0618 (1/4–20 Tap Subprogram)	
M99	

The only changes to the drilling subprogram are the tool numbers, the drilling cycle, and the next subprogram call.

The last subprogram is the tapping cycle. Again, it is very similar, but there are no additional subprograms to initiate.

O0618 (1/4–20 Tap Subprogram)
G0 G90 G54 X1. Y–1.
T3 M6
G43 H3 Z1. M3 S500
G84 Z–.65 R.1 F25.
G80
M99

Notice that this subprogram is canceling canned cycles with the G80 command before it leaves the subprogram. After the G80 command, the M99 function will return the subprogram back to the previous subprogram (O0617), where the controller will read the M99 function and continue to progress back to the main program. At the end of the main program, there is a return to machine home and the M30 function ends the program.

This may not seem like a more efficient way to program for a single tapped hole, and it probably is not, but the ability to adapt this series of programs to a common machining operation can be very beneficial. As the configuration changes, the subprograms are easily modified.

Figure 26-13 shows a part with a pattern of 16 holes oriented in a straight line. For this part, the main program stays the same and the subprograms will have a looped incremental move to complete all 16 holes. The subprogram can be arranged in a few different ways, but the following is one possible solution.

O0616 (Spot Drill Subprogram)	
G0 G90 G54 X1. Y–1.	
T1 M6	
G43 H1 Z1. M3 S2250	
G81 Z–.140 R.1 F5.	
G91 X1. L15 (K15)	L address used to loop on a Haas controller; K address used to loop on a Fanuc controller
G80	
M98 P0617	
M99	

Notice that the only line added is the G91 line. Everything else can remain the same. This can be used for holes in the Y axis direction or even an angled hole pattern. The machine controller is in incremental positioning mode (G91) when this subprogram ends, but the next subprogram commands absolute positioning mode (G90) in the first line. It is acceptable to add the G90 command after the G80 command for safety if desired.

The same technique can be used for multiple parts held on a CNC machine table or fixture while using multiple work offsets. The difference would be moving the work offset designations into the main program and out of the subprograms. For example, if four of the parts shown in **Figure 26-13** were made in one operation, the main program would look similar to the following.

O0615 (Main Program to Drill and Tap 1/4–20 Holes)	
(Subprograms are O0616, O0617, and O0618)	
G20	
G0 G17 G40 G49 G80	
G91 G28 Z0.	
G90 G54	First part
M98 P0616	
G90 G55	Second part

Continued

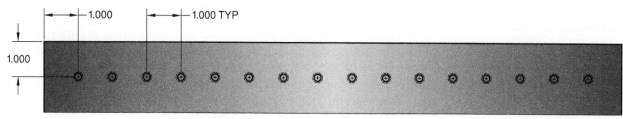

Figure 26-13. A part with a straight-line hole pattern made up of 16 holes spaced equally.

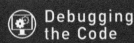

Debugging the Code

Programming Modes

Notice the main program is in absolute positioning mode (G90) to command work offsets. When the subprogram is entered, absolute positioning is still active. The machine moves to the first position at X1. Y–1. and the G81 cycle is initiated. It is not until the following line that incremental positioning mode (G91) is programmed to perform the looping cycle. After the cycle is completed, the G90 command is re-entered before the controller enters the next subprogram. There are many things happening rapidly, so keep track of which programming modes are active at every step.

M98 P0616	
G90 G56	Third part
M98 P0616	
G90 G57	Fourth part
M98 P0616	
G91 G28 Z0.	
M30	

The subprograms are the same, except the **G0 G90 G54** code is removed so the work coordinates are not overwritten from the main program. The subprograms can be written as follows.

O0616 (Spot Drill Subprogram)
T1 M6
X1. Y–1.
G43 H1 Z1. M3 S2250
G81 Z–.140 R.1 F5.
G91 X1. L15 (K15)
G80
G90
M98 P0617
M99

26.7 Contouring Subprograms

The subprogram examples in this chapter have been based on hole machining operations. These are common applications for subprograms. However, there are many more applications, and they are only limited to the programmer's creativity.

Remember that the best application for a subprogram is any repetitive function or toolpath. **Figure 26-14** shows an example of a part that can use a subprogram for an operation other than drilling. This is a rectangular part with a raised boss feature. The boss has rounded and angled sides. The outside contour of the boss needs to be machined in multiple steps. The work origin for this part is established at the top-left corner of the part.

The best practice for machining contours is to send the tool as deep as possible in the Z axis, and then make smaller stepovers in the X or Y axis direction. This will allow for increased tool flute engagement and higher feed rates. The boss feature on this part is .500″ deep, so the tool will be programmed to Z–.495 to allow for a finish pass, and a 1/2″ end mill will be used. The programmer wants to make five passes with a distance of .100″ between each pass. Cutter compensation will be used, and the path will look like **Figure 26-15**.

The key to using subprograms in this application is cutter compensation. Instead of calculating all the positions around the part and keeping the tool away at the right distance, cutter compensation will be activated in the main program, and the final contour shape will be programmed in the subprogram. The main program is as follows.

Goodheart-Willcox Publisher

Figure 26-14. A part with a raised boss feature machined in a contouring operation by making repetitive cutting passes with an end mill.

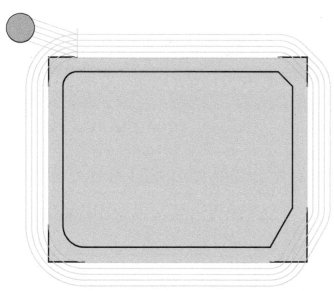

Figure 26-15. Simulated cutting path for the contouring operation used to create the raised boss.

```
O2706 (Main Program)
(T1 | 1/2 FLAT END MILL | H1 |)
G20
G0 G17 G40 G49 G80 G90
T1 M6
G0 G90 G54 X-.5 Y.5 S5500 M3
G43 H1 Z2. M8
Z.2
G41 D1 X.5 Y-.25 F110.
M98 P2716 (Subprogram)
G41 D2 X.5 Y-.25
M98 P2716 (Subprogram)
G41 D3 X.5 Y-.25
M98 P2716 (Subprogram)
G41 D4 X.5 Y-.25
M98 P2716 (Subprogram)
G41 D5 X.5 Y-.25
M98 P2716 (Subprogram)
G91 G28 Z0.
M30
```

In this example, cutter compensation is initiated in the main program, before the subprogram. During setup, the D offset values are recorded on the tool offset page as follows:

- D1 = .900
- D2 = .800
- D3 = .700
- D4 = .600
- D5 = .500 (The actual cutter size)

The subprogram will be used to control the contour coordinates and the Z axis position of the tool.

```
O2716 (Subprogram)
G1 Z-.495 F50.
X4. F110.
X4.25 Y-.5
Y-2.5806
X3.8635 Y-3.25
X.625
G2 X.25 Y-2.875 I0. J.375
G1 Y-.5
G2 X.5 Y-.25 I.25 J0.
G1 G40 Y.25
Z.2
G0 Z2.
X-.5 Y.5
M99
```

In the subprogram, the tool is already positioned in the X and Y axis locations to begin with the correct cutter compensation offset value initiated. Not only does this reduce programming time, but it simplifies editing. A change to the subprogram will be reflected in all of the passes, and an adjustment to the cutter compensation amounts will change the depths of cuts.

Figure 26-16 shows another part that serves as an example of using a subprogram for a contouring operation. This time, instead of altering the X or Y position with the subprogram, the Z axis will be used for multiple depth cuts. This is a square part with a circular groove that is .375″ (3/8″) wide with a 2″ diameter centerline. The .500″ depth makes it difficult to complete the cut in one or two depth cuts, so multiple steps will be required.

To create the best possible groove, a tool one size smaller than the 3/8″ groove (5/16″) will be used to rough the groove to depth, and then a finishing tool can be used to complete each wall of the groove. A straight plunge into a piece of material with a flat end mill is always the worst-case scenario. The second-worst scenario is making the full diameter cut with no relief, or no way for the chips to evacuate. For this part, both of those scenarios will be happening. For this example, the tool will plunge .050″ deep into the groove, make a full 360° circular cut, and repeat that motion 10 times for the .500″ depth cut. The main program will position the tool in the starting location and then the subprogram will make only the motion necessary for the groove machining. The main program is as follows.

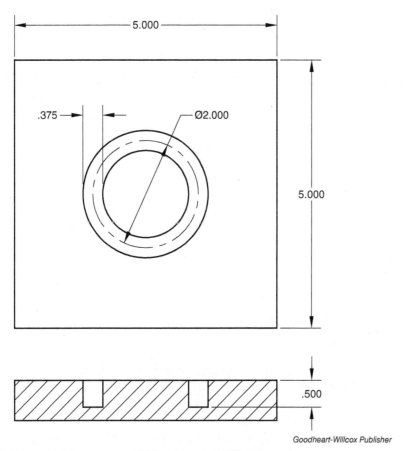

Figure 26-16. A part with a groove .375″ wide and .500″ deep.

O2707 (Main Program)	
(T1 | 5/16 FLAT END MILL)	
G20	
G0 G17 G40 G49 G80 G90	
T1 M6	
G0 G90 G54 X3.5 Y–2.5 S3600 M3	Starting location
G43 H1 Z.2 M8	
G1 Z0. F25.	
M98 P2717 L10 (Subprogram)	
G0 G90 Z1.	
G91 G28 Z0.	
T2 M6	Proceed to finishing tool
...	
...	
...	
M30	

The following is the subprogram (O2717).

O2717 (Subprogram)	
G91 G1 Z−.05 F5.	Incremental plunge amount into material
G3 I-1.0 F10.	Full circle move for 2″ diameter groove
M99	

This is a very simple subprogram—the end mill just plunges down in the Z axis and completes a cut in a full circle. If it is determined that the cut is too deep or too shallow, it is easily edited. Just ensure that the plunge amount in the Z axis is multiplied by the number of passes to cut the desired depth (.050 × 10 repeat cuts in this case).

26.8 Using Subprograms in Multiaxis Machining

Often, multiaxis machining is thought of as a very complicated process where all of the axes of the machine are moving together and the tool is chasing around the material in all directions. But, most multiaxis machining processes consist of rotating the material to a desired location and then performing standard machining operations. If a single operation, or toolpath, is repetitive between rotational positions, it can be a good candidate for the use of subprograms.

One of the most common parts machined on a five-axis mill is the impeller. See **Figure 26-17**. An *impeller* is a rotating device made up of a series of vanes that force fluid or air outward to create centrifugal force. There are many different types of impellers and their uses vary, but generally, all the vanes have the same shape or form. This can be an opportunity to use a subprogram to program a single vane and then rotate that program to multiple locations.

In this type of programming, either separate work offsets can be used for each rotation, or the same work offset can be used with varying rotations. The following program uses multiple work offsets.

From the Programmer

Subprogram Applications

The possibilities for subprograms and layered subprograms are almost endless, depending on the application and creativity of the programmer. Continue to look for repetitive operations and determine if there is a time savings in programming. It is even possible to turn processes such as tool changes into a subprogram. Startup blocks can also be programmed as subprograms. In the following chapters, probing operations and operations using macro variables are discussed. These processes are suitable for subprograms as well.

Dmitry Kalinovsky/Shutterstock.com

Figure 26-17. Impeller vanes are common features produced on multiaxis machines.

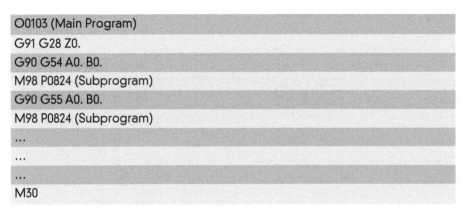

In this program, the orientation of the A axis and B axis is stored in the work offset and separate work offsets are used for each rotation. In the following program, the orientation of the A axis and B axis is set to zero in the G54 work offset and rotation is programmed from that location. The main program will initiate rotation and the subprograms will repeat machining for one vane.

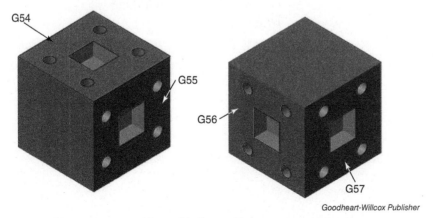

Another type of operation performed on a multiaxis machine is machining mirrored patterns on alternate sides, or faces, of a part. See **Figure 26-18**. These patterned features are handled the same way as previously discussed, either by using different work offsets or shifting back to the main program for the rotational value. Neither method is "better" or more efficient. They are just different techniques used to attack the same problem.

Goodheart-Willcox Publisher

Figure 26-18. On parts produced in multiaxis machining, patterned features on different faces can be programmed by using multiple work offsets or rotating the machine axes in conjunction with the use of subprograms.

Chapter 26 Review

Summary

- The ability to use different work offsets with subprograms can substantially reduce cycle times and improve machining efficiency.
- Depending on the controller type, most modern CNC machines have many more work offsets besides the G54–G59 designations. Consult the programming manual for the controller in use for specific work offset designations.
- Using work offsets with subprograms is an alternative to programming in incremental positioning mode. Programming in absolute positioning mode will reduce programming errors.
- In lathe programming, work offsets can be used for machining multiple parts from a single piece of bar stock. This reduces cycle times and simplifies production of multiple parts.
- Often in CNC milling, multiple parts are produced simultaneously using multiple vises or a fixture. In these cases, using subprograms and multiple work offsets will reduce setup time and improve efficiency.
- The process of nesting a subprogram in another subprogram is called layering.
- In multiaxis machining, repetitive operations on work planes oriented at different angles can be programmed using subprograms.

Review Questions

Answer the following questions using the information provided in this chapter.

Know and Understand

1. Which of the following G-code commands is *not* used to designate a work offset?
   A. G54
   B. G32
   C. G115
   D. G54.1 P110

2. *True or False?* A work offset must be defined inside a subprogram for it to be initiated.

3. When using subprograms, which work offset has to be used to identify the work origin for the material?
   A. G54
   B. G54.1
   C. G55
   D. Any work offset designator can be used.

4. The M98 function tells the machine to go to ____.
   A. the main program
   B. a subprogram
   C. the previous program
   D. machine home

5. Which of the following is a possible benefit to using work offset designators with subprograms?
   A. Faster setup.
   B. Easier operation.
   C. Simplified programming.
   D. All of the above.

6. *True or False?* Coordinates for work offset locations are entered on the tool offset page of the controller.

7. On a CNC lathe, if work offsets are not available, the _____ command can be used for multiple part programming.
   A. G32
   B. G90
   C. G10
   D. G91

8. What is the primary reason to have the G90 command and work offset initiated from the main program?
   A. It allows using the same subprogram for all work offset positions.
   B. It allows using incremental moves in a subprogram.
   C. It must be used to make the subprogram work.
   D. None of the above.

9. *True or False?* A subprogram that is called by another subprogram is referred to as a layered subprogram.

10. When is it advantageous to use subprograms in multiaxis machining?
    A. When machining patterned features on different faces.
    B. When programming identical toolpaths on different angular work planes.
    C. When programming time can be minimized.
    D. All of the above.

11. *True or False?* In a multiaxis machining program with subprograms, separate work offsets can be used to rotate the work plane, or one work offset can be utilized and A and B values can be used to define rotational coordinates.

12. What are the primary benefits of using subprograms with work offsets?
    A. Faster setups and more efficient programming.
    B. Faster machining times.
    C. More accurate machining.
    D. All of the above.

Apply and Analyze

1. Referring to the drawing and the blanks provided below, give the correct codes that are missing for the main program. The main program is O3565 and the subprogram is O2244. The G54 work offset is used to initially set up the workpiece material location.

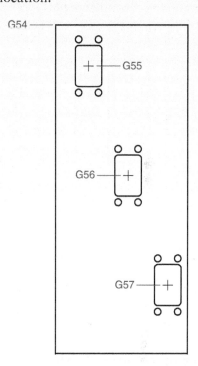

| _____ (Main Program) |
| G91 G28 Z0. |
| G90 G54 X0. Y0. |
| _____ |
| _____ _____ (Subprogram) |
| G90 _____ |
| _____ _____ (Subprogram) |
| G90 _____ |
| _____ _____ (Subprogram) |
| G91 G28 Z0. |
| G91 G28 Y0. |
| _____ |

2. Referring to the drawing of the lathe setup and the blanks provided below, give the correct codes that are missing for the subprogram and main program. The subprogram is O2244 and the main program is O5255. The G54 work offset is used for the first part.

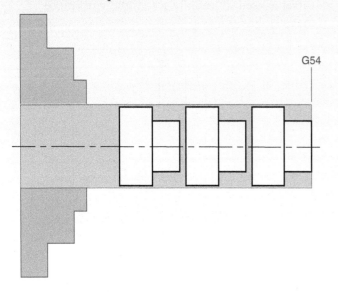

Subprogram

____ (Subprogram)
T101
G97 S437 M3
G0 1.75 Z0. M8
G50 S3600
G96 S200
G71 U.1 R.1
G71 P100 Q110 U.02 W.01 F.01
N100 G0 X.2 S200
G1 X1.125
Z-.625
X1.75
N110 Z-1.375
G0 Z0.
G70 P100 Q110
G0 Z0.
M9
G28 U0. W0. M5
T100
M01
(OD CUTOFF TOOL)
T0202
G97 S500 M3
G0 G54 X2.15 Z-1.5 M8
G1 X1.95 F.0025
X-.02
G0 X1.95
M9
G28 U0. W0. M5
T0200
____

Main Program

____ (Main Program)
G28 U0. W0.
____
____ ____ (Subprogram)
____
____ ____ (Subprogram)
____
____ ____ (Subprogram)
G28 U0. W0.
M30

Critical Thinking

1. As you progress through your career and gain experience with different types of machines, it will become apparent that there are some basic machining principles that always apply. Using subprograms with work offsets is one way to apply these principles more effectively. Describe a specific subprogramming example that uses work offsets and explain how it can be applied in a machine shop to improve efficiency and production.

2. Explain why CNC machine builders have added work offsets to their controllers. How has this feature simplified processes for CNC machinists and programmers?

3. One of the decisions the programmer often makes in the planning process is how the workpiece will be secured for machining. How can subprogramming influence this decision? How does this affect the way the programmer approaches the parts being machined?

27 Probing for Work Offsets

Chapter Outline

27.1 Introduction
27.2 Bore Probing Cycle
27.3 Boss Probing Cycle
27.4 Rectangular Pocket Probing Cycle
27.5 Rectangular Block Probing Cycle
27.6 Pocket X Axis Probing Cycle
27.7 Pocket Y Axis Probing Cycle
27.8 Web X Axis Probing Cycle
27.9 Web Y Axis Probing Cycle
27.10 Outside Corner Probing Cycle
27.11 Inside Corner Probing Cycle
27.12 Single Surface Probing Cycle

Learning Objectives

After completing this chapter, you will be able to:

- Describe the features of a probing system used in establishing work offsets.
- Identify the correct probing cycle for setting a work offset.
- Explain how to complete the information required in a probing cycle to perform a part probe.

Key Terms

probe
probing system
Wireless Intuitive Probing System (WIPS)
work coordinate system (WCS)
work offset

27.1 Introduction

Many chapters in this text have discussed the use of work offsets in programming and machining. A *work offset* defines the distance from the machine home position to the part zero location on the workpiece. A work offset establishes the *work coordinate system (WCS)*. All CNC programs are written from a part zero location on the material. It is the starting point, or reference point, of the program. Establishing this point is a critical part of the machining process. Old machinists used to refer to this point as the "bomb sight."

There are a number of methods to establish a work offset during machine setup. As discussed in the mill section of this text, these methods include using an edge finder, indicator, or hard stop. The key point to understand is that this position is vital to the successful creation of the part. If the work offset is off by just a few thousandths of an inch, it shifts the entire machining toolpath on the material. The most accurate and repeatable method to set work offsets is to use an electronic probing system. A *probing system* is a hardware addition to a CNC machine that provides accurate measurement feedback. A *probe* is a precision measurement instrument that contacts a surface of a part to record its position. This chapter will explore the features of a probing system and the process used to set work offsets.

A number of enhancements have been made in probing equipment and user interfaces on the machine to improve the simplicity of probing cycles. Most controllers have a similar system format that allows the setup person to use an electronic probe to quickly find the work offset zero location, but this text is going to focus on the Haas controller system. Haas uses the *Wireless Intuitive Probing System (WIPS)* in conjunction with its Visual Quick Code (VQC) programming language to create a seamless, efficient method to set work offsets on the controller.

There are some differences in the appearance of the user interface from older controller types to the latest controllers, but the selection of routines and the basic commands are the same. It is important to note that this is a "semi-automated process," meaning that the probe will have to be manually located at a part feature or surface and at the appropriate position in the Z axis, but the probing cycle uses a macro program to actually perform the probing. Haas Automation, in conjunction with Renishaw, has developed a series of macro programs that operate in the background of the controller. These programs are 9000 series programs and are not normally visible to the operator unless made visible with a parameter setting. It is not critical for the operator or programmer to know these programs, but it is important to note that these 9000 series programs should not be altered in any way or they will not properly operate.

Probing cycles used to set work offsets on a Haas controller are discussed in the following sections. These cycles simplify the setup process and can be leveraged to the setup person's advantage. To use these cycles, the machine must be equipped with a Renishaw spindle probe and the appropriate software and macros must be present in the machine controller. The specific location in the controller where these cycles are found will differ between controllers. Refer to the operator's manual for the controller to locate the VQC pages for work offsets.

From the Programmer

Parameter Entry

Each probing cycle has a different set of unknown parameters that must be set before the cycle is started. Become familiar with these parameters and the minimum requirements to complete each probing cycle. Once the correct probing cycle is determined, position the probe manually to its correct starting position and enter the parameter data. If the data is not correct, the cycle will fail and the incorrect data will need to be corrected.

27.2 Bore Probing Cycle

The bore probing cycle will probe in four directions to accurately measure the diameter of an existing hole and set the work offset at the hole's center location. The probe should be centered in the existing bore before starting the cycle. The centerline of the probe's ruby sphere should be below the top of the part surface for an accurate measurement. See **Figure 27-1**.

Information for the probing cycle is entered on the Bore Setup page. See **Figure 27-2**. The visual display on the page shows a representation of where the probe should be positioned prior to the cycle (near the approximate center). The number in red in the top-left corner (54) identifies the work offset being set. The up and down cursor arrow keys can be used to move through the available work offset numbers. The Approximate Diameter text box is used to enter the approximate diameter of the existing hole. This allows the machine to know the approximate size of the hole. The X Adjust and Y Adjust text boxes can be used to adjust the center location found by the probe. If necessary, use the appropriate text box to enter a value to shift the work offset zero location in either or both axis directions. Once the probe is in position and the proper settings are made, press the cycle start button to start the cycle.

From the Programmer

Probe Position

For each probing cycle, you will be asked to position the probe near the center of the part or near a surface or corner of the part. This position does not need to be that close. Depending on some parameter settings, as long as the probe is within about 1/4″ of the correct position, everything will be fine. The probe will search for the surface and the machine will alarm out if the probe is too far away or too close.

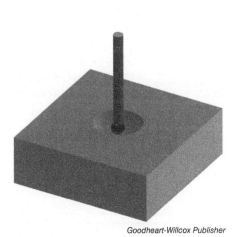

Goodheart-Willcox Publisher

Figure 27-1. To use the bore probing cycle, the centerline of the probe's sphere must be below the top of the part being measured.

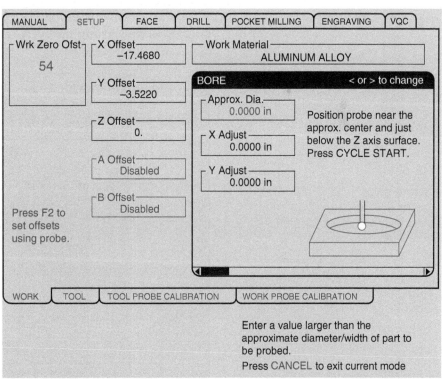

Goodheart-Willcox Publisher

Figure 27-2. The Bore Setup page is used to enter parameters for the bore probing cycle, including the approximate bore diameter and X and Y axis shift amounts, if necessary.

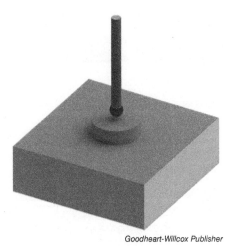

Figure 27-3. To use the boss probing cycle, the centerline of the probe's sphere must be above the top of the part being measured.

27.3 Boss Probing Cycle

The boss probing cycle will probe in four directions to accurately measure the diameter of an existing boss and set the work offset at the boss's center location. The probe should be centered above the existing boss before starting the cycle. The centerline of the probe's ruby sphere should be safely above the top of the part surface for an accurate measurement. See **Figure 27-3**.

Information for the probing cycle is entered on the Boss Setup page. See **Figure 27-4**. The visual display on the page shows a representation of where the probe should be positioned prior to the cycle (near the approximate center). The number in red in the top-left corner (54) identifies the work offset being set. The Approximate Diameter text box is used to enter the approximate diameter of the existing boss. The Incremental Z text box is used to enter the incremental distance from the current location in the Z axis to the probe depth. The probe will travel outside the diameter size set in the Approximate Diameter text box and then down in the Z axis before it measures. If necessary, the X Adjust and Y Adjust text boxes can be used to adjust the center location found by the probe. Once the probe is in position and the proper settings are made, press the cycle start button to start the cycle.

27.4 Rectangular Pocket Probing Cycle

The rectangular pocket probing cycle will probe in four directions to accurately locate the center of a rectangular-shaped pocket and set the work offset at the center location. The probe should be centered in the existing pocket before starting the cycle. The centerline of the probe's ruby sphere

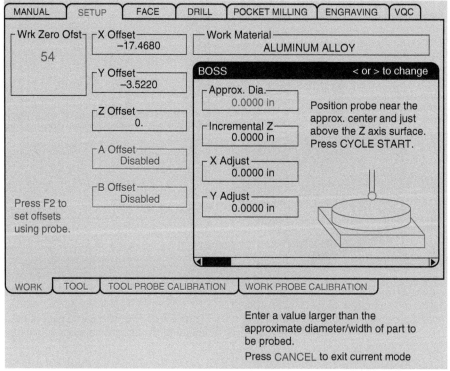

Figure 27-4. The Boss Setup page is used to enter parameters for the boss probing cycle.

should be below the top of the part surface for an accurate measurement. See **Figure 27-5**.

Information for the probing cycle is entered on the Rectangle Pocket page. See **Figure 27-6**. The visual display on the page shows a representation of where the probe should be positioned prior to the cycle (near the approximate center). The number in red in the top-left corner (54) identifies the work offset being set. The X Length text box is used to enter the approximate length of the pocket in the X axis direction. The Y Width text box is used to enter the approximate width of the pocket in the Y axis direction. The values entered in these text boxes allow the machine to know the approximate size of the pocket. If necessary, the X Adjust and Y Adjust text boxes can be used to adjust the center location found by the probe. Once the probe is in position and the proper settings are made, press the cycle start button to start the cycle.

27.5 Rectangular Block Probing Cycle

The rectangular block probing cycle will probe in four directions to accurately locate the center of a rectangular-shaped block and set the work offset at the center location. The probe should be centered above the existing block before starting the cycle. The centerline of the probe's ruby sphere should be safely above the top of the part surface for an accurate measurement. See **Figure 27-7**. The part shown has a boss feature, but this probing cycle can be used for the outside of a rectangular block of any shape.

Information for the probing cycle is entered on the Rectangle Block page. See **Figure 27-8**. The number in red in the top-left corner (54)

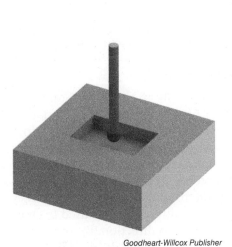

Goodheart-Willcox Publisher

Figure 27-5. To use the rectangular pocket probing cycle, the centerline of the probe's sphere must be below the top of the part being measured.

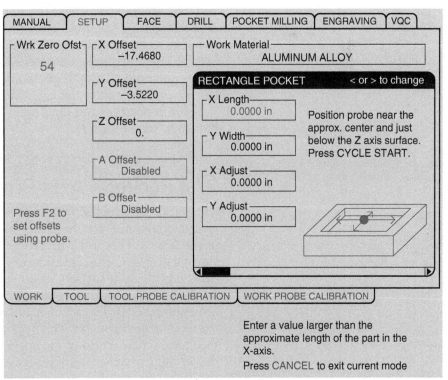

Goodheart-Willcox Publisher

Figure 27-6. The Rectangle Pocket page is used to enter parameters for the rectangular pocket probing cycle.

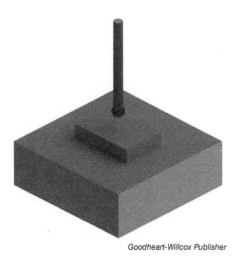

Figure 27-7. To use the rectangular block probing cycle, the centerline of the probe's sphere must be above the top of the part being measured.

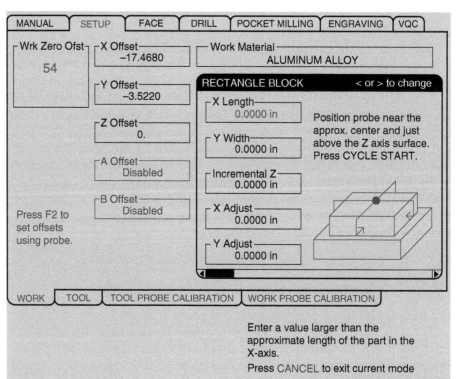

Figure 27-8. The Rectangle Block page is used to enter parameters for the rectangular block probing cycle.

identifies the work offset being set. The X Length text box is used to enter the approximate length of the block in the X axis direction. The Y Width text box is used to enter the approximate width of the block in the Y axis direction. The values entered in these text boxes allow the machine to know the approximate size of the block. The Incremental Z text box is used to enter the incremental distance from the current location in the Z axis to the probe depth. If necessary, the X Adjust and Y Adjust text boxes can be used to adjust the center location found by the probe. Once the probe is in position and the proper settings are made, press the cycle start button to start the cycle.

27.6 Pocket X Axis Probing Cycle

The pocket X axis probing cycle will probe only in the X axis direction to accurately measure the center location of a rectangular-shaped pocket in the X axis. The probe should be centered in the existing pocket before starting the cycle. The centerline of the probe's ruby sphere should be below the top of the part surface for an accurate measurement. See **Figure 27-9**.

Information for the probing cycle is entered on the Pocket X Axis page. See **Figure 27-10**. The number in red in the top-left corner (54) identifies the work offset being set. The Approximate Width text box is used to enter the approximate width of the pocket in the X axis direction. Be aware that this probing cycle will only set the X axis coordinate for the work offset. If necessary, the X Adjust text box can be used to adjust the center location found by the probe. Once the probe is in position and the proper settings are made, press the cycle start button to start the cycle.

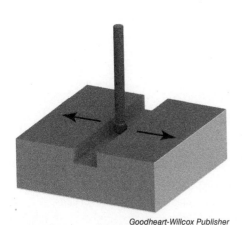

Figure 27-9. To use the pocket X axis probing cycle, the centerline of the probe's sphere must be below the top of the part being measured.

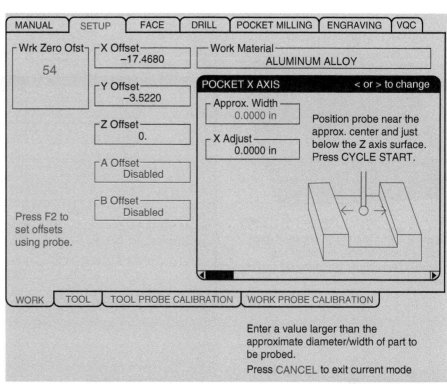

Figure 27-10. The Pocket X Axis page is used to enter parameters for the pocket X axis probing cycle.

27.7 Pocket Y Axis Probing Cycle

The pocket Y axis probing cycle will probe only in the Y axis direction to accurately measure the center location of a rectangular-shaped pocket in the Y axis. The probe should be centered in the existing pocket before starting the cycle. The centerline of the probe's ruby sphere should be below the top of the part surface for an accurate measurement. See **Figure 27-11**.

Information for the probing cycle is entered on the Pocket Y Axis page. See **Figure 27-12**. The number in red in the top-left corner (54) identifies the work offset being set. The Approximate Width text box is used to enter the approximate width of the pocket in the Y axis direction. Be aware that this probing cycle will only set the Y axis coordinate for the work offset. If necessary, the Y Adjust text box can be used to adjust the center location found by the probe. Once the probe is in position and the proper settings are made, press the cycle start button to start the cycle.

27.8 Web X Axis Probing Cycle

The web X axis probing cycle will probe only in the X axis direction to accurately measure the center location of a rectangular-shaped boss feature in the X axis. The probe should be centered above the top surface before starting the cycle. The centerline of the probe's ruby sphere should be safely above the top of the part surface for an accurate measurement. This probing cycle will only set the X axis coordinate for the work offset. See **Figure 27-13**.

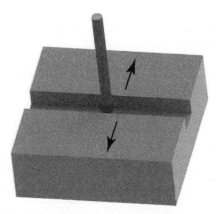

Figure 27-11. To use the pocket Y axis probing cycle, the centerline of the probe's sphere must be below the top of the part being measured.

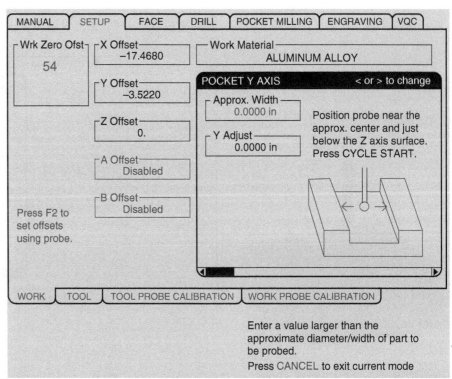

Figure 27-12. The Pocket Y Axis page is used to enter parameters for the pocket Y axis probing cycle.

Information for the probing cycle is entered on the Web X Axis page. See **Figure 27-14**. The number in red in the top-left corner (54) identifies the work offset being set. The Approximate Width text box is used to enter the approximate width of the web in the X axis direction. If necessary, the X Adjust text box can be used to adjust the center location found by the probe. The Incremental Z text box is used to enter the incremental distance the probe will travel from the current location to the probe depth in the Z axis. Once the probe is in position and the proper settings are made, press the cycle start button to start the cycle.

27.9 Web Y Axis Probing Cycle

The web Y axis probing cycle will probe only in the Y axis direction to accurately measure the center location of a rectangular-shaped boss feature in the Y axis. The probe should be centered above the top surface before starting the cycle. The centerline of the probe's ruby sphere should be safely above the top of the part surface for an accurate measurement. This probing cycle will only set the Y axis coordinate for the work offset. See **Figure 27-15**.

Information for the probing cycle is entered on the Web Y Axis page. See **Figure 27-16**. The number in red in the top-left corner (54) identifies the work offset being set. The Approximate Width text box is used to enter the approximate width of the web in the Y axis direction. If necessary, the Y Adjust text box can be used to adjust the center location found by the probe. The Incremental Z text box is used to enter the incremental distance the probe will travel from the current location to the probe depth in the Z axis. Once the probe is in position and the proper settings are made, press the cycle start button to start the cycle.

Chapter 27 Probing for Work Offsets 597

Figure 27-13. To use the web X axis probing cycle, the centerline of the probe's sphere must be above the top of the part being measured.

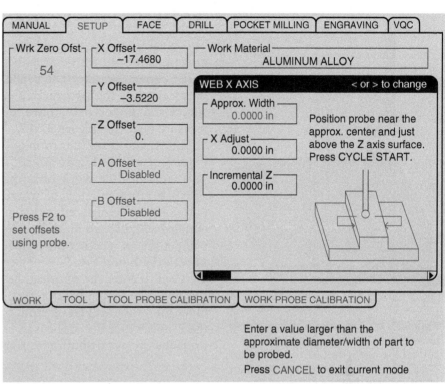

Figure 27-14. The Web X Axis page is used to enter parameters for the web X axis probing cycle.

Figure 27-15. To use the web Y axis probing cycle, the centerline of the probe's sphere must be above the top of the part being measured.

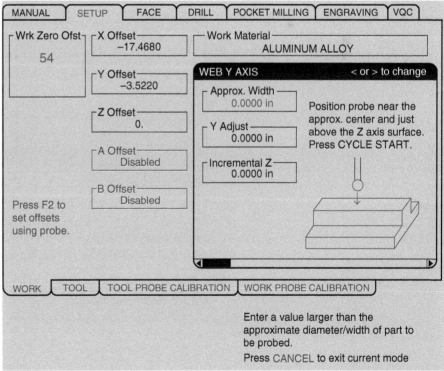

Figure 27-16. The Web Y Axis page is used to enter parameters for the web Y axis probing cycle.

27.10 Outside Corner Probing Cycle

The outside corner probing cycle will probe in both the X and Y axis directions to locate an external corner of a block and set the work offset at the corner location. The probe should be located above the part at the intersection of the two surface edges. The centerline of the probe's ruby sphere should be safely above the top of the part surface for an accurate measurement. See **Figure 27-17**.

Information for the probing cycle is entered on the Outside Corner Finding page. See **Figure 27-18**. The number in red in the top-left corner (54) identifies the work offset being set. The visual display on the page shows a representation of where the probe should be positioned. Each part corner is designated with a number. The number corresponding to the corner being probed must be entered in the Corner text box. This number will determine whether the probe moves in a positive or negative direction in both the X and Y axes. The Incremental X and Incremental Y text boxes are used to enter the distance the probe will travel along each axis before probing. The Incremental Z text box is used to enter the incremental distance the probe will travel from the current location to the probe depth in the Z axis. Once the probe is in position and the proper settings are made, press the cycle start button to start the cycle.

Goodheart-Willcox Publisher

Figure 27-17. The outside corner probing cycle is used to probe a corner location and set the work offset coordinates at the designated corner. The centerline of the probe's sphere must be above the top of the part being measured.

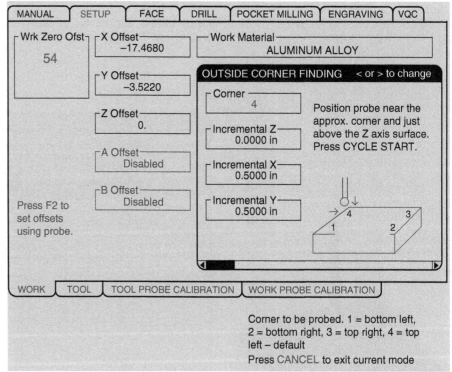

Goodheart-Willcox Publisher

Figure 27-18. The Outside Corner Finding page is used to enter parameters for the outside corner probing cycle.

27.11 Inside Corner Probing Cycle

The inside corner probing cycle will probe in both the X and Y axis directions to locate an inside corner of a block and set the work offset at the corner location. The probe should be located above the part at the intersection of the two surface edges. The centerline of the probe's ruby sphere should be safely above the top of the part surface for an accurate measurement. See **Figure 27-19**.

Information for the probing cycle is entered on the Inside Corner Finding page. See **Figure 27-20**. The number in red in the top-left corner (54) identifies the work offset being set. The visual display on the page shows a representation of where the probe should be positioned. Each part corner is designated with a number. The number corresponding to the corner being probed must be entered in the Corner text box. This number will determine whether the probe moves in a positive or negative direction in both the X and Y axes. The Incremental X, Incremental Y, and Incremental Z text box settings are the same as those used in the outside corner probing cycle. Once the probe is in position and the proper settings are made, press the cycle start button to start the cycle.

Goodheart-Willcox Publisher

Figure 27-19. The inside corner probing cycle is used to probe a corner location and set the work offset coordinates at the designated corner. The centerline of the probe's sphere must be above the top of the part being measured.

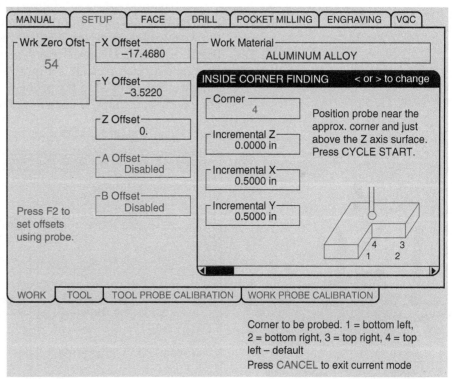

Goodheart-Willcox Publisher

Figure 27-20. The Inside Corner Finding page is used to enter parameters for the inside corner probing cycle.

> **From the Programmer**
>
> **Identifying Probing Cycles**
>
> Selecting the correct probing cycle for the desired work offset and type of part feature is critical. Some of the probing cycles will only locate one axis position and it may require two different cycles to set the X and Y axis coordinates. Being familiar with the workpiece, programming requirements, and setup procedure will help in identifying the correct probing cycle.

27.12 Single Surface Probing Cycle

The single surface probing cycle will probe in the X, Y, or Z axis direction to locate a surface of a workpiece. This is probably the most common probing cycle used, but it can only be used to set one axis coordinate setting for a work offset at a time. The starting point of the probe should be located at a safe distance near the surface being probed and should be at the appropriate Z axis height. See **Figure 27-21**.

Information for the probing cycle is entered on the Single Surface page. See **Figure 27-22**. The number in red in the top-left corner (54) identifies the work offset being set. Next to the visual display are the X Distance, Y Distance, and Z Distance text boxes. Only one text box can be used for the single surface cycle. It is important to establish the correct axis distance and direction (+ or –) so the probe will know the correct direction of travel. For example, if the X Distance text box entry is .500, the probe will travel

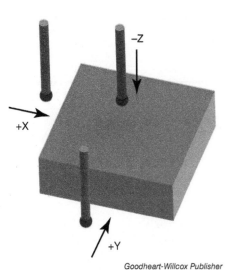

Goodheart-Willcox Publisher

Figure 27-21. The single surface probing cycle is used to probe a part surface and set the coordinate for the corresponding axis in the work offset. The starting point of the probe must be at a safe distance away from the part being measured.

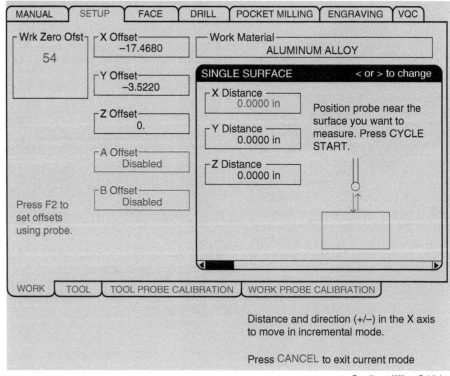

Goodheart-Willcox Publisher

Figure 27-22. The Single Surface page is used to enter parameters for the single surface probing cycle.

1/2″ in the positive X axis direction to probe for the part surface. If the X Distance text box entry is −.500, the probe will travel in a negative X axis direction.

Probing in the Z axis direction will post a negative Z axis coordinate for the specified work offset on the work offset page. Probing a surface in the Z axis direction is usually only done if an electronic tool setter probe is used to establish tool offsets. Be cautious when setting the Z axis coordinate for a work offset. If tools are touched off on the top of the part to set tool offsets, and the Z axis coordinate setting in the work offset does not account for the tool offset measurement, it will cause a catastrophic collision.

 From the Programmer

Benefits of Probing Systems

Investing in a machine with a probing system will result in an immediate and substantial return. Probing is an extremely fast and accurate method for establishing work offsets. Probing reduces the amount of time the machine is idle and helps eliminate errors in measurement. The measuring accuracy of a properly calibrated and utilized probing system is approximately .00005″ (50 millionths of an inch). Not all machines will have a probing system, but those that do will allow for considerable time savings and efficiency in part setup.

Chapter 27 Review

Summary

- The process of setting work offsets is greatly enhanced through the use of a probing system. Probing greatly simplifies the machine setup process and provides the most accurate and repeatable method for establishing the work coordinate system.
- The bore probing cycle is used to measure the diameter of an existing hole and set the work offset at the hole's center location.
- The boss probing cycle is used to measure the diameter of an existing boss and set the work offset at the boss's center location.
- The rectangular pocket probing cycle is used to locate the center of a rectangular-shaped pocket and set the work offset at the center location.
- The rectangular block probing cycle is used to locate the center of a rectangular-shaped block and set the work offset at the center location.
- The pocket X axis probing cycle is used to measure the center location of a pocket in the X axis. This cycle will set the X axis coordinate for the work offset. The pocket Y axis probing cycle is used to measure the center location of a pocket in the Y axis. This cycle will set the Y axis coordinate for the work offset.
- The web X axis probing cycle is used to measure the center location of a boss feature in the X axis. This cycle will set the X axis coordinate for the work offset. The web Y axis probing cycle is used to measure the center location of a boss feature in the Y axis. This cycle will set the Y axis coordinate for the work offset.
- The outside corner probing cycle is used to locate an external corner of a block and set the work offset at the corner location.
- The inside corner probing cycle is used to locate an inside corner of a block and set the work offset at the corner location.
- The single surface probing cycle is used to probe a part surface and set the coordinate for the corresponding axis in the work offset.

Review Questions

Answer the following questions using the information provided in this chapter.

Know and Understand

1. *True or False?* A probing system uses macro programs and a spindle probe installed on the machine to perform probing routines.

2. Which of the following probing cycles is used to set the coordinates of the work offset to the center of an existing hole?
   A. Boss probing cycle
   B. Single surface probing cycle
   C. Outside corner probing cycle
   D. Bore probing cycle

3. Which of the following probing cycles measures the center location of a raised circular feature and sets the coordinates of the work offset to the center of the feature?
   A. Boss probing cycle
   B. Single surface probing cycle
   C. Outside corner probing cycle
   D. Bore probing cycle

4. *True or False?* To use the rectangular block probing cycle, the starting point of the probe should be below the top of the block.

5. *True or False?* The pocket X axis probing cycle probes in the X axis direction to measure the center location of a rectangular-shaped boss feature.

6. The _____ text box is used to enter the distance the probe will travel from the current location to the probe depth when setting parameters for the web X axis probing cycle.
   A. Approximate Width
   B. X Adjust
   C. Y Adjust
   D. Incremental Z

7. Where should the probe be located when using the outside corner probing cycle?
   A. Above the part at the center of the part.
   B. Below the part at the center of the part.
   C. Above the part at the intersection of the two surface edges.
   D. Below the part at the intersection of the two surface edges.
8. *True or False?* The number representing the corner being probed must be entered in the Corner text box when using the inside corner probing cycle.
9. *True or False?* The single surface probing cycle is used to set the X and Y axis coordinate settings for the specified work offset in a single cycle.
10. Which of the following probing cycles should be used to probe a square boss feature and set the X and Y work offset coordinates to the center of the feature?
    A. Boss probing cycle
    B. Single surface probing cycle
    C. Outside corner probing cycle
    D. Rectangular block probing cycle

Apply and Analyze

1. Where should the probe be positioned prior to using the outside corner probing cycle? Explain the parameters that are entered when using the cycle.
2. Where should the probe be positioned prior to using the single surface probing cycle? Explain the parameters that are entered when using the cycle.
3. Explain how the X and Y coordinates found by the probe can be adjusted automatically by entering parameters when using the boss probing cycle.

Critical Thinking

1. As discussed in this chapter, defining the part zero location on the workpiece is a vitally important step. What are some of the errors that will occur in machining as a result of a misaligned work coordinate system?
2. What are the primary benefits of using a probing system to define work offsets in work setup? How does a probing system improve production, reduce costs, and lead to better quality in parts?
3. Today's technology has given us the ability to use a global positioning system (GPS) in a handheld device or in our automobiles. In the past, there were less accurate ways to locate starting positions and destination points when traveling. How is probing technology similar to a GPS system? How does probing help make machinists more efficient?

28 Probing Inside of the Program

Chapter Outline

28.1 Introduction
28.2 Using the Probe as a Tool
28.3 Starting Probe Position and Protected Moves
28.4 Single Surface Measurement
28.5 Web/Pocket Measurement
28.6 Bore/Boss Measurement

Learning Objectives

After completing this chapter, you will be able to:

- Explain how a probing cycle is used inside a main program to establish work offsets.
- Turn the probe on and off in a program using the P9832 and P9833 macros.
- Use the P9810 macro for protected movements inside the machine's work envelope.
- Use the P9811 probing cycle to measure a single surface and record work offset coordinates.
- Use the P9812 probing cycle to measure a web or pocket and record work offset coordinates.
- Use the P9814 probing cycle to measure a hole feature or a boss and record work offset coordinates.

Key Terms

macro
protected move
variable

Chapter opening photo credit: Pixel B/Shutterstock.com

28.1 Introduction

In Chapter 27, you learned how a probing cycle can be used to establish the part zero location in a work coordinate system. This improves the efficiency of the setup process and gives the setup person and operator a tremendous advantage. However, the process of setting parameters for a probing cycle at the machine does fall somewhat short in fully automating the machining process. As discussed in Chapter 27, on a Haas controller, the Wireless Intuitive Probing System (WIPS) allows the operator to define parameters for a probing cycle and run the cycle. To use this system, the operator must change the machine's work mode to MDI (manual data input), go through different screen pages, manually position the probe in the right location, input the correct parameters, and run the probing cycle. A good operator can do this in under a minute and it is not that difficult. But, what if it were possible to build this process into the program and automatically probe every part to establish an accurate work offset? What if the operator just pressed the cycle start button and the probe came out to measure the part, the probing cycle only took 5–10 seconds to accomplish, and the controller kept right on running the part program? Now that is efficient!

As discussed in Chapter 27, probing cycles are actually programs written with protected numbers in the 9000 series. Normally these programs are hidden from the program list screen, but they can be made visible by changing a parameter setting. For example, the single surface probing cycle is program number O9811. By knowing which programs perform these cycles, and knowing how to add cycle parameters, these programs can be included as subprograms in part programs. However, before a probing cycle can be used in a main program, or in a subprogram, there are some initial items that need to be put in place.

28.2 Using the Probe as a Tool

If the probe is going to be called in the program to perform a probing operation, it must be designated as a tool, assigned a tool offset, and programmed similarly to a standard tool. Be aware that the probe head and stylus are precise and fragile instruments. The probe is a highly accurate measuring device and needs to be treated as such. It is best practice to assign a probe to a machine and leave the probe in the same tool number location. Taking the probe in and out of a machine, or switching it between machines, will require the probing system to be recalibrated and can compromise accuracy. For the examples in this chapter, the probe will be installed in the last available tool number location (#25), and it will remain in that location.

In the following example, a simple single surface probing cycle will be programmed. This cycle will be accomplished in the main program. Later in this chapter, probing cycles will be used in subprograms to speed up the programming process. The following program will start just like a "normal" milling program, calling tool #25 to the spindle and positioning it in place.

> **From the Programmer**
>
> **Initial Work Offset Setting**
>
> Before this program can be run, there must be a "rough" G54 work offset designation. The machine has to have some idea of where part zero is located. One way to do this, when touching off the probe tool on top of the material, is to move the center of the probe over the desired G54 work offset origin and set the work offset coordinates. This does not need to be exact, because you are going to use the probe to set the exact work offset location.

O1968 (Milling Program with Probing)
G20 G17 G90 G80 G40
T25 M6
G0 G90 G54 X–1. Y–1.
G43 H25 Z2.

The program is initiated, a startup block is added, and tool #25 (the probe) is put in the spindle. Note that there is no M3 function to turn on the spindle. The spindle should never be turned on when the probe is in the spindle. The next step is moving the probe to a safe distance away from the part in the X and Y axes. See **Figure 28-1**. When selecting this position, be aware of any workholding that might interfere with the probe movement. At this point, the probe has not been turned on, so if it hits a clamp, a vise, or the material, it will severely damage the probe. The last step in this portion of the program is to initiate the tool length offset for the probe with the G43 command.

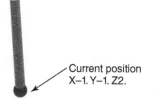
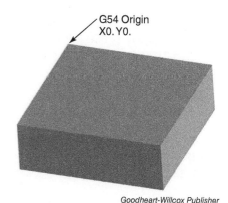

Goodheart-Willcox Publisher

Figure 28-1. A probe is programmed in the same manner as other tools. A tool length offset must be assigned to the probe and the probe must be positioned at a safe location prior to being used in a probing cycle.

28.3 Starting Probe Position and Protected Moves

Now that the probe is in a safe position in the X and Y axes, and safely above the Z0. plane, the controller needs to turn on the probing head, and then move safely into a probing position. These actions will require a special G-code command not previously discussed in this text—the G65 command. The G65 command is used to call a custom macro. As discussed in Chapter 27, all probing cycles are macro subprograms. A *macro* is a custom subroutine that defines one or more actions performed by the macro when called by a program. In addition to initiating probing cycles, macros are used for actions such as turning the probe on and off and commanding positioning moves by the probe. Each block in the program referencing a macro must begin with the G65 command. This command is accompanied by a P address and value designating the number of the macro program. This is similar to calling a subprogram with the M98 function, but programs stored as macros are called with the G65 command.

At this point in the program, the probe is safely positioned above the part in the Z axis (Z2.). Before the probe is moved to a starting position, it must be turned on. This is done with the macro program P9832. This macro will orient the spindle and turn on the probe's electronics. A green light will appear on the probing head when the probe is on. In the following program, the probe is powered on and then moved to the starting position.

O1968 (Milling Program with Probing)
G20 G17 G90 G80 G40
T25 M6
G0 G90 G54 X–1. Y–1.
G43 H25 Z2.
G65 P9832
G65 P9810 Z–.5

Debugging the Code

Feed Rate

The P9810 macro moves the probe in feed mode, not rapid mode. This means the speed of movement can be adjusted. For example, the line can be programmed as **G65 P9810 Z–.5 F100**. The specified feed rate is modal to the P9810 macro, so once set, it stays at the same feed rate until it is altered. This should be a pretty fast feed, but determine the rate you are comfortable with and leave it set.

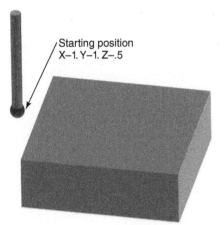

Goodheart-Willcox Publisher

Figure 28-2. Before the probe is moved to the starting position, it is turned on. Then, a protected move is used to position the probe safely away from the part.

In the last block, the P9810 macro is used to command a special type of move called a *protected move*. This move will allow the probe to be driven to a specified coordinate in a safe mode. During a protected move, if the probe encounters an obstruction, it will stop moving and the machine will alarm out before the probe stem breaks or a collision occurs. The result of the move to the probe's starting position with the P9810 macro is shown in **Figure 28-2**. This is the most important macro in a probing cycle. It is critical to use the P9810 macro whenever the probe is being moved and is not in a measurement cycle.

28.4 Single Surface Measurement

The probe is now in the correct X, Y, and Z starting position. It is time to send the probe to the part surface to measure and set the appropriate work offset coordinate, G54 X0. in this case. The next block in the program will command the probe to search for the part material in the positive X axis direction. Once triggered, the probe will make a small retract and re-probe for accuracy.

O1968 (Milling Program with Probing)
G20 G17 G90 G80 G40
T25 M6
G0 G90 G54 X–1. Y–1.
G43 H25 Z2.
G65 P9832
G65 P9810 Z–.5
G65 P9811 X0. S1.

The P9811 macro is the single surface measurement probing cycle. The cycle will initiate movement in the X axis direction because the specified coordinate is X0. The controller will record the location of the material that the probe finds as the X axis zero coordinate position. This is important information because the work offset setting could be X–.02, meaning that a tool could be programmed to machine at X0. and it would make a .020″ cut on that edge.

The designation following the X0. coordinate is S1. This is a reference to work coordinate offset 1, or the G54 work offset. This is how work offsets were registered on the very first CNC machines. The designation S2. references work coordinate offset 2, or the G55 work offset, and so forth. The S designator and other letter codes used in probing cycles are called *inputs*. The value following an input letter defines data used in the cycle or a setting to be updated by the controller. The following are common input letters in probing cycles.

- **S.** Designates the work offset to be updated by the cycle. The designation S1. is a reference to work coordinate offset 1, or the G54 work offset. The designation S2. is a reference to work coordinate offset 2, or the G55 work offset, and so on.
- **T.** Designates the tool offset to be updated by the cycle. By adding a tool number, the measurement made by the probe can be used to adjust the tool offset setting.

- **H.** Designates the required tolerance for the feature. The related value is used by the machine to determine whether the feature is within tolerance when the feature is probed. This allows the probe to measure for acceptance or rejection of the part feature. This is discussed in more detail in Chapter 29.
- **W.** Used to print or output inspection data. With certain controllers or external hardware, the physical data, such as the feature size and position, can be output to third-party software and recorded in an inspection report.

Thus far, the probe has measured the X axis location (using P9811) and recorded the G54 work offset X axis location (using S1.). It is worth noting that without the S1. code, the probe will still position itself, find the X axis edge, and even store that data in a macro variable on the controller. However, it will not update the G54 X axis work offset coordinate. The S1. designation is critical to writing the X axis positional data to the work offset page.

With the X axis position recorded, the program will make additional safe moves to relocate the probe stylus to measure the Y axis location. This location will be recorded as the work offset Y axis coordinate. Then, the probe will be moved to a safe location and the P9833 macro will be used to turn off the probe. The completed program is as follows.

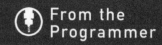

From the Programmer

Macros and Variables

Macros are covered in more detail in Chapter 30 of this text. Variables serve a key role in macros. A *variable* is a placeholder used to store and access data for programming purposes. The probing cycles covered in this chapter are used to output measurement data to macro variables. For example, the single surface probing cycle writes measured values to global variables based on what is measured in the cycle. The variable data can be referenced to perform actions in macros, as discussed in Chapter 30.

O1968 (Milling Program with Probing)	
G20 G17 G90 G80 G40	
T25 M6	
G0 G90 G54 X–1. Y–1.	
G43 H25 Z2.	
G65 P9832	
G65 P9810 Z–.5	
G65 P9811 X0. S1.	
G65 P9810 Z.5	Safe move to a +Z position above part
G65 P9810 X1. Y1.	Safe move to an XY location before probing
G65 P9810 Z–.5	Safe move to a –Z location before probing
G65 P9811 Y0. S1.	Measure Y axis location and set G54 work offset Y0. location
G65 P9810 Z1.	Safe move to clear top of part
G65 P9833	Turn off probe
G0 G91 G28 Z0.	Send machine home in Z axis

The next line in the program could be a call for the first tool used to machine, the M30 function if the program is only setting the work offset X and Y coordinates, or the M99 function if this program is used as a subprogram. This would be a great application for a subprogram. If it is common or standard practice to use a vise with the top-left corner of the material as the part zero location, this program could be stored as a probing subprogram in a main machining program to set the work offset coordinates for every part.

28.5 Web/Pocket Measurement

The P9812 measurement macro can be used to measure a web or pocket feature and establish work offset coordinates for the feature. By specifying movement in the X or Y axis, this cycle can be used to measure any web or pocket in the X or Y axis direction. **Figure 28-3** shows a part with a pocket feature. To complete the probing measurement for this feature, the X or Y axis probing direction must be entered along with the nominal width of the pocket. The following probing cycle is used to measure the pocket feature in the X axis direction. For this example, the pocket is 1″ wide.

O2804 (Pocket Probing in X Axis Direction)	
G20 G17 G90 G80 G40	
T25 M6	
G0 G90 G54 X0. Y0.	Position probe at center of pocket
G43 H25 Z2.	
G65 P9832	Turn on probe
G65 P9810 Z–.3	Safe Z axis move to below top of part
G65 P9812 X1. S1.	Measure 1″ wide pocket in X axis direction; set G54 work offset X0. location
G65 P9810 Z.5	Safe move to a Z position above part
G65 P9833	Turn off probe
G0 G91 G28 Z0.	Send machine home in Z axis
M99	Return to main program

The following probing cycle is used to measure the pocket feature in the Y axis direction.

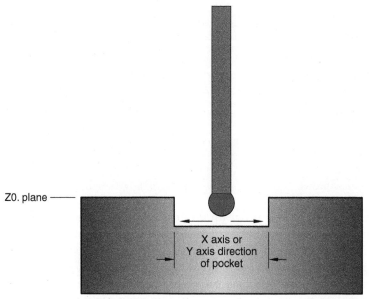

Goodheart-Willcox Publisher

Figure 28-3. The P9812 measurement macro can be used to measure a pocket feature in the X axis or Y axis direction. The probing direction and nominal width of the feature are specified in the probing cycle.

O2804 (Pocket Probing in Y Axis Direction)	
G20 G17 G90 G80 G40	
T25 M6	
G0 G90 G54 X0. Y0.	Position probe at center of pocket
G43 H25 Z2.	
G65 P9832	Turn on probe
G65 P9810 Z-.3	Safe Z axis move to below top of part
G65 P9812 Y1. S1.	Measure 1" wide pocket in Y axis direction; set G54 work offset Y0. location
G65 P9810 Z.5	Safe move to a Z position above part
G65 P9833	Turn off probe
G0 G91 G28 Z0.	Send machine home in Z axis
M99	Return to main program

These two programs will each complete a measurement cycle for a pocket in a single axis. To probe a closed pocket in the X and Y axes, repeat the P9812 cycle in two consecutive lines, one in the X axis direction and one in the Y axis direction. **Figure 28-4** shows a pocket 2" wide in the X axis and 1.5" wide in the Y axis. The following program is used to measure the pocket feature in each direction and set the X axis and Y axis work offset coordinates to the center of the pocket.

O2805 (Closed Pocket Probing)	
G20 G17 G90 G80 G40	
T25 M6	
G0 G90 G54 X0. Y0.	Position probe at center of pocket
G43 H25 Z2.	
G65 P9832	Turn on probe
G65 P9810 Z-.3	Safe Z axis move to below top of part
G65 P9812 X2. S1.	Measure 2" wide pocket in X axis direction; set G54 work offset X0. location

Continued

Goodheart-Willcox Publisher

Figure 28-4. A closed pocket can be measured in two axis directions by using two separate probing cycles for the X and Y axis measurements.

G65 P9812 Y1.5 S1.	Measure 1.5″ wide pocket in Y axis direction; set G54 work offset Y0. location
G65 P9810 Z.5	Safe move to a Z position above part
G65 P9833	Turn off probe
G0 G91 G28 Z0.	Send machine home in Z axis
M99	Return to main program

Creating a probing cycle program for a web feature is very similar. An additional Z axis move is needed to drive the probe tip below the top of the part and the R address is used to specify the approach distance on the web. The Z axis move is an absolute positioning move in the negative Z axis direction. The R address value specifies an incremental amount of clearance on each side of the web. See **Figure 28-5**. The following probing cycle is used to measure a 1″ web feature in the X axis direction and set the X axis work offset coordinate to the center of the pocket.

O2806 (Web Probing in X Axis Direction)	
G20 G17 G90 G80 G40	
T25 M6	
G0 G90 G54 X0. Y0.	Position probe at center of web
G43 H25 Z2.	
G65 P9832	Turn on probe
G65 P9810 Z.3	Safe Z axis move to above top of part
G65 P9812 X1. Z–.5 R.25 S1.	Measure 1″ wide web in X axis direction; probe at Z–.5 with .25″ clearance

Continued

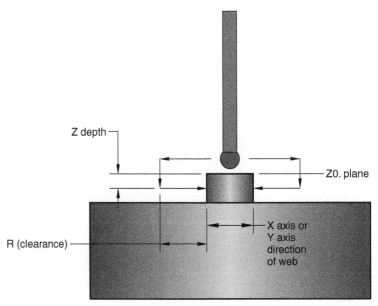

Goodheart-Willcox Publisher

Figure 28-5. To use a web measurement probing cycle, the probe is positioned above the top surface before starting the cycle. The cycle requires an axis direction to designate the direction of measurement and a Z axis move to the probing depth. The R address is used to specify a clearance distance for the probe to travel before moving to the probing depth.

G65 P9810 Z.5	Safe move to a Z position above part
G65 P9833	Turn off probe
G0 G91 G28 Z0.	Send machine home in Z axis
M99	Return to main program

The following probing cycle is used to measure a 1″ web feature in the Y axis direction.

O2806 (Web Probing in Y Axis Direction)	
G20 G17 G90 G80 G40	
T25 M6	
G0 G90 G54 X0. Y0.	Position probe at center of web
G43 H25 Z2.	
G65 P9832	Turn on probe
G65 P9810 Z.3	Safe Z axis move to above top of part
G65 P9812 Y1. Z−.5 R.25 S1.	Measure 1″ wide web in Y axis direction; probe at Z−.5 with .25″ clearance
G65 P9810 Z.5	Safe move to a Z position above part
G65 P9833	Turn off probe
G0 G91 G28 Z0.	Send machine home in Z axis
M99	Return to main program

These two programs will each complete a measurement cycle for a web in a single axis. To probe a rectangular block in the X and Y axis directions, repeat the P9812 cycle in two consecutive lines, one in the X axis direction and one in the Y axis direction. **Figure 28-6** shows a part with a rectangular boss 2″ wide in the X axis direction and 1.5″ wide in the Y axis direction. The following program measures the block in each direction and sets the X axis and Y axis work offset coordinates to the center of the block.

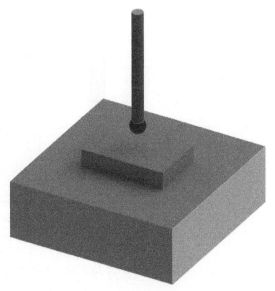

Goodheart-Willcox Publisher

Figure 28-6. A rectangular boss can be measured in two axis directions by using two separate web measurement probing cycles for the X and Y axis measurements.

O2807 (Web Probing for a Block)	
G20 G17 G90 G80 G40	
T25 M6	
G0 G90 G54 X0. Y0.	Position probe at center of web
G43 H25 Z2.	
G65 P9832	Turn on probe
G65 P9810 Z.3	Safe Z axis move to above top of part
G65 P9812 X2. Z-.5 R.25 S1.	Measure 2″ wide block in X axis direction; probe at Z-.5 with .25″ clearance
G65 P9812 Y1.5 Z-.5 R.25 S1.	Measure 1.5″ wide block in Y axis direction; probe at Z-.5 with .25″ clearance
G65 P9810 Z.5	Safe move to a Z position above part
G65 P9833	Turn off probe
G0 G91 G28 Z0.	Send machine home in Z axis
M99	Return to main program

28.6 Bore/Boss Measurement

The P9814 measurement macro can be used to measure a bore or boss feature and establish work offset coordinates for the feature. The P9814 cycle will probe a round feature, either internally or externally, in four directions and calculate the feature center. The D address is used to enter the nominal diameter of the round feature. To correctly measure a hole feature, the probe is moved below the top face before the cycle begins. See **Figure 28-7**.

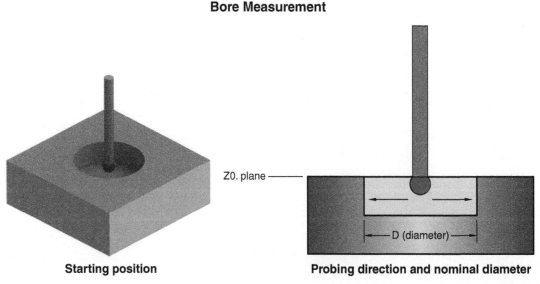

Figure 28-7. The P9814 measurement macro can be used to measure a bore or boss feature. The nominal diameter of the feature is specified in the probing cycle. To measure a hole feature, the probe is positioned below the top face before starting the cycle.

The following cycle is used to measure a hole feature and set the X axis and Y axis work offset coordinates to the center of the hole.

O2808 (Bore Probing)	
G20 G17 G90 G80 G40	
T25 M6	
G0 G90 G54 X0. Y0.	Position probe at center of bore
G43 H25 Z2.	
G65 P9832	Turn on probe
G65 P9810 Z-.5	Safe Z axis move to below top of part
G65 P9814 D1.5 S1.	Measure 1.5" diameter bore; set G54 work offset X0. Y0. location
G65 P9810 Z.5	Safe move to a Z position above part
G65 P9833	Turn off probe
G0 G91 G28 Z0.	Send machine home in Z axis
M99	Return to main program

In each of the previously discussed probing cycles, the assumption was that the work offset zero location was set to the location of the feature that was being probed. For example, in the bore probing cycle, the probe was moved to X0. Y0. and the S1. designation was used to set the coordinates for the G54 work offset at the center location. However, it is common for other surfaces to be machined in a secondary operation after the 1.5" diameter bore has been completed. It is possible to measure the completed bore and establish the work offset at a different location. In this case, the desired G54 work offset location is the top-left corner of the part. The top-left corner is located 1.5" away from the bore center in both the X and Y axis directions. See **Figure 28-8**.

For this probing cycle, the "rough" G54 work offset location is set at the top-left corner of the part material. The following program is used to

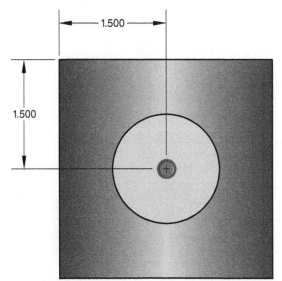

Goodheart-Willcox Publisher

Figure 28-8. The bore is measured after it is machined and the location found by the probe is referenced to establish the work offset. The work offset is established at the top-left corner of the part for a secondary operation. This location is exactly 1.5" away from the measured hole center in the X and Y axis directions.

command the controller to move to the X1.5 Y–1.5 position, probe the 1.5″ bore, and find its center. Because the machine is told that it is located at X1.5 Y–1.5, when the bore center is located, the controller will make the calculations to adjust the G54 work offset to the top-left corner of the part material. This location is exactly 1.5″ away from the measured hole center in the X and Y axis directions.

O2809 (Bore Probing)	
G20 G17 G90 G80 G40	
T25 M6	
G0 G90 G54 X1.5 Y–1.5	Position probe at center of bore
G43 H25 Z2.	
G65 P9832	Turn on probe
G65 P9810 Z–.5	Safe Z axis move to below top of part
G65 P9814 D1.5 S1.	Measure 1.5″ diameter bore; set G54 work offset X0. Y0. location at top-left corner of part
G65 P9810 Z.5	Safe move to a Z position above part
G65 P9833	Turn off probe
G0 G91 G28 Z0.	Send machine home in Z axis
M99	Return to main program

All of the probing cycles can be used in a similar manner to measure a feature and then shift the work offset coordinates to an alternate position.

A similar process is used to create a probing cycle program for a boss feature. An additional Z axis move is needed to drive the probe tip below the top of the part and the R address is used to specify the clearance distance from the outside surface. See **Figure 28-9**. The following cycle is used to measure a 1.5″ boss and set the X axis and Y axis work offset coordinates to the center of the feature.

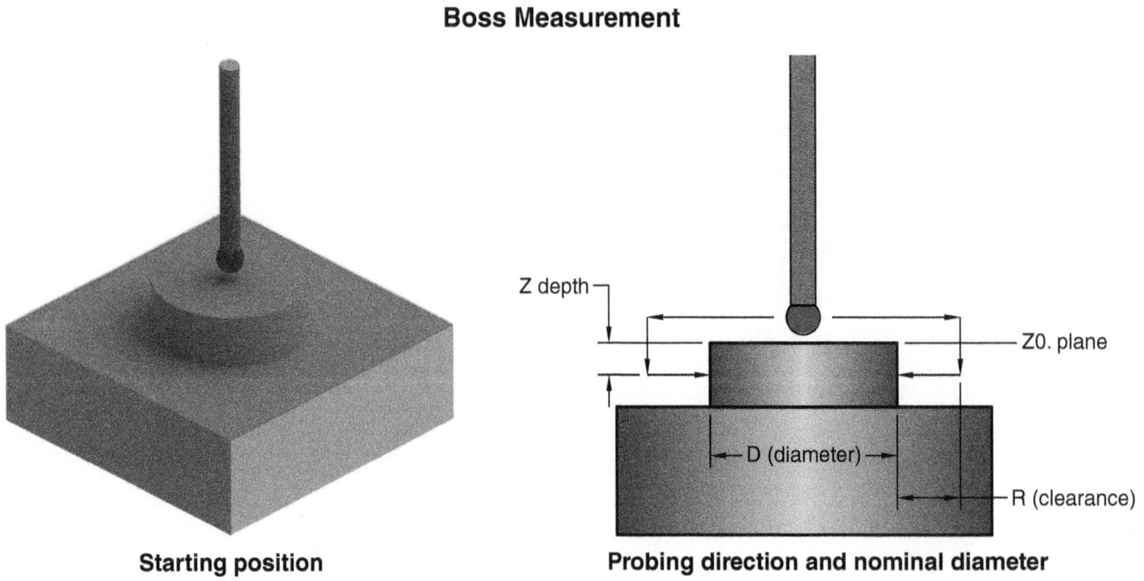

Boss Measurement

Starting position

Probing direction and nominal diameter

Goodheart-Willcox Publisher

Figure 28-9. To measure a boss using the P9814 measurement macro, the probe is positioned above the top surface before starting the cycle. The cycle requires the nominal diameter of the feature and a Z axis move to the probing depth. The R address is used to specify a clearance distance for the probe to travel before moving to the probing depth.

O2810 (Boss Probing)	
G20 G17 G90 G80 G40	
T25 M6	
G0 G90 G54 X0. Y0.	Position probe at center of boss
G43 H25 Z2.	
G65 P9832	Turn on probe
G65 P9810 Z.3	Safe Z axis move to above top of part
G65 P9814 D1.5 Z-.3 R.25 S1.	Measure 1.5" diameter boss; probe at Z-.3 with .25" clearance; set G54 work offset X0. Y0. location
G65 P9810 Z.5	Safe move to a Z position above part
G65 P9833	Turn off probe
G0 G91 G28 Z0.	Send machine home in Z axis
M99	Return to main program

 From the Programmer

Using Probing Cycles in Subprograms

Probing cycles allow the programmer to measure any standard feature and establish work offset coordinates with great speed and accuracy. The probing cycle programs in this chapter were completed with the M99 function. This methodology is used to store probing cycles as subprograms. In each case, the probing routine is used to measure a feature and set the work offset coordinates to the designated location. If these subprograms are carefully written and called in a main program, it will only take one line of programming and minor edits for feature sizes to establish all of the required work offsets.

Chapter 28 Review

Summary

- Probing cycles can be included inside main programs to simplify the setup process. The ability to measure features and set work offset coordinates with a probing system greatly improves the efficiency of the machining process.
- Probing cycles are macro subprograms. Macros are also used for actions such as turning the probe on and off and commanding positioning moves by the probe.
- Before the probe can be used in a probing cycle, it must be turned on with the P9832 macro. After completion of the probing cycle, and when the probe is safely away from any material interference, the probe is shut down with the P9833 macro.
- The probe is a highly precise and fragile measuring instrument. The P9810 macro is used to safely move the unit around the part material and any other obstructions.
- The P9811 probing cycle allows for measurement of a single surface in the X, Y, or Z axis and establishment of work offset coordinates.
- The P9812 probing cycle can be used to measure a web or pocket feature in a single axis and establish work offset coordinates.
- The P9814 probing cycle can be used to measure a bore or boss feature and establish work offset coordinates.
- When measuring a feature in a probing cycle, the work offset coordinates can be set to the location of the feature, or the work offset can be shifted to an alternate location by specifying the location in the program.

Review Questions

Answer the following questions using the information provided in this chapter.

Know and Understand

1. *True or False?* Before a probing cycle is initiated in a program, the M3 function must be used to turn on the spindle.

2. Which command is used to call a custom macro?
   A. G54
   B. G55
   C. G65
   D. G70

3. *True or False?* When a macro is referenced in a program, the P address designates the number of the macro program.

4. The ____ macro is used to turn on the probe.
   A. P9810
   B. P9811
   C. P9832
   D. P9833

5. The ____ macro is used to designate a protected move for the probe.
   A. P9810
   B. P9811
   C. P9832
   D. P9833

6. The ____ macro allows for measurement of a single surface and establishment of work offset coordinates.
   A. P9810
   B. P9811
   C. P9832
   D. P9833

7. Which of the following designations references the G54 work offset in a probing cycle?
   A. S1.
   B. S2.
   C. S3.
   D. S4.

8. *True or False?* The P9812 measurement macro is used to probe a web or pocket feature.

9. Which of the following macros is used to probe a rectangular block feature?
   A. P9810
   B. P9812
   C. P9814
   D. P9832

10. *True or False?* To use a boss probing cycle, the probe must be positioned below the top surface of the part before starting the cycle.

Apply and Analyze

1. Referring to the blanks provided below, give the correct codes that are missing for the single surface probing cycle. This cycle is used to measure in the X axis direction and set the X axis zero coordinate position for the G54 work offset.

O2812
G20 G17 G90 G80 G40
T25 _____
G0 G90 G54 X–1.5 Y–1.5
_____ H25 Z2.
G65 _____
G65 P9810 Z–.5
G65 _____ X0. _____
G65 _____ Z1.
G65 _____
G0 G91 G28 Z0.

2. Referring to the blanks provided below, give the correct codes that are missing for the web probing cycle. This cycle is a subprogram used to probe a block shape measuring 3" in the X axis direction and 1" in the Y axis direction. The cycle sets the work offset coordinates for the G54 work offset and allows for .500" of clearance on each side of the block when moving to the probing depth.

O2813
G20 G17 G90 G80 G40
T25 _____
G0 G90 G54 X0. Y0.
_____ H25 Z2.
G65 _____
G65 _____ Z.3
G65 _____ X3. Z–.5 _____ _____
G65 _____ _____ Z–.5 _____ _____
G65 P9810 Z.5
G65 _____
G0 G91 G28 Z0.

3. Referring to the blanks provided below, give the correct codes that are missing for the bore probing cycle. This cycle is a subprogram used to measure a bore 2.5" in diameter and set the work offset coordinates for the G54 work offset.

O2814
G20 G17 G90 G80 G40
T25 _____
G0 G90 G54 X0. Y0.
_____ H25 Z2.
G65 _____
G65 P9810 Z–.5
G65 _____ _____ _____
G65 _____ Z.5
G65 _____
G0 G91 G28 Z0.

Critical Thinking

1. There are multiple ways to locate and set work offset coordinates during machine setup. What are the benefits of using probing cycles directly in a program? Consider factors such as part quality and cost savings.

2. External components such as electronic probes have greatly altered the way parts are machined. What are some other types of external components that are not standard on CNC machines and how can they be used to improve efficiency?

3. As discussed in this chapter, it is possible to create subprograms to store probing cycles. Normally, a CNC programmer is responsible for multiple machines or work cells in a machine shop. What are some ways that programming can be standardized and simplified throughout an entire company? Explain how standardization can make a financial impact by improving existing processes.

29 On Machine Verification

Chapter Outline

29.1 Introduction

29.2 Pros and Cons of On Machine Verification

29.3 Tool Offset Adjustment and Tolerance Verification

    29.3.1 Single Surface Measurement

    29.3.2 Web/Pocket Measurement

    29.3.3 Bore/Boss Measurement

    29.3.4 Measuring Multiple Features

Learning Objectives

After completing this chapter, you will be able to:

- Explain how to use a probing program to measure machined features, adjust tool offsets, and verify that features meet tolerance requirements.
- Use the P9811 probing cycle in a program for feature measurement.
- Use the P9812 probing cycle in a program to measure a web or pocket.
- Use the P9814 probing cycle in a program to measure a bore or boss.
- Explain how to use probing cycles to measure multiple features on a part.

Key Terms

backlash

geometric dimensioning and tolerancing (GD&T)

on machine verification

true position

29.1 Introduction

The last two chapters of this text have discussed probing and probe programming. These chapters covered methods used to probe features and set work offset coordinates. In Chapter 28, you learned how macro subprograms can be used in a main program to execute probing cycles and automate the process of setting work offsets. These same macro subprograms can be used to physically measure machined features in a CNC machine and provide a pass/fail result to the operator. This process is commonly called on machine verification, or OMV. *On machine verification* refers to checking and verifying features during the machining process to confirm a part meets design requirements. This process is used to check features after a machining operation is complete, but prior to removing the part from the machine.

29.2 Pros and Cons of On Machine Verification

Before exploring how features can be measured for verification in a program, it is worth discussing the benefits and potential disadvantages of measuring parts on the machine with an electronic probe. At first glance, it seems like an obvious decision to make sure that all the parts that have been machined are in tolerance before taking them out of the machine. For many machine shops, this is the preferred method. So what could possibly be the harm in this method?

Throughout this textbook, there has been an emphasis on the importance of efficiency. Efficiency does not just refer to the speed of a certain operation or part cycle time. Efficiency pertains to all of the processes that occur in a manufacturing environment. It is a measure of how processes are performed to reduce production time, mistakes, and the complexity of the job. An efficient machining process takes into account every aspect of production, including part verification. Taking a part out of the machine and measuring it is risky, because if the part is out of tolerance, it will be difficult to reinstall in the exact same location and re-machine the part. However, if the machining process is consistent, it saves cycle time to inspect the piece while another piece is running. By comparison, inspecting the part on the machine with a probe will add significant time to the overall manufacturing time. If the parts are being machined correctly, there is no need for 100% inspection. There may be parts with single features that have close tolerances requiring full inspection, and this may be a suitable application for on machine verification.

Another potential concern in using on machine verification is what is commonly referred to as "checking your own homework." To put it in simple terms, the CNC machine is being asked to verify what it completed. If the work offset coordinates were set incorrectly and are off by .005″, for example, the probe will measure the location of a feature with that .005″ error. If the machine has a mechanical movement error, or *backlash*, that same error will be seen with the movement of the probe. *Backlash* is mechanical play that occurs between the lead screw and mating nut in a screw assembly when a machine axis changes direction. In the process of machining a part, it is possible for a part that is out of tolerance to be measured as acceptable by the probe. To account for this, probe inspection is often performed at the machine, but at least some secondary inspection will be required. Do not assume that because the probe inspected the part, it is the final product that meets the customer's specifications.

When properly utilized, however, a probing system can be a game changer. Since the beginning of manufacturing, company owners and manufacturing engineers have known that inspection is considered a *non-value added* necessity. Non-value added means that the company is not getting paid for a service. No customer will pay an invoice that charges for inspection. Why? Because there is a fundamental assumption that the customer is paying for parts that are made to the print requirements—not for the machinist's time to check if the parts are made correctly. This is why it is so important to make sure the machining process will create a quality product. By adding a probe inspection to check features controlled by tight tolerances, the machinist can be assured that the parts are acceptable and the process is under control.

29.3 Tool Offset Adjustment and Tolerance Verification

The following sections cover probing cycles used in programs to measure and verify features. These are similar to the probing cycles that were covered in Chapter 28, but different parameter entries will be used to adjust tool offsets and verify machined features. The difference in application is that the feature being probed has already been machined.

The probing cycles covered in this chapter represent only a partial list of the cycles available to complete on machine verification. However, these probing cycles are the most frequently used and are sufficient to complete most probing applications. Probing cycles used for on machine verification can be stored as subprograms or they can be placed in the main program. Because these cycles are used to measure features at specific locations on the part after a machining operation, they are generally written directly into the main program, and not used as subprograms.

29.3.1 Single Surface Measurement

The P9811 single surface measurement probing cycle is used to measure a single surface in the X, Y, or Z axis. **Figure 29-1** shows a part that will be measured with the probe in the X axis direction. The probe will be turned on and moved to the starting position, and then the probing cycle will be initiated.

O2901 (Milling Program with Probing)	
G20 G17 G90 G80 G40	
T25 M6	
G0 G90 G54 X–1. Y–1.	
G43 H25 Z2.	
G65 P9832	Turn on probe
G65 P9810 Z–.5	Safe move to Z–.5 before probing
G65 P9811 X0. T1	Probe to X0. and update tool offset for tool #1

Before this part was machined, there was a coordinate for the G54 work offset X0. location stored in the machine controller. The part surface was machined to remove material in the X axis. The probing cycle is being used to measure the machined location. The final block in the program includes the T1 tool offset designation. It is important to note

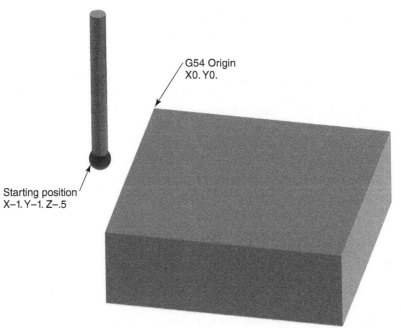

Figure 29-1. Using the P9811 measurement macro to measure the surface of the part in the X axis direction after the surface has been machined.

that the G65 P9811 X0. block can stand alone in the program without the tool offset designation. In that case, the probe will just measure the location and record the X axis coordinate value under a macro variable in the machine controller. The probe simply finds the part edge and records its location. The additional entries in the program are what force the controller to make a calculation with the measurement and do something. When the work offset designation S1. is added, a coordinate location is entered into the G54 work offset. When the T1 tool offset designation is added, the diameter offset setting of the tool is altered. For this to be successful, the tool must be milling with cutter compensation turned on. See **Figure 29-2.** In this operation, the G41 command is used to initiate cutter compensation left.

The probing cycle program referenced the tool offset designated T1. The T number must reflect the correct tool offset to be altered. If the probe measures the edge at X.005, it will alter the tool diameter offset (or radial offset) designated with the G41 D1 offset accordingly.

Now, the same part can be measured again to check if the part is in tolerance and meets the print requirements. The part dimension in the X axis direction is 3.000 ±.005. See **Figure 29-3.** An additional probing cycle will be used to inspect this dimension. The H address is used in the cycle to designate the tolerance. The lines of code for this cycle can be added to the existing program.

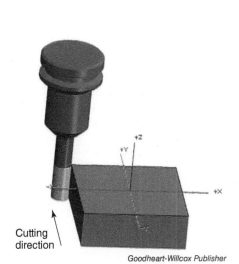

Figure 29-2. The side of the part is milled to remove material in the X axis direction. Cutter compensation is active. In this case, the G41 command is used to offset the tool to the left of the programmed cutting path and the tool diameter offset setting is referenced in the machining program.

Debugging the Code

Tool Offset Adjustment

Remember that the material was cut, the surface was measured with a probe, and the tool diameter offset setting was adjusted, but the part is still not to size. For the tool offset setting to be utilized, the machining cycle has to be run again.

O2901 (Milling Program with Probing)
G20 G17 G90 G80 G40
T25 M6
G0 G90 G54 X–1. Y–1.
G43 H25 Z2.

Continued

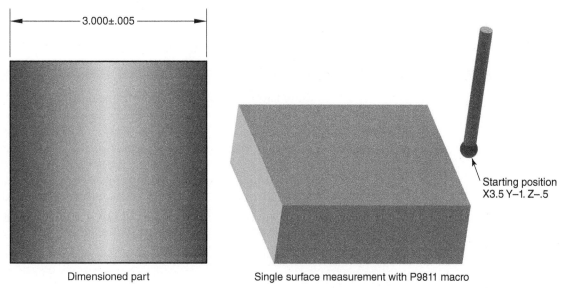

Figure 29-3. A single surface probing cycle is used to measure the surface on the right end of the part to verify that the part meets the specified tolerance on the print. The tolerance designation H.005 is used by the controller to calculate whether the feature meets the requirement.

G65 P9832	Turn on probe
G65 P9810 Z−.5	Safe move to Z−.5
G65 P9811 X0. T1	Probe to X0. and update tool offset for tool #1
G65 P9810 Z.5	Safe move to Z.5
G65 P9810 X3.5	Safe move to X3.5
G65 P9810 Z−.5	Safe move to Z−.5
G65 P9811 X3.0 H.005	Measure the 3.0 dimension with a tolerance of ±.005
G65 P9810 Z.5	Safe move to Z.5
G65 P9833	Turn off probe

Notice that all of the moves to reposition the probe to X3.5 Z−.5 are performed with the P9810 macro to protect the probing head. The X3.5 Z−.5 position is the starting position for the probing cycle to measure the surface on the right end of the part. After correct positioning, the probe is commanded to go to X3.0 and the tolerance designation H.005 is specified. The probing software only allows for a ± dimensional tolerance, so this must be programmed accordingly. For example, if the part is dimensioned as 3.000 +.010/−.000, the P9811 macro block would be programmed to X3.005 H.005. If the part is not in tolerance, the machine will stop and display an *Out of Tolerance* message.

29.3.2 Web/Pocket Measurement

The P9812 measurement macro is used to measure a web or pocket feature. A web or pocket feature can be probed to adjust a tool offset setting for machining or to verify that tolerance requirements are met. It is important to understand that while this discussion is focused on programming the CNC machine, these operations must be considered through the mind-set of the inspector and not just the machinist. Depending on the tolerance

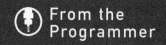

From the Programmer

Single Surface Z Axis Measurement

A single surface measurement probing cycle used to probe a surface in the Z axis produces a different result when using the T address. If a surface is probed in the Z axis direction and a tool offset is designated with the T address, it will adjust the corresponding tool length offset, not the diameter offset. This is often done with a face milling tool, but it can be accomplished with any tool. It is also possible to probe multiple part depths along the Z axis, but those depths are calculated relative to the probe's tool length offset.

requirements and part shape, more than one probing cycle may be needed to ensure that the part is made to the print specifications. **Figure 29-4** shows a part with a pocket being measured in the Y axis direction. Note that a web or pocket measurement can be made in the X axis direction by changing the direction of movement by the probe.

The first dimension that should catch the attention of the programmer is the .500″ slot width dimension. The slot feature is controlled by a tolerance of ±.001″. This is a very close tolerance for a slot spanning 3″ of material. Positioning the probe in the center of the slot and taking one measurement might not allow for keeping the slot feature in tolerance. Adjusting the tool diameter offset can be accomplished with a single probing cycle, but in order to check the slot tolerance, multiple measurements should be taken. See **Figure 29-5**. The following is a single program that includes probing operations for each measurement.

O2905 (Pocket Probing in Y Axis Direction)	
G20 G17 G90 G80 G40	
T25 M6	
G0 G90 G54 X1.5 Y–1.5	Position probe at center of pocket
G43 H25 Z2.	
G65 P9832	Turn on probe
G65 P9810 Z–.3	Safe Z axis move to below top of part
G65 P9812 Y.5 T5	Measure .500″ wide pocket in Y axis direction; update tool offset for tool #5
G65 P9810 X.5	Safe move to left side of pocket
G65 P9812 Y.5 H.001	Measure .500″ wide pocket with a tolerance of ±.001
G65 P9810 X2.5	Safe move to right side of pocket
G65 P9812 Y.5 H.001	Measure .500″ wide pocket with a tolerance of ±.001
G65 P9810 Z.5	Safe move to a Z position above part
G65 P9833	Turn off probe

After the probe is safely positioned inside the slot at X1.5 Y–1.5, the P9812 macro is commanded and the T5 tool offset designation is specified.

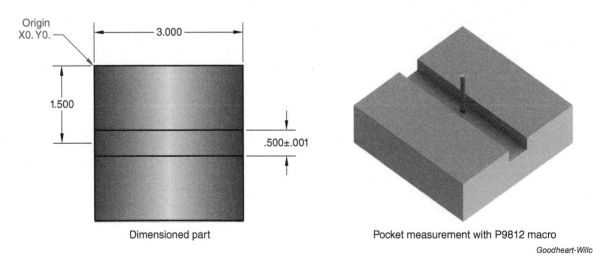

Dimensioned part

Pocket measurement with P9812 macro

Goodheart-Willcox Publisher

Figure 29-4. Using the P9812 measurement macro to measure a pocket in the Y axis direction after it has been machined.

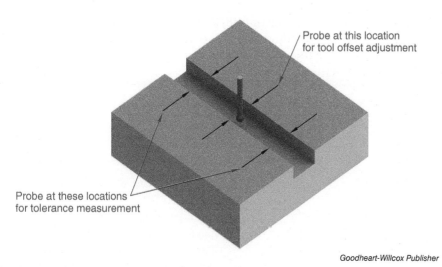

Figure 29-5. A pocket measurement probing cycle is performed to measure the center location of the slot and adjust the tool offset setting for machining. Two additional probing cycles are performed at opposite ends of the slot to check the ±.001″ tolerance for the slot dimension.

This will update the tool diameter offset (or radial offset) on the tool offset page for tool #5. It does not make sense to measure the feature and adjust this in multiple locations, because doing so would update the offset in every probing cycle and only the last measured position would be relevant. Next, the probe is safely positioned on one end of the slot and the feature is measured against the ±.001″ tolerance. After this cycle, the probe is safely moved to the opposite end of the slot and the feature is re-measured against the ±.001″ tolerance. Measuring in multiple locations verifies the slot is straight and consistent throughout. Depending on the programmer's comfort level, the slot could be measured as many times as required to ensure a quality product.

The same technique can be used for a .500″ wide web feature that protrudes in the Z axis direction. See **Figure 29-6**. The width dimension for

Figure 29-6. The probe is positioned to measure the center location of the web and adjust the tool offset setting. Two additional probing cycles are performed at opposite ends of the web to check the ±.001″ tolerance for the web dimension.

the web has a close tolerance of ±.001″. In the following program, the first probing cycle is used to adjust the tool offset and two more probing cycles are used to check the tolerance. An R address value is used in each cycle to step around the feature safely.

O2906 (Web Probing in Y Axis Direction)	
G20 G17 G90 G80 G40	
T25 M6	
G0 G90 G54 X1.5 Y−1.5	Position probe at center of web
G43 H25 Z2.	
G65 P9832	Turn on probe
G65 P9810 Z.3	Safe Z axis move to above top of part
G65 P9812 Y.5 R.25 Z−.3 T5	Measure .500″ wide web in Y axis direction; probe at Z−.3 with .25″ clearance; update tool offset for tool #5
G65 P9810 X.5	Safe move to left side of web
G65 P9812 Y.5 R.25 Z−.3 H.001	Measure .500″ wide web with a tolerance of ±.001
G65 P9810 X2.5	Safe move to right side of web
G65 P9812 Y.5 R.25 Z−.3 H.001	Measure .500″ wide web with a tolerance of ±.001
G65 P9810 Z.5	Safe move to a Z position above part
G65 P9833	Turn off probe

29.3.3 Bore/Boss Measurement

Bores with close tolerances are a very common requirement in machining. Measuring hole features with conventional tools presents certain challenges. There are multiple tools for measuring holes, including pin gages, bore gages, inside diameter micrometers, and telescoping gages. However, a better method is to use a probing system. A full contact probe measures quickly and produces very accurate results. This section covers probing methods used to measure hole size and location.

Figure 29-7 shows a part with a hole machined on the top face. The hole has a limit dimension of 1.499–1.501. In the following program, the P9814 measurement macro is used to measure the hole diameter at .500″ deep and adjust the tool diameter offset setting. Note that the T address in this cycle only works when using an end mill with cutter compensation active and programming the tool motion with circular interpolation commands.

O2907 (Bore Probing)	
G20 G17 G90 G80 G40	
T25 M6	
G0 G90 G54 X1.5 Y−1.5	Position probe at center of bore
G43 H25 Z2.	
G65 P9832	Turn on probe
G65 P9810 Z−.5	Safe Z axis move to below top of part
G65 P9814 D1.5 T5	Measure 1.5″ bore; update tool offset for tool #5
G65 P9810 Z.5	Safe move to a Z position above part
G65 P9833	Turn off probe

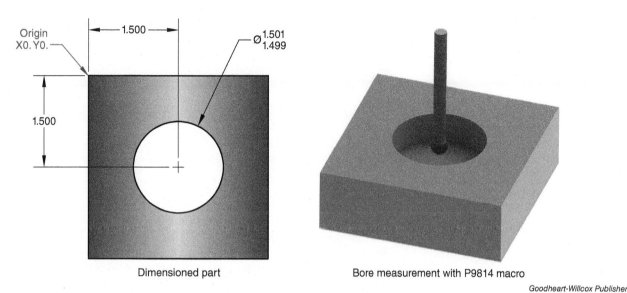

Figure 29-7. Using the P9814 measurement macro to measure a bore after it has been machined.

The following program is used to re-measure the bore and check the tolerance for the 1.499–1.501 dimension using the H address. The tolerance for this dimension can be expressed as a plus-minus tolerance using the value ±.001.

O29071 (Bore Probing)	
G20 G17 G90 G80 G40	
T25 M6	
G0 G90 G54 X1.5 Y–1.5	Position probe at center of bore
G43 H25 Z2.	
G65 P9832	Turn on probe
G65 P9810 Z–.5	Safe Z axis move to below top of part
G65 P9814 D1.5 H.001	Measure 1.5" bore with a tolerance of ±.001
G65 P9810 Z.5	Safe move to a Z position above part
G65 P9833	Turn off probe

If the part is not in tolerance, the machine will stop and display an *Out of Tolerance* message.

For deeper holes, it will be important to measure the diameter at multiple depths. Hole features can have out-of-roundness conditions or taper throughout the length of the hole. The customer will expect the hole feature to meet the print requirements at all locations in the hole. Consider a 1.5" diameter hole that is 1.5" deep. The following program is used to probe at three depths to measure the hole diameter and verify that it is in tolerance.

O29072 (Bore Probing)	
G20 G17 G90 G80 G40	
T25 M6	
G0 G90 G54 X1.5 Y–1.5	Position probe at center of bore
G43 H25 Z2.	
G65 P9832	Turn on probe

Continued

G65 P9810 Z–.3	Safe Z axis move to below top of part
G65 P9814 D1.5 H.001	Measure 1.5″ bore with a tolerance of ±.001
G65 P9810 Z–.75	Safe Z axis move to second probe depth
G65 P9814 D1.5 H.001	
G65 P9810 Z–1.2	Safe Z axis move to third probe depth
G65 P9814 D1.5 H.001	
G65 P9810 Z.5	Safe move to a Z position above part
G65 P9833	Turn off probe

The P9814 macro probes in four directions to measure the hole diameter. The four measured locations are at the 3 o'clock, 6 o'clock, 9 o'clock, and 12 o'clock positions, relative to the X axis. This cannot be altered in the macro. There will be instances when the part features will not allow for measuring at these locations. In such instances, a three-point inspection can be used. The P9823 measurement macro allows for three-point bore or boss measurement.

Figure 29-8 shows a part with a keyway slot oriented at the 6 o'clock position. Using the P9814 macro to measure the slot will generate a probe alarm, or false probe measurements. In this case, the P9823 macro can be used to measure the bore at three locations. In the macro program, the three probing locations are specified as angular locations relative to the X axis. Just as with other angular calculations in CNC programming, 0° is oriented at the 3 o'clock position. See **Figure 29-9**.

Because the probing cycle will measure at three locations, it is normal practice to measure the locations at 120° apart. This is not mandatory, but recommended for best results. The cycle in the following program will measure at 60°, 180°, and 300° to avoid the keyway. The angles are specified with the A, B, and C addresses.

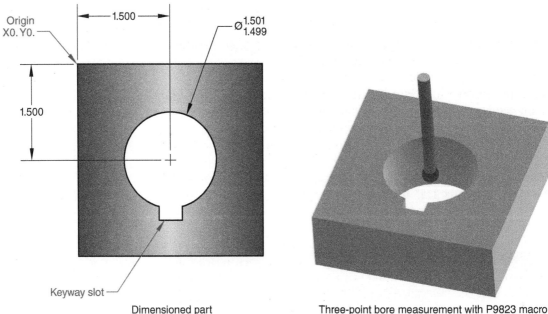

Dimensioned part

Three-point bore measurement with P9823 macro

Goodheart-Willcox Publisher

Figure 29-8. The P9823 measurement macro is used to measure a bore or boss at three locations. This allows the probe to measure the bore shown at locations away from the keyway slot. The three probing locations are specified as angular locations relative to the X axis.

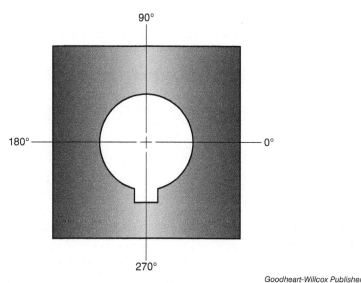

Figure 29-9. Angular positions used in programming probing locations. The 0° position is oriented at 3 o'clock.

O2909 (3 Point Bore Probing)	
G20 G17 G90 G80 G40	
T25 M6	
G0 G90 G54 X1.5 Y–1.5	Position probe at center of bore
G43 H25 Z2.	
G65 P9832	Turn on probe
G65 P9810 Z–.5	Safe Z axis move to below top of part
G65 P9823 A60. B180. C300. D1.5 H.001	Probe at 60°, 180°, and 300° positions
G65 P9810 Z.5	Safe move to a Z position above part
G65 P9833	Turn off probe

In the P9823 block, the A address value represents the first probing angle from the bore center measured from the X axis. The B address value represents the second angle and the C address value represents the third angle. These angular movements can be in any order and at any angle.

In the previous examples, the tolerance designations used to measure features represented tolerances applied to control size dimensions, such as the limit dimension in **Figure 29-7**. Tolerances are also used to control the allowable variation of a feature from a specified location. *True position* is the theoretically exact location of a feature of size. *True position* is a term used in geometric dimensioning and tolerancing (GD&T). *Geometric dimensioning and tolerancing (GD&T)* is a system that uses geometric tolerances to define allowable variations of form, orientation, location, profile, and runout. In the GD&T system, a positional tolerance can be used to control the allowable variation of a feature relative to a specified position. When applied to a hole, a positional tolerance establishes a cylindrical tolerance zone centered at the true position. The axis of the feature must be located within the tolerance zone. The size of the tolerance zone for a circular feature is specified on a print with a diameter value, indicating that the zone is cylindrical (round). The tolerance zone represents the maximum allowable error of the feature in the X and Y axis directions.

A bore or boss feature can be measured in a probing cycle to verify that it meets the requirements of a positional tolerance. In the program, the M address is used to specify the tolerance value. Referring to the part in **Figure 29-7**, there is a very close tolerance used to control the size of the hole. There is also a .006″ positional tolerance requirement for the hole. The machined hole must meet both requirements. The following program is used to verify the .006″ diameter positional tolerance is met.

O29091 (Bore Probing)	
G20 G17 G90 G80 G40	
T25 M6	
G0 G90 G54 X1.5 Y–1.5	Position probe at center of bore
G43 H25 Z2.	
G65 P9832	Turn on probe
G65 P9810 Z–.5	Safe Z axis move to below top of part
G65 P9814 D1.5 M.006	Measure 1.5″ bore with a true position tolerance of .006
G65 P9810 Z.5	Safe move to a Z position above part
G65 P9833	Turn off probe

A machined boss feature can also be measured using the P9814 or P9823 macro. The part shown in **Figure 29-10** has a 1.5″ boss located at X1.5 Y1.5. The work origin for this part is located at the bottom-left corner on the top face. In the following program, the P9814 macro is used to perform a four-point inspection. An R address value is used to step around the feature safely in the negative Z axis direction.

O2910 (4 Point Boss Probing)	
G20 G17 G90 G80 G40	
T25 M6	
G0 G90 G54 X1.5 Y1.5	Position probe at center of boss
G43 H25 Z2.	

Continued

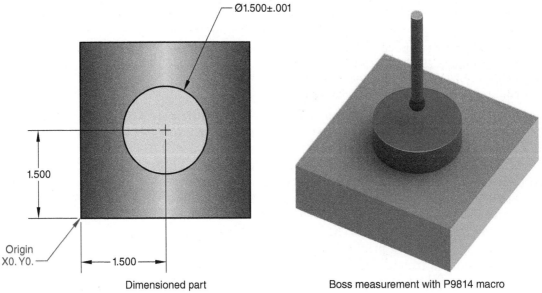

Figure 29-10. Using the P9814 measurement macro to measure a boss after it has been machined.

G65 P9832	Turn on probe
G65 P9810 Z.3	Safe Z axis move to above top of part
G65 P9814 D1.5 Z−.3 R.25 H.001	Measure 1.5″ diameter boss with a tolerance of ±.001; probe at Z−.3 with .25″ clearance
G65 P9810 Z.5	Safe move to a Z position above part
G65 P9833	Turn off probe

The P9823 macro is used in the following program to perform a three-point inspection.

O29101 (3 Point Boss Probing)	
G20 G17 G90 G80 G40	
T25 M6	
G0 G90 G54 X1.5 Y1.5	Position probe at center of boss
G43 H25 Z2.	
G65 P9832	Turn on probe
G65 P9810 Z.3	Safe Z axis move to above top of part
G65 P9823 A0. B120. C240. D1.5 Z−.3 R.25 H.001	Measure 1.5″ diameter boss with a tolerance of ±.001; probe at Z−.3 with .25″ clearance
G65 P9810 Z.5	Safe move to a Z position above part
G65 P9833	Turn off probe

29.3.4 Measuring Multiple Features

The probing cycles discussed in this chapter have been presented individually for explanatory purposes. However, the real power of measuring with a probe is achieved when an entire part is verified on the machine. **Figure 29-11** shows a part with multiple features. These features would normally be verified by adding probing cycles to the end of the program, after machining, but they could be measured after each machining operation, depending on the efficiency desired. For this part, consider all tolerances to be ±.003″. In **Figure 29-12**, the probing locations are numbered in order of execution. The following program is used to measure each feature in order.

O0326 (Full Inspection Probing)	
G20 G17 G90 G80 G40	
T25 M6	
G0 G90 G54 X1. Y1.25	Position probe at center of bore
G43 H25 Z2.	
G65 P9832	Turn on probe
G65 P9810 Z−.3	Safe Z axis move to below top of part
G65 P9814 D.875 H.003	Measure at Point 1
G65 P9810 Z.5	Safe move to a Z position above part
G65 P9810 X4. Y2.	Position probe at center of bore

Continued

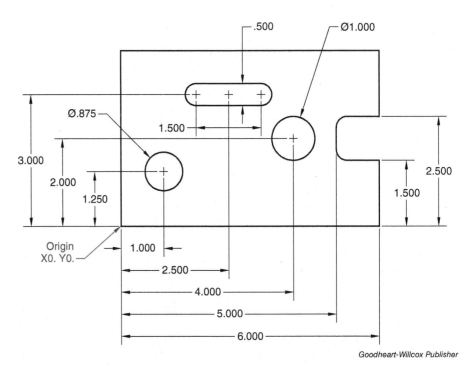

Goodheart-Willcox Publisher

Figure 29-11. A part with multiple machined features. Each feature is measured in a probing cycle to verify it meets tolerance requirements.

G65 P9810 Z-.3	Safe Z axis move to below top of part
G65 P9814 D1. H.003	Measure at Point 2
G65 P9810 Z.5	Safe move to a Z position above part
G65 P9810 X2.5 Y3.	Position probe at center of slot
G65 P9810 Z-.3	Safe Z axis move to below top of part
G65 P9812 Y.5 H.003	Measure at Point 3
G65 P9810 Z.5	Safe move to a Z position above part
G65 P9810 X5.5 Y1.8	Safe move to XY location before probing
G65 P9810 Z-.3	Safe move to a -Z location before probing
G65 P9811 Y1.5 H.003	Measure at Point 4
G65 P9810 X5.5 Y2.2	Safe move to XY location before probing
G65 P9811 Y2.5 H.003	Measure at Point 5
G65 P9810 X5.5 Y2.	Safe move to XY location before probing
G65 P9811 X5.0 H.003	Measure at Point 6
G65 P9810 Z.5	Safe move to a Z position above part
G65 P9833	Turn off probe

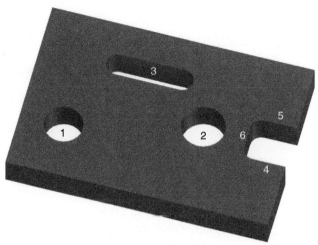

Goodheart-Willcox Publisher

Figure 29-12. Numbered locations identifying the order in which probing cycles are performed to measure and verify features after machining.

If these probing cycles were added at the end of the program, the machine would be sent home in the Z axis and the M30 function would be used to rewind the program.

From the Programmer

Advancing Probing Technology

Probe measuring cycles can be used to eliminate unnecessary inspection and improve part quality and consistency. Probing technology has advanced tremendously and there are now options for high-speed scanning and contactless measuring devices. These devices will soon be standard equipment on CNC machines. Becoming proficient with probing cycles will make the programmer a valuable member of any manufacturing team.

Chapter 29 Review

Summary

- On machine verification is the process of measuring part features after the machining operation is complete, but prior to removing the part from the machine. On machine verification can be performed with high accuracy and speed using probing cycles.
- The P9811 macro can be used to measure a single part surface. The T address can be used in the program to alter the tool diameter offset setting. By using the H address with a tolerance value, a surface can be measured to verify it is in tolerance.
- The P9812 macro can be used to measure a slot, pocket, or web. The probing cycle can be used to adjust the tool offset or verify the feature is in tolerance. The probe will travel in the X or Y axis direction, depending on the designation in the P9812 block.
- The P9814 macro can be used to measure a bore or boss. The tool offset can be adjusted if the hole is machined using an end mill with cutter compensation. The M address can be used to verify the feature meets the requirements of a positional tolerance.

Review Questions

Answer the following questions using the information provided in this chapter.

Know and Understand

1. *True or False?* On machine verification (OMV) refers to checking and verifying features on a part prior to removing the part from the machine.
2. *True or False?* A probing cycle used for on machine verification can be stored as a subprogram or written directly into the main program.
3. Which address code is used to adjust a tool diameter offset setting when measuring a feature in a probing cycle?
   A. D
   B. P
   C. S
   D. T
4. Which address code is used to specify a tolerance value to be verified when measuring a feature in a probing cycle?
   A. F
   B. H
   C. S
   D. T
5. The _____ probing cycle is used to measure a web or pocket feature.
   A. P9810
   B. P9811
   C. P9812
   D. P9814
6. The _____ probing cycle is used to measure a bore or boss feature.
   A. P9810
   B. P9811
   C. P9812
   D. P9814
7. The _____ probing cycle address is used to specify a tolerance value when measuring a feature to verify it meets the requirements of a positional tolerance.
   A. H
   B. M
   C. S
   D. T
8. The true _____ of a feature of size is the theoretically exact location of the feature.
   A. dimension
   B. position
   C. tolerance
   D. variation
9. *True or False?* The P9822 probing cycle is used to measure a bore or boss feature at three locations.
10. *True or False?* Probing cycles used to check multiple machined features can be added at the end of a main program or after each machining operation.

Apply and Analyze

1. Referring to the blanks provided below, give the correct codes that are missing for the single surface probing cycle. This cycle is used to measure a surface in the X axis direction and adjust the tool offset for tool #5.

   O2912
   G20 G17 G90 G80 G40
   T25 _____
   G0 G90 G54 X–1.5 Y–1.5
   G43 _____ Z2.
   G65 P9832
   G65 _____ Z–.5
   G65 P9811 X0. _____

2. Referring to the blanks provided below, give the correct codes that are missing for the pocket probing cycle. This cycle is used to probe a .375″ wide slot with a tolerance of .001″ in the Y axis direction.

   O2913
   G20 G17 G90 G80 G40
   T25 _____
   G0 G90 G54 X3.5 Y1.5
   G43 _____ Z2.
   G65 _____
   G65 _____ Z–.3
   G65 P9812 _____ _____
   G65 _____ Z.5
   G65 _____
   G0 G91 G28 Z0.

3. Referring to the blanks provided below, give the correct codes that are missing for the bore probing cycle. This cycle is used to probe a .750″ diameter bore with a tolerance of .003″.

   O2914
   G20 G17 G90 G80 G40
   T25 _____
   G0 G90 G54 X2.25 Y1.75
   G43 _____ Z2.
   G65 _____
   G65 _____ Z–.5
   G65 _____ _____ _____
   G65 _____ Z.5
   G65 P9833

Critical Thinking

1. On machine verification provides instant feedback to the operator, but slows down the manufacturing time. What types of parts would benefit from OMV? What are the impacts to consider if the production run is increased from five parts to 5000 parts?

2. Adding new technology to machines always takes some effort to convince management and the shop owner. How does adding probes to existing machines benefit the company? Consider inspection costs compared to the costs of a probing system, along with the costs of machined parts that are out of tolerance and longer cycle times.

3. Probing is not new technology, but it has evolved rapidly. What are the latest advances in full contact probing? What other technologies are being developed besides contact probing? Are these technologies readily available and currently in use on modern CNC machines?

30 Macros

Chapter Outline

30.1 Introduction
30.2 Using Variables in Macros
30.3 Local Variables
30.4 Global Variables
30.5 System Variables
    30.5.1 Current Machine Position Variables
    30.5.2 Work Offset Variables
    30.5.3 Tool Offset Variables
    30.5.4 Programmable Alarm Message Variable
30.6 Macro Programming Example

Learning Objectives

After completing this chapter, you will be able to:

- Explain the purpose of macros and how they can be utilized.
- Identify local variables, global variables, and system variables and describe how they are used in macro programming.
- Explain how to set work offset coordinates with system variables.
- Explain how to use a system variable to alter a tool offset setting.

Key Terms

global variables
local variables
macro
system variables
variable

30.1 Introduction

This chapter serves as a general introduction to macros, variables, and macro programming techniques. Macro subprograms used in probe programming were covered in Chapters 27–29. The purpose of this chapter is to make the programmer aware of the additional applications of macros and to explore the often under-utilized power of the CNC machine controller. A *macro* is a custom subroutine that defines one or more actions performed by the macro. In general terms, a macro is used to control a program using external data. There are many types of applications for macros, and they are only limited to the creativity and skill of the programmer. Before going further, it is important to note that macros can be used to alter machine parameters. Changes to parameters as a result of misguided or misinformed programming can cause issues such as corrupting the controller software, requiring a full reinstall of all machine parameters. For this reason, this chapter will focus on basic applications of macros.

The term *macro* has a very broad meaning. For example, macros are used in almost every program without us even being aware. The M6 tool change function is a macro. In the infant stages of CNC machines (during the 1970s and early 1980s), programming a tool change looked similar to the following.

G91 G28 Z0. M8	
M5	Spindle stop
M19	Spindle orient
M6	

The machine had to be told to return to machine zero, turn off coolant, turn off the spindle, and orient the spindle. It is taken for granted today that the M6 function will complete all of these tasks. This is because the M6 function is a macro program that commands the machine controller to perform the tasks in the background, without the operator seeing it happen.

Macros reduce programming time and make setup and machining processes more efficient. Often, macros are used to automate common operations on the machine. As discussed in previous chapters, in probe programming, macros are used for tasks such as turning the probe on and off and performing a specific probing cycle. Macros can also be used to automate setup tasks and machining operations such as setting work offset coordinates and drilling holes. For example, a custom drilling routine can be stored and called in a program whenever a pattern of holes is required on a part. Macro variables can be used in the program as needed to machine a certain number of holes at the required depth and location.

To start with, it is important to understand what is happening in the CNC machine controller. *CNC* stands for computer numerical control, meaning that the machine's computer is *numerically* controlled. A CNC machine is driven by numeric inputs. Most of these inputs are stored in the parameters section of the controller, and then there is another set of inputs in the variables section (or macro variables section) of the controller. Variables serve a key function in macro programming. A *variable* is a placeholder used to store and access data. Variables give macros flexibility by allowing data to be stored as values in a temporary location. Values stored by variables can then be used to perform actions in a program.

The variables section of the controller contains values that the machine uses, and they are generally locked in place. For example, the variable for the G54 work offset X axis coordinate is variable #5221 on a Haas controller. When the work offset coordinates are set on the work offset page, the X axis position entered by the operator is stored in variable #5221. The operator may think that when the G54 X axis position is referenced in a program, the controller looks at the work offset page to find that location, but it actually looks at variable #5221. There is another set of variables that are categorized as *open* and can be used by the programmer. For example, the operator can go to MDI (manual data input), input #101 = 1, and press the Enter key. This input assigns the value 1 to the user-defined variable #101, so by looking at the variable page, the operator will see that variable #101 now has a 1 in the value column.

30.2 Using Variables in Macros

Every controller manufacturer, whether it is Fanuc, Haas, Mazak, Okuma, or others, will define variable numbers specific to that manufacturer's controller. In addition, many machine builders define certain variables for operation of their machine, even if they use an off-the-shelf controller (such as a Fanuc controller). Therefore, there is no standard variable list for all machines. The only way to verify what each variable number controls is to consult your machine's operating manual. In this text, only the Haas variable lists will be used. The following sections discuss variables that are stored for machine use and variables that can be leveraged to the programmer's benefit. There are three different types of variables: local variables, global variables, and system variables. Each type has a specific purpose and is identified in a range of numbers on the controller.

30.3 Local Variables

Local variables are variables defined for temporary use in a macro subprogram called with the G65 command. Local variables are numbered in the controller as #1–#33. When variables are used in a program or in any subprogram, they are expressed with the # symbol. Local variables are defined by entering a letter address and the desired variable value on the G65 command block. The variable values are then used in the macro subprogram called by the G65 command. Each variable value is assigned to a variable number corresponding to the letter address. In other words, on a G65 macro command line, the *A* address will input data into variable #1, the *B* address will input data into variable #2, the *C* address will input data into variable #3, and so on. Local variables are passed to the macro subprogram and are used *locally* by the macro subprogram. Local variables are not passed to other macro programs, so be aware that local variables #1–#33 are only for use in the macro called by the G65 command.

30.4 Global Variables

Global variables are variables that are intended for general use in macro programming. There are approximately 500 global variables in the Haas controller. Global variables are not used by any other function of the machine and are wide open to use by the programmer. **Figure 30-1** shows the variables defined as global variables. There are roughly 400 usable variables and 100 non-usable variables. The non-usable variables are used for calibration and measurement data in probing routines. Variables #550–#599 store the probe calibration data. These variables should be avoided. If these were changed, the probe would not provide accurate data, or would need recalibration. Variables #150–#199 store probing measurement data, such as the X, Y, or Z location of a single surface measurement, and the measured bore diameter in a bore probing cycle. These variables should also be avoided. The usable global variables are completely open to use by themselves or in conjunction with system variables.

As previously discussed, in MDI mode, it is possible to manually enter a line of code that simply states #100 = 1. This entry assigns a value of 1 to variable #100. In the controller, variable #100 can be looked at, and a *1* will appear in the value column for the variable. At this point, the value *1* is essentially serving no purpose. But, if a few lines are written in a program, it can serve a purpose. The following program is an example.

```
O3001
#100 = 1
#101 = 2
#102 = #100 + #101
M30
```

This program states that variable #100 has a value of *1*, variable #101 has a value of *2*, and variable #102 is equal to the value of variable #100 plus the value of variable #101. It is easy to see that the value of variable #102 is *3*. Although this example is very simple, it shows how variables can be used to perform calculations or represent values in a program. For example, as discussed in previous chapters, operations can be repeated by specifying a loop count. The K or L address is used to create loops, depending on the controller type. The number of loops can be programmed using a variable instead of a fixed number. In the following program, an incremental move is looped to repeat a drilling cycle.

Usable Global Variables

Variables	Description
#100–#149	General purpose variables
#500–#549	General purpose variables
#600–#699	General purpose variables
#800–#999	General purpose variables

Non-usable Global Variables

Variables	Description
#150–#199	Probe values
#550–#599	Probe calibration data

Goodheart-Willcox Publisher

Figure 30-1. Usable and non-usable global variables. The usable variables are available for general use. The non-usable variables store probing calibration and measurement data and should be avoided.

```
O3002
#102 = 10                          Set variable #102
T1 M6
G90 G54 X1. Y-1. M3 S3000
G43 H1 Z1. M8
G81 G98 R.1 Z-.1 F10.
G91 X1. L#102                      L address references variable #102
```

In this example, the amount of loops is controlled by the value set in variable #102. The value (10) is established inside the program. If this part had multiple rows of drilled holes along the X axis, with a different number of holes to be drilled in each row, variables could be set for each row of holes. Variables can be used to create increasing counts, as shown in the following example.

```
#125 = #125 + 1
#126 = #125
```

To begin the cycle, variable #125 will have a value of 0. The program then tells the controller that variable #125 is equal to variable #125 plus 1, changing the value of variable #125 to 1. The last step is to set the value of variable #126 to the same value as #125, which is currently 1. The next time the program is executed, the value of variable #125 is still 1, and the program states that variable #125 equals the current setting plus 1, making it 2, and so on. Every time the program sees this sequence, it will add a value of 1 to the value of variable #126. Variable #126 can then be utilized for operations such as engraving serial numbers and counting the number of cycles completed.

The use of global variables is not limited to basic addition functions. Other mathematical functions can be used to calculate variable values. For example, trigonometric functions can be used to calculate values for angles. The following examples show the programming format used with standard mathematical functions. Brackets are used instead of parentheses in expressions to separate operations where needed.

Subtraction

```
#105 = 10.7
#106 = 25.2 - #105
```

Multiplication

```
#110 = 3.5
#111 = 2.6 * #110
```

Division

```
#115 = 12
#116 = #115/3.14
```

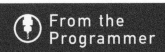

From the Programmer

Using Global Variables

There is virtually no limit to the number of applications for programming with global variables. Global variables can be used in programs or subprograms to automate functions, perform math calculations, and represent standalone values to simplify programming. The topic of macro programming is so extensive that there will be separate machine manuals listing all of the macro variables and how to use them. Macro programming with global variables is advanced-level programming and will take considerable effort and experience to master.

Trigonometric Functions

#125 = SIN[38]	Calculates the sine of 38° or .6157
#126 = 35	Sets variable #126 to the known angle of a triangle
#127 = COS[#126]	Calculates the cosine of 35° set in variable #126 or .8192
#128 = TAN[32]	Calculates the tangent of 32° or .6249

Additional Functions

#135 = SQRT[25]	Calculates the square root of 25 (5) and sets variable #135 to that value
ABS[–12.5]	Calculates the absolute value of –12.5

The ABS function is the absolute function. It will calculate a positive value for a given number. Some mathematical operations will produce a negative result and will require a conversion to positive output. The input **ABS[–12.5]** will return the positive value 12.5.

Boolean functions and binary functions can also be used in calculations. A Boolean function compares two values to determine whether the expression is true or false. A binary function compares two conditions to determine whether the expression is true or false. Boolean functions include the following.

- EQ = *Equal to*
- NE = *Not equal to*
- GT = *Greater than*
- LT = *Less than*
- GE = *Greater than or equal to*
- LE = *Less than or equal to*

Binary functions include the following.

- AND = *And* function
- OR = *Or* function

30.5 System Variables

System variables store values that represent the current operating conditions and modes in the controller. System variables are the largest group of variables in the controller. They allow the programmer to access a variety of controller conditions and functions, including tool offset and wear offset settings, work coordinates, timers, alarms, and even axis load values. System variable values can be used in programming to change the way the controller functions. When a system variable is read by the controller, the variable value can alter the machine's operation. However, some system variables are defined as *read only*. This means the system variable can be seen, but not altered. The key thing to know about system variables is that they can be used in conjunction with global variables to perform operations or calculations. The following sections discuss some of the categories of system variables and the data the variables contain. Later in this chapter, a macro programming example will show how global variables and system variables can be used in conjunction to measure parts, adjust tool offsets, and display a message on the machine telling the operator if the part is acceptable.

30.5.1 Current Machine Position Variables

The system variables in the #5021–#5026 series can be used to read or record the machine's current coordinate position in any axis. See **Figure 30-2**. If a position needs to be recorded, the movement can be stopped and the corresponding system variable can be transferred to a global variable. For example, the input #101 = #5021 records the machine's current X axis position to variable #101.

30.5.2 Work Offset Variables

The system variables in the #5221–#5326 series store the work offset coordinate values for the G54–G59 work offsets. See **Figure 30-3**.

The following macro program will machine a part using the G54 work offset set by the probe, then machine a second part that is located exactly 2″ in the positive X axis direction from the first part. The probe sets the work coordinate system on the first part, then the work origin is shifted in the X axis to machine the next part.

...	Program machines first part
#105 = #5221	The original G54 work offset X axis coordinate position is stored in variable #105
#5221 = #5221 + 2	The G54 work offset X axis coordinate value is shifted +2″
...	Program machines second part
#5221 = #105	The original G54 work offset X axis coordinate position is restored from variable #105

Another use for the work offset system variables is to pre-establish the work offset coordinates, specifically in the X and Y axes. Many machine shops will use fixtures that are precisely located on the machine table. When a new fixture is added, or a new part is mounted onto the fixture, the work offset position will be the same every time. It saves considerable time in setup to add the work offset coordinates to the program. For

> **From the Programmer**
>
> **Reading Variables**
>
> Use caution when referencing system variables and ensure you are using the correct variable number for the intended purpose. Each variable represents a specific condition or behavior in the controller. Incorrect use can result in damage to the part or machine.

Current Machine Position System Variables

Variable	Description
#5021	X axis position
#5022	Y axis position
#5023	Z axis position
#5024	A axis position
#5025	B axis position
#5026	C axis position

Goodheart-Willcox Publisher

Figure 30-2. System variables that store the machine's current position in each axis.

Work Offset System Variables

Variable	Work Offset	Description
#5221	G54	X axis coordinate
#5222	G54	Y axis coordinate
#5223	G54	Z axis coordinate
#5224	G54	A axis coordinate
#5225	G54	B axis coordinate
#5226	G54	C axis coordinate
#5241	G55	X axis coordinate
#5242	G55	Y axis coordinate
#5243	G55	Z axis coordinate
#5244	G55	A axis coordinate
#5245	G55	B axis coordinate
#5246	G55	C axis coordinate
#5261	G56	X axis coordinate
#5262	G56	Y axis coordinate
#5263	G56	Z axis coordinate
#5264	G56	A axis coordinate
#5265	G56	B axis coordinate
#5266	G56	C axis coordinate
#5281	G57	X axis coordinate
#5282	G57	Y axis coordinate
#5283	G57	Z axis coordinate
#5284	G57	A axis coordinate
#5285	G57	B axis coordinate
#5286	G57	C axis coordinate
#5301	G58	X axis coordinate
#5302	G58	Y axis coordinate
#5303	G58	Z axis coordinate
#5304	G58	A axis coordinate
#5305	G58	B axis coordinate
#5306	G58	C axis coordinate
#5321	G59	X axis coordinate
#5322	G59	Y axis coordinate
#5323	G59	Z axis coordinate
#5324	G59	A axis coordinate
#5325	G59	B axis coordinate
#5326	G59	C axis coordinate

Goodheart-Willcox Publisher

Figure 30-3. System variables that store the work offset coordinate values for the G54–G59 work offsets.

example, if a new fixture is installed, and the G54 work offset coordinates are going to be X–12.25 and Y–7.5, the program can start by writing these values to the appropriate work offset system variables.

```
O3002
#5221 = –12.25
#5222 = –7.5
...
```

This technique allows the operator to load the fixture and press the cycle start button without having to establish the work offset coordinates.

30.5.3 Tool Offset Variables

The system variables in the #2001–#2800 series store values for tool offset settings. See **Figure 30-4**.

Variables #2001–#2200 store the tool length offset position for each tool. This is the offset used with the G43 command. The value for tool #1 is stored in variable #2001, the value for tool #2 is stored in variable #2002, and so on.

Variables #2201–#2400 store the tool length offset wear value for each tool. The value for tool #1 is stored in variable #2201, the value for tool #2 is stored in variable #2202, and so on.

Variables #2401–#2600 store the tool diameter offset for each tool. This is the offset used with the G41 and G42 commands. The value for tool #1 is stored in variable #2401, the value for tool #2 is stored in variable #2402, and so on. The tool diameter offset variables can be used to adjust stored tool diameter offsets after probing features for size, as discussed later in this chapter.

Variables #2601–#2800 store the tool diameter offset wear value for each tool. The value for tool #1 is stored in variable #2601, the value for tool #2 is stored in variable #2602, and so on.

Debugging the Code

Tool Offset Settings
As discussed in Chapter 29, probing cycles use the T address to adjust tool diameter offset settings. When the T address is commanded, the controller uses the tool offset system variables to alter the designated tool diameter offset.

30.5.4 Programmable Alarm Message Variable

Custom alarms can be programmed to stop the machine when an error in machining occurs or a certain condition is detected by the controller. Alarms can be programmed using variable #3000. When a programmed alarm is generated, it is similar to the default alarms on the machine. To program an alarm, a value from 1 to 999 is assigned to variable #3000. The following format is used.

```
#3000 = 15 (PART OUT OF TOLERANCE)
```

Tool Offset System Variables

Variables	Description
#2001–#2200	Tool length offsets (tools 1–200)
#2201–#2400	Tool length offset wear values (tools 1–200)
#2401–#2600	Tool diameter offsets (tools 1–200)
#2601–#2800	Tool diameter offset wear values (tools 1–200)

Goodheart-Willcox Publisher

Figure 30-4. System variables that store values for tool length offsets and tool diameter offsets.

Any message can be added inside the parentheses. When an alarm is generated, the message flashes on the display to alert the operator. The value stored by the variable (in this case, 15) is added to 1000 and the resulting value represents the alarm number. When the alarm is generated, the machine stops and the program must be reset by the operator to continue the machining process.

30.6 Macro Programming Example

Now that you have an understanding of the different variable types, the following example shows how global variables and system variables can be leveraged in machining a basic part. In this example, a 1/2″ end mill (tool #5) is being used to finish machine a bore. The bore has a diameter dimensioned as 1.500–1.501. This is a very close tolerance bore. It is not going to be efficient to stop after each part is machined and use a conventional measuring device to verify the size. A smart programmer is going to use machining subprograms, a probing cycle subprogram, and variables to verify the size and adjust the tool diameter offset wear as needed. The following shows the main program and subprograms for producing the part.

O3005
M98 P3006 (Subprogram to rough bore, Tool #4)
M98 P3007 (Subprogram to finish bore, Tool #5)
G103 P1 (Block look-ahead function)
M98 P3008 (Subprogram to probe bore with G65 P9814 cycle)
#105 = 1.500 (Bore target diameter)
IF [#188 GT [#105 + .001]] GOTO3000 (Alarm)
#106 = #105 – #188 (Target diameter as probed)
#2605 = #2605 – #106 (Adjust tool wear offset)
G103
M30
N3000
#3000 = 15 (Bore Oversize)

There are many things happening in this program. The following explains the individual blocks in more detail.

- **M98 P3006.** This block calls the machining subprogram that uses tool #4 to rough out the bore diameter.
- **M98 P3007.** This block calls the machining subprogram that uses tool #5 to finish the 1.500–1.501 bore diameter.
- **G103 P1.** The G103 command limits the controller's look-ahead function. Normally, the CNC machine controller looks ahead and reads multiple blocks of code in the program. This enables the machine to change directions rapidly without pausing or hesitating. The more powerful the controller, the further it can look ahead in the program. In this case, however, the programmer does not want the controller looking ahead because there is a macro subprogram used for probing and the program contains conditional statements. The P address specifies the number of blocks the controller looks ahead.

- **M98 P3008.** This block calls the probing cycle subprogram to probe the bore diameter and verify the size. In the subprogram, the G65 P9814 D1.5 block is used to measure the bore and store the data points in the probing data measurement variables (variables #150–#199). Remember that these are *non-usable* global variables. This means that the variables store data, but they should not be altered by the programmer. The probed diameter is written to variable #188.
- **#105 = 1.500.** This block defines a global variable (variable #105). The 1.500 target diameter will be stored in this variable for use with a system variable in calculations later in the program.
- **IF [#188 GT [#105 + .001]] GOTO3000.** This block begins with the IF function and contains the GT Boolean function. The IF function is used to evaluate a conditional statement and represents a branch point in the program. The conditional statement is a logical expression. When combined with the IF function, it tells the controller what to do. In this case, the expression tells the controller how to proceed if the part is the correct size, or if it is out of tolerance. The GT function references variable #188, which stores the probed diameter. This is where the *actual* measured size of the bore is stored. When read by the controller, this block states that "If the actual size (variable #188) is greater than (GT) 1.500 (variable #105) plus .001, go to (GOTO) line 3000." The entry GOTO is the GOTO function. If this statement is true, the controller goes to line 3000 (N3000) and an alarm is generated. If the probe measures the bore at 1.502, for example, the conditional statement is true and the program jumps to the designated block number (N3000). The next block that follows defines the alarm.
  - **#3000 = 15 (Bore Oversize).** This will stop the machine and display the message "Bore Oversize" at the bottom of the controller screen. This means the part is out of tolerance and cannot be re-machined. If the conditional statement is not true, meaning the probed bore size is not greater than 1.501 (#105 + .001), the program will continue to the next block.
- **#106 = #105 – #188.** This block will calculate the difference between the target diameter of 1.500 (variable #105) and the probed diameter (variable #188) and write that value to variable #106. For example, if the probed diameter is 1.498, the calculation is as follows.

$$\#106 = 1.500 - 1.498$$

$$\#106 = .002$$

- **#2605 = #2605 – #106.** Variable #2605 stores the tool diameter offset wear value for tool #5. This expression states that variable #2605 = variable #2605 (the existing value) minus the calculated value stored in variable #106, or #2605 = 0 – .002, assuming the initial tool diameter offset wear value was 0. The operator should be able to verify that the tool wear offset for tool #5 changes and is set to a –.002 value.
- **G103.** This line cancels the block look-ahead limiting function.

The finishing program can be run again to recut the bore to the final size with the correct tool diameter offset wear value.

From the Programmer

Applying Macros

This is only one example of how measurement data can be leveraged to assist in machining parts. However, it is easy to see that with more knowledge and practice, macros can be applied in many other ways to automate common functions using data stored in the machine. Macros are another example of using creativity and programming skill to elevate efficiency.

Chapter 30 Review

Summary

- The term *macro* has a very broad meaning. In general terms, a macro is used to control a program using external data.
- Local variables are variables defined for temporary use in a macro subprogram called with the G65 command. Global variables are open to the programmer to use for general purposes. System variables are specifically assigned by the controller. System variables are usable by the programmer unless they are defined as read only.
- Work offset coordinates defined for the G54–G59 work offsets are stored in system variables in the #5221–#5326 series. Work offset coordinates can be directly input into these variables in the program, or the variable values can be referenced in calculations.
- Tool offset settings are stored in system variables in the #2001–#2800 series. Tool offset values can be directly input into these variables in the program, or the variable values can be referenced in calculations.

Review Questions

Answer the following questions using the information provided in this chapter.

Know and Understand

1. Which of the following is an application of macro programming?
   A. P9814 probing cycle
   B. M6 tool change function
   C. Programmable machine alarms
   D. All of the above.

2. *True or False?* System variables are numbered in the controller as #1–#33.

3. *True or False?* Global variables are general purpose variables used in macro programming.

4. Which of the following variables stores the Y axis coordinate position of the G54 work offset?
   A. #5221
   B. #5222
   C. #5223
   D. #5224

5. Which of the following variables stores the tool length offset position for tool #1?
   A. #2001
   B. #2201
   C. #2401
   D. #2601

6. Which of the following variables is a local variable?
   A. #10
   B. #100
   C. #150
   D. #5025

7. The ____ function is an example of a Boolean function.
   A. ABS
   B. IF
   C. EQ
   D. SIN

8. Alarms can be programmed using variable ____.
   A. #100
   B. #500
   C. #2200
   D. #3000

Apply and Analyze

1. Explain the purpose of local variables, global variables, and system variables. How do they differ in application?

2. Write an expression that stores the value 10.0 in variable #110. Write a second expression that calculates the difference between 20.0 and the value of variable #110 and records the value to variable #115.

3. Explain the difference between usable and non-usable global variables. In which number ranges are usable global variables designated?

Critical Thinking

1. Macro programming might seem like an entirely different way to program, but it really is not. It is another way to use data that is already available in the controller. How does the use of variables compare to the use of other types of codes in CNC programming?

2. Macro programming can reduce the length of programs and make them easier to edit. What are some examples of places in programs where macros can be used for greater efficiency? Consider parts with repetitive features and families of similar parts.

3. Probe programming is a form of macro programming. In probing cycles, measured values are written to global variables. Global variable outputs defined in Renishaw probing cycles are listed in the Renishaw probe programming manual, which can be found online. Referring to the outputs in the manual, select a specific type of feature and explain how a probed measurement could be verified in a program. Explain how macro variables could be used to produce the feature to meet tolerance requirements.

31 Additional Tips and Tricks

Chapter Outline

31.1 Introduction
31.2 Verifying the Program
31.3 Establishing Tool Offsets
31.4 Running the First Part
31.5 CAM Programming
31.6 Lathe Taper
31.7 Second-to-Last Cut and Finish Pass
31.8 Programming Feed Rates for Arcs
31.9 Subprogramming Applications
31.10 Metric Programming
31.11 Scaling
31.12 Mirror Imaging

Learning Objectives

After completing this chapter, you will be able to:

- Explain processes used to test programs before production machining begins.
- Describe techniques used to adjust tool offsets to verify a program and control critical features.
- Identify the cause of shaft taper in turning operations and explain how it is resolved.
- Calculate feed rates for corner radius features in milling operations.
- Explain how to use the G51 command to scale a toolpath.
- Explain how to use the G101 command to mirror a toolpath.

Key Terms

chatter
computer-aided manufacturing (CAM)
dry run
graphics mode
work coordinate system (WCS)

31.1 Introduction

CNC programmers, setup personnel, and machine operators are highly skilled and innovative workers. People in machine shops are almost exclusively responsible for the advances in CNC machines, tooling, and workholding. Think about that for a minute. When a CNC machinist has an idea to make a job easier, that person can design it, machine it, and then implement the idea. When machinists need a special tool for a job, they can make it themselves. Some of the best machines and accessories have been developed in machine shops to meet a specific need in making products or to improve efficiency. This chapter explores some of the "tricks of the trade" that will help in making quality parts as efficiently as possible.

31.2 Verifying the Program

After the stock is loaded, the tools are set, and the work coordinates are established, the operator wants to push the big green button and make some parts. But before a tool is driven into the material, the program should be verified. The term *verified* means that the actual code is checked for errors. For example, are there two M-codes on the same line? Is cutter compensation being turned on and off correctly? Are there any other syntax errors or codes and commands that the machine does not recognize?

There are a couple of different methods to verify the program. The first method, a ***dry run***, is a test run of the program to check for correct machine motion and accuracy without cutting stock. Dry run mode is a special machine function that is activated with a button or switch and should be available on every machine on the market. It allows the operator to have full control over the rapid move rates and feed rates in a program. All feed rates will be overridden and controlled by the operator using the manual override switches. The work plane is shifted so that it is safely above the part and the operator can slow down rapid movements and accelerate feed moves, if desired, to ensure that the axis movements and program syntax are correct.

The second method is to test a program in ***graphics mode***. Graphics mode is available on almost all machines built after 2010. Depending on the machine and controller type, this function is normally found on the control panel. Graphics mode can be used to run a very basic two-dimensional visual representation of the program. The program is executed in full rapid movement and should only take a few seconds to run. Although this is only a very basic simulation of what the tool is going to do, it will generate alarms if the code is incorrect.

31.3 Establishing Tool Offsets

There are many techniques that can be used to establish tool offsets to improve part quality and machining efficiency. Some of these techniques can be applied as part of the machine setup and others can be applied to a machining operation to improve part accuracy.

After a work setup is complete, and before running the first part, a test run can be made by offsetting the tools in a positive Z axis direction to keep them above the material at a safe distance. This allows the operator to see the toolpaths in real time without cutting the stock. This can be done with each individual tool, but it is much simpler to offset the *work coordinate system (WCS)* so the entire work plane is shifted in the positive Z axis direction. This works for both mills and lathes and can be done by offsetting the programmed work offset. It is important to make a note of the offset amount, so it can be removed after the test run. Use a common, round number for the Z axis offset, such as Z2.0 or Z4.0. After the program is tested, then remove that offset by applying an identical value in the negative Z axis direction. Make sure that the Z axis offset amount is more than the deepest Z axis movement in the program to prevent a tool from engaging the material.

When machining on a lathe, it is always best practice to offset the tools in the positive X axis direction before running the first part or after changing an insert. This will leave the part "bigger" than programmed and allow the operator to recut to the desired finish dimension. The best practice, depending on part size and shape, is to make the offset value at least .010" in the X axis direction. If too much material is left to machine, it could cause the tool to make a cut heavy enough to influence the final dimension. If the offset is too small, the part could be undersized, or a second pass will only allow for a couple thousandths of an inch to cut, causing the tool to push away and not machine cleanly.

In CNC milling, it is important to consider tool diameter offset settings for operations that use cutter compensation. Tool diameter offsets can be adjusted to test a program. By making the tool diameter offset larger, the tool will stay away from the final toolpath, allowing a recut if needed. For example, when a .500 diameter end mill is programmed using cutter compensation, the tool diameter offset can be set to .510, leaving material on the finished part. Setting the diameter offset to a larger value makes the controller think the tool is larger, keeping it away from the finish size. Make sure to record the diameter offset used so that it can be removed after testing the program.

These are all techniques used for part setups. There is another technique applied for tool offsets that is specifically for use in programming. Normally, tool numbers correspond to the same number designations used for stored tool offsets on the machine and specified offsets in the program. For example, tool #5 usually has a tool length offset programmed as H5 and a tool diameter offset programmed as D5. But using the same numbers for each offset is not required. In some cases, it might be beneficial to have different offsets. **Figure 31-1** shows a part with three features that can all

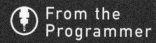

From the Programmer

Tool Offset Adjustment

If a machinist is cutting a part on a lathe in a test run and the final desired diameter on the part is 1.000", a preferred practice is to offset the tool by .010". After the cycle is completed, the machinist will measure the diameter with a micrometer and offset the difference from the measurement to the exact 1.000" dimension. For example, if the part measures 1.008" after the test run, offset the tool .008". Do not try to recut at .004", or stay on the high side of the tolerance. Measure the diameter once, offset the tool to the middle value of the tolerance, and re-run. This provides the best chance for success.

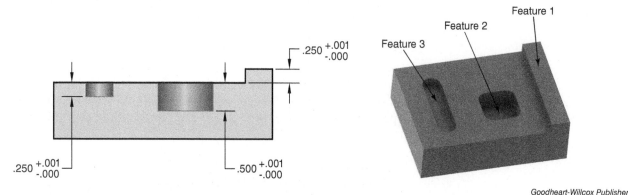

Goodheart-Willcox Publisher

Figure 31-1. A part with three separate features to be milled using the same tool. Different tool length offset values can be assigned to control the machined depth of each feature to meet tolerance requirements.

be finished to the correct Z axis depth with a .500 diameter end mill. The end mill is designated as tool #5. It is beneficial to use a single tool in this case, because the dimensions are established off the same surface (the top surface). However, because the features all have very close tolerances, it could be difficult to keep all three dimensions to the middle value (mean) of the tolerance.

In this case, it could be beneficial to machine the features with different tool length offset values. In the program, the G43 command is used with the H address to designate tool length offsets. For example, the raised boss (feature 1) could be programmed using the G43 H5 tool offset designation. The closed pocket (feature 2) could be programmed using the G43 H15 tool offset designation. The slot (feature 3) could be programmed using the G43 H25 tool offset designation. This allows the programmer to control the machined depth of each feature independently. The example part in **Figure 31-1** is a milled part. This technique can also be used in lathe operations when turning parts that have faces or diameters with close tolerances.

31.4 Running the First Part

Running the first part, especially in a new program, can be stressful. However, this process can be made less stressful by just slowing down the machine and following a couple of good practices. The first step is to make sure that there is an M01 code at every tool change in the program. This allows the operator to make an optional stop at each tool change. Before the program is tested, the optional stop function on the machine should be turned on using the Optional Stop button or switch. Next, reduce the rapid move rate by using the rapid override keys on the control panel. This is probably the most important action to take when machining the first part. The machine will default to 100% rapid on startup. Reduce this percentage to 25% or to 5%, if warranted. The program will almost always start with rapid positioning, so give the operator enough time to stop the machine if needed.

The next step is to place the machine in single block operational mode. This is done by pressing the Single Block key on the control panel. This tells the controller to execute only one line of the program for each time the cycle start button is pushed. Next, navigate to the *Distance to Go* display screen on the controller. This display screen will indicate to the operator how much farther the machine is going to travel to execute the current line of code. Once the machine is in single block mode, with the rapid move rate reduced, and the *Distance to Go* display screen active, press the cycle start button. Once the machine begins traveling toward the material, the machine can be stopped by using the Feed Hold button, and the distance to the part can be estimated on the *Distance to Go* display screen. It is a matter of preference, but once the machine is at the initial cutting depth, single block mode can be turned off. The program will run until the next tool change, where the optional stop function will halt the program. Repeat this process for each tool in the program.

31.5 CAM Programming

Computer-aided manufacturing (CAM) is the process of using computer software to define machining operations and generate CNC part programs. CAM software is the standard for machine shops operating in the modern machining environment. A CAM system allows a programmer to construct part geometry, select tooling, define toolpaths, and specify machine parameters and functions used in manufacturing a part. There are numerous benefits to programming in a CAM system. Understanding how G-codes and M-codes work in the controller and how programs are written is an absolute necessity to become a great programmer, but do not be a stubborn "hand programmer." A CAM system gives the programmer much more control of machining toolpaths and makes the programming process more efficient. The tools of a CAM system make the process of machining a very complex surface a simple operation. Knowing how to operate a CAM system is a skill that every modern CNC programmer must have.

When creating a program on a CAM system, take the same approach that is used when creating a program "by hand." Take the time to select the appropriate tools, toolholders, and part material, and define any other required parameters that are controllable in the CAM system. See **Figure 31-2**. Investing additional time in preparation will reap benefits in the final program produced. There is nothing more frustrating for an operator than getting a mediocre program from the programmer. If any changes are made at the machine, for any reason, go back to the CAM file and update the programmed toolpaths for future use.

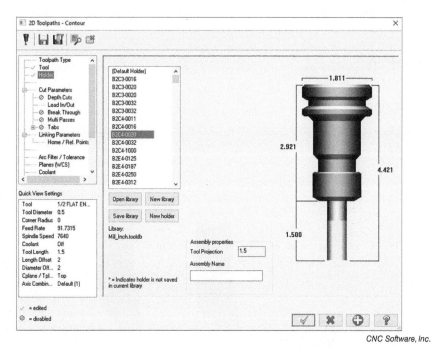

CNC Software, Inc.

Figure 31-2. A CAM software system allows the programmer to select the cutting tools and toolholders used in machining. Custom tooling can be created by defining the dimensions and other properties of the tool.

31.6 Lathe Taper

One of the common operations on a lathe is turning, where material is removed from the outside diameter of a shaft over an extended length. This creates a fulcrum effect where the material diameter closest to the part workholding is more rigid than the material diameter that is farthest from the workholding. When the cutting tool engages the material farthest from the workholding, the tool is pushed away from the workpiece by cutter pressure at the cutting edge. Conversely, the closer the cutter travels to the part workholding, the less it will deflect. This effect causes some taper in the shaft diameter and is referred to as tool deflection. The amount of taper produced will depend on the shaft diameter and length.

Whenever this type of machining is performed on a lathe, it will be necessary to check the machined diameter in multiple locations along the shaft. In **Figure 31-3**, the shaft is 1.500″ diameter and 3″ long. The programming code used for the finish pass is as follows.

```
G1 X1.5 F.01
Z-3.0
```

After the part is machined, the dimensional inspection shows the right-hand side of the shaft is 1.502″ diameter, the middle is 1.501″ diameter, and the diameter next to the shoulder is 1.500″. There is a .002″ taper in the shaft. A simple alteration can resolve this issue. Taper can be programmed out of the stock by adjusting the beginning and ending X axis coordinates in the program. Placing an X axis move on the same line as the Z axis move should remove or drastically reduce the taper.

First option

```
G1 X1.5 F.01
X1.502 Z-3.0
```

Second option

```
G1 X1.498 F.01
X1.500 Z-3.0
```

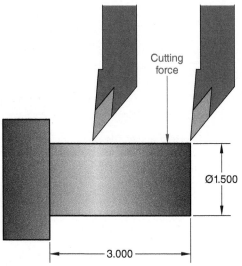

Goodheart-Willcox Publisher

Figure 31-3. In a turning operation performed on a shaft, tool deflection can be caused by cutter pressure at the unsupported end of the shaft, resulting in taper in the shaft diameter.

Either one of these options will force the tool deeper on the right-hand side of the shaft and gradually taper to the Z–3. position.

For machining operations on longer shafts, a rotating live center is often mounted on a tailstock to provide longitudinal support. This helps eliminate tool chatter and prevent tool failure when the tool enters the cut. *Chatter* is unwanted tool vibration caused by excessive or insufficient tool engagement or inadequate tool support. Chatter can damage the tool and result in a rough surface finish. Using a live center secures the end of the shaft, but this can cause another condition where tool deflection occurs at the center of the part and the diameter increases in size in the middle. See **Figure 31-4**. The center of the shaft is the farthest point from the two workholding locations. In this case, the part can be programmed in a similar manner to the previous example to produce a straight part. In this example, the shaft is 1.000″ diameter and 5″ long. To reduce or remove an increase in diameter of .002″ in the center, the tool movement is programmed as follows.

```
G1 X1.0
X.998 Z–2.5
X1.0 Z–5.0
```

The programmed movement may require more or less taper depending on the material type and size of the cutter nose radius.

31.7 Second-to-Last Cut and Finish Pass

There is an old adage in machining that goes back to manual machining: The finish pass is not as important as the second-to-last cut. Machinists have known for decades that the finish pass will never hold the required dimensional tolerance if it is not consistent. To be consistent, it must be made correctly every time. Often, operators alter finish passes by adjusting cutter offsets to "chase" a finished part size. There are many different variables to consider in machining a part to a near perfect finish size, but focus on the cut right before the finish pass. Do not make the mistake of trying to leave .002″ or .003″ of material for the finish pass. The problem with that approach is that the cutter needs to be able to cut, and if there is not enough material, the tool will rub against the workpiece. The other mistake is to think that leaving more material is a better idea. If the workpiece has .100″ of stock for the final pass, the tool will deflect, or push away, and leave a taper or rough surface finish. For lathe operations, the recommended practice is to leave the cutter nose radius size on both sides

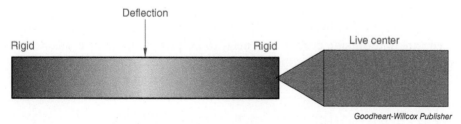

Goodheart-Willcox Publisher

Figure 31-4. A tailstock with a live center is often used to provide support for long workpieces on a lathe. Tool deflection can occur at the center of the part, resulting in a larger diameter in the middle.

of the workpiece to finish. In other words, a tool with a .015″ cutter nose radius needs .030″ of material for the finish pass. In a milling operation, about .010″ to .020″ of material on a wall should be adequate. Of course, these numbers can change, but if you are having issues with the final cut, the best strategy to resolve the issue is to leave the correct amount of material with the second-to-last cut.

31.8 Programming Feed Rates for Arcs

More often than not, a programmer will use one feed rate for the entire machining pass in a milling operation. On parts with rounded features, this often results in tool chatter, or a rough surface finish, in both internal and external radii. The reason for this is that the machine is being told to move in at least two axes at the same time to create a radius. This can produce a faceted surface finish. Circular moves require a different feed rate than linear moves. Outside radii should be cut at a faster rate than linear surfaces, and inside radii should be cut at a slower rate than linear surfaces.

The following formula is used to calculate the feed rate for an outside corner radius.

$$OF = \frac{F \times (R + r)}{R}$$

In this formula, OF is the outside corner radius feed rate, F is the linear feed rate, R is the part radius, and r is the tool radius. If a 1/2″ diameter end mill is being used to machine a .250″ outside corner radius and the linear feed rate is 30 inches per minute (ipm), the following is the feed rate calculation for the part radius.

$$OF = \frac{30 \times (.250 + .250)}{.250}$$

$$OF = 60 \text{ ipm}$$

The following formula is used to calculate the feed rate for an inside corner radius.

$$IF = \frac{F \times (R - r)}{R}$$

In this formula, IF is the inside corner radius feed rate, F is the linear feed rate, R is the part radius, and r is the tool radius. If a 1/2″ diameter end mill is being used to machine a .625″ inside corner radius and the linear feed rate is 15 inches per minute (ipm), the following is the feed rate calculation for the part radius.

$$IF = \frac{15 \times (.625 - .250)}{.625}$$

$$IF = 9 \text{ ipm}$$

These formulas are a recommended starting point, but some minor adjustments may be needed to produce the desired results. The key point to remember is that inside corner arcs need to be programmed at a slower feed rate and outside corner arcs need to be programmed at a faster feed rate for the same surface conditions.

31.9 Subprogramming Applications

The previous chapters in this text discussed a variety of subprogramming applications. Subprogramming is an under-utilized tool that can be used in many different programming scenarios. Subprograms help reduce programming time and can help eliminate programming errors. Most programs used in probing applications, such as probing for work offset coordinates and part verification, can be stored as subprograms in the controller for easy access through the main program. Create a list of these subprograms for quick reference when writing a new program. Once this is done, start adding macro variables to the subprograms or main programs to make them more efficient and editable. For example, the G65 P9814 bore/boss probing cycle requires a D address value to identify the bore or boss diameter to be measured. Make the D address value a variable by programming a block such as **G65 P9814 D#125 S1.** In the beginning of the program, the D value can be defined with the block **#125 = 1.5 (Bore diameter).** This enables the cycle to be programmed once and allows for quick editing and programming.

Always look for any opportunity to use a subprogram. Any part that has a repeating pattern of features can benefit from a subprogram. Families of parts that are similar in design but vary in size or pattern location are good candidates for subprograms. The machine operators and setup personnel will begin to recognize these standard subprograms and this will make their jobs easier.

31.10 Metric Programming

Programming in metric units can allow for specifying offsets and writing code to a higher degree of accuracy. This is a programming "trick" to meet very close tolerances, but it is important to note that it might not work on most CNC machines. The following explains the reasoning behind this strategy and why it might not work.

Throughout this text, all programs have been completed in inch mode using the G20 command. Most CNC machine shops in the United States are still working from inch-based prints and are writing programs primarily in inch units. In inch mode, most CNC machines can read values in increments up to one ten thousandths of an inch, or four decimal places (.0001″). This allows for very accurate positioning. However, there are some machining applications that require a tighter tolerance.

If a program is written in metric units using the G21 command, the controller will allow input up to three decimal places, or .001 mm. The conversion from millimeters to inches is:

$$1 \text{ mm} = .03937″$$

If .001 mm is converted to inches, it equals .000039″, or roughly four one hundred thousandths of an inch. This allows the operator to offset tools to a metric value of .001 mm, which is about 2 1/2 times smaller than the smallest programmable inch increment (.0001″) when converted. To put it in simpler terms, instead of specifying a .0001″ offset, it is possible to specify a .00004″ offset.

However, there is a reason why an offset that small might not work. There are very few machines on the market that can position to that level of precision. Although a value equivalent to .00004″ can be stored in the offset setting, can the machine position itself, in a repeatable fashion, that accurately? In most cases, it cannot. But some machines can, and it is worth the consideration to try if needed.

31.11 Scaling

There are additional specialty functions and cycles that are not common in CNC programming, but they can be useful if applied correctly. One of these is the G51 scaling option. This allows the programmer to scale a toolpath in a positive or negative direction. Scaling is another application that can work hand-in-hand with subprogramming.

Figure 31-5 shows a part with a set of three square-shaped features. The squares increase in size from 1″ to 2″ to 3″. The part can be machined by scaling a milling toolpath defined in a single subprogram. In the following program, the first subprogram is used to machine the 1″ square feature without scaling. Then, the G51 command is used to set the scale factor for the 2″ square feature and the subprogram is called again. This sequence is repeated to machine the 3″ square feature.

T1 M6	
G0 G90 G54 X0. Y0.	
G43 H1 Z1. M3 S3500	
M98 P3101	Subprogram for 1″ square machining
G51 X0. Y0. P2	X0. Y0. is the center point of the scaling routine; P2 specifies 2X scale factor
M98 P3101	Same machining subprogram
G51 X0. Y0. P3	X0. Y0. is the center point of the scaling routine; P3 specifies 3X scale factor
M98 P3101	Same machining subprogram
G91 G28 Z0.	
M30	

The G51 block specifies the center point for scaling in the X and Y axes. In this example, the Z axis positions are programmed in the subprogram, but it is possible to also scale the Z axis by adding a Z axis coordinate to the G51 block. The P address is used to specify the scale factor. This value enlarges or reduces the size of the existing toolpath. The programmed tool length offsets and any tool diameter offsets will not be

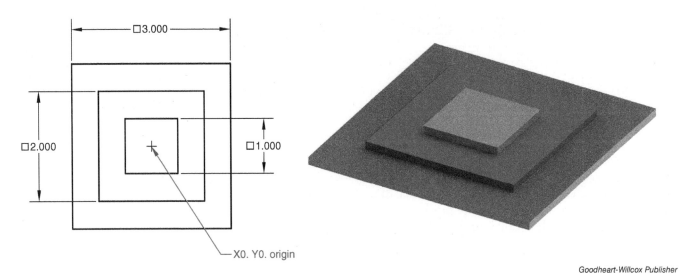

Figure 31-5. A part with three square-shaped features. The 2″ and 3″ features can be machined by scaling the toolpath used to machine the 1″ feature.

affected by the scaling routine. The G51 command can be used with any toolpath, including cutting paths for drill grids, bolt hole circles, contours, and pockets. If used appropriately, scaling can be a useful tool.

31.12 Mirror Imaging

Mirror imaging is a function used to mirror, or flip, a toolpath about an axis in the work coordinate system (WCS). This function is seldom used, but it allows the programmer to simplify programming for symmetrical parts. The G101 command is used to activate mirror imaging. This command is modal and will stay active until the G100 command is used to disable mirror imaging. It is possible to mirror a machining toolpath about the X axis, Y axis, or both axes at the same time. It is also possible to mirror a toolpath about the Z axis, but this is not a common function.

Figure 31-6 shows a series of operations used to machine parts in the four quadrants of the WCS. As discussed earlier in this text, the four quadrants in the XY work plane are numbered 1–4 in a counterclockwise direction. The first part in **Figure 31-6** is machined in the first quadrant relative to the X0. Y0. origin. The remaining parts are machined by mirroring the toolpath to the other quadrants. The toolpath movements are defined in a single subprogram. After the first part is machined, mirror imaging is activated and the toolpath is mirrored about the Y axis to machine the second part. The third part is machined by mirroring the toolpath about the X axis. The fourth part is machined by mirroring the toolpath about

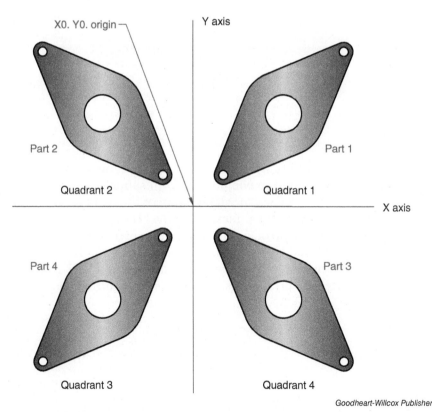

Goodheart-Willcox Publisher

Figure 31-6. Using mirror imaging to machine a part with a symmetrical shape. The part can be machined in one quadrant of the WCS and then the toolpath can be repeated by mirroring it about the appropriate axis.

the X and Y axes. The same subprogram is used to machine each part. It is important to note that the X0. Y0. origin is the controlling point about which toolpath motion is mirrored. When the G101 command is initiated, it is followed by the mirror axis designation defining the mirroring direction.

T1 M06	
G0 G90 G54 X.5 Y.5	Rapid to first position
S5000 M3	
G43 H1 Z.1	
(Begin Cutting Blocks)	
M98 P3107	Subprogram for Part 1
G0 Z.1	Rapid retract above part
G101 Y0.	Turn on mirror imaging for Y axis
X.5 Y.5	Rapid to first position
M98 P3107	Subprogram for Part 2
G100 Y0.	Turn off mirror imaging for Y axis
G0 Z.1	
X.5 Y.5	
G101 X0.	Turn on mirror imaging for X axis
X.5 Y.5	
M98 P3107	Subprogram for Part 3
G100 X0.	Turn off mirror imaging for X axis
G0 Z.1	
X.5 Y.5	
G101 X0. Y0.	Turn on mirror imaging for X axis and Y axis
X.5 Y.5	
M98 P3107	Subprogram for Part 4
G100 X0. Y0.	Turn off mirror imaging for X axis and Y axis
G91 G28 Z0.	
M30 (End program)	

As shown in this example, mirror imaging allows for repeatable machining and is best used in conjunction with subprograms. When programming a toolpath to be mirrored, be aware of the work origin location and the direction of the X and Y axes. When the G101 command is used to mirror a toolpath, the controller will reverse the direction of motion programmed in the G02 and G03 circular interpolation commands and adjust the cutter compensation mode defined with the G41 and G42 commands. This may influence the surface finish of the part and cutting conditions.

Chapter 31 Review

Summary

- A programmer should implement a process to verify a CNC program before it is executed. A program can be checked for syntax errors or incorrect codes by testing it in dry run mode or graphics mode.
- In both milling and lathe operations, individual tools can be assigned multiple tool offsets. This gives the programmer more flexibility and allows features to be controlled independently.
- In turning operations, taper can occur on the outside diameter of a shaft when cutting pressure forces the tool away from the material at the point farthest from the workholding. Taper can be programmed out of the stock by adjusting the beginning and ending X axis diameter positions in the program.
- When milling an inside corner radius or an outside corner radius, it is important to alter feed rates to maintain surface finish. Inside corner radii are programmed at slower feed rates, and outside corner radii are programmed at faster feed rates.
- Programming in metric units can allow for specifying offsets and writing code to a higher degree of accuracy. This strategy may be limited by the positioning abilities of the machine.
- The G51 command can be used to scale the size of an existing toolpath. This allows for repetitive machining of similar features that vary in size.
- The G101 command can be used to mirror a toolpath. This is useful for symmetrical parts and is best suited for use in subprograms.

Review Questions

Answer the following questions using the information provided in this chapter.

Know and Understand

1. *True or False?* When adjusting a tool diameter offset to test a milling program, the offset should be made smaller so that the tool will stay away from the final toolpath.
2. *True or False?* In a turning operation, a cutting tool will have less deflection from the material as it travels closer to the part workholding.
3. *True or False?* When testing a program at the machine, the rapid movement should be set to a faster rate by using the rapid override keys.
4. Which of the following commands is used to scale an existing toolpath?
   A. G51
   B. G52
   C. G53
   D. G54
5. Which of the following commands is used to activate mirror imaging?
   A. G98
   B. G99
   C. G100
   D. G101

Apply and Analyze

1. Explain the steps that are typically taken to check the program and prepare the machine when running the first part.
2. Using a 5/8″ diameter end mill, what is the feed rate for a .250″ outside corner radius if the linear feed rate is 20 ipm?
3. What is the purpose of mirror imaging? How does this function lend itself to subprogramming?

Critical Thinking

1. Everyone uses "shortcuts" in everyday life to save time. This is a great strategy for practical reasons, as long as quality does not suffer. One of the primary reasons for writing a subprogram is to shorten the length of a program. What are some of the factors to think about when creating subprograms to maintain the accuracy and safety of the operation?
2. Why is it important to check a program before it is executed? How does this make machining more efficient? What are some of the errors that could occur if the program or machine setup is not correct?
3. In programs that use repetitive machining functions, such as scaling and mirror imaging, why is it more efficient to use subprograms? Consider not only the programming time, but also the machine setup and operation. How do subprograms make editing easier?

Reference Section

The following pages contain tables and charts that will be useful as reference in a variety of areas. To make locating information easier, the material in this section is listed below, along with the page number.

- Rules for Determining Speeds and Feeds 667
- Feeds and Speeds for HSS Drills, Reamers, and Taps 668
- Decimal Equivalents: Number-Size Drills 669
- Decimal Equivalents: Letter-Size Drills 669
- 60° V-Type Thread Dimensions: Fractional Sizes 670–671
- 60° V-Type Thread Dimensions: Metric Sizes 672
- Tap Drill Sizes 673–674
- Conversion Table: US Customary to SI Metric 675
- Conversion Table: SI Metric to US Customary 676
- EIA and AIA National Codes for CNC Programming 677–679

Rules for Determining Speeds and Feeds

To Find	Having	Rule	Formula
Speed of cutter in feet per minute (fpm)	Diameter of cutter and revolutions per minute	Diameter of cutter (in inches), multiplied by 3.1416 (π), multiplied by revolutions per minute, divided by 12	$fpm = \dfrac{\pi D \times rpm}{12}$
Speed of cutter in meters per minute (mpm)	Diameter of cutter and revolutions per minute	Diameter of cutter, multiplied by 3.1416 (π), multiplied by revolutions per minute, divided by 1000	$mpm = \dfrac{D(mm) \times \pi \times rpm}{1000}$
Revolutions per minute (rpm)	Feet per minute and diameter of cutter	Feet per minute, multiplied by 12, divided by circumference of cutter (πD)	$rpm = \dfrac{fpm \times 12}{\pi D}$
Revolutions per minute (rpm)	Meters per minute and diameter of cutter in millimeters (mm)	Meters per minute, multiplied by 1000, divided by circumference of cutter (πD)	$rpm = \dfrac{mpm \times 1000}{\pi D}$
Feed per revolution (FR)	Feed per minute and revolutions per minute	Feed per minute, divided by revolutions per minute	$FR = \dfrac{F}{rpm}$
Feed per tooth per revolution (ftr)	Feed per minute and number of teeth in cutter	Feed per minute (in inches or millimeters), divided by number of teeth in cutter × revolutions per minute	$ftr = \dfrac{F}{T \times rpm}$
Feed per minute (F)	Feed per tooth per revolution, number of teeth in cutter, and RPM	Feed per tooth per revolution, multiplied by number of teeth in cutter, multiplied by revolutions per minute	$F = ftr \times T \times rpm$
Feed per minute (F)	Feed per revolution and revolutions per minute	Feed per revolution, multiplied by revolutions per minute	$F = FR \times rpm$
Number of teeth per minute (TM)	Number of teeth in cutter and revolutions per minute	Number of teeth in cutter, multiplied by revolutions per minute	$TM = T \times rpm$

rpm = Revolutions per minute
T = Teeth in cutter
D = Diameter of cutter
π = 3.1416 (pi)
fpm = Speed of cutter in feet per minute

TM = Teeth per minute
F = Feed per minute
FR = Feed per revolution
ftr = Feed per tooth per revolution
mpm = Speed of cutter in meters per minute

Goodheart-Willcox Publisher

Feeds and Speeds for HSS Drills, Reamers, and Taps

Material	Brinell	Drills Speed (sfm)	Point	Feed	Reamers Speed (sfm)	Feed	Taps (sfm) Threads per Inch 3–7 1/2	8–15	16–24	25–up
Aluminum	99–101	200–250	118°	M	150–160	M	50	100	150	200
Aluminum bronze	170–187	60	118°	M	40–45	M	12	25	45	60
Bakelite	...	80	60-90°	M	50–60	M	50	100	150	200
Brass	192–202	200–250	118°	H	150–160	H	50	100	150	200
Bronze, common	166–183	200–250	118°	H	150–160	H	40	80	100	150
Bronze, phosphor, 1/2 hard	187–202	175–180	118°	M	130–140	M	25	40	50	80
Bronze, phosphor, soft	149–163	200–250	118°	H	150–160	H	40	80	100	150
Cast iron, soft	126	140–150	90°	H	100–110	H	30	60	90	140
Cast iron, medium soft	196	80–110	118°	M	50–65	M	25	40	50	80
Cast iron, hard	293–302	45–50	118°	L	67–75	L	10	20	30	40
Cast iron, chilled*	402	15	150°	L	8–10	L	5	5	10	10
Cast steel	286–302	40–50*	118°	L	70–75	L	20	30	40	50
Celluloid	...	100	90°	M	75–80	M	50	100	150	200
Copper	80–85	70	100°	L	45–55	L	40	80	100	150
Drop forgings (steel)	170–196	60	118°	M	40–45	M	12	25	45	60
Duralumin	90–104	200	118°	M	150–160	M	50	100	150	200
Everdur	179–207	60	118°	L	40–45	L	20	30	40	50
Machinery steel	170–196	110	118°	H	67–75	H	35	50	60	85
Magnet steel, soft	241–302	35–40	118°	M	20–25	M	20	40	50	75
Magnet steel, hard*	321–512	15	150°	L	10	L	5	10	15	25
Manganese steel, 7% – 13%	187–217	15	150°	L	10	L	15	20	25	30
Manganese copper, 30% Mn.*	134	15	150°	L	10–12	L	...	...	...	...
Malleable iron	112–126	85–90	118°	H	...	H	20	30	40	50
Mild steel, .20 –.30 C	170–202	110–120	118°	H	75–85	H	40	55	70	90
Molybdenum steel	196–235	55	125°	M	35–45	M	20	30	35	45
Monel metal	149–170	50	118°	M	35–38	M	8	10	15	20
Nickel, pure*	187–202	75	118°	L	40	L	25	40	50	80
Nickel steel, 3 1/2%	196–241	60	118°	L	40–45	L	8	10	15	20
Rubber, hard	...	100	60-90°	L	70–80	L	50	100	150	200
Screw stock, C.R.	170–196	110	118°	H	75	H	20	30	40	50
Spring steel	402	20	150°	L	12–15	L	10	10	15	15
Stainless steel	146–149	50	118°	M	30	M	8	10	15	20
Stainless steel, C.R.*	460–477	20	118°	L	15	L	8	10	15	20
Steel, .40 to .50 C	170–196	80	118°	M	8–10	M	20	30	40	50
Tool, SAE, and forging steel	149	75	118°	H	35–40	H	25	35	45	55
Tool, SAE, and forging steel	241	50	125°	M	12	M	15	15	25	25
Tool, SAE, and forging steel*	402	15	150°	L	10	L	8	10	15	20
Zinc alloy	112–126	200–250	118°	M	150–175	M	50	100	150	200

*Use specially constructed heavy-duty drills.

Note: Carbon steel tools should be run at speeds 40% to 50% of those recommended for high-speed steel.

Spiral point taps may be run at speeds 15% to 20% faster than regular taps.

Goodheart-Willcox Publisher

Decimal Equivalents: Number-Size Drills

Drill	Size of Drill in Inches	Drill	Size of Drill in Inches	Drill	Size of Drill in Inches	Drill	Size of Drill in Inches
1	.2280	21	.1590	41	.0960	61	.0390
2	.2210	22	.1570	42	.0935	62	.0380
3	.2130	23	.1540	43	.0890	63	.0370
4	.2090	24	.1520	44	.0860	64	.0360
5	.2055	25	.1495	45	.0820	65	.0350
6	.2040	26	.1470	46	.0810	66	.0330
7	.2010	27	.1440	47	.0785	67	.0320
8	.1990	28	.1405	48	.0760	68	.0310
9	.1960	29	.1360	49	.0730	69	.0292
10	.1935	30	.1285	50	.0700	70	.0280
11	.1910	31	.1200	51	.0670	71	.0260
12	.1890	32	.1160	52	.0635	72	.0250
13	.1850	33	.1130	53	.0595	73	.0240
14	.1820	34	.1110	54	.0550	74	.0225
15	.1800	35	.1100	55	.0520	75	.0210
16	.1770	36	.1065	56	.0465	76	.0200
17	.1730	37	.1040	57	.0430	77	.0180
18	.1695	38	.1015	58	.0420	78	.0160
19	.1660	39	.0995	59	.0410	79	.0145
20	.1610	40	.0980	60	.0400	80	.0135

Goodheart-Willcox Publisher

Decimal Equivalents: Letter-Size Drills

Drill	Size of Drill in Inches	Drill	Size of Drill in Inches	Drill	Size of Drill in Inches	Drill	Size of Drill in Inches
A	0.234	H	0.266	O	0.316	V	0.377
B	0.238	I	0.272	P	0.323	W	0.386
C	0.242	J	0.277	Q	0.332	X	0.397
D	0.246	K	0.281	R	0.339	Y	0.404
E	0.250	L	0.290	S	0.348	Z	0.413
F	0.257	M	0.295	T	0.358		
G	0.261	N	0.302	U	0.368		

Goodheart-Willcox Publisher

Copyright Goodheart-Willcox Co., Inc.

60° V-Type Thread Dimensions: Fractional Sizes
National Special Thread Series

Nominal Size	Threads per Inch	Major Diameter (Inches)	Minor Diameter (Inches)	Pitch Diameter (Inches)	Tap Drill for 75% Thread†	Clearance Drill Size*
1/16"	64	.0625	.0422	.0524	3/64"	51
5/64"	60	.0781	.0563	.0673	1/16"	45
3/32"	48	.0938	.0667	.0803	49	40
7/64"	48	.1094	.0823	.0959	43	32
1/8"	32	.1250	.0844	.1047	3/32"	29
9/64"	40	.1406	.1081	.1244	32	24
5/32"	32	.1563	.1157	.1360	1/8"	19
5/32"	36	.1563	.1202	.1382	30	19
11/64"	32	.1719	.1313	.1516	9/64"	14
3/16"	24	.1875	.1334	.1604	26	8
3/16"	32	.1875	.1469	.1672	22	8
13/64"	24	.2031	.1490	.1760	20	3
7/32"	24	.2188	.1646	.1917	16	1
7/32"	32	.2188	.1782	.1985	12	1
15/64"	24	.2344	.1806	.2073	10	1/4"
1/4"	24	.2500	.1959	.2229	4	17/64"
1/4"	27	.2500	.2019	.2260	3	17/64"
1/4"	32	.2500	.2094	.2297	7/32"	17/64"
5/16"	20	.3125	.2476	.2800	17/64"	21/64"
5/16"	27	.3125	.2644	.2884	J	21/64"
5/16"	32	.3125	.2719	.2922	9/32"	21/64"
3/8"	20	.3750	.3100	.3425	21/64"	25/64"
3/8"	27	.3750	.3269	.3509	R	25/64"

†Refer to tables "Decimal Equivalents: Number-Size Drills" and "Decimal Equivalents: Letter-Size Drills."
*Clearance drill makes hole with standard clearance for diameter of nominal size.
**Standard spark plug size.

Goodheart-Willcox Publisher

(continued)

60° V-Type Thread Dimensions: Fractional Sizes *(continued)*
National Special Thread Series

Nominal Size	Threads per Inch	Major Diameter (Inches)	Minor Diameter (Inches)	Pitch Diameter (Inches)	Tap Drill for 75% Thread †	Clearance Drill Size*
7/16"	24	.4375	.3834	.4104	X	29/64"
7/16"	27	.4375	.3894	.4134	Y	29/64"
1/2"	12	.5000	.3918	.4459	27/64"	33/64"
1/2"	24	.5000	.4459	.4729	29/64"	33/64"
1/2"	27	.5000	.4519	.4759	15/32"	33/64"
9/16"	27	.5625	.5144	.5384	17/32"	37/64"
5/8"	12	.6250	.5168	.5709	35/64"	41/64"
5/8"	27	.6250	.5769	.6009	19/32"	41/64"
11/16"	11	.6875	.5694	.6285	19/32"	45/64"
11/16"	16	.6875	.6063	.6469	5/8"	45/64"
3/4"	12	.7500	.6418	.6959	43/64"	49/64"
3/4"	27	.7500	.7019	.7259	23/32"	49/64"
13/16"	10	.8125	.6826	.7476	23/32"	53/64"
7/8"	12	.8750	.7668	.8209	51/64"	57/64"
7/8"	18**	.8750	.8028	.8389	53/64"	57/64"
7/8"	27	.8750	.8269	.8509	27/32"	57/64"
15/16"	9	.9375	.7932	.8654	53/64"	61/64"
1"	12	1.0000	.8918	.9459	59/64"	1 1/64"
1"	27	1.0000	.9519	.9759	31/32"	1 1/64"
1 5/8"	5 1/2	1.6250	1.3888	1.5069	1 29/64"	1 41/64"
1 7/8"	5	1.8750	1.6152	1.7451	1 11/16"	1 57/64"
2 1/8"	4 1/2	2.1250	1.8363	1.9807	1 29/32"	2 5/32"
2 3/8"	4	2.3750	2.0502	2.2126	2 1/8"	2 13/32"

†Refer to tables "Decimal Equivalents: Number-Size Drills" and "Decimal Equivalents: Letter-Size Drills."
*Clearance drill makes hole with standard clearance for diameter of nominal size.
**Standard spark plug size.

Goodheart-Willcox Publisher

60° V-Type Thread Dimensions: Metric Sizes
International Standard

Major Diameter (mm)	Pitch (mm)	Minor Diameter (mm)	Pitch Diameter (mm)	Tap Drill for 75% Thread (mm)	Tap Drill for 75% Thread† (No. or Inches)	Clearance Drill Size*
2.0	.40	1.48	1.740	1.6	1/16"	41
2.3	.40	1.78	2.040	1.9	48	36
2.6	.45	2.02	2.308	2.1	45	31
3.0	.50	2.35	2.675	2.5	40	29
3.5	.60	2.72	3.110	2.9	33	23
4.0	.70	3.09	3.545	3.3	30	16
4.5	.75	3.53	4.013	3.75	26	10
5.0	.80	3.96	4.480	4.2	19	3
5.5	.90	4.33	4.915	4.6	14	15/64"
6.0	1.00	4.70	5.350	5.0	9	1/4"
7.0	1.00	5.70	6.350	6.0	15/64"	19/64"
8.0	1.25	6.38	7.188	6.8	H	11/32"
9.0	1.25	7.38	8.188	7.8	5/16"	3/8"
10.0	1.50	8.05	9.026	8.6	R	27/64"
11.0	1.50	9.05	10.026	9.6	V	29/64"
12.0	1.75	9.73	10.863	10.5	Z	1/2"
14.0**	1.25	12.38	13.188	13.0	33/64"	9/16"
14.0	2.00	11.40	12.701	12.0	15/32"	9/16"
16.0	2.00	13.40	14.701	14.0	35/64"	21/32"
18.0	1.50	16.05	17.026	16.5	41/64"	47/64"
18.0	2.50	14.75	16.376	15.5	39/64"	47/64"
20.0	2.50	16.75	18.376	17.5	11/16"	13/16"
22.0	2.50	18.75	20.376	19.5	49/64"	57/64"
24.0	3.00	20.10	22.051	21.0	53/64"	31/32"
27.0	3.00	23.10	25.051	24.0	15/16"	13/32"
30.0	3.50	25.45	27.727	26.5	1 3/64"	1 13/64"
33.0	3.50	28.45	30.727	29.5	1 11/64"	1 21/64"
36.0	4.00	30.80	33.402	32.0	1 17/64"	1 7/16"
39.0	4.00	33.80	36.402	35.0	1 3/8"	1 9/16"
42.0	4.50	36.15	39.077	37.0	1 29/64"	1 43/64"
45.0	4.50	39.15	42.077	40.0	1 37/64"	1 13/16"
48.0	5.00	41.50	44.752	43.0	1 11/16"	1 29/64"

†Refer to tables "Decimal Equivalents: Number-Size Drills" and "Decimal Equivalents: Letter-Size Drills."

*Clearance drill makes hole with standard clearance for diameter of nominal size.

**Standard spark plug size.

Goodheart-Willcox Publisher

Tap Drill Sizes
Probable Percentage of Full Thread Produced in Tapped Hole Using Stock Sizes of Drill

Tap	Tap Drill	Decimal Equivalent of Tap Drill	Theoretical % of Thread	Probable Oversize (Mean)	Probable Hole Size	Percentage of Thread	Tap	Tap Drill	Decimal Equivalent of Tap Drill	Theoretical % of Thread	Probable Oversize (Mean)	Probable Hole Size	Percentage of Thread
0–80	56	.0465	83	.0015	.0480	74	8–32	29	.1360	69	.0029	.1389	62
	3/64	.0469	81	.0015	.0484	71		28	.1405	58	.0029	.1434	51
1–64	54	.0550	89	.0015	.0565	81	8–36	29	.1360	78	.0029	.1389	70
	53	.0595	67	.0015	.0610	59		28	.1405	68	.0029	.1434	57
1–72	53	.0595	75	.0015	.0610	67		9/64	.1406	68	.0029	.1435	57
	1/16	.0625	58	.0015	.0640	50	10–24	27	.1440	85	.0032	.1472	79
2–56	51	.0670	82	.0017	.0687	74		26	.1470	79	.0032	.1502	74
	50	.0700	69	.0017	.0717	62		25	.1495	75	.0032	.1527	69
	49	.0730	56	.0017	.0747	49		24	.1520	70	.0032	.1552	64
2–64	50	.0700	79	.0017	.0717	70		23	.1540	67	.0032	.1572	61
	49	.0730	64	.0017	.0747	56		5/32	.1563	62	.0032	.1595	56
3–48	48	.0760	85	.0019	.0779	78		22	.1570	61	.0032	.1602	55
	5/64	.0781	77	.0019	.0800	70	10–32	5/32	.1563	83	.0032	.1595	75
	47	.0785	76	.0019	.0804	69		22	.1570	81	.0032	.1602	73
	46	.0810	67	.0019	.0829	60		21	.1590	76	.0032	.1622	68
	45	.0820	63	.0019	.0839	56		20	.1610	71	.0032	.1642	64
3–56	46	.0810	78	.0019	.0829	69		19	.1660	59	.0032	.1692	51
	45	.0820	73	.0019	.0839	65	12–24	11/64	.1719	82	.0035	.1754	75
	44	.0860	56	.0019	.0879	48		17	.1730	79	.0035	.1765	73
4–40	44	.0860	80	.0020	.0880	74		16	.1770	72	.0035	.1805	66
	43	.0890	71	.0020	.0910	65		15	.1800	67	.0035	.1835	60
	42	.0935	57	.0020	.0955	51		14	.1820	63	.0035	.1855	56
	3/32	.0938	56	.0020	.0958	50	12–28	16	.1770	84	.0035	.1805	77
4–48	42	.0935	68	.0020	.0955	61		15	.1800	78	.0035	.1835	70
	3/32	.0938	68	.0020	.0958	60		14	.1820	73	.0035	.1855	66
	41	.0960	59	.0020	.0980	52		13	.1850	67	.0035	.1885	59
5–40	40	.0980	83	.0023	.1003	76		3/16	.1875	61	.0035	.1910	54
	39	.0995	79	.0023	.1018	71	1/4–20	9	.1960	83	.0038	.1998	77
	38	.1015	72	.0023	.1038	65		8	.1990	79	.0038	.2028	73
	37	.1040	65	.0023	.1063	58		7	.2010	75	.0038	.2048	70
5–44	38	.1015	79	.0023	.1038	72		13/64	.2031	72	.0038	.2069	66
	37	.1040	71	.0023	.1063	63		6	.2040	71	.0038	.2078	65
	36	.1065	63	.0023	.1088	55		5	.2055	69	.0038	.2093	63
6–32	37	.1040	84	.0023	.1063	78		4	.2090	63	.0038	.2128	57
	36	.1065	78	.0026	.1091	71	1/4–28	3	.2130	80	.0038	.2168	72
	7/64	.1094	70	.0026	.1120	64		7/32	.2188	67	.0038	.2226	59
	35	.1100	69	.0026	.1126	63		2	.2210	63	.0038	.2248	55
	34	.1110	67	.0026	.1136	60	5/16–18	F	.2570	77	.0038	.2608	72
	33	.1130	62	.0026	.1156	55		G	.2610	71	.0041	.2651	66
6–40	34	.1110	83	.0026	.1136	75		17/64	.2656	65	.0041	.2697	59
	33	.1130	77	.0026	.1156	69		H	.2660	64	.0041	.2701	59
	32	.1160	68	.0026	.1186	60							

(continued)

Tap Drill Sizes (continued)
Probable Percentage of Full Thread Produced in Tapped Hole Using Stock Sizes of Drill

Tap	Tap Drill	Decimal Equivalent of Tap Drill	Theoretical % of Thread	Probable Oversize (Mean)	Probable Hole Size	Percentage of Thread	Tap	Tap Drill	Decimal Equivalent of Tap Drill	Theoretical % of Thread	Probable Oversize (Mean)	Probable Hole Size	Percentage of Thread
5/16–24	H	.2660	86	.0041	.2701	78	1″–14	59/64	.9219	84	.0060	.9279	78
	I	.2720	75	.0041	.2761	67		15/16	.9375	67	.0060	.9435	61
	J	.2770	66	.0041	.2811	58	1 1/8–7	31/32	.9688	84	.0062	.9750	81
3/8–16	5/16	.3125	77	.0044	.3169	72		63/64	.9844	76	.0067	.9911	72
	O	.3160	73	.0044	.3204	68		1″	1.0000	67	.0070	1.0070	64
	P	.3230	64	.0044	.3274	59		1 1/64	1.0156	59	.0070	1.0226	55
3/8–24	21/64	.3281	87	.0044	.3325	79	1 1/8–12	1 1/32	1.0313	87	.0071	1.0384	80
	Q	.3320	79	.0044	.3364	71		1 3/64	1.0469	72	.0072	1.0541	66
	R	.3390	67	.0044	.3434	58							
7/16–14	T	.3580	86	.0046	.3626	81	**Taper Pipe**				**Straight Pipe**		
	23/64	.3594	84	.0046	.3640	79							
	U	.3680	75	.0046	.3726	70	Thread	Drill			Thread	Drill	
	3/8	.3750	67	.0046	.3796	62	1/8–27	R			1/8–27	S	
	V	.3770	65	.0046	.3816	60	1/4–18	7/16			1/4–18	29/64	
7/16–20	W	.3860	79	.0046	.3906	72	3/8–18	37/64			3/8–18	19/32	
	25/64	.3906	72	.0046	.3952	65	1/2–14	23/32			1/2–14	47/64	
	X	.3970	62	.0046	.4016	55	3/4–14	59/64			3/4–14	15/16	
1/2–13	27/64	.4219	78	.0047	.4266	73	1–11 1/2	1 5/32			1–11 1/2	1 3/16	
	7/16	.4375	63	.0047	.4422	58	1 1/4–11 1/2	1 1/2			1 1/4–11 1/2	1 33/64	
1/2–20	29/64	.4531	72	.0047	.4578	65	1 1/2–11 1/2	1 47/64			1 1/2–11 1/2	1 3/4	
9/16–12	15/32	.4688	87	.0048	.4736	82	2–11 1/2	2 7/32			2–11 1/2	2 7/32	
	31/64	.4844	72	.0048	.4892	68	2 1/2–8	2 5/8			2 1/2–8	2 21/32	
9/16–18	1/2	.5000	87	.0048	.5048	80	3–8	3 1/4			3–8	3 9/32	
	33/64	.5156	65	.0048	.5204	58	3 1/2–8	3 3/4			3 1/2–8	3 25/32	
5/8–11	17/32	.5313	79	.0049	.5362	75	4–8	4 1/4			4–8	4 9/32	
	35/64	.5469	66	.0049	.5518	62							
5/8–18	9/16	.5625	87	.0049	.5674	80							
	37/64	.5781	65	.0049	.5831	58							
3/4–10	41/64	.6406	84	.0050	.6456	80							
	21/32	.6563	72	.0050	.6613	68							
3/4–16	11/16	.6875	77	.0050	.6925	71							
7/8–9	49/64	.7656	76	.0052	.7708	72							
	25/32	.7812	65	.0052	.7864	61							
7/8–14	51/64	.7969	84	.0052	.8021	79							
	13/16	.8125	67	.0052	.8177	62							
1″–8	55/64	.8594	87	.0059	.8653	83							
	7/8	.8750	77	.0059	.8809	73							
	57/64	.8906	67	.0059	.8965	64							
	29/32	.9063	58	.0059	.9122	54							
1″–12	29/32	.9063	87	.0060	.9123	81							
	59/64	.9219	72	.0060	.9279	67							
	15/16	.9375	58	.0060	.9435	52							

Standard Tool Co.

Conversion Table: US Customary to SI Metric

When You Know:	Multiply By:		To Find:
	Very Accurate	Approximate	
Length			
inches	* 25.4		millimeters
inches	* 2.54		centimeters
feet	* 0.3048		meters
feet	* 30.48		centimeters
yards	* 0.9144	0.9	meters
miles	* 1.609344	1.6	kilometers
Weight			
grains	15.43236	15.4	grams
ounces	* 28.349523125	28.0	grams
ounces	* 0.028349523125	0.028	kilograms
pounds	* 0.45359237	0.45	kilograms
short ton	* 0.90718474	0.9	tonnes
Volume			
teaspoons		5.0	milliliters
tablespoons		15.0	milliliters
fluid ounces	29.57353	30.0	milliliters
cups		0.24	liters
pints	* 0.473176473	0.47	liters
quarts	* 0.946352946	0.95	liters
gallons	* 3.785411784	3.8	liters
cubic inches	* 0.016387064	0.02	liters
cubic feet	* 0.028316846592	0.03	cubic meters
cubic yards	* 0.764554857984	0.76	cubic meters
Area			
square inches	* 6.4516	6.5	square centimeters
square feet	* 0.09290304	0.09	square meters
square yards	* 0.83612736	0.8	square meters
square miles		2.6	square kilometers
acres	* 0.40468564224	0.4	hectares
Temperature			
Fahrenheit	*	5/9 (after subtracting 32)	Celsius
Density			
pounds per cubic feet	1.602×10	16	kilograms per cubic meter
Force			
ounces (F)	2.780×10^{-1}		newtons
pounds (F)	4.448×10^{-3}		kilonewtons
kips	4.448		kilonewtons
Stress			
pounds/square inch (psi)	6.895×10^{-3}		megapascals
kips/square inch (ksi)	6.895		megapascals
Torque			
ounce-inches	7.062×10^3		newton-meters
pound-inches	1.130×10^{-1}		newton-meters
pound-feet	1.356		newton-meters

* = Exact

Conversion Table: SI Metric to US Customary

When You Know:	Multiply By:		To Find:
	Very Accurate	Approximate	
Length			
millimeters	0.0393701	0.04	inches
centimeters	0.3937008	0.4	inches
meters	3.280840	3.3	feet
meters	1.093613	1.1	yards
kilometers	0.621371	0.6	miles
Weight			
grains	0.00228571	0.0023	ounces
grams	0.03527396	0.035	ounces
kilograms	2.204623	2.2	pounds
tonnes	1.1023113	1.1	short tons
Volume			
milliliters		0.2	teaspoons
milliliters	0.06667	0.067	tablespoons
milliliters	0.03381402	0.03	fluid ounces
liters	61.02374	61.024	cubic inches
liters	2.113376	2.1	pints
liters	1.056688	1.06	quarts
liters	0.26417205	0.26	gallons
liters	0.03531467	0.035	cubic feet
cubic meters	61023.74	61023.7	cubic inches
cubic meters	35.31467	35.0	cubic feet
cubic meters	1.3079506	1.3	cubic yards
cubic meters	264.17205	264.0	gallons
Area			
square centimeters	0.1550003	0.16	square inches
square centimeters	0.00107639	0.001	square feet
square meters	10.76391	10.8	square feet
square meters	1.195990	1.2	square yards
square kilometers		0.4	square miles
hectares	2.471054	2.5	acres
Temperature			
Celsius	*9/5 (then add 32)		Fahrenheit

* = Exact

Goodheart-Willcox Publisher

EIA and AIA National Codes for CNC Programming
Preparatory (G) Functions

G word	Explanation
G00	Denotes a rapid traverse rate for point-to-point positioning.
G01	Describes linear interpolation blocks; reserved for contouring.
G02, G03	Used with circular interpolation.
G04	Sets a calculated time delay during which there is no machine motion (dwell).
G05, G07	Unassigned by the EIA. May be used at the discretion of the machine tool or system builder. Could also be standardized at a future date.
G06	Used with parabolic interpolation.
G08	Acceleration code. Causes the machine, assuming it is capable, to accelerate at a smooth exponential rate.
G09	Deceleration code. Causes the machine, assuming it is capable, to decelerate at a smooth exponential rate.
G10-G12	Normally unassigned for CNC systems. Used with some hard-wired systems to express blocks of abnormal dimensions.
G13-G16	Direct the control system to operate on a particular set of axes.
G17-G19	Identify or select a coordinate plane for such functions as circular interpolation or cutter compensation.
G20-G32	Unassigned according to EIA standards. May be assigned by the control system or machine tool builder.
G33-G35	Selected for machines equipped with thread-cutting capabilities (generally referring to lathes). G33 is used when a constant lead is sought, G34 is used when a constantly increasing lead is required, and G35 is used to designate a constantly decreasing lead.
G36-G39	Unassigned.
G40	Terminates any cutter compensation.
G41	Activates cutter compensation in which the cutter is on the left side of the work surface (relative to the direction of the cutter motion).
G42	Activates cutter compensation in which the cutter is on the right side of the work surface.
G43, G44	Used with cutter offset to adjust for the difference between the actual and programmed cutter radii or diameters. G43 refers to an inside corner, and G44 refers to an outside corner.
G45-G49	Unassigned.
G50-G59	Reserved for adaptive control.
G60-G69	Unassigned.
G70	Selects inch programming.
G71	Selects metric programming.
G72	Selects three-dimensional CW circular interpolation.
G73	Selects three-dimensional CCW circular interpolation.
G74	Cancels multiquadrant circular interpolation.
G75	Activates multiquadrant circular interpolation.
G76-G79	Unassigned.
G80	Cancel cycle.
G81	Activates drill, or spot drill, cycle.
G82	Activates drill with a dwell.
G83	Activates intermittent, or deep-hole, drilling.
G84	Activates tapping cycle.
G85-G89	Activates boring cycles.

(continued)

EIA and AIA National Codes for CNC Programming (continued)
Preparatory (G) Functions (continued)

G word	Explanation
G90	Selects absolute input. Input data is to be in absolute dimensional form.
G91	Selects incremental input. Input data is to be in incremental form.
G92	Preloads registers to desired values (for example, preloads axis position registers).
G93	Sets inverse time feed rate.
G94	Sets inches (or millimeters) per minute feed rate.
G95	Sets inches (or millimeters) per revolution feed rate.
G97	Sets spindle speed in revolutions per minute.
G98, G99	Unassigned.

Miscellaneous (M) Functions

M word	Explanation
M00	Program stop. Operator must cycle start in order to continue with the remainder of the program.
M01	Optional stop. Acted upon only when the operator has previously signaled for this command by pushing a button. When the control system senses the M01 code, machine will automatically stop.
M02	End of program. Stops the machine after completion of all commands in the block. May include rewinding of tape.
M03	Starts spindle rotation in a clockwise direction.
M04	Starts spindle rotation in a counterclockwise direction.
M05	Spindle stop.
M06	Executes the change of a tool (or tools) manually or automatically.
M07	Turns coolant on (flood).
M08	Turns coolant on (mist).
M09	Turns coolant off.
M10	Activates automatic clamping of the machine slides, workpiece, fixture, spindle, etc.
M11	Deactivates automatic clamping.
M12	Inhibiting code used to synchronize multiple set of axes, such as a four-axis lathe that has two independently operated heads or slides.
M13	Combines simultaneous clockwise spindle motion and coolant on.
M14	Combines simultaneous counterclockwise spindle motion and coolant on.
M15	Sets rapid traverse or feed motion in the + direction.
M16	Sets rapid traverse or feed motion in the − direction.
M17, M18	Unassigned.
M19	Oriented spindle stop. Stops spindle at a predetermined angular position.
M20-M29	Unassigned.
M30	End of data. Used to reset control and/or machine.
M31	Interlock bypass. Temporarily circumvents a normally provided interlock.
M32-M39	Unassigned.
M40-M46	Signals gear changes if required at the machine; otherwise, unassigned.
M47	Continues program execution from the start of the program, unless inhibited by an interlock signal.
M48	Cancels M49.
M49	Deactivates a manual spindle or feed-override and returns to the programmed value.

(continued)

EIA and AIA National Codes for CNC Programming *(continued)*
Miscellaneous (M) Functions *(continued)*

M word	Explanation
M50–M57	Unassigned.
M58	Cancels M59.
M59	Holds the rpm constant at its value.
M60–M99	Unassigned.

Other Address Characters

Address character	Explanation
A	Angular dimension about the X axis.
B	Angular dimension about the Y axis.
C	Angular dimension about the Z axis.
D	Can be used for an angular dimension around a special axis, for a third feed function, or for tool offset.
E	Used for angular dimension around a special axis or for a second feed function.
H	Fixture offset.
I, J, K	Centerpoint coordinates for circular interpolation.
L	Not used.
O	Used on some N/C controls in place of the customary sequence number word address N.
P	Third rapid traverse code—tertiary motion dimension parallel to the X axis.
Q	Second rapid traverse code—tertiary motion dimension parallel to the Y axis.
R	First rapid traverse code—tertiary motion dimension parallel to the Z axis (or to the radius) for constant surface speed calculation.
U	Secondary motion dimension parallel to the X axis.
V	Secondary motion dimension parallel to the Y axis.
W	Secondary motion dimension parallel to the Z axis.

Goodheart-Willcox Publisher

Glossary

3 + 1 milling. The process of positioning a fourth axis at a fixed position and completing three-axis milling operations. (2)

A

absolute positioning. A positioning mode in which coordinate values are measured from the coordinate system origin. (2, 3, 15)

acute angle. Angle that measures less than 90°. (1, 14)

address. A single-letter character that defines what a machine should do with the numerical data that follows. A letter precedes the number in a word. (3, 5, 16, 17)

address code. A letter code that is coupled with a number to create a word. The address code defines what the machine controller should do with the numerical data that follows. (5, 17)

angle. A measure of the rotational distance between two intersecting lines or line segments from their intersection point, usually given in degrees. (1, 14)

arc. A portion of a circle. (1, 14)

axial drilling. A lathe live tooling operation in which the travel of the tool is perpendicular to the part face and parallel to the Z axis direction. (20, 23)

B

backlash. Mechanical play that occurs between the lead screw and mating nut in a screw assembly when a machine axis changes direction. (29)

ball screw. A long threaded shaft mated with a ball nut in a ball screw assembly, used to produce linear motion through rotational motion. (22)

bar feeder. A separate unit from a CNC lathe that holds bar stock and automatically loads it into the lathe for machining. (21, 25)

bar puller. A programmable lathe tool mounted to a turret that grips bar stock and pulls it to a specific length through the machine spindle. (21, 25)

bell mouth. A tapered hole shape that is larger at the top and smaller toward the bottom. (10)

blind hole. A hole that does not extend through the part completely. (10)

block. A single word or a series of words forming a complete line of CNC code. (3, 5, 16)

bolt circle. A theoretical circle on which the center points of holes lie in a circular pattern of holes. (1, 10)

boring. A machining operation used to enlarge an existing hole to a finish size. (10, 15, 20)

boss. A protruding feature that is part of an existing solid part. (11)

bull nose end mill. An end mill with a rounded edge used to produce a radius transition between surfaces. (6)

burr. A rough raised edge left on material from a machining operation. (8)

C

canned cycle. An abbreviated machining cycle that automates complex repetitive commands and functions, designed to reduce program length and simplify editing. (5, 10, 16, 19, 20)

Cartesian coordinate system. System of point specification that locates each point in a plane with a pair of alphanumeric coordinates. (2, 15)

chamfer. A beveled edge machined to relieve a sharp corner. (8, 14, 21)

chatter. Unwanted tool vibration caused by excessive or insufficient tool engagement or inadequate tool support. (31)

chip load. The actual thickness of a chip being cut or the depth of each cutting edge as it passes through a material. (1)

chuck. A workholding device that secures a workpiece for machining. (18)

circle. A closed plane curve that is an equal distance at all points from its center point. (1, 14)

climb milling. Milling operation in which the cutting tool rotates in the same direction as the feed direction of the material. (8)

closed pocket. A pocket encased on all sides by the part. (9)

closing statement. The ending lines of a program that prepare the machine to run the next part in a safe manner. (7)

Note: The number in parentheses following each definition indicates the chapter in which the term can be found.

comment line. A block of code consisting of text enclosed by parentheses, used by the programmer to communicate important information to the setup person and operator. Text appearing in parentheses represents a comment and is not read by the machine. (7)

complementary angles. Two angles that add up to 90°. (1, 14)

computer-aided manufacturing (CAM). The process of using computer software to define machining operations and generate CNC part programs. (31)

contouring. A machining operation that follows a joined path or single piece of geometry to cut material. (8, 19)

conventional milling. Milling operation in which the cutting tool rotates against the feed direction of the material. (8)

cosine. For a given angle, the ratio of the adjacent side to the hypotenuse of a right triangle. (1, 14)

counterbore. An enlarged cylindrical feature at the top of a hole that provides a flat surface for a fastener to sit flush below the part face. (10, 20)

countersink. An enlarged conical feature at the top of a hole used to create clearance and allow the top of a fastener to sit flush with the part face. (10, 20)

crash. An accident that occurs when the tool or spindle collides with the machine, material, or workholding. (13)

curved line. An arc, partial circle, or spline. (1, 14)

cutter compensation. A programmed offset from the center line of the cutting tool to the tool's edge along a cutting path. (8)

D

denominator. The number on the bottom of a fraction that indicates the whole number quantity into which the parts are divided. (1, 14)

depth of cut. The depth along the Z axis the tool is engaged in the material. Also called *axial engagement*. (9)

design intent. The specification of how a part is to function, defined by dimensions and notes that describe critical features and manufacturing requirements. (6, 18)

dial indicator. A precision measuring instrument with a dial face used to check for alignment in machine setup and to inspect manufactured parts. (13)

diameter. The distance across a circle through the center point. (1, 14)

dovetail slot. A slot that has two angled sides and a flat bottom. (9)

dry run. A test run of the program performed to check for correct machine motion and accuracy without cutting stock. (31)

dwell. A function that pauses machine axis motion for a specified length of time without affecting other machine functions. (5)

E

edge finder. A cylindrical alignment tool used to establish the position of a workpiece in relation to the machine. During part setup, the tool is brought into contact with a side of the part to establish coordinates for the work coordinate system. (12, 13)

end mill. A cutting tool used in milling for roughing, finishing, slotting, contouring, and profiling. (6)

end-of-block (EOB). The end of a single line of code in a program defined by a special character, usually a semicolon. (5)

equilateral triangle. Triangle with all equal sides and all equal angles. (1, 14)

F

face mill: A cutting tool designed for machining the top face of a part. (11)

face milling: Machining the top surface of a part with the face of the cutting tool parallel to the surface being cut. (11)

facing: Machining the face of a part to remove any excess material or to establish a consistent part length. (11, 19)

fillet. An internal radial edge that blends adjacent surfaces. (14)

fixture. A custom workholding device used to position and secure an irregular-shaped workpiece for machining. (6)

flood coolant. Liquid coolant used to flood the cutting area during machining. (4, 17)

flute. A recessed groove along a cutting edge allowing for chip removal. (1)

fly cutter: A single-point cutting tool used in facing operations to make shallow cuts and produce a flat surface with a very fine finish. (11)

full four-axis milling. Milling in which all of the primary axes are moving while one additional axis is utilized simultaneously. (2)

G

G-code. A program address identifying a preparatory command. (3, 16)

geometric dimensioning and tolerancing (GD&T). A system that uses geometric tolerances to define allowable variations of form, orientation, location, profile, and runout. (29)

geometric shape. In mathematics, a two-dimensional object, such as a circle, arc, triangle, polygon, or quadrilateral; or a three-dimensional object, such as a sphere, pyramid, cube, or polyhedron. (1, 14)

global variables. Variables that are intended for general use in macro programming. (30)

graphics mode. A viewing mode on a CNC machine used to verify proper operation of a program. (31)

grooving. An operation that typically refers to machining a narrow three-sided cavity. (21)

H

helical entry. A method of cutting tool entry in which the tool enters the material in a helical motion and movement occurs simultaneously along three axes. (9)

helix. A three-dimensional spiral. (9)

hypotenuse. The longest side of a right triangle, which is always located across from the right angle. (1, 14)

I

I and K method. In lathe programming, an arc programming method in which incremental distances from the tool's starting point to the arc center are used as parameters in defining radial movement. (19)

I, J, and K method. An arc programming method in which incremental distances from the tool's starting point to the arc center are used as parameters in defining radial movement. (8)

impeller. A rotating device made up of a series of vanes that force fluid or air outward to create centrifugal force. (26)

incremental positioning. A positioning mode in which coordinate values are measured from the current point to the next point. (2, 3, 15)

insert. A replaceable cutting tool with a polygonal or round shape that is fastened to a matching toolholder. (18)

island: A protruding standalone feature located inside of a pocket. (11)

isosceles triangle. Triangle with two equal sides and two equal angles. (1, 14)

L

lathe. A metal cutting machine that holds and rotates the workpiece while the cutting tool removes material. (14)

layering. The process of nesting subprograms in other subprograms. (26)

lead. The distance the thread advances along the axis of the cylinder in one revolution of the part. (22)

line. Continuous, straight, one-dimensional geometric element with no end. (1, 14)

linear interpolation. A method of determining a straight-line distance by calculating intermediate points between a start point and end point. (3, 16)

line segment. Part of a line with a definite beginning and end. (1, 14)

live tool. A driven tool used to perform drilling or milling operations on a lathe while the workpiece remains in orientation with the main spindle. (15, 20, 23)

live tooling lathe. A machine equipped with powered tools that can perform various operations while the workpiece remains in orientation with the main spindle. (20, 23)

local variables. Variables defined for temporary use in a macro subprogram called with the G65 command. (30)

M

machine home. The machine position at which the X, Y, and Z axes are at their furthest limits. (2, 13, 15)

machine setup. The process of mounting the work in the correct position, setting the work coordinates for the part, and establishing tool length and diameter offsets. (13)

macro. A custom subroutine that defines one or more actions performed by the macro. (28, 30)

main program. Initial, overarching program that provides direction to the machine and may contain subprograms. (25)

major diameter. The largest diameter of the thread measured from one crest to the other crest. (14, 22)

M-code. A program address identifying a machine control or program control command. (17)

mill stop. A positioning tool with an end stop for locating a part to be mounted in a vise, used to align parts quickly and precisely in the same position. (12)

mill-turn machine. A CNC machine that offers full turning capability along with full milling capability in one machine. (20, 23)

minor diameter. The smallest diameter of the thread measured from one root to the other root. (14, 22)

miscellaneous function. A machine control function or programming function designated with the program address *M*. (4, 17)

mist coolant. A mix of fluid and air sprayed at the cutting area during machining. (4, 17)

modal command. A programmed command that stays on until the mode is canceled or until a subsequent command changes the machine's condition. (3, 16)

multiaxis machine. A machine with more axes of travel than a standard machine. (20)

multi-start threads. Threads made up of two or more intertwined threads running parallel to one another. (22)

N

nominal size. A designated value used for general identification purposes. (14, 22)

nonmodal command. A programmed command that is only active in the block in which it appears and terminates as soon as the function is complete. (3, 16)

numerator. The number on top of a fraction that indicates the number of parts in the fraction. (1, 14)

O

obtuse angle. Angle that measures more than 90°. (1, 14)

on machine verification. The process of checking and verifying features with the part in the machine to confirm the part meets design requirements. (29)

opening statement. The beginning lines that make up a program, used to cancel any existing modal commands that may be in effect and place the machine in the desired starting condition. (7)

open pocket. A pocket enclosed on three sides, but open on the fourth side. (9)

order of operations. The sequence in which machining operations will occur to produce a part. (6, 18)

origin. Zero point on a line or coordinate system. (2, 15)

P

parallelogram. A quadrilateral in which the opposite sides are both parallel and equal. (1)

parting off. A process in which a machined part is separated from the rest of the bar stock held in a lathe chuck or collet after it is completed. (21)

peck drilling. A technique for incrementally drilling a hole in which the tool drives down to a partial depth, fully retracts to remove chips and allow coolant to enter, and then re-enters the hole at a deeper depth. Used whenever the depth of a hole is four times greater, or more, than the diameter of the drill. (10, 20)

pitch. The distance from a point on one thread to another point at the same location on the next thread, measured parallel to the axis of the cylinder. (14, 22)

pitch diameter. The theoretical diameter measured at the point where the width of the ridge is equal to the width of the groove. (22)

pocket. An internal cavity with a flat or open bottom. (9)

point-to-point programming. Programming method that refers to driving the machine through a series of connected coordinates. (8, 10, 19)

polar coordinates. Coordinates that reference a linear distance and an angle. (2)

polygon. A two-dimensional shape with straight sides. (1)

preparatory command. A programmed command that prepares the machine controller, or presets the machine, into a specific state of operation. (3, 16)

probe. A precision measurement instrument that contacts a surface of a part to record its position. (27)

probing system. A hardware addition to a CNC machine that provides accurate measurement feedback. (27)

program. A series of commands and designated functions specified in combination with axis positions to control the machine's movement and operation. (25)

program body. The section of a program that contains all of the movements and commands that direct a machine through specific tools to create a part according to the print. (7)

program loop. A program function that repeats a program after it has completed. (4, 17)

program number. The number identifying a program in the opening block, consisting of the O address and a four- or five-digit number. (5, 17)

program rewind. A program function that ends a program and returns it to the beginning so it can be restarted. (4, 17)

protected move. A safe move used to protect a spindle probe. If the probe encounters an obstruction, it will stop moving and the machine will alarm out before the probe stem breaks or a collision occurs. (28)

Pythagorean theorem. Mathematical property of right triangles that states that the square of one side plus the square of the other side is equal to the square of the hypotenuse or longest side. (1, 14)

Q

quadrilateral. A four-sided, two-dimensional polygon. (1)

R

radial drilling. A lathe live tooling operation in which the travel of the tool is perpendicular to the part diameter and parallel to the X axis direction. (20, 23)

radius. The distance from the center point to the edge of a circle. (1, 14)

radius method. An arc programming method in which the arc radius is used to define radial movement. Also called the *R method*. (8, 19)

ramp entry. A method of cutting tool entry in which the tool enters the material on an angle. (9)

reamer. A cutting tool used to enlarge a previously drilled hole to a finished size. (6, 20)

reaming. An operation used to enlarge an existing hole to a final size and produce a smooth finish. (6, 20)

rectangle. A quadrilateral with two pairs of equal sides and four right (90°) angles. (1)

rhombus. A quadrilateral with four equal sides but no right (90°) angles. (4)

right angle. Angle that measures exactly 90°. (1, 14)

right triangle. Triangle with one 90° angle. (1, 14)

rotary. A rotating unit that can be mounted to a three-axis machining center, used to establish a fourth axis for rotation of the workpiece.

round. An external radial edge that blends adjacent surfaces. (14)

S

scalene triangle. Triangle with no equal sides or angles. (1, 14)

setup. Phase in CNC machining that involves setting up the workholding for the part, setting up the required tooling, and making a test run of the program. (12, 24)

setup person. The individual responsible for setting up the part and tooling for machining operations. (12)

setup sheet. A document that shows detailed information for setting up a part to be machined. (12, 24)

sine. For a given angle, the ratio of the opposite side to the hypotenuse of a right triangle. (1, 14)

single-point threading. A process in which a formed insert makes a series of passes to create the desired thread. (22)

slot. An internal cavity with straight sides and open or closed ends. (9)

spindle synchronization. In threading canned cycles on a lathe, a process in which the spindle rotation is oriented to the tool feed rate so that each cut enters at the same rotational position of the material. (22)

spot drill. A drilling tool used to start holes in CNC applications. (6, 10, 20)

spotface. A shallow bearing surface normally used to provide a smooth flat spot on a casting or forging. (10, 20)

spring pass. A final pass that removes any material that remains as a result of tool deflection on the initial cutting pass. (10, 20)

square. A quadrilateral with four equal sides and four right (90°) angles. (1)

startup block. A block in the opening statement of a program that cancels any machine conditions left from the previous program or establishes a new starting condition for the current program. (3, 7, 16)

stepover amount. The amount of material being cut by the side of the end mill. Also called *radial engagement*. (9)

straight angle. Angle that measures exactly 180°. (1, 14)

subprogram. Secondary program that can be called within a main program to perform specific tasks. (25)

supplementary angles. Two angles that add up to 180°. (1, 14)

system variables. Variables used to store values that represent the current operating conditions and modes in the controller. (30)

T

tailstock. A manually operated or programmable device that provides longitudinal support for a workpiece. (18)

tangent. For a given angle, the ratio of the opposite side to the adjacent side of a right triangle. (1, 14)

tangent line. A line that touches an arc or circle at exactly one point. (1, 14)

taper. A diameter that changes in size uniformly along its length. (14)

tapped hole. A hole with internal threads. (10)

tapping. The process of forming internal threads in a drilled hole with a tap. (6)

thread. A helical groove machined along a cylindrical or conical surface. (22)

thread angle. The included angle between the sides of the thread. (22)

thread depth. The perpendicular distance between the major diameter and minor diameter of the thread. Also called *thread height*. (14, 22)

thread form. The basic profile shape of the thread. (22)

threads per inch (TPI). The total number of complete revolutions of the thread in exactly 1″. (22)

through hole. A hole that completely extends through the material. (10)

tool length offset. A measurement used to compensate for different lengths of tools used in machining. The tool length offset represents the distance the tool travels from the machine home position to the work coordinate system origin. (13)

tool nose radius compensation. A programmed offset from the center of the cutting tool to the tool's edge along a cutting path. (19)

tool offset. A measurement used to compensate for different lengths of tools used in machining. A tool offset represents the distance the tool travels from machine home to the part origin. (21, 24)

tool operation. In a program, a section of the program body defining all of the motions and actions performed while one tool is held in the spindle. (7)

triangle. A polygon that has three sides that form three angles. (1, 14)

trigonometry. A branch of mathematics that deals with the relationships between the sides and angles of triangles and with functions of angles. (1, 14)

true position. The theoretically exact location of a feature of size. (29)

trunnion. A multiaxis attachment mounted directly onto a machine to allow for rotation in one or two directions beyond standard three-axis machining. (2)

T-slot. A T-shaped slot used to mount work in machining. (9)

turning. A machining operation that removes material from a rotating workpiece. (14)

turret. A tool holding device that secures tools for cutting and stores tools when not in use. A turret serves as an automatic tool changer on a CNC turning center. (17)

V

variable. A placeholder used to store and access data for programming purposes. (28, 30)

W

Wireless Intuitive Probing System (WIPS). Probing system equipment and software used to automate part setup and inspection tasks on a CNC machine. (27)

word. A letter and number grouping used to execute a command in a program. (3, 5, 16, 17)

work coordinate system (WCS). The system of point specification that defines where the zero point of the part physically sits within the work envelope of the machine. (13, 26, 27, 31)

work envelope. The space defining the range of a machine's normal operation. (2, 15, 26)

workholding. Any device that is used to secure a workpiece against the forces of machining. (6, 18)

work offset. A coordinate setting that establishes the location of the work origin relative to the machine home position. (3, 12, 13, 15, 24, 26, 27)

work origin. The coordinate point representing the X0. Y0. location on a part. (2)

work shift. A programmed shift of the active work coordinate system to a different location for machining. (24, 25)

Z

zero return. The movement of the machine to its machine home position, the point at which the X, Y, and Z axes are set at zero. (7, 13)

Index

3 + 1 milling, 35

A
absolute positioning, 31–32, 54, 323
acute angle, 11, 297
address, 46, 80, 336, 352
address codes, 78–91, 367–372
    A (rotary axis movement—X axis), 81
    B (rotary axis movement—Y axis), 82
    B (tailstock movement), 368
    C (rotary axis movement and automatic chamfering), 368
    C (rotary axis movement—Z axis), 82
    D (depth of cut), 369
    D (tool diameter offset), 82
    E (precision feed rate), 369
    F (feed rate), 82, 369–370
    G (preparatory command), 83
    H (tool length offset), 83–84
    I (arc center location), 84
    J (arc center location), 84
    K (arc center location), 85
    L (loop count), 85–86
    M (miscellaneous function), 86
    N (block number), 86, 370
    O (program number), 86–87, 370
    P (program number and dwell), 87
    P and Q (cycle starting block and cycle ending block), 370
    Q (repeat depth), 87–88
    R (rapid plane position and arc radius), 88
    S (spindle speed), 88, 371
    T (tool number), 88–89, 371–372
    U and W (alternate axis designations), 372
    U, V, and W (alternate axis designations), 89–90
    X (axis movement), 90
    Y (axis movement), 91
    Z (axis movement), 91
angles, 11–12, 296–297
    complementary, 12, 297
    supplementary, 12, 297
angular moves, calculating, 134–137
arc, 6, 293
axial drilling, 442, 507

B
backlash, 622
ball screw, 482
bar feeder, 473, 557
bar puller, 473, 557
bell mouth, 210
blind hole, 184
block, 46, 80, 336
bolt circles, 16–19, 207
bore/boss measurement, 614–617, 628–633
bore probing cycle, 591
boring, 197, 324, 438
boss, 227
boss probing cycle, 592
bull nose end mill, 102
burr, 145

C
CAM software programs, 516–519, 656–657
canned cycles, 86, 192–203, 345, 408–425, 434
    G70 (finishing cycle), 415–417
    G71 (roughing cycle), 412–415, 418–421
    G73 (high-speed peck drilling cycle), 194–195
    G73 (pattern repeating cycle), 421–424
    G74 (left-hand tapping cycle), 196–197
    G76 (precision boring cycle), 197
    G80 (cancel canned cycles), 198
    G81 (standard drilling cycle), 198
    G82 (spot drilling cycle), 198–199
    G83 (peck drilling cycle), 199–200
    G84 (right-hand tapping cycle), 200–201
    G85 (boring cycle), 201–203
    G94 (facing cycle), 424–425
    G98 and G99 (initial plane and return plane), 193–194
Cartesian coordinate system, 26, 316–328
C axis coordinate programming, 326–328
chamfer, 145, 302, 461–463
chamfering, 145–147
chatter, 659
chip load, 20
chuck, 381–383
circle, 6–8, 293
climb milling, 128
closed pocket, 154
closed slots, 169–171
closing statement, 119–121
comment line, 114–116
complementary angles, 12, 297
computer-aided manufacturing (CAM), 656–657
contouring (mill programming), 124–147, 398
    calculating angular moves, 134–137
    calculating radial moves, 138–145
    chamfering, 145–147
    cutter compensation, 128–134
    point-to-point programming, 126–128
    subprograms, 578–582
contour programming (lathe programming), 396–425
    angular moves, 406–408
    canned cycles, 408–425
    inside diameter features, 399
    point-to-point programming, 398–399
    radial moves, 402–406
    tool nose radius compensation, 400–402
conventional milling, 128
coolant functions, 69–72, 360–364
coordinate systems, 24–39, 316–328
    absolute and incremental positioning, 31–32, 323
    machine home and work origin, 32–33, 325–326
    number line, 26, 316
    polar coordinates, 38–39
    three-dimensional, 28–31
    two-dimensional, 27–28

cosine, 13, 298
counterbore, 189, 435
countersink, 189, 435
crash, 262
cutter compensation, 128–134

D

denominator, 4, 291
depth of cut, 163
design intent, 97, 379
dial indicator, 264
diameter, 7, 293
dovetail slot, 177–178
dry run, 654
dwell, 87

E

edge finder, 241, 266
end mill, 98
end-of-block (EOB), 80
entry strategy, 156–160
equilateral triangle, 9, 294

F

face mill, 220
face milling, 220
facing, 220–227, 424
    creating program, 226–227
    program strategy, 223–225
    tools, 220–223
feed rates, 20, 309
    arcs, 660
fillet, 293
five-axis machines, 36–38
fixture, 100
flood coolant, 69, 360
flute, 20
fly cutter, 220
four-axis machines, 33–35
fractions, converting to decimals, 4–6, 290–292
full four-axis milling, 35
full radius grooves, 470–472

G

G-codes, 44–56, 334–346
    G00 (rapid positioning), 50–51, 340
    G01 (linear interpolation), 51–52, 340–341
    G02 and G03 (circular interpolation), 52, 342–343
    G17 (XY plane), 507
    G18 (XZ plane), 507
    G19 (YZ plane), 507
    G20 and G21 (measurement modes), 52–53, 343
    G28 (machine zero return), 343–344
    G32 single pass threading cycle, 485–487
    G40, G41, and G42 (cutter compensation), 55–56
    G40, G41, and G42 (tool nose radius compensation), 344–345
    G43 (tool length offset), 118
    G51 (scaling), 662–663
    G53 (machine coordinate system), 117
    G54 (work offset), 56
    G65 (custom macro), 607–608, 641
    G70 (finishing cycle), 345, 415–417
    G71 (roughing cycle), 345, 412–415, 418–421
    G72 (roughing cycle), 345–346
    G73 (roughing cycle), 345–346
    G74 (left-hand tapping cycle), 196–197
    G75 (grooving cycle), 463–466
    G76 (multiple pass threading cycle), 488–491
    G76 (precision boring cycle), 197
    G80 (cancel canned cycles), 198
    G81 (standard drilling cycle), 198, 434–435
    G82 (spot drilling cycle), 198–199, 435–436
    G83 (peck drilling cycle), 199–200, 436–437
    G84 (right-hand tapping cycle), 200–201, 437–438
    G85 (boring cycle), 201–202, 438–439
    G86 (boring cycle), 202–203
    G90 and G91 (absolute and incremental positioning), 53–55
    G92 thread cutting cycle, 491–492
    G94 (facing cycle), 424–425
    G98 and G99 (initial plane and return plane), 193–194
    G101 (mirror imaging), 663–664
    G241 (radial drilling cycle), 508–510
    G242 (radial spot drilling cycle), 510
    G243 (radial peck drilling cycle), 510–511
    G245 (radial boring cycle), 511
    G246 (radial boring cycle), 512
    live tooling G-codes, 506–507
    using in programs, 46, 336–337
geometric dimensioning and tolerancing (GD&T), 631
geometric shapes, 6–11, 292–296
    circles, 6–8, 293
    lines, 11, 296
    polygons, 8–10
    triangles, 293–296
global variables, 641–644
graphics mode, 654
grooves, 454–472
    full radius, 470–472
    G75 grooving cycle, 463–466
    programming chamfers, 461–463
    program planning, 454–458
    straight wall, 458–461
    tapered wall, 466–470
grooving, 454
grooving programs, 454–458
    groove depth, 457
    tool offsets, 455–456
    groove width and shape, 456–457

H

helical entry, 159–160
helical milling, 210
helix, 159
hole drilling cycles, 432–438
    G80 (cancel canned cycles), 434
    G81 (standard drilling and spot drilling cycle), 434–435
    G82 (spot drilling and counterboring cycle), 435–436
    G83 (peck drilling cycle), 436–437
    G84 (right-hand tapping cycle), 437–438
hole machining, 430–445
    drilling cycles, 432–439
    live tooling, 439–445
hole machining operations, 182–213
    canned cycles, 192–203
    helical milling, 210
    hole patterns, 203–209
    hole shapes and types, 184–190
    machining holes, 190–192
    thread milling, 210–213
hole patterns, 203–209
    angled, 205–207
    bolt circles, 207–209
    straight line, 204–205
hypotenuse, 9, 294

I

I and K method, 404–406
I, J, and K method, 139–141
impeller, 582
incremental positioning, 31–32, 54, 323
information exchange, 240–241
insert, 385
inside corner probing cycle, 599
inside diameter contour programming, 399
island, 227
island machining, 227–233
    creating program, 231–233
    strategy, 227–231
isosceles triangle, 9, 294

L

lathe programming
    address codes, 367–372
    canned cycles, 408–425
    contour programming, 396–425
    hole machining, 430–445
    program planning, 376–391
    subprogramming, 554–558, 569–571
    turning speeds and feeds, 308–309
lathe setup, 524–539
    setup sheet formats, 534–537
    setup sheets, 533–534
    tool installation, 526–527
    tool offsets, 527–531
lathe taper, 657–659
lead, 482
line, 11, 296
linear interpolation, 51–52, 340
live tool, 326, 440, 505
live tooling, 439–445, 502–519
    CAM software programming, 516–519
    G-codes, 506–507
    M-codes, 506
    radial hole machining canned cycles, 507–512
    slot milling, 512–516
    special considerations, 505–507
live tooling lathe, 504
local variables, 641

M

machine functions
    coolant functions, 69–72, 360–364
    spindle functions, 72–74, 364–367
    see also M-codes
machine home, 32–33, 264, 325–326
machine setup, 260–281
    running first piece, 278–280
    setting workholding, 262–264
    tool diameter offsets, 274–278
    tool length offsets, 268–274
    work coordinates, 264–268
macros, 607, 638–649
    global variables, 641–644
    local variables, 641
    programming example, 648–649
    system variables, 644–648
    using variables, 641
magnetic chucks, 101
main program, 546
major diameter, 306, 482
math calculations, 2–20, 288–309
    angles, 11–12, 296–297
    bolt circles, 16–19
    converting fractions to decimals, 4–6, 290–292
    geometric shapes, 6–11, 292–296
    milling speeds and feeds, 19–20
    precision, 5–6, 291–292
    tapers, 303–306
    thread measurements, 306–308
    trigonometry, 12–16, 298–303
    turning speeds and feeds, 308–309
M-codes, 60–74, 350–372
    live tooling, 506
    M00 (program stop), 64–65, 355
    M01 (optional stop), 66, 356–357
    M02, M30, and M99 (program end), 66–68, 357–359
    M03 (spindle forward), 72–73
    M04 (spindle reverse), 73–74
    M05 (spindle stop), 74
    M06 (tool change), 117–118
    M07 (mist coolant), 69
    M08 (flood coolant), 70–71
    M09 (coolant off), 71–72
    machine functions, 68–74, 359–367
    program functions, 64–68, 354–359
    program stop, 64–66, 355
metric programming, 661
mill program format, 110–121
    opening statement, 113–117
    program body, 117–120
    program closing statement, 120–121
mill programming
    address codes, 78–91
    program format, 110–121
    program planning, 94–107
    subprogramming, 547–554
mill stop, 241
mill-turn machine, 504
minor diameter, 307, 482
mirror imaging, 663–664
miscellaneous functions, 62, 352
    see also M-codes
mist coolant, 69, 360
modal command, 46, 336
multiaxis machine, 439
multiaxis machining, 582–583
multi-start threads, 496

N

nominal size, 308, 483
nonmodal command, 50, 340
number line, 26, 316
numerator, 4, 291

O

obtuse angle, 11, 297
on machine verification, 620–635
    bore/boss measurement, 628–633
    measuring multiple features, 634–635
    single surface measurement, 623–625
    tool offset adjustment and tolerance verification, 623–635
    web/pocket measurement, 625–628
opening statement, 113–117
    comment lines, 114–116
    startup commands, 116–117
open pocket, 154
open slots, 167–169
order of operations, 103–107, 389–391
origin, 26, 316
outside corner probing cycle, 598

P

parallelogram, 10
parting off, 473–475
part origin, 325–326
part workholding, 99–101, 383–385
    chucks, 381–383
    fixtures, 100
    lathe collet chuck, 383
    magnetic chucks, 101
    tailstock, 383–385
    vacuum tables, 100
    vises, 99–100
peck drilling, 192, 433
pitch, 308, 482
pitch diameter, 482

pocket finishing, 166–167
pocket milling, 154–166
    depth of cut and stepover amount, 163–166
    entry strategy, 156–160
pocket probing cycles, 594–595
pocket roughing, 161–163
pockets, 154–167
point-to-point programming, 126–128, 190, 398–399
polar coordinates, 38–39
polygon, 8–10
preparatory commands, 46, 336
    see also G-codes
print review, 97–99, 379–381
probe, 590
probing, 588–601, 604–617
    bore/boss measurement, 614–617
    bore probing cycle, 591
    boss probing cycle, 592
    inside corner probing cycle, 599
    outside corner probing cycle, 598
    pocket X axis probing cycle, 594
    pocket Y axis probing cycle, 595
    protected moves, 607–608
    rectangular block probing cycle, 593–594
    rectangular pocket probing cycle, 592–593
    single surface measurement, 608–609
    single surface probing cycle, 600–601
    starting probe position, 607–608
    web/pocket measurement, 610–614
    web X axis probing cycle, 595–596
    web Y axis probing cycle, 596–597
probing system, 590
production efficiency, 245–254
program, 546
program body, 117–120
program closing statement, 120–121
program loop, 68, 359
program number, 86, 370
program planning, 94–107, 376–391
    order of operations, 103–107, 389–391
    part workholding, 99–101, 383–385
    print review, 97–99, 379–381
    tool selection, 101–102, 385–388
program rewind, 67, 358
protected macro, 608
Pythagorean theorem, 10, 295

Q
quadrilateral, 10

R
radial drilling, 443, 507
radial hole machining canned cycles, 507–512
    G241 (radial drilling cycle), 508–510
    G242 (radial spot drilling cycle), 510
    G243 (radial peck drilling cycle), 510–511
    G245 (radial boring cycle), 511–512
radial move calculation, 138–145
    I, J, and K method, 139–141
    radius method, 141–145
radial moves, 402–406
    I and K method, 404–406
    radius method, 402–404
radius, 7, 293
radius method, 141–145, 402–404
ramp entry, 157–159
reamer, 98, 438
reaming, 438
rectangle, 10
rectangular block probing cycle, 593–594
rectangular pocket probing cycle, 592–593
rhombus, 10
right angle, 11, 296
right triangle, 9, 294
rotary, 33
round, 293

S
scalene triangle, 8, 294
scaling, 662–663
setup, 240, 526
setup person, 240
setup sheets, 238–254, 533–534
    formats, 243–245, 534–537
    header, 242, 533
    information exchange, 240–241
    production efficiency, 245–254
    rough stock, 242, 534
    setup and operation, 240–241
    special instructions, 242, 534
    tool data, 242, 534
    workholding and fixturing, 242, 534
sine, 13, 298
single-point threading, 484
single surface measurement, 608–609, 623–625
single surface probing cycle, 600–601
slot milling, 512–516
slots, 167–178
    closed slots, 169–171
    dovetail slots, 177–178
    open slots, 167–169
    T-slots, 172–176
speeds and feeds, 19–20, 308–309
spindle functions, 72–74, 364–367
spindle synchronization, 485
spot drill, 98, 186, 435
spotface, 190, 435
spring pass, 201, 438
square, 10
startup block, 56, 113, 346
startup commands, 116–117
stepover amount, 164
straight angle, 12, 297
straight wall grooves, 458–461
subprograms, 546, 562–583
    applications, 660–661
    contouring, 578–582
    layering, 574–578
    multiaxis machining, 582–583
    multiple parts and fixtures, 572–573
    work offsets, 564–569
supplementary angles, 12, 297
system variables, 644–648
    current machine position variables, 645
    programmable alarm message variable, 647–648
    tool offset variables, 647
    work offset variables, 645–647

T
tailstock, 383–385
tangent, 13, 298
tangent line, 11, 296
tapered threads, 493–495
tapered wall grooves, 466–470
tapers, 303–306
tapped hole, 186
tapping, 103
thread, 482
thread angle, 482
thread depth, 307, 482
thread form, 482

threading, 480–497
    G32 single pass threading cycle, 485–487
    G76 multiple pass threading cycle, 488–491
    G92 thread cutting cycle, 491–492
    multi-start threads, 496–497
    programming threads, 484–492
    tapered threads, 493–495
    terminology, 482–484
thread measurements, 306–308
    major diameter and minor diameter, 306–307
    thread depth, 307–308
    thread pitch, 308
thread milling, 210–213
threads per inch (TPI), 483
three-dimensional coordinate systems, 28–31
through hole, 186
tool diameter offset, 274–278
tool installation, 526–527
tool length offset, 268–274
tool nose radius compensation, 400–402
tool offsets, 268–278, 455–456, 527–531, 654–656
    establishing, 654–656
    tool diameter offset, 274–278, 530–531
    tool length offset, 268–274, 528–530
tool operation, 117
tool selection, 101–102, 385–388
triangle, 8, 293–296
trigonometry, 12–16, 298–303
true position, 631
trunnion, 37
T-slot, 172–176
turret, 364
two-axis coordinate systems, 317–325
    absolute and incremental positioning, 323
    diameter programming, 320–323
    inside diameter features, 324
two-dimensional coordinate systems, 27–28

V

vacuum tables, 100
variable, 609, 640
vises, 99–100

W

web/pocket measurement, 610–614, 625–628
web probing cycles, 595–597
Wireless Intuitive Probing System (WIPS), 590
word, 46, 80, 336, 352
work coordinates, 264–268
    locating on materials, 265–267
    setting on machine control, 267–268
work coordinate system (WCS), 264, 564, 590, 655
work envelope, 32, 325, 572
workholding, 99–101, 262–264, 383–385
work offset, 56, 241, 264, 326, 530–533, 564–569, 590
work origin, 32–33
work shift, 530, 556

Z

zero return, 116, 265